M000248255

ELEMENTS OF PHOTONICS

Volume I

WILEY SERIES IN PURE AND APPLIED OPTICS

Founded by Stanley S. Ballard, University of Florida

EDITOR: Bahaa E.A. Saleh, Boston University

BEISER · *Holographic Scanning*
BERGER-SCHUNN · *Practical Color Measurement*
BOYD · *Radiometry and The Detection of Optical Radiation*
BUCK · *Fundamentals of Optical Fibers*
CATHEY · *Optical Information Processing and Holography*
CHUANG · *Physics of Optoelectronic Devices*
DELONE AND KRAINOV · *Fundamentals of Nonlinear Optics of Atomic Gases*
DERENIAK AND BOREMAN · *Infrared Detectors and Systems*
DERENIAK AND CROWE · *Optical Radiation Detectors*
DE VANY · *Master Optical Techniques*
GASKILL · *Linear Systems, Fourier Transform, and Optics*
GOODMAN · *Statistical Optics*
HOBBS · *Building Electro-Optical Systems: Making It All Work*
HUDSON · *Infrared System Engineering*
IIZUKA · *Elements of Photonics Volume I: In Free Space and Special Media*
IIZUKA · *Elements of Photonics Volume II: For Fiber and Integrated Optics*
JUDD AND WYSZECKI · *Color in Business, Science, and Industry*. Third Edition
KAFRI AND GLATT · *The Physics of Moire Metrology*
KAROW · *Fabrication Methods for Precision Optics*
KLEIN AND FURTAK · *Optics*, Second Edition
MALACARA · *Optical Shop Testing*, Second Edition
MILONNI AND EBERLY · *Lasers*
NASSAU · *The Physics and Chemistry of Color: The Fifteen Causes of Color*, Second Edition
NIETO-VESPERINAS · *Scattering and Diffraction in Physical Optics*
O'SHEA · *Elements of Modern Optical Design*
SALEH AND TEICH · *Fundamentals of Photonics*
SCHUBERT AND WILHELMI · *Nonlinear Optics and Quantum Electronics*
SHEN · *The Principles of Nonlinear Optics*
UDD · *Fiber Optic Sensors: An Introduction for Engineers and Scientists*
UDD · *Fiber Optic Smart Structures*
VANDERLUGT · *Optical Signal Processing*
VEST · *Holographic Interferometry*
VINCENT · *Fundamentals of Infrared Detector Operation and Testing*
WILLIAMS AND BECKLUND · *Introduction to the Optical Transfer Function*
WYSZECKI AND STILES · *Color Science: Concepts and Methods, Quantitative Data and Formulae*, Second Edition
XU AND STROUD · *Acousto-Optic Devices*
YAMAMOTO · *Coherence, Amplification, and Quantum Effects in Semiconductor Lasers*
YARIV AND YEH · *Optical Waves in Crystals*
YEH · *Optical Waves in Layered Media*
YEH · *Introduction to Photorefractive Nonlinear Optics*
YEH AND GU · *Optics of Liquid Crystal Displays*

ELEMENTS OF PHOTONICS

Volume I
In Free Space and Special Media

Keigo Iizuka

University of Toronto

WILEY-INTERSCIENCE
A JOHN WILEY & SONS, INC., PUBLICATION

This book is printed on acid-free paper. ⊗

Copyright © 2002 by John Wiley & Sons, New York, N.Y. All rights reserved.

Published simultaneously in Canada.

No part of this publication may be reproduced, stored in a retrieval system or transmitted in any form or by any means, electronic, mechanical, photocopying, recording, scanning or otherwise, except as permitted under Sections 107 or 108 of the 1976 United States Copyright Act, without either the prior written permission of the Publisher, or authorization through payment of the appropriate per-copy fee to the Copyright Clearance Center, 222 Rosewood Drive, Danvers, MA 01923, (508) 750-8400, fax (508) 750-4744. Requests to the Publisher for permission should be addressed to the Permissions Department, John Wiley & Sons, Inc., 605 Third Avenue, New York, NY 10158-0012, (212) 850-6011, fax (212) 850-6008, E-Mail: PERMREQ@WILEY.COM.

For ordering and customer service, call 1-800-CALL-WILEY.

Library of Congress Cataloging-in-Publication Data:

Iizuka, Keigo, 1931–
  Elements of photonics / Keigo Iizuka.
    p. cm. — (Wiley series in pure and applied optics)
  "A Wiley-Interscience publication."
  Includes indexes.
  Contents: v. 1. In free space and special media — v. 2. For fiber and integrated optics.
  ISBN 0-471-83938-8 (v. 1 : alk. paper) — ISBN 0-471-40815-8 (v. 2 : alk. paper)
    1. Photonics. 2. Integrated optics. 3. Fiber optics. I. Title. II. Series.
  TA1520 .I35 2002
  621.3–dc21   98-015244

Printed in the United States of America.

10 9 8 7 6 5 4 3 2 1

Kuro, starling dear,
nature's gentle companion
from start to finish

CONTENTS

Volume I

8   Phase Conjugate Optics

Appendix A   Derivation of the Fresnel–Kirchhoff Diffraction Formula from the Rayleigh–Sommerfeld Diffraction Formula

Appendix B   Why the Analytic Signal Method is Not Applicable to the Nonlinear System

Appendix C   Derivation of P_{NL}

Answers to Problems

Index

CONTENTS

Volume II

12   Detecting Light   **796**

PREFACE

After visiting leading optics laboratories for the purpose of producing the educational video *Fiber Optic Labs from Around the World* for the Institute of Electrical and Electronics Engineers (IEEE), I soon realized there was a short supply of photonics textbooks to accommodate the growing demand for photonics engineers and evolving fiber-optic products. This textbook was written to help fill this need.

From my teaching experiences at Harvard University and the University of Toronto, I learned a great deal about what students want in a textbook. For instance, students hate messy mathematical expressions that hide the physical meaning. They want explanations that start from the very basics, yet maintain simplicity and succinctness. Most students do not have a lot of time to spend reading and looking up references, so they value a well-organized text with everything at their fingertips. Furthermore, a textbook with a generous allotment of numerical examples helps them better understand the material and gives them greater confidence in tackling challenging problem sets. This book was written with the student in mind.

The book amalgamates fundamentals with applications and is appropriate as a text for a fourth year undergraduate course or first year graduate course. Students need not have a previous knowledge of optics, but college physics and mathematics are prerequisites.

Elements of Photonics is comprised of two volumes. Even though cohesiveness between the two volumes is maintained, each volume can be used as a stand-alone textbook.

Volume I is devoted to topics that apply to propagation in free space and special media such as anisotropic crystals. Chapter 1 begins with a description of Fourier optics, which is used throughout the book, followed by applications of Fourier optics such as the properties of lenses, optical image processing, and holography.

Chapter 2 deals with evanescent waves, which are the basis of diffraction unlimited optical microscopes whose power of resolution is far shorter than a wavelength of light.

Chapter 3 covers the Gaussian beam, which is the mode of propagation in free-space optical communication. Topics include Bessel beams characterized by an unusually long focal length, optical tweezers useful for manipulating microbiological objects like DNA, and laser cooling leading to noise-free spectroscopy.

Chapter 4 explains how light propagates in anisotropic media. Such a study is important because many electrooptic and acoustooptic crystals used for integrated optics are anisotropic. Only through this knowledge can one properly design integrated optics devices.

Chapter 5 comprehensively treats external field effects, such as the electrooptic effect, elastooptic effect, magnetooptic effect, and photorefractive effect. The treatment includes solid as well as liquid crystals and explains how these effects are applied to such integrated optics devices as switches, modulators, deflectors, tunable filters, tunable resonators, optical amplifiers, spatial light modulators, and liquid crystal television.

Chapter 6 deals with the state of polarization of light. Basic optical phenomena such as reflection, refraction, and deflection all depend on the state of polarization of the light. Ways of converting light to the desired state of polarization from an arbitrary state of polarization are explained.

Chapter 7 explains methods of constructing and using the Poincaré sphere. The Poincaré sphere is an elegant tool for describing and solving polarization problems in the optics laboratory.

Chapter 8 covers the phase conjugate wave. The major application is for optical image processing. For example, the phase conjugate wave can correct the phasefront distorted during propagation through a disturbing medium such as the atmosphere. It can also be used for reshaping the light pulse distorted due to a long transmission distance inside the optical fiber.

Volume II is devoted to topics that apply to fiber and integrated optics.

Chapter 9 explains how a lightwave propagates through a planar optical guide, which is the foundation of integrated optics. The concept of propagation modes is fully explored. Cases for multilayer optical guides are also included.

Chapter 10 is an extension of Chapter 9 and describes how to design a rectangular optical guide that confines the light two dimensionally in the x and y directions. Various types of rectangular optical guides used for integrated optics are compared. Electrode configurations needed for applying the electric field in the desired direction are also summarized.

Chapter 11 presents optical fibers, which are the key components in optical communication systems. Important considerations in the choice of optical fibers are attenuation during transmission and dispersion causing distortion of the light pulse. Such special-purpose optical fibers as the dispersion-shifted fiber, polarization-preserving fiber, diffraction grating imprinted fiber, and dual-mode fiber are described. Methods of cabling, splicing, and connecting multifiber cables are also touched on.

Chapter 12 contains a description of light detectors for laboratory as well as communication uses. Mechanisms for converting the information conveyed by photons into their electronic counterparts are introduced. Various detectors, such as the photomultiplier tube, the photodiode, and the avalanche photodiode, and various detection methods, such as direct detection, coherent detection, homodyne detection, and detection by stimulated Brillouin scattering, are described and their performance is compared for the proper choice in a given situation.

Chapter 13 begins with a brief review of relevant topics in quantum electronics, followed by an in-depth look at optical amplifiers. The optical amplifier has revolutionized the process of pulse regeneration in fiber-optic communication systems. The chapter compares two types of optical amplifier: the semiconductor optical amplifier and the erbium-doped fiber amplifier. Knowledge gained from the operation of a single fiber amplifier is applied to the analysis of concatenated fiber amplifiers.

Chapter 14 is devoted to lasers, which is a natural extension of the preceding chapter on optical amplifiers. The chapter begins with an overview of different types of lasers, followed by an in-depth treatment of semiconductor lasers, which are the preferred light sources for most fiber-optic communication systems. The basic relationship among the laser structure, materials, and operational characteristics are clarified. The ability to tune the laser wavelength, which is indispensible to the wavelength division multiplexing of the communication system, is addressed. The quantum well, quantum wire, and quantum dot laser diodes that have low threshold current and hence a high upper limit on the modulation frequency are also included. The erbium-doped or Raman fiber lasers that are simple in structure and easy to install in an optical fiber system are also explained.

In Chapter 15, an introduction to the nonlinear (Kerr) effect is presented. Optical devices based on the Kerr effect are controlled by photons and can respond much faster than those controlled by electrons. The chapter also provides the mechanism of formation of a soliton wave. A light pulse that propagates in an optical fiber spreads due to the dispersion effect of the fiber, but as the intensity of the pulse is increased, the nonlinear effect of the fiber starts to generate a movement directed toward the center of the light pulse. When these two counteracting movements are balanced, a soliton wave pulse that can propagate distortion-free over thousands of kilometers is formed. The attraction of distortion-free pulse propagation is that it can greatly reduce, or even eliminate, the need for pulse regenerators (repeaters) in long-haul fiber-optic communication systems.

Chapter 16 interweaves the design skills developed throughout the book with realistic problems in fiber-optic communication systems.

The problems at the end of each chapter are an integral part of the book and supplement the explanations in the text.

As a photonics textbook, each volume would be sufficient for a two-semester course. If time is really limited, Chapter 16 alone can serve as a crash course in fiber-optic communication systems and will give the student a good initiation to the subject.

For those who would like to specialize in optics, I highly recommend reading through each volume, carefully and repeatedly. Each chapter will widen your horizon of optics that much more. You will be amazed to discover how many new applications are born by adding a touch of imagination to a fundamental concept.

This two-volume work has been a long time in the making. I applaud Beatrice Shube, and George Telecki and Rosalyn Farkas of John Wiley & Sons for their superhuman patience. Sections of the manuscript went through several iterations of being written, erased, and then rewritten. As painstaking as this process was, the quality of the manuscript steadily improved with each rewrite.

I am very grateful to Professor Joseph W. Goodman of Stanford University who first suggested I publish my rough lecture notes in book form.

I am indebted especially to Mary Jean Giliberto, who spent countless hours proofreading the text, smoothing the grammatical glitches, and checking equations and numerical examples for completeness and accuracy. I greatly valued her comments and perspective during our many marathon discussions. This book was very much a partnership, in which she played a key role.

I would like to express my gratitude to Dr. Yi Fan Li, who provided much input to Chapter 15 on nonlinear optics, and Professor Subbarayan Pasupathy of the University of Toronto and Professor Alfred Wong of the University of California, Los Angeles, who critically read one of the appendixes. Frankie Wing Kei Cheng has double-checked the equations and calculations of the entire book.

I would also like to acknowledge the following students, who went through the manuscript very critically and helped to refine it: Claudio Aversa, Hany H. Loka, Benjamin Wai Chan, Soo Guan Teh, Rob James, Christopher K. L. Wah, and Megumi Iizuka.

Lena Wong's part in typing the entire manuscript should not be underestimated. I also owe my gratidue to Linda Espeut for retyping the original one-volume manuscript into the current two-volume manuscript. I wish to express my heartfelt thanks to my wife, Yoko, and children, Nozomi, Izumi, Megumi, and Ayumi, for their kind sacrifices. Ayumi Iizuka assisted in designing the cover of the book.

KEIGO IIZUKA

University of Toronto

ELEMENTS OF PHOTONICS

Volume I

1

FOURIER OPTICS: CONCEPTS AND APPLICATIONS

Welcome to the exciting field of photonics. Chapter 1 is a quick tour of Fourier optics, a vital foundation for the chapters that follow. The branch of optics that can be analyzed by means of the Fourier transform is known as Fourier optics. The presentation of this subject is a condensed version of several texts on the subject [1–8]. This chapter starts with expressions for plane waves and a collection of special functions that are often used in photonics. The rest of the chapter is devoted to problems that can nicely be solved by Fourier optics, including various diffraction patterns, thin lenses, optical signal processing, spatial filters, and holography. A more rigorous derivation of the diffraction equations is also added at the end.

1.1 PLANE WAVES AND SPATIAL FREQUENCY

The representation of plane waves is introduced first, followed by a discussion of spatial frequency.

1.1.1 Plane Waves

The expression for a plane wave propagating in an arbitrary direction when observed at an arbitrary point in space will be derived. Let's first restrict ourselves to a two-dimensional (2D) vacuum medium such as shown in Fig. 1.1a.

Let a plane wave observed at the origin at time t be expressed as

$$\mathbf{E}(0, 0, t) = \mathbf{E}_0(0, 0)e^{-j\omega t} \tag{1.1}$$

where the vector $\mathbf{E}_0(0, 0)$ represents the amplitude and direction of polarization, and ω represents the angular frequency. In this book the sign convention of $e^{-j\omega t}$ rather than $e^{j\omega t}$ is used. The direction of propagation associated with the sign convention is discussed in the boxed note.

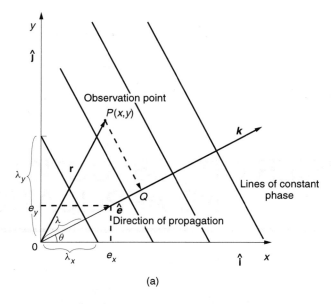

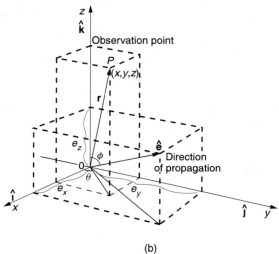

Figure 1.1 Expression for a plane wave. (a) A plane wave propagating in the $\hat{\mathbf{e}}$ direction and observed at $P(x, y)$ in 2D space. (b) The position vector $\mathbf{r}$ and direction $\hat{\mathbf{e}}$ of propagation of a plane wave in 3D space.

The direction of propagation is expressed by the unit vector

$$\hat{\mathbf{e}} = e_x\hat{\mathbf{i}} + e_y\hat{\mathbf{j}} \tag{1.2}$$

where

$$e_x = \cos\theta, \qquad e_y = \sin\theta \tag{1.3}$$

and $\hat{\mathbf{i}}$ and $\hat{\mathbf{j}}$ are unit vectors in the x and y directions, respectively.

There are two equally valid conventions for expressing the time dependence of the electric wave; they are $e^{j\omega t}$ and $e^{-j\omega t}$. Depending on the choice of convention, the wave expression is different.

Let's take the expression

$$E = \mathrm{Re}\left[E_0 e^{j(-\omega t + \beta z)}\right] = E_0 \cos(-\omega t + \beta z)$$

as an example.

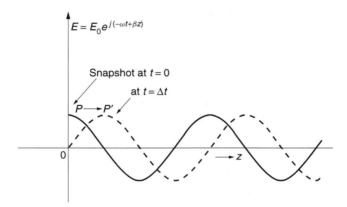

$E = E_0 e^{j(-\omega t + \beta z)}$ is a forward wave.

The peak occurs when the term within the cosine bracket equals zero. Therefore, when a snapshot of the wave is taken at $t = 0$, the location of the peak is at $z = 0$. Now, another snapshot is taken at Δt seconds later. The position Δz of the peak is again where the value of the bracketed term is zero. The peak has moved to a positive new location at $\Delta z = (\omega/\beta)\Delta t$. Thus, this equation represents a forward wave with the phase velocity

$$v_p = \omega/\beta$$

On the other hand, as time increases with

$$E = E_0 \cos(\omega t + \beta z)$$

the peak moves toward the negative z direction, $\Delta z = -(\omega/\beta)\Delta t$, and this represents the backward wave.

The rule is, whenever the signs of t and z are different, as in $E = E_0 e^{j(-\omega t + \beta z)}$ or $E = E_0 e^{j(\omega t - \beta z)}$, the waves are forward waves; and whenever the signs are the same, as in $E = E_0 e^{j(\omega t + \beta z)}$ or $E = E_0 e^{j(-\omega t - \beta z)}$, the waves are backward waves. In this book the convention of $e^{-j\omega t}$ is used, unless otherwise stated, because the forward wave $E = E_0 e^{j(-\omega t + kz)}$ has a positive sign on the z.

Let us now observe this wave from the observation point $P(x, y)$, which is connected to the origin by the position vector $\mathbf{r}$,

$$\mathbf{r} = x\hat{\mathbf{i}} + y\hat{\mathbf{j}} \tag{1.4}$$

$\overline{0Q}$ is the distance between the origin and the projection of point P in the direction of $\hat{e}$ as shown in Fig. 1.1a. If the field travels at the speed of light, the field observed at point P is that which has left the origin $\overline{0Q}/c$ seconds ago. $\overline{0Q}$ can be represented by the scalar product $\hat{e} \cdot \mathbf{r}$. Thus, the field $E(x, y)$ observed at P is

$$\mathbf{E}(x, y, t) = \mathbf{E}_0(0, 0)e^{-j\omega(t - \hat{e} \cdot \mathbf{r}/c)} \tag{1.5}$$

Equation (1.5) is rewritten as

$$\mathbf{E}(x, y, t) = \mathbf{E}_0(0, 0)e^{-j\omega t + j\mathbf{k} \cdot \mathbf{r}} \tag{1.6}$$

where

$$\mathbf{k} = \frac{\omega}{c}\hat{e} = \frac{2\pi}{\lambda}\hat{e} \tag{1.7}$$

and where λ is the wavelength in vacuum. $\mathbf{k}$ is called the *vector propagation constant*. Using Eq. (1.2) the vector propagation constant $\mathbf{k}$ can be expanded in Cartesian coordinates as

$$\mathbf{k} = \frac{2\pi}{\lambda}\cos\theta\,\hat{i} + \frac{2\pi}{\lambda}\sin\theta\,\hat{j} \tag{1.8}$$

Equation (1.6) is the expression for a plane wave propagating at a speed of c in the $\hat{e}$ direction observed at point P. So far, we have assumed a refractive index equal to 1. Inside a linear medium with refractive index n, the frequency does not change, but the wavelength is reduced to λ/n and the speed of propagation is slowed down to a velocity of $v = c/n$. Hence, the propagation constant becomes $n\mathbf{k}$.

1.1.2 Spatial Frequency

Next, the vector propagation constant $\mathbf{k}$ will be rewritten in terms of spatial frequency. It is important to note that spatial frequency is different from temporal frequency. Temporal frequency f is defined as the number of wavelengths that pass through a particular point in space per unit time, whereas spatial frequency f_s is defined as the number of wavelengths in a unit of distance:

$$f_s = \frac{1}{\lambda} \tag{1.9}$$

The most popular unit of f_s is lines/mm or lines/m. In the field of spectroscopy, f_s is called the wavenumber with units of cm^{-1}.

The relationship between the spatial and temporal frequencies of a plane wave is

$$f = cf_s \tag{1.10}$$

Equation (1.10) was obtained by comparing Eq. (1.9) with the temporal frequency

$$f = \frac{c}{\lambda} \tag{1.11}$$

The spatial frequency depends on the direction in which the unit distance is taken.

Referring to Fig. 1.1a, we see that λ_x and λ_y are

$$\lambda_x = \frac{\lambda}{\cos\theta}$$
$$\lambda_y = \frac{\lambda}{\sin\theta}$$

(1.12)

The spatial frequencies in these directions are

$$f_x = \frac{1}{\lambda_x} = e_x f_s$$
$$f_y = \frac{1}{\lambda_y} = e_y f_s$$

(1.13)

$$f_s = \frac{1}{\lambda}$$

where Eq. (1.3) was used.

Earlier, we derived the propagation vector $\mathbf{k}$ to be

$$\mathbf{k} = 2\pi\left(\frac{\cos\theta}{\lambda}\mathbf{i} + \frac{\sin\theta}{\lambda}\mathbf{j}\right)$$

in Eq. (1.8). This can now be expressed more elegantly using Eq. (1.12) as

$$\mathbf{k} = 2\pi\left(\frac{1}{\lambda_x}\hat{\mathbf{i}} + \frac{1}{\lambda_y}\hat{\mathbf{j}}\right) = 2\pi(f_x\hat{\mathbf{i}} + f_y\hat{\mathbf{j}})$$

where, to repeat, f_x and f_y are spatial frequencies. From this, it follows that $\mathbf{k} \cdot \mathbf{r}$ is expressed as

$$\mathbf{k} \cdot \mathbf{r} = 2\pi f_x x + 2\pi f_y y$$

(1.14)

From Eq. (1.12), we obtain

$$\left(\frac{1}{\lambda}\right)^2 = \left(\frac{1}{\lambda_x}\right)^2 + \left(\frac{1}{\lambda_z}\right)^2$$

which can be rewritten as

$$f_s^2 = f_x^2 + f_y^2$$

(1.15)

Finally, the two-dimensional expression will be extended into three dimensions by adding a unit vector $\hat{\mathbf{k}}$ in the z direction. The direction of propagation $\hat{\mathbf{e}}$ becomes

$$\hat{\mathbf{e}} = e_x\hat{\mathbf{i}} + e_y\hat{\mathbf{j}} + e_z\hat{\mathbf{k}}$$

(1.16)

With the coordinates shown in Fig. 1.1b, the components of the unit vector $\hat{\mathbf{e}}$ are

$$e_x = \sin\phi\cos\theta$$
$$e_y = \sin\phi\sin\theta \qquad (1.17)$$
$$e_z = \cos\phi$$

With $\phi \neq 90°$ in Eq. (1.17), the expression corresponding to Eq. (1.8) is

$$\mathbf{k} = 2\pi\frac{\cos\theta}{\lambda}\sin\phi\,\hat{\mathbf{i}} + 2\pi\frac{\sin\theta}{\lambda}\sin\phi\,\hat{\mathbf{j}} + 2\pi\frac{\cos\phi}{\lambda}\hat{\mathbf{k}} \qquad (1.18)$$

and

$$\lambda_x = \frac{\lambda}{\cos\theta\sin\phi}$$
$$\lambda_y = \frac{\lambda}{\sin\theta\sin\phi} \qquad (1.19)$$
$$\lambda_z = \frac{\lambda}{\cos\phi}$$

and in terms of spatial frequencies,

$$f_x = \frac{1}{\lambda_x} = f_s e_x$$
$$f_y = \frac{1}{\lambda_y} = f_s e_y$$
$$f_z = \frac{1}{\lambda_z} = f_s e_z \qquad (1.20)$$
$$f_s = \frac{1}{\lambda}$$

The three-dimensional (3D) position vector $\mathbf{r}$ is

$$\mathbf{r} = x\hat{\mathbf{i}} + y\hat{\mathbf{j}} + z\hat{\mathbf{k}} \qquad (1.21)$$

Similar to the two-dimensional case, $\mathbf{E}$ is expressed as

$$\mathbf{E}(x, y, z, t) = \mathbf{E}(0, 0, 0)e^{-j\omega t + j\mathbf{k}\cdot\mathbf{r}} \qquad (1.22)$$

where

$$\mathbf{k}\cdot\mathbf{r} = 2\pi f_x x + 2\pi f_y y + 2\pi f_z z \qquad (1.23)$$

and where

$$f_s^2 = f_x^2 + f_y^2 + f_z^2 \qquad (1.24)$$

Example 1.1 For a plane wave propagating in the direction

$$\theta = 30°, \qquad \phi = 45°$$

find the expression for the field when observed at point $P(1, 2, 3) \times 10^{-6}$ m. The free-space wavelength is $\lambda = 1.55$ μm. The vector $\mathbf{E}_0$ representing the polarization is $\hat{\mathbf{i}} + 2\hat{\mathbf{j}} - 1.86\hat{\mathbf{k}}$.

Solution From the value $\mathbf{k} \cdot \mathbf{r}$ in Eq. (1.23) and from Eqs. (1.19) and (1.20), the x component is

$$k_x x = 2\pi f_x x$$

$$= \frac{2\pi}{\lambda} e_x x$$

$$= \frac{2\pi}{\lambda} \sin \phi \cos \theta \cdot x$$

$$= 2.48$$

Similarly, the y and z components are

$$k_y y = 2.86$$

$$k_z z = 8.60$$

and

$$\omega = \frac{2\pi c}{\lambda} = 1.22 \times 10^{15} \text{ rad/s}$$

From Eq. (1.22), the expression for $\mathbf{E}$ is

$$\mathbf{E}[(1, 2, 3) \times 10^{-6}, t] = (\hat{\mathbf{i}} + 2\hat{\mathbf{j}} - 1.86\hat{\mathbf{k}})e^{j13.94 - j1.22 \times 10^{15} t} \qquad \square$$

Example 1.2 A plane wave propagating in a given medium is expressed as

$$\mathbf{E}(x, y, z, t) = \mathbf{E}_0 e^{j(2x + 3y + 4z) \times 10^6 - j10^{15} t} \qquad (1.25)$$

(a) Find the *unit vector* for the direction of propagation.
(b) What are the *values of θ and ϕ* that characterize the direction of propagation?
(c) Find the refractive index of the medium.
(d) Find the vector expression of $\mathbf{E}_0$ of Eq. (1.25), assuming that $\mathbf{E}_0$ is polarized in the $x = x_1$ plane and the amplitude is 5.0.

Solution

(a) The direction of the vector parallel to the propagation direction is $2\mathbf{i} + 3\mathbf{j} + 4\mathbf{k}$. The unit vector is found by normalizing this vector. The result is

$$\hat{\mathbf{e}} = \frac{1}{\sqrt{29}}(2\hat{\mathbf{i}} + 3\hat{\mathbf{j}} + 4\hat{\mathbf{k}})$$

and therefore

$$\hat{\mathbf{e}} = 0.37\hat{\mathbf{i}} + 0.56\hat{\mathbf{j}} + 0.74\hat{\mathbf{k}}$$

(b) The direction of the unit vector in terms of θ and ϕ can be solved for by using Eq. (1.17):

$$\theta = 56.3°$$

$$\phi = 42.0°$$

(c) In order to determine the refractive index n, the velocity of the wave needs to be determined since $n = c/v$. The velocity is expressed by the product of the temporal frequency and the wavelength or $v = f\lambda$. The first part of the exponential term of Eq. (1.25) contains the wavelength information whereas the second term contains the temporal frequency information.

To solve for the wavelength, the spatial frequency f_s is first obtained from its components f_x, f_y, and f_z. Explicitly,

$$\mathbf{k} \cdot \mathbf{r} = 2\pi(f_x x + f_y y + f_z z) = (2x + 3y + 4z) \times 10^6$$

where

$$f_x = \frac{2 \times 10^6}{2\pi} \text{ lines/m}$$

$$f_y = \frac{3 \times 10^6}{2\pi} \text{ lines/m}$$

$$f_z = \frac{4 \times 10^6}{2\pi} \text{ lines/m}$$

and hence $f_s = 0.86 \times 10^6$ lines/m from the sum of the squares.

The inverse of the spatial frequency is the wavelength and is equal to

$$\lambda = 1.17 \times 10^{-6} \text{ m}$$

From the second part of the exponential term in Eq. (1.25) we now obtain the temporal frequency, which is

$$f = \frac{10^{15}}{2\pi} = 1.59 \times 10^{14} \text{ Hz}$$

Hence, the phase velocity v is

$$v = \lambda f = 1.86 \times 10^8 \text{ m/s}$$

Finally, from the velocity we obtain the refractive index

$$n = \frac{c}{v} = 1.61$$

(d) Let the amplitude vector be

$$\mathbf{E}_0 = (a\hat{\mathbf{i}} + b\hat{\mathbf{j}} + c\hat{\mathbf{k}})$$

With a plane wave, the direction of polarization is perpendicular to the direction of propagation. Thus, their dot product should be zero:

$$\mathbf{E}_0 \cdot \hat{\mathbf{e}} = 0$$

Since $\mathbf{E}_0$ is polarized in the $x = x_1$ plane, it follows that

$$a = 0$$

and

$$e_y b + e_z c = 0$$

From Eq. (1.25), and the given magnitude of 5, a pair of equations are obtained that can be solved for b and c, namely,

$$3b + 4c = 0$$
$$b^2 + c^2 = 5^2$$

Hence, the vector expression of $\mathbf{E}_0$ is

$$\mathbf{E}_0 = \pm(4\hat{\mathbf{j}} - 3\hat{\mathbf{k}})$$ □

1.2 FOURIER TRANSFORM AND DIFFRACTION PATTERNS IN RECTANGULAR COORDINATES

Referring to Fig. 1.2a, the field distribution from a source is observed on a screen. The field distribution on the screen is called a diffraction pattern.

We will demonstrate that the diffraction pattern can be elegantly expressed by the Fourier transform of the source. Let $E(x_i, y_i)$ represent the field at point P on the screen placed a distance z_i away from the source field $E(x_0, y_0)$. The distributed source $E(x_0, y_0)$ is considered as an ensemble of point sources. Each point source radiates a spherical wave. The field at the observation point P is comprised of contributions from an ensemble of fields radiated from all the point sources. The contribution of the point source located at (x_0, y_0) to point P at (x_i, y_i) is

$$dE(x_i, y_i) = \frac{e^{jkr}}{r} E(x_0, y_0) dx_0 \, dy_0 \qquad (1.26)$$

where $E(x_0, y_0)$ is the magnitude of the point source located at (x_0, y_0) and r is the distance between (x_0, y_0) and (x_i, y_i). The distance r is expressed as

$$r = \sqrt{z_i^2 + (x_i - x_0)^2 + (y_i - y_0)^2} \qquad (1.27)$$

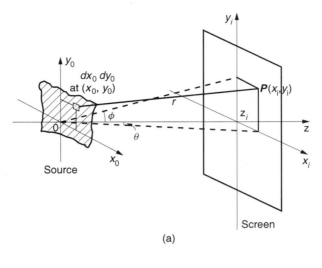

(a)

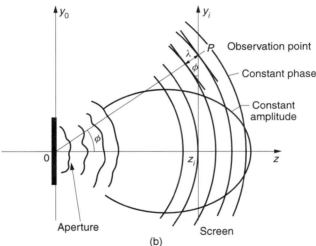

(b)

Figure 1.2 Field distribution from a source observed on a screen. (a) Geometry. (b) Distribution of the field in the (y_i, z_i) plane.

The contribution of the spherical waves from all the point sources to $E(x_i, y_i)$ is

$$E(x_i, y_i) = K \iint \frac{e^{jkr}}{r} E(x_0, y_0) dx_0 \, dy_0 \tag{1.28}$$

This equation is known as the Fresnel–Kirchhoff diffraction formula. The amplitude of the diffracted field is inversely proportional to its wavelength and is expressed as

$$K = \frac{1}{j\lambda} \tag{1.29}$$

The constant K will be derived later in Appendix A of Volume I.

If the point of observation is far enough away or in the vicinity of the z axis (paraxial),

$$z_i^2 \gg (x_i - x_0)^2 + (y_i - y_0)^2 \tag{1.30}$$

then the distance r can be simplified by the binomial expansion as

$$r \doteq z_i \left(1 + \frac{(x_i - x_0)^2 + (y_i - y_0)^2}{2z_i^2} \right) \tag{1.31}$$

which can be rewritten as

$$r \cong z_i + \frac{x_i^2 + y_i^2}{2z_i} - \frac{x_i x_0 + y_i y_0}{z_i} + \frac{x_0^2 + y_0^2}{2z_i} \tag{1.32}$$

The region of z_i for which the approximate expression Eq. (1.32) is valid is called the Fresnel region or the near-field region. As the distance is further increased in the z direction, the last term in Eq. (1.32) becomes negligible for the finite size of the source. This region of z_i is called the Fraunhofer region or far-field region. In this chapter we are concerned about the far-field. In the far-field region, the approximation for r is

$$r = z_i + \frac{x_i^2 + y_i^2}{2z_i} - \frac{x_i x_0 + y_i y_0}{z_i} \tag{1.33}$$

By substituting this approximation into the exponential term of the Fresnel–Kirchhoff diffraction formula, Eq. (1.28), the field becomes

$$E(x_i, y_i) = \frac{1}{j\lambda z_i} e^{jk[z_i + (x_i^2 + y_i^2)/2z_i]} \iint_{-\infty}^{\infty} E(x_0, y_0) e^{-j2\pi(f_x x_0 + f_y y_0)} \, dx_0 \, dy_0$$

$$\text{with} \quad f_x = \frac{x_i}{\lambda z_i}, \qquad f_y = \frac{y_i}{\lambda z_i} \tag{1.34}$$

We recognize that the integral is the two-dimensional Fourier transform of the field in the x, y domain into the f_x, f_y domain:

$$\mathcal{F}\{g(x, y)\} = \iint_{-\infty}^{\infty} g(x, y) e^{-j2\pi(f_x x + f_y y)} \, dx \, dy \tag{1.35}$$

Or in mathematical terms, the diffraction pattern is

$$E(x_i, y_i) = \frac{1}{j\lambda z_i} e^{jk[z_i + (x_i^2 + y_i^2)/2z_i]} \mathcal{F}\{E(x_0, y_0)\}_{f_x = x_i/\lambda z_i, \ f_y = y_i/\lambda z_i} \tag{1.36}$$

where $\mathcal{F}$ denotes the Fourier transform.

In short, *the Fraunhofer diffraction pattern is the Fourier transform of the source field.*

Sometimes, the angular distribution rather than the planar distribution is desired. For this case, the azimuth angle ϕ and the elevation angle θ, measured with respect to the center of the source field, are approximated as $\sin\theta \doteq x_i/z_i$ and $\sin\phi \doteq y_i/z_i$. Hence, the values in Eq. (1.34) are

$$f_x = \frac{\sin\theta}{\lambda}, \qquad f_y = \frac{\sin\phi}{\lambda} \qquad (1.37)$$

The branch of optics that can be analyzed by means of the Fourier transform is categorized as Fourier optics. First, the physical meaning of f_x and f_y in Eq. (1.34) will be explored. For simplicity, only the distribution in the (y_i, z_i) plane will be considered. Figure 1.2b shows a typical phase and amplitude distribution of the field diffracted from an aperture source whose dimensions are much smaller than the distance to the screen.

In the region far from the aperture, the phase distribution is more like that of a spherical wave. With the source placed at the origin, the phase front along the y_i axis near $y_i = 0$ is always parallel to the y_i axis. In the vicinity of this point, there is no variation in the phase of the field with respect to y_i. Hence, the field has zero spatial frequency at $y_i = 0$. (The variation of the field amplitude with y_i is normally much slower than that of the phase.) As shown in Fig. 1.2b, the change in the variation of phase increases as the point of observation P moves along the y_i axis, such that, eventually, the wavelength λ_{y_i} measured along the y_i axis will approach the wavelength of free space.

Mathematically, λ_{y_i} at the observation point $P(y_i, z_i)$ is, from the geometry in Fig. 1.2b,

$$\lambda_{y_i} = \frac{\lambda}{\sin\phi}$$

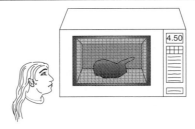

Microwave oven door shields us from microwaves but not from light.

How is it that a microwave oven door shields us from microwaves but not from light waves, allowing us to see our food as it cooks? The answer is found by examining the Fresnel–Kirchhoff diffraction formula applied to the mesh of our microwave oven door. The mesh is equivalent to a series of equally spaced apertures. Hence, this causes diffraction to occur, and the Fresnel–Kirchhoff equation, Eq. (1.34), is used to calculate the diffracted field.

Let us calculate the amplitude of the microwave and light wave field observed at the center: $x_i = y_i = 0$. The integral in Eq. (1.34) is the same for both microwaves and light, but as shown by the factor K of Eq. (1.29), the amplitude of the field is inversely proportional to the wavelength. Thus, the amplitude of the microwaves is about 10,000 times smaller than that of light. Thank goodness for the presence of K.

The field located at P, therefore, has the spatial frequency of

$$f_{y_i} = \frac{\sin \phi}{\lambda} \cong \frac{y_i}{z_i \lambda}$$

If we wish to construct a spatial frequency filter to pick selectively a particular f_{y_i} spatial frequency, we can place an opaque screen in the (x_i, y_i) plane and poke a hole in it at a particular location. The location of this hole for a desired spatial frequency can be calculated by rearranging the above equation to give

$$y_i = f_{y_i} \lambda z_i$$

The usefulness of Eq. (1.36) extends throughout the electromagnetic spectrum. For example, the inverse of this calculation is used in X-ray crystallography. By knowing the X-ray diffraction pattern of a crystal, the structure of the crystal is found by its inverse Fourier transform. The same is true with radio astronomy. By probing $E(x_i, y_i)$, the radio radiation pattern of a star, the structure can be analyzed by the inverse Fourier transform in a similar manner. Yet another application is to use this relationship to find the radiation pattern of an antenna (Problem 1.4). The antenna radiation pattern for a given current distribution $I(x_0, y_0)$ can be obtained [9]. With a few simple substitutions, $E(x_0, y_0)$ is replaced by $\frac{1}{2}\eta I(x_0, y_0) \sin \theta$, where $\eta = 120\pi$ is the intrinsic impedance of free space, and $\sin \theta$ is necessary to convert E_z to E_θ, since antenna theory expresses its patterns in terms of E_θ.

Before closing this section, we will demonstrate that the Fresnel field or the *near* field can also be expressed in terms of the Fourier transform. As previously stated, the Fresnel region is valid when r is approximated with the addition of the quadratic phase factor term of the source in Eq. (1.32). By similarly substituting this value of r into the Fresnel–Kirchhoff diffraction formula, the field in the Fresnel region becomes

$$E(x_i, y_i) = \frac{1}{j\lambda z_i} e^{jk[z_i + (x_i^2 + y_i^2)/2z_i]} \mathcal{F}\left\{ \underbrace{E(x_0, y_0)}_{\text{Input}} \underbrace{e^{jk(x_0^2 + y_0^2)/2z_i}}_{\substack{\text{Part of the} \\ \text{point spread} \\ \text{function}}} \right\}_{f_x = x_i/\lambda z_i, \, f_y = y_i/\lambda z_i}$$

$$(1.38)$$

Alternatively, the Fresnel field can be expressed elegantly as a convolution of two terms explained as follows. Recall that the approximation of r, the distance to the screen, is from Eq. (1.31)

$$r = z_i \left(1 + \frac{(x_i - x_0)^2 + (y_i - y_0)^2}{2z_i^2} \right)$$

When r is directly substituted into the Fresnel–Kirchhoff diffraction formula, Eq. (1.28), the result is

$$E(x_i, y_i) = j\frac{1}{\lambda z_i} \iint E(x_0, y_0)\{e^{jkz_i[1 + (x_i - x_0)^2/2z_i^2 + (y_i - y_0)^2/2zi^2]}\}\, dx_0\, dy_0$$

The expression in the curly brackets is in the form of $f(x_i - x_0, y_i - y_0)$. From this observation, we note that the above expression takes on the form of a convolution:

$$g(x) * f(x) = \int g(\xi) f(x - \xi) d\xi$$

Thus,

$$E(x_i, y_i) = E_0(x_i, y_i) * f_{z_i}(x_i, y_i) \tag{1.39}$$

where

$$f_{z_i}(x_i, y_i) = \frac{1}{j\lambda z_i} e^{jk[z_i + (x_i^2 + y_i^2)/2z_i]} \tag{1.40}$$

The function $f_{z_i}(x_i, y_i)$ is called the point spread function (or impulse response function of free space) and is identical with the field at (x_i, y_i, z_i) when a point source is placed at the origin of the source coordinates. Whether Eq. (1.38) or (1.39) is used, the results are the same.

The Fourier transform F_{z_i} of the point spread function in Eq. (1.40) is

$$F_{z_i} = e^{jkz_i - j\pi\lambda z_i (f_x^2 + f_y^2)} \tag{1.41}$$

The derivation of Eq. (1.41) is found in the boxed note.

The Fourier transform F_{z_i} of the point spread function f_{z_i}, given by Eq. (1.41), will be derived. Let's start with the easy to remember Fourier transform [10]

$$\mathcal{F}\{e^{-\pi x^2}\} = e^{-\pi f^2}$$

The Fourier transform of this function is the original function itself.
First, the x_i component of Eq. (1.40) is rewritten as

$$e^{jkx_i^2/2z_i} = e^{[-\pi(x_i/\sqrt{j\lambda z_i})^2]}$$

The similarity theorem of the Fourier transform is

$$\mathcal{F}\{g(\alpha x)\} = \frac{1}{\alpha} G\left(\frac{f_x}{\alpha}\right) \tag{1.42}$$

where G is the Fourier transform of $g(x)$. If Eq. (1.42) is used, the Fourier transform of the x_i component is

$$\mathcal{F}\{e^{[-\pi(x_i/\sqrt{j\lambda z_i})^2]}\} = \sqrt{j\lambda z_i} \, e^{-j\pi\lambda z_i f_x^2}$$

For the two-dimensional case, the Fourier transform is

$$\mathcal{F}\{e^{jk(x_i^2 + y_i^2)/2z_i}\} = j\lambda z_i e^{-j\pi\lambda z_i(f_x^2 + f_y^2)} \tag{1.43}$$

Hence, the Fourier transform of the point spread function is

$$\mathcal{F}\left\{\frac{1}{j\lambda z_i} e^{jk[z_i + (x_i^2 + y_i^2)/2z_i]}\right\} = e^{jkz_i - j\pi\lambda z_i(f_x^2 + f_y^2)} \tag{1.44}$$

Example 1.3   A Fresnel diffraction pattern observed at a certain distance away from the source is identical with that obtained by a series of diffractions taking place successively from one fictitious plane after another up to the screen. This phenomenon is related to Huygens' principle. Prove this using the example of dividing the distance z_i into d_1 and d_2.

Solution   Let the source function be $E_0(x, y)$. The distance z_i to the screen is arbitrarily divided into two, d_1 and d_2, with $z_i = d_1 + d_2$ as shown in Fig. 1.3. The diffraction pattern on a fictitious screen at distance d_1 will be used as the input for another diffraction pattern on a screen at an additional distance d_2. This result will be compared with that obtained when a single diffraction pattern impinges directly from the input onto the screen at distance $z_i = d_1 + d_2$.

From Eq. (1.39), the diffraction pattern of $E_0(x, y)$ on the fictitious screen at d_1 is

$$E_1(x, y) = E_0(x, y) * f_{d_1}(x, y) \tag{1.45}$$

The diffraction pattern of $E_1(x, y)$ on the screen at an additional distance d_2 is

$$E_2(x_i, y_i) = E_0(x_i, y_i) * f_{d_1}(x_i, y_i) * f_{d_2}(x_i, y_i) \tag{1.46}$$

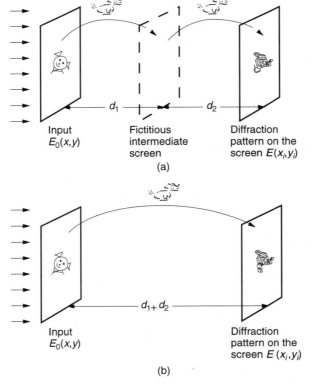

Figure 1.3   Comparison of diffraction patterns with and without a fictitious screen. (a) Two-step diffraction. (b) One-step diffraction.

The Fourier transform and inverse Fourier transform are successively performed to make use of the product rule of Fourier transforms. The result is

$$E_2(x_i, y_i) = \mathcal{F}^{-1}\{\varepsilon(f_x, f_y) \cdot F_{d_1}(f_x, f_y) \cdot F_{d_2}(f_x, f_y)\} \tag{1.47}$$

From Eq. (1.41),

$$E_2(x_i, y_i) = \mathcal{F}^{-1}\{\varepsilon(f_x, f_y)e^{jk(d_1+d_2)}e^{-j\pi\lambda(d_1+d_2)(f_x^2+f_y^2)}\} \tag{1.48}$$

where $\varepsilon(f_x, f_y)$ is the Fourier transform of $E_0(x, y)$. From Eq. (1.39), the one-step calculation of the diffraction pattern on the screen over the distance z_i is

$$E_2(x_i, y_i) = E_0(x_i, y_i) * f_{d_1+d_2}(x_i, y_i) \tag{1.49}$$

Thus, the result of the single diffraction is identical with successive diffractions. This fact conforms with Huygens' principle that a wave propagates by creating new wavefronts from the pattern of the old wavefront.   □

1.3   FOURIER TRANSFORM IN CYLINDRICAL COORDINATES

Photographic plates are rectangular, but most optical components like lenses, retarders, and apertures are cylindrically symmetric. The relationships between rectangular spatial coordinates (x, y) and spatial frequency coordinates (f_x, f_y) and cylindrical spatial coordinates (r, θ) and spatial frequency coordinates (ρ, ϕ) are

$$
\begin{array}{ll}
x = r\cos\theta & f_x = \rho\cos\phi \\[4pt]
y = r\sin\theta & f_y = \rho\sin\phi \\[4pt]
dx\,dy = r\,dr\,d\theta & df_x\,df_y = \rho\,d\rho\,d\phi
\end{array} \tag{1.50}
$$

This relationship is illustrated in Fig. 1.4. The two-dimensional Fourier transform in rectangular coordinates

$$G(f_x, f_y) = \iint_{-\infty}^{\infty} g(x, y)e^{-j2\pi(f_x x + f_y y)}\,dx\,dy \tag{1.51}$$

is converted into cylindrical coordinates using the relationships of Eq. (1.50) as

$$G(\rho, \phi) = \int_0^{\infty}\int_0^{2\pi} g(r, \theta)e^{-j2\pi\rho r\cos(\theta-\phi)}r\,dr\,d\theta \tag{1.52}$$

In order to simplify the calculation of the double integral, $g(r, \theta)$ is first separated into functions of r and θ. Since $g(r, \theta)$ is periodic with respect to θ with period 2π, it can be expanded into a Fourier series as

$$g(r, \theta) = \sum_{n=-\infty}^{\infty} a_n e^{j2\pi(n/T)\theta} \tag{1.53}$$

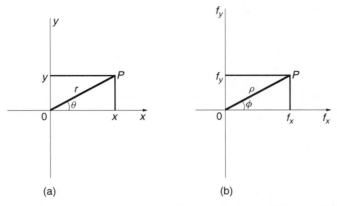

Figure 1.4 Change of coordinates from rectangular to cylindrical. (a) Space domain. (b) Spatial frequency domain.

where the coefficient a_n is

$$a_n = \frac{1}{T} \int_{-T/2}^{T/2} g(r, \theta) e^{-j2\pi(n/T)\theta} \, d\theta \tag{1.54}$$

Substituting 2π for the period T gives

$$g(r, \theta) = \sum_{n=-\infty}^{\infty} g_n(r) e^{jn\theta} \tag{1.55}$$

where

$$g_n(r) = \frac{1}{2\pi} \int_{-\pi}^{\pi} g(r, \theta) e^{-jn\theta} \, d\theta \tag{1.56}$$

Inserting Eq. (1.55) into (1.52) gives

$$G(\rho, \phi) = \sum_{n=-\infty}^{\infty} \int_{0}^{\infty} r \, dr \int_{0}^{2\pi} g_n(r) e^{jn\theta - j2\pi\rho r \cos(\theta-\phi)} \, d\theta \tag{1.57}$$

The integral with respect to θ can be expressed in terms of the Bessel function of the first kind of nth order [10] as

$$J_n(z) = \frac{1}{2\pi} \int_{\alpha}^{2\pi+\alpha} e^{j(n\beta - z\sin\beta)} \, d\beta \tag{1.58}$$

Noting that

$$\cos(\theta - \phi) = \sin(\theta - \phi + \pi/2) \tag{1.59}$$

and letting

$$\beta = \theta - \phi + \pi/2 \tag{1.60}$$

and inserting Eqs. (1.58), (1.59), and (1.60) into (1.52) finally gives

$$G(\rho, \phi) = \sum_{n=-\infty}^{\infty} (-j)^n e^{jn\phi} 2\pi \int_0^{\infty} r g_n(r) J_n(2\pi\rho r)\, dr \qquad (1.61)$$

where $g_n(r)$ is given by Eq. (1.56). Conversely, the inverse Fourier transform is given by

$$g(r, \theta) = \sum_{n=-\infty}^{\infty} (j)^n e^{jn\theta} 2\pi \int_0^{\infty} \rho G_n(\rho) J_n(2\pi\rho r)\, d\rho \qquad (1.62)$$

where

$$G_n(\rho) = \frac{1}{2\pi} \int_{-\pi}^{\pi} G(\rho, \phi) e^{-jn\phi}\, d\phi \qquad (1.63)$$

When $n \neq 0$, the Fourier transform in cylindrical coordinates is called the Fourier–Hankel transform of the nth order. When $n = 0$, it is called the Fourier–Bessel transform and is written as $B\{g(r)\} = G(\rho)$, and $B^{-1}\{G(\rho)\} = g(r)$.

For the special case where there is no θ dependence, such as a circular aperture, then

$$g(r, \theta) = g(r) \qquad (1.64)$$

and Eq. (1.56) becomes

$$\begin{aligned}
g_n(r) &= \frac{g(r)}{2\pi} \int_{-\pi}^{\pi} e^{-jn\theta}\, d\theta \\
&= \frac{g(r)}{2\pi} \left[\frac{e^{-jn\theta}}{-jn} \right]_{-\pi}^{\pi} \\
&= \begin{cases} g(r) & n = 0 \\ 0 & n \neq 0 \end{cases}
\end{aligned} \qquad (1.65)$$

Terms with nonzero n disappear from Eqs. (1.61) and (1.62), and these equations simplify to

$$G(\rho) = 2\pi \int_0^{\infty} r g(r) J_0(2\pi\rho r)\, dr \qquad (1.66)$$

$$g(r) = 2\pi \int_0^{\infty} \rho G(\rho) J_0(2\pi\rho r)\, d\rho \qquad (1.67)$$

which are, as mentioned above, the Fourier–Bessel transform $B\{g(r)\}$ and its inverse $B^{-1}\{G(\rho)\}$.

Example 1.4 Find the Fourier–Hankel transform of a circular aperture with a one-sixth section obstruction as shown in Fig. 1.5.

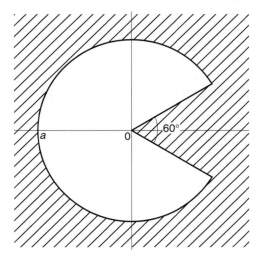

Figure 1.5   Circular aperture with obstruction.

Solution $G(\rho, \phi) = $ whole circle $-$ wedge portion. The wedge portion is expanded into a Fourier series as

$$g_n(r) = \frac{1}{2\pi} \int_{-\pi/6}^{\pi/6} g(r, \theta) e^{-jn\theta} \, d\theta \tag{1.68}$$

$$= \frac{1}{2\pi} \left[\frac{e^{-jn\theta}}{-jn} \right]_{-\pi/6}^{\pi/6}$$

$$g_n(r) = \begin{cases} \dfrac{1}{n\pi} \sin \dfrac{n\pi}{6} & n \neq 0 \\ \dfrac{1}{6} & n = 0 \end{cases} \tag{1.69}$$

$$G(\rho, \phi) = 2\pi \int_0^a r J_0(2\pi\rho r) \, dr$$

$$- 2 \sum_{\substack{n = -\infty \\ \text{except } n = 0}}^{\infty} (-j)^n \frac{e^{jn\phi}}{n} \sin\left(\frac{n\pi}{6}\right) \int_0^a r J_n(2\pi\rho r) \, dr$$

$$- \frac{\pi}{3} \int_0^a r J_0(2\pi\rho r) \, dr \tag{1.70}$$

$$G(\rho, \phi) = \frac{5\pi}{6} \frac{a}{\rho} J_1(2\pi\rho a) - 2 \sum_{\substack{n = -\infty \\ \text{except } n = 0}}^{\infty} (-j)^n \frac{e^{jn\phi}}{n} \sin\left(\frac{n\pi}{6}\right) \int_0^a r J_n(2\pi\rho r) \, dr \tag{1.71}$$

$\square$

Fourier transforms performed in either rectangular or cylindrical coordinates provide the same results, but one is usually more convenient than the other.

Now that the significance of the Fourier transform has been demonstrated, the next section describes representations of the source shape function and their Fourier transforms.

1.4 SPECIAL FUNCTIONS IN PHOTONICS AND THEIR FOURIER TRANSFORMS

1.4.1 Rectangle Function

An aperture function of unit width can be represented by the rectangle function. The rectangle function $\Pi(x)$ shown on the left in Fig. 1.6a is defined as

$$\Pi(x) = \begin{cases} 1 & |x| \leq \frac{1}{2} \\ 0 & |x| > \frac{1}{2} \end{cases} \tag{1.72}$$

The Fourier transform of the rectangle function is

$$\mathcal{F}\{\Pi(x)\} = \int_{-1/2}^{1/2} e^{-j2\pi f x} \, dx$$

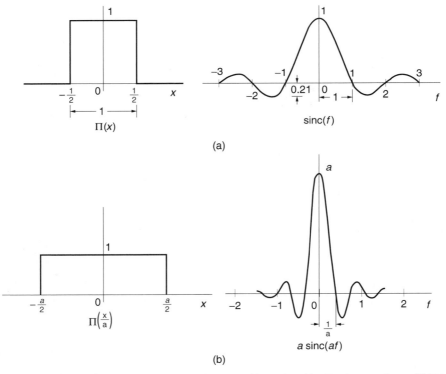

(a)

(b)

Figure 1.6 The rectangular function $\Pi(x)$ and $\Pi(x/a)$. (a) $\Pi(x)$ and its Fourier transform. (b) $\Pi(x/a)$ and its Fourier transform.

$$= \left[\frac{e^{-j2\pi fx}}{-j2\pi f} \right]_{-1/2}^{1/2}$$

$$= \frac{\sin \pi f}{\pi f} \tag{1.73}$$

The right-hand side of Eq. (1.73) is the sinc function, defined as

$$\text{sinc}(f) = \frac{\sin \pi f}{\pi f} \tag{1.74}$$

As indicated in Fig. 1.6a, the main lobe of $\text{sinc}(f)$ has a width of 2, as measured along the f axis, and attains unit height at $f = 0$. Side lobes decrease in height as $|f|$ increases. The extrema of $\text{sinc}(f)$ occur when f is near an odd multiple of $\frac{1}{2}$, and the zeros of $\text{sinc}(f)$ are located at every integer. The height of the first side lobe is approximately 0.21 that of the main lobe. Now, let us extend the definition of Eq. (1.72) to an aperture with width a. The rectangle function for such an aperture is

$$\Pi\left(\frac{x}{a}\right) = \begin{cases} 1 & |x| \le a/2 \\ 0 & |x| > a/2 \end{cases} \tag{1.75}$$

The Fourier transform is

$$\mathcal{F}\left\{\Pi\left(\frac{x}{a}\right)\right\} = \int_{-a/2}^{a/2} e^{-j2\pi fx} \, dx \tag{1.76}$$

$$= a \, \text{sinc}(af)$$

Equation (1.76) certainly can be derived directly using the similarity theorem of the Fourier transform, Eq. (1.42). As indicated in Fig. 1.6b, $\text{sinc}(af)$ also has its main lobe at $f = 0$, but its height is a. The width of the main lobe is now $2/a$. Side lobes with decaying amplitudes appear with extrema near odd multiples of $1/2a$ and zeros at integral multiples of $1/a$. The ratio of the height of the first side lobe to the main lobe still remains at 0.21.

1.4.2 Triangle Function

The triangle function is defined as

$$\Lambda(x) = \begin{cases} 1 - |x| & |x| \le 1 \\ 0 & |x| > 1 \end{cases} \tag{1.77}$$

and is shown on the left in Fig. 1.7. Unlike $\Pi(x)$, the width of the base is 2. The fact that the triangle function can be generated from the convolution of two rectangular functions makes the calculation of the Fourier transform simple:

$$\Lambda(x) = \Pi(x) * \Pi(x) \tag{1.78}$$

and

$$\mathcal{F}\{\Lambda(x)\} = \text{sinc}^2(f) \tag{1.79}$$

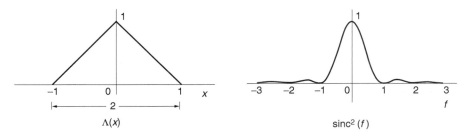

Figure 1.7   The triangle function $\Lambda(x)$ and its Fourier transform.

The triangle function with base width $2a$ can be expressed as the convolution of two rectangular functions as

$$\Lambda\left(\frac{x}{a}\right) = \frac{1}{a}\Pi\left(\frac{x}{a}\right) * \Pi\left(\frac{x}{a}\right) \tag{1.80}$$

Since the convolution of $\Pi(x/a)$ at $f = 0$ is a, the factor $1/a$ is necessary in Eq. (1.80) to make $\Lambda(x/a)$ unity at $f = 0$. The Fourier transform of Eq. (1.80) is

$$\mathcal{F}\left\{\Lambda\left(\frac{x}{a}\right)\right\} = a\,\text{sinc}^2(af) \tag{1.81}$$

The graph of $\Lambda(x)$ with its Fourier transform is shown in Fig. 1.7. Compared to $\text{sinc}(f)$, the side lobes of $\text{sinc}^2(f)$ are significantly lower in height. An input aperture having a $\Lambda(x_0)$ distribution is used to reduce the side lobes in its diffraction pattern. This technique is called *apodizing* to reduce the side lobes. Apodal means a creature without legs, as, for example, eels or whales. Apodization of a lens is performed by darkening the lens toward the edge of the lens. Apodization of a radiation pattern from an antenna array is achieved by reducing the element antenna current toward the edge of the array [9].

The intensity pattern $I(x_i, y_i)$ of the diffracted field is expressed as

$$I(x_i, y_i) = E(x_i, y_i)E^*(x_i, y_i) \tag{1.82}$$

The field intensity patterns of the diffraction from a normal and an apodized slit are obtained by inserting Eqs. (1.76) and (1.79) into Eqs. (1.36) and (1.82) as

$$I_s(x_i, y_i) = \left(\frac{a}{\lambda z_i}\right)^2 \text{sinc}^2\left(\frac{a}{\lambda z_i}x_i\right)\delta^2(y_i) \tag{1.83}$$

$$I_{as}(x_i, y_i) = \left(\frac{a}{\lambda z_i}\right)^2 \text{sinc}^4\left(\frac{a}{\lambda z_i}x_i\right)\delta^2(y_i) \tag{1.84}$$

where δ is the delta function (see Section 1.4.5). The field intensity, which is EE^*, is not the same as the power intensity, which is $1/\eta|E|^2$ (see Section 2.3.1).

Examples of a normal and an apodized slit are shown in Fig. 1.8a, and the corresponding intensity patterns of the diffraction are compared in Fig. 1.8b. The reduction of the side lobe levels by the apodization is clearly demonstrated.

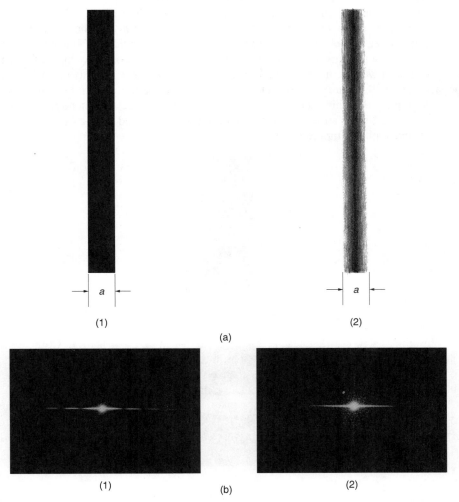

a

(1)   (2)

(a)

(1)   (b)   (2)

Figure 1.8. Comparison of (1) a normal slit and (2) an apodized slit. (a) Geometry of the apertures. The dark portions are the openings. (b) Diffraction patterns.

Example 1.5 Find the Fraunhofer diffraction pattern of a rectangular aperture with dimensions $a \times l$.

Solution The input function is given by

$$E(x_0, y_0) = \Pi\left(\frac{x_0}{a}\right) \Pi\left(\frac{y_0}{l}\right) \tag{1.85}$$

The field $E(x_i, y_i)$ in the $z = z_i$ plane is found from Eqs. (1.36) and (1.85):

$$E(x_i, y_i) = \frac{al}{j\lambda z_i} e^{jk[z_i + (x_i^2 + y_i^2)/2z_i]} \operatorname{sinc}\left(a\frac{x_i}{\lambda z_i}\right) \operatorname{sinc}\left(l\frac{y_i}{\lambda z_i}\right) \tag{1.86}$$

It should be noted that x_0 and y_0 are independent variables. Equation (1.86) is not the convolution of the two sinc functions but the product of the two.

The aperture and a photograph of its diffraction pattern are shown in Fig. 1.9. An important feature of the diffraction pattern is that the width of the main lobe is narrowed with widening of the aperture. To remember the concept, think of a water hose. The narrower the nozzle is pinched, the wider the water is sprayed. The smaller the structure of the source (or object), the wider the radiation pattern becomes. X-ray crystallography uses this fact very wisely to analyze molecular structure. The X-ray pattern scattered from an angstrom-sized structure is enlarged enough to be recorded by an ordinary photographic plate. □

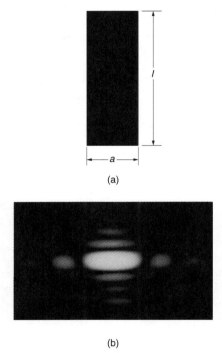

(a)

(b)

Figure 1.9 A rectangular aperture and its far-field diffraction pattern. (a) Geometry. The dark portion is the opening. (b) Diffraction pattern.

A firefighter knows that a narrower nozzle expands the beam. In the same way, a narrower source produces a larger diffraction pattern.

1.4.3 Sign and Step Functions

The sign function, sgn x, is positive unity for positive values of x and negative unity for negative values of x:

$$\text{sgn}(x) = \begin{cases} 1 & x > 0 \\ 0 & x = 0 \\ -1 & x < 0 \end{cases} \tag{1.87}$$

It is used to express a phase reversal at $x = 0$. The Fourier transform of Eq. (1.87) is

$$\mathcal{F}\{\text{sgn}(x)\} = \lim_{\alpha \to 0} \left[\int_{-\infty}^{0} (-1)e^{-j2\pi fx + \alpha x}\, dx + \int_{0}^{\infty} e^{-j2\pi fx - \alpha x}\, dx \right] \tag{1.88}$$

The presence of α is necessary to perform the integral. After integration, α is reduced to zero. The result is

$$\mathcal{F}\{\text{sgn}(x)\} = \frac{1}{j\pi f} \tag{1.89}$$

The step function, $H(x)$, is immediately generated from the sign function:

$$H(x) = \tfrac{1}{2}[1 + \text{sgn}(x)] \tag{1.90}$$

The step function is used to mask one-half of a plane. Its Fourier transform is

$$\mathcal{F}\{H(x)\} = \frac{1}{2}\left(\delta(f) + \frac{1}{j\pi f} \right) \tag{1.91}$$

where δ represents the delta function and is explained in Section 1.4.5.

Figure 1.10 shows the geometry of the step function and its diffraction pattern. It is worth noting that even though the aperture does not have symmetry with respect to $x_0 = 0$, the intensity pattern of the diffraction has a symmetry with respect to the edge at $x_0 = 0$. The streak pattern is always perpendicular to the direction of the edge, and its intensity decreases monotonically with distance away from the edge.

1.4.4 Circle Function

In order to describe a circular aperture, the circle function shown in Fig. 1.11a is defined as

$$\text{circ}(r) = \begin{cases} 1 & r \leq 1 \\ 0 & r > 1 \end{cases} \tag{1.92}$$

The Fourier transform of the circle function is found from Eq. (1.66):

$$B\{\text{circ}(r)\} = 2\pi \int_{0}^{1} rJ_0(2\pi\rho r)\, dr \tag{1.93}$$

The Bessel function has the property

$$x^n J_{n-1}(x) = \frac{d}{dx}[x^n J_n(x)] \tag{1.94}$$

(a)

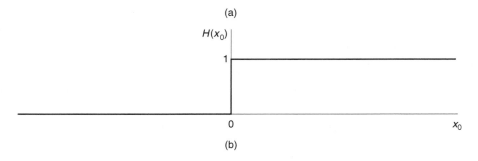

(b)

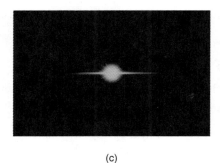

(c)

Figure 1.10 Diffraction from a semi-infinite screen. (a) Semi-infinite screen. (b) Step function. (c) Diffraction pattern.

Substituting $n = 1$ in Eq. (1.94) and integrating both sides of the equation gives

$$\int x J_0(x)\,dx = x J_1(x) \tag{1.95}$$

Using $x = 2\pi\rho r$, Eq. (1.93) becomes

$$B\{\operatorname{circ}(r)\} = 2\pi \frac{1}{(2\pi\rho)^2} \int_0^{2\pi\rho} x J_0(x)\,dx$$

$$= \frac{2\pi}{(2\pi\rho)^2}[x J_1(x)]_0^{2\pi\rho}$$

$$B\{\operatorname{circ}(r)\} = \frac{1}{\rho} J_1(2\pi\rho) \tag{1.96}$$

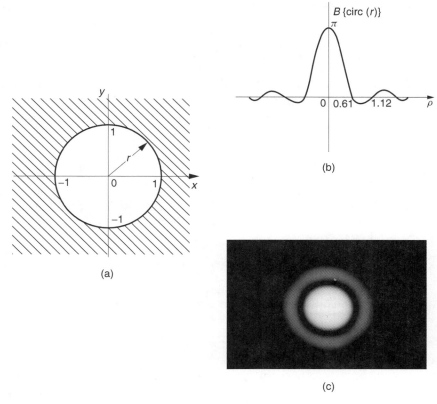

Figure 1.11 The circle function. (a) Geometry. (b) Fourier transform. (c) Diffraction pattern.

The graph of Eq. (1.96) is shown in Fig. 1.11b. The similarity theorem for the Fourier–Bessel transform does not follow Eq. (1.42); that is,

$$B\{g(ar)\} = \frac{1}{a^2}G\left(\frac{\rho}{a}\right) \tag{1.97}$$

and

$$B\left\{\text{circ}\left(\frac{r}{a}\right)\right\} = \frac{a}{\rho}J_1(2\pi\rho a) \tag{1.98}$$

The circle function and its Fourier transform are shown in Figs. 1.11a and 1.11b, respectively. The intensity pattern of the diffraction from the circular aperture with radius a is

$$I(r_i) = \left(\frac{a^2}{\lambda z_i}\right)^2 \left(\frac{J_1(2\pi a\rho)}{a\rho}\right)^2 \tag{1.99}$$

with

$$\rho = \frac{r_i}{\lambda z_i}$$

which can be rewritten as

$$I(r_i) = \left(\frac{ka^2}{z_i}\right)^2 \left(\frac{J_1(kar_i/z_i)}{kar_i/z_i}\right)^2 \tag{1.100}$$

where r_i is the radial coordinate in the plane of the diffraction pattern. This diffraction pattern was first derived by Sir George Biddell Airy and is referred to as the Airy pattern. The diffraction photograph is shown in Fig. 1.11c.

1.4.5 Delta Function

The delta function $\delta(x)$ is conveniently made to represent a point source [10]. Its amplitude is confined within a minute range of $x = \pm\varepsilon$, while it is zero outside this range, as shown in Fig. 1.12a. The amplitude grows to infinity as ε shrinks to zero, but in such a way that the area enclosed by the curve is always unity. In the limit $\varepsilon \to 0$, $\delta(0) \to \infty$, while satisfying

$$\int_{-\varepsilon}^{\varepsilon} \delta(x)\,dx = 1 \tag{1.101}$$

The delta function is most often used in an integral form:

$$\int_{-\infty}^{\infty} f(x)\delta(x-a)\,dx = f(a) \tag{1.102}$$

The integrand is shown in Fig. 1.12b. The region where the product is nonzero is only at $x = a \pm \varepsilon$. Consequently, $f(x)$ in this region can be approximated by the constant

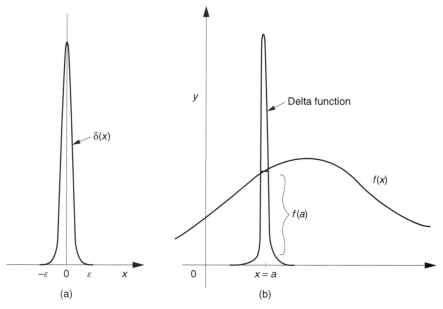

Figure 1.12 Diagram showing $\int f(x)\delta(x-a)\,dx = f(a)$. (a) Delta function. (b) Integral including the delta function.

value $f(a)$, and the constant $f(a)$ can be brought outside the integral. The integral of the delta function is unity and Eq. (1.102) holds. Equation (1.102) is used to sample (or sift) the value of $f(x)$ at $x = a$. This is called the sifting property of the delta function.

The Fourier transform of the delta function is

$$\mathcal{F}\{\delta(x)\} = \int_{-\infty}^{\infty} e^{-j2\pi fx}\delta(x)\,dx \qquad (1.103)$$

Since the delta function samples the value of $e^{-j2\pi fx}$ at $x = 0$, Eq. (1.103) becomes

$$\mathcal{F}\{\delta(x)\} = 1 \qquad (1.104)$$

Next, the inverse Fourier transform of $\delta(f)$ is considered. The inverse Fourier transform uses $e^{j2\pi fx}$ instead of $e^{-j2\pi fx}$ on the right-hand side of Eq. (1.103), and therefore

$$\mathcal{F}^{-1}\{\delta(f)\} = 1$$

Taking the Fourier transform of both sides gives

$$\mathcal{F}\{1\} = \int_{-\infty}^{\infty} e^{-j2\pi fx}\,dx = \delta(f) \qquad (1.105)$$

Next, the value of $\delta(bx)$ will be expressed in terms of $\delta(x)$:

$$\int_{-\varepsilon}^{\varepsilon} f(x)\delta(bx)\,dx = \frac{1}{b}\int_{-b\varepsilon}^{b\varepsilon} f\left(\frac{y}{b}\right)\delta(y)\,dy \qquad (1.106)$$

Thus, Eq. (1.106) becomes

$$\int_{-\varepsilon}^{\varepsilon} f(x)\delta(bx)\,dx = \frac{1}{b}f(0) \qquad (1.107)$$

Equation (1.102) with $a = 0$ gives

$$\int_{-\varepsilon}^{\varepsilon} f(x)\delta(x)\,dx = f(0) \qquad (1.108)$$

A comparison of Eqs. (1.107) and (1.108) leads to

$$\delta(bx) = \frac{\delta(x)}{|b|} \qquad (1.109)$$

The absolute value is placed in the denominator of Eq. (1.109) because the result is the same for $-b$ and $+b$. Another property of the delta function is that its convolution with a function is the function itself; namely,

$$f(x) * \delta(x - a) = f(x - a) \qquad (1.110)$$

because

$$\int_{-\infty}^{\infty} f(\eta)\delta(x - \eta - a)\,d\eta = f(x - a) \qquad (1.111)$$

1.4.6 Shah Function (Impulse Train Function)

An array of equally spaced delta functions, such as shown in Fig. 1.13a, is called the shah function (or comb function), and it is denoted by $\text{Ш}(x)$. Mathematically, the shah function is represented as

$$\text{Ш}(x) = \sum_{n=-\infty}^{\infty} \delta(x - n) \qquad (1.112)$$

Two major applications of $\text{Ш}(x)$ are the following:

1. The generation of a sampled function. The sampled function $g_s(x)$ is an array of delta functions whose envelope is proportional to $g(x)$, such as shown in

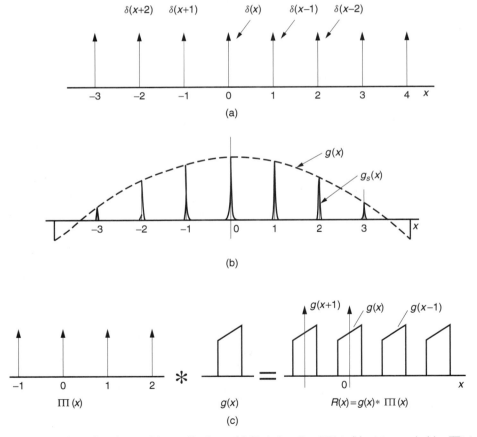

Figure 1.13 Shah function and its applications. (a) Shah function $\text{Ш}(x)$. (b) $g(x)$ sampled by $\text{Ш}(x)$. (c) Step and repeat function $R(x)$.

Fig. 1.13b. The sampled function can be generated by simply multiplying $g(x)$ by the shah function:

$$g_s(x) = g(x) \text{Ш}(x) \tag{1.113}$$

A typical use for $g_s(x)$ is in optical signal processing, as, for example, processing a newspaper photograph that is made of closely sampled points.

2. The generation of a step and repeat function. The step and repeat function, such as that shown in Fig. 1.13c, can be generated by convolving $g(x)$ with the shah function:

$$R(x) = g(x) * \text{Ш}(x) \tag{1.114}$$

From the definition of convolution, $R(x)$ is

$$R(x) = \int_{-\infty}^{\infty} g(\tau)\text{Ш}(x - \tau)\, d\tau$$

$$= \int_{-\infty}^{\infty} g(\tau) \sum_{n=-\infty}^{\infty} \delta(x - \tau - n)\, d\tau \tag{1.115}$$

Using Eq. (1.102), Eq. (1.115) becomes

$$R(x) = \sum_{n=-\infty}^{\infty} g(x - n) \tag{1.116}$$

which steps and repeats $g(x)$ at a unit interval.

The shah function that steps and repeats at an interval other than the unit interval warrants special attention. The step and repeat function at an interval of a is expressed as

$$g(x) * \text{Ш}\left(\frac{x}{a}\right) = \int_{-\infty}^{\infty} g(\tau) \sum_{-\infty}^{\infty} \delta\left(\frac{x - \tau - an}{a}\right) d\tau \tag{1.117}$$

From Eq. (1.109), Eq. (1.117) becomes

$$g(x) * \text{Ш}\left(\frac{x}{a}\right) = a \sum_{n=-\infty}^{\infty} g(x - an) \tag{1.118}$$

Thus, the step and repeat function at an interval of a is

$$R(x) = \frac{1}{a} g(x) * \text{Ш}\left(\frac{x}{a}\right) \tag{1.119}$$

The existence of the factor $1/a$ should be noted.

Next, the Fourier transform of the shah function will be derived. Because the shah function is a periodic function, it can be expanded into a Fourier series with a period of unity as

$$\text{III}(x) = \sum_{n=-\infty}^{\infty} a_n e^{j2\pi(n/T)x} \tag{1.120}$$

where

$$a_n = \frac{1}{T} \int_{-T/2}^{T/2} \delta(x) e^{-j2\pi(n/T)x} \, dx = 1$$

and where T is the period of the delta functions and is unity. Thus, Eq. (1.120) becomes

$$\text{III}(x) = \sum_{n=-\infty}^{\infty} e^{j2\pi nx} \tag{1.121}$$

Whenever x is an integer, $\text{III}(x)$ becomes infinite, so that Eq. (1.121) constitutes an array of delta functions spaced by unity. Equation (1.121) is an alternate expression for the shah function. Thus, using Eqs. (1.35), (1.105), and (1.121), the Fourier transform of the shah function becomes

$$\mathcal{F}\{\text{III}(x)\} = \sum_{n=-\infty}^{\infty} \delta(f-n) \equiv \text{III}(f) \tag{1.122a}$$

Similarly,

$$\mathcal{F}\{\text{III}(x/a)\} = a\text{III}(af) \tag{1.122b}$$

$\text{III}(x)$ is a very special function in that the Fourier transform is the same as the function itself.

1.4.7   Diffraction from an Infinite Array of Similar Apertures with Regular Spacing

Making use of the shah function, the diffraction pattern will be calculated for a one-dimensional array of slits such as shown in Fig. 1.14a(1). The slits are identical rectangle functions with width a, and the slits are equally spaced with period b, thereby forming a step and repeat function. The transmittance of this step and repeat function is expressed using Eqs. (1.75) and (1.119) as

$$E(x_0, y_0) = \frac{1}{b} \Pi\left(\frac{x_0}{a}\right) * \text{III}\left(\frac{x_0}{b}\right) \tag{1.123}$$

The diffraction pattern is given by Eq. (1.34) as

$$E(x_i, y_i) = a \, \text{sinc}\left(\frac{a}{\lambda z_i} x_i\right) \cdot \text{III}\left(\frac{b}{\lambda z_i} x_i\right) \delta\left(\frac{y_i}{\lambda z_i}\right) \tag{1.124}$$

where the quadratic phase factor will be suppressed in this section.

A photograph of the diffraction pattern is shown in Fig. 1.14b(1). The pattern consists of an array of bright spikes with spacing $\lambda z_i/b$. For this particular array, the ratio between the slit width a and the period b is $b/a = 5$. This means that the fifth spike overlaps with the first null of $\text{sinc}(af_x)$ and the intensity of the fifth spike is faint.

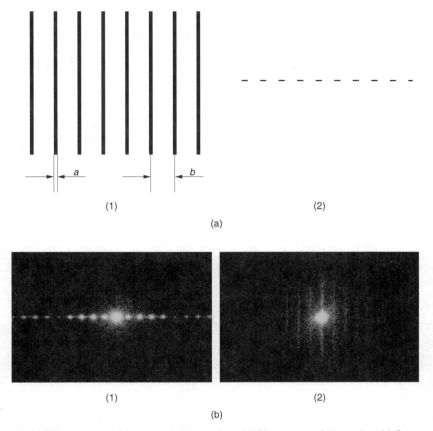

Figure 1.14 Comparison of (1) an array of long slits and (2) an array of short slits. (a) Geometry of the apertures. The dark portions are the openings. (b) Diffraction patterns.

Figure 1.14b(2) was included in order to demonstrate how the dimension in f_y of the diffraction pattern stretches as the height of the slits y_0 is shortened, while keeping other parameters unchanged.

Next, a two-dimensional array will be formed out of rectangular apertures with dimensions $a \times l$, with period b in the x direction and m in the y direction, as shown in Fig. 1.15a(2).

The aperture is represented by

$$E(x_0, y_0) = \frac{1}{bm} \left[\Pi \left(\frac{x_0}{a} \right) * \amalg \left(\frac{x_0}{b} \right) \right] \left[\Pi \left(\frac{y_0}{l} \right) * \amalg \left(\frac{y_0}{m} \right) \right] \tag{1.125}$$

The diffraction pattern is given by

$$E(x_i, y_i) = \frac{al}{j\lambda z_i} \text{ sinc} \left(\frac{a}{\lambda z_i} x_i \right) \text{ sinc} \left(\frac{l}{\lambda z_i} y_i \right) \amalg \left(\frac{b}{\lambda z_i} x_i \right) \amalg \left(\frac{m}{\lambda z_i} y_i \right) \tag{1.126}$$

The photograph of the diffraction pattern in Fig. 1.15b(2) shows a grid of bright spikes. The spacing of the spikes is $\lambda z_i / b$ in the x_i direction and $\lambda z_i / m$ in the y_i direction. Note that the brightness of the spikes is not uniform and the overall distribution of the bright spikes is similar to that of the single rectanglar aperture shown in Fig. 1.15b(1).

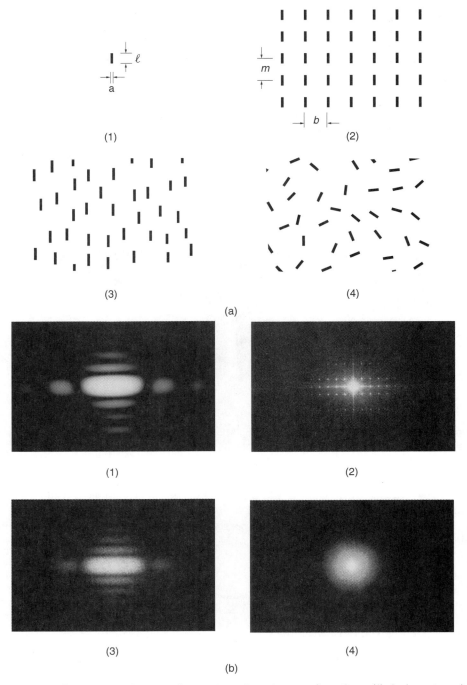

Figure 1.15 Comparison of rectangular apertures in various configurations: (1) single rectangular aperture, (2) rectangular apertures in a grid pattern, (3) rectangular apertures with random position in the vertical orientation, and (4) randomly arranged rectangular apertures. (a) Geometry. (b) Diffraction patterns.

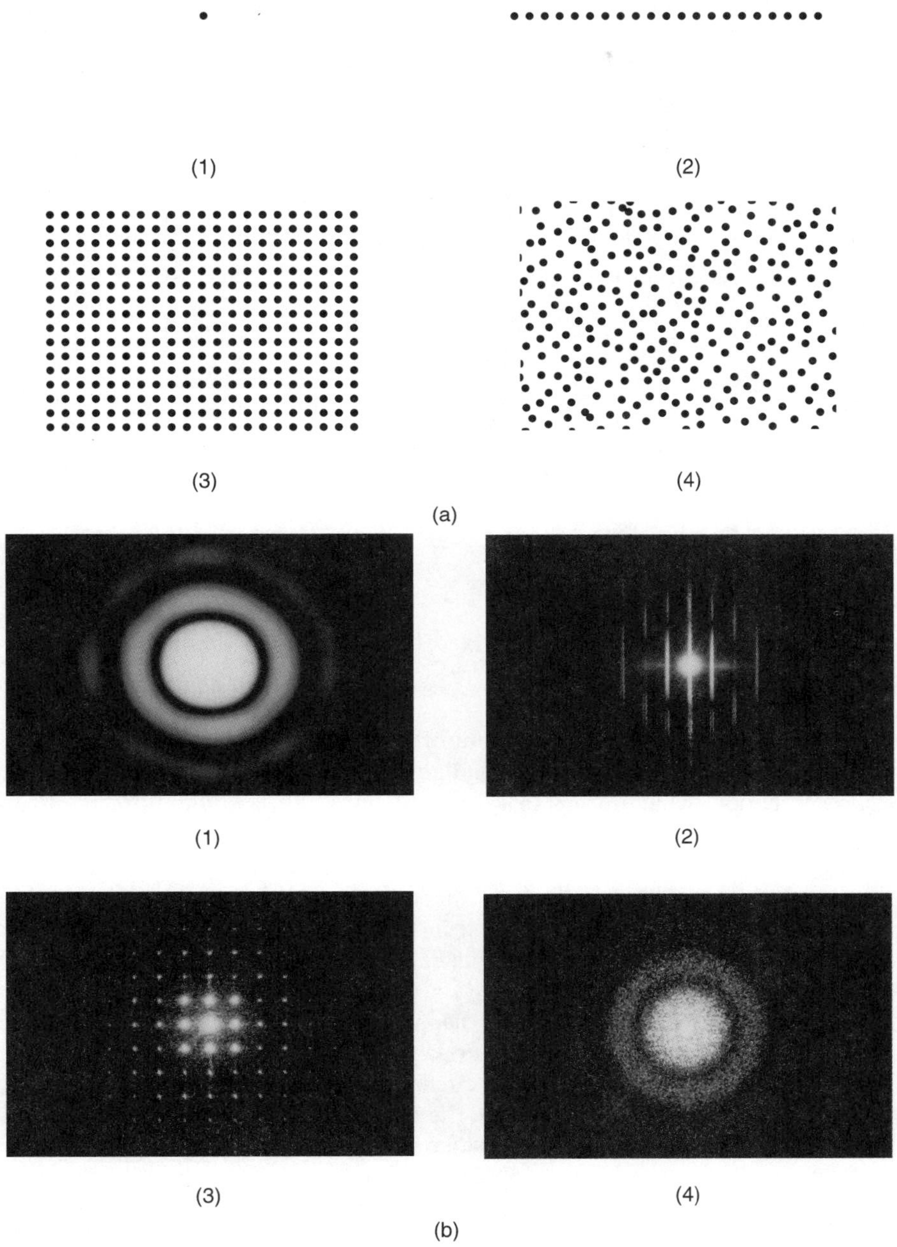

(1)　　　　　　　　(2)

(3)　　　　　　　　(4)

(a)

(1)　　　　　　　　(2)

(3)　　　　　　　　(4)

(b)

Figure 1.16 Circular apertures in various configurations: (1) single circular aperture, (2) circular apertures in a linear array, (3) circular apertures in a grid pattern, and (4) randomly arranged circular apertures. (a) Geometry. (b) Diffraction patterns.

1.4.8 Diffraction from an Infinite Array of Similar Apertures with Irregular Spacing

In the configuration shown in Fig. 1.15a(3), the elements of the array are randomly arranged, but without any rotation of the individual elements. For simplicity, consider a one-dimensional array with random spacing. If each rectangular aperture is translated in the x_0 direction by a random distance b_j from the origin, then the aperture distribution is represented by

$$E(x_0, y_0) = \sum_{j=0}^{N-1} \Pi \left(\frac{x_0 - b_j}{a} \right) \Pi \left(\frac{y_0}{l} \right) \tag{1.127}$$

Using the shift theorem of the Fourier transform

$$\mathcal{F}\{g(x - a)\} = e^{-j2\pi f a} G(f) \tag{1.128}$$

and

$$I(x_i, y_i) = E(x_i, y_i) \times E^*(x_i, y_i)$$

the intensity distribution of the diffraction pattern is

$$I(x_i, y_i) = E(x_i, y_i)(1 + e^{-j2\pi b_1 f_x} + e^{-j2\pi b_2 f_x} + \cdots)$$

$$\times E^*(x_i, y_i)(1 + e^{j2\pi b_1 f_x} + e^{j2\pi b_2 f_x} + \cdots)$$

$$= |E(x_i, y_i)|^2 \left(N + 2 \sum_{j=1}^{N-1} \sum_{k=1}^{N-1} \cos 2f_x(b_j - b_k) \right) \tag{1.129}$$

where $|E(x_i, y_i)|^2$ is the diffraction pattern of a single rectangular aperture.

Since b_j and b_k are random, the second term is a superposition of cosine functions of a random period, which means that the resultant is zero and the diffraction pattern becomes $N|E(x_i, y_i)|^2$. The intensity distribution of the diffraction is the same as that of the single rectangular aperture but with N times the intensity, as shown in Fig. 1.15b(3).

Finally, the rectangular apertures are randomized with respect to rotation as well as translation. The diffraction photograph in Fig. 1.15b(4) looks like one that would have been obtained by rotating the single rectangular aperture diffraction pattern in Fig. 1.15b(1) about its center.

Figure 1.16a and 1.16b show what happens when the rectangular apertures in Fig. 1.15a are replaced by circular apertures.

The diffraction pattern of the single circular aperture in Fig.1.16a(1) is shown in Fig. 1.16b(1). When this circular aperture is arranged in a one-dimensional array, as shown in Fig. 1.16a(2), the diffraction pattern is made up of an array of vertical lines (shah function) as shown in Fig. 1.16b(2). The brightness of the lines is not uniform, but the overall brightness distribution resembles the diffraction pattern of the single circular aperture shown in Fig. 1.16a(1).

Next, the circular apertures are arranged in grid form with period b in both the x_0 and y_0 directions as indicated in Fig. 1.16a(3). A photograph of the diffraction pattern is shown in Fig. 1.16b(3). The grid pattern of the diffraction looks like the pattern that would be obtained by the product of the pattern of an array in the x_0 direction and the pattern of an array in the y_0 direction.

As in the case of the rectangular element apertures shown in Fig. 1.15b(2), the overall pattern of the brightness of the spikes in Fig. 1.16b(3) has a similar distribution to the diffraction pattern of a circular element aperture.

In Fig. 1.16a(4) the circular apertures are arranged in a random manner. As with the rectangular aperture, the photograph of the diffraction pattern in Fig. 1.16b(4) resembles that of a single circular aperture but with N times the intensity.

The speckle patterns are due to the finite number of element apertures and decrease with an increase in the number of elements.

1.4.9   Diffraction from a Finite Array

So far the dimensions of the array have been assumed to be infinite. In this section, the effect of the finiteness of the array will be explained. An example of a two-dimensional finite size rectangular aperture array is shown in Fig. 1.17a. The element apertures have a size of $a \times l$ and are spaced b and m apart in the x_0 and y_0 directions, respectively.

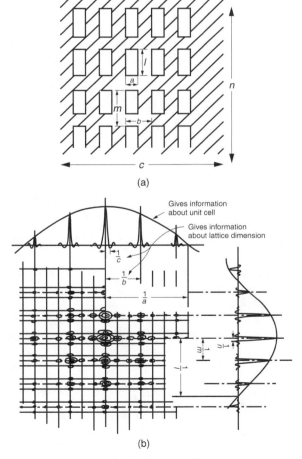

Figure 1.17   Diffraction from an array of finite size. (a) Two-dimensional array of windows with external dimensions $c \times n$. (b) Diffraction pattern of (a).

The extent of the array is limited to $c \times n$ in the x_0 and y_0 directions. The expression for the transmittance is

$$E(x_0, y_0) = \frac{1}{bm} \left[\Pi\left(\frac{x_0}{a}\right) * \text{Ш}\left(\frac{x_0}{b}\right) \right] \Pi\left(\frac{x_0}{c}\right)$$
$$\times \left[\Pi\left(\frac{y_0}{l}\right) * \text{Ш}\left(\frac{y_0}{m}\right) \right] \Pi\left(\frac{y_0}{n}\right) \tag{1.130}$$

The step and repeat functions in the square brackets are truncated by the rectangle functions $\Pi(x_0/c)$ and $\Pi(y_0/n)$.

The Fourier transform of the transmittance pattern is obtained from Eqs. (1.76) and (1.122b):

$$\mathcal{F}\{E_0(x_0, y_0)\} = acln[\ \text{sinc}(a f_x)\text{Ш}(b f_x)] * \text{sinc}(c f_x)$$
$$\times [\ \text{sinc}(l f_y)\text{Ш}(m f_y)] * \text{sinc}(n f_y) \tag{1.131}$$

The order of the dimensions in the transmittance pattern is

$$a < b < c$$

where a is the width of the window, b is the spacing between windows, and c is the overall dimension. In the diffraction pattern in Fig. 1.17b, the order of the dimensions is inverted and reversed; namely,

$$\frac{1}{c} < \frac{1}{b} < \frac{1}{a} \tag{1.132}$$

where $1/c$ is the size of an individual spike, $1/b$ is the spacing between spots, and $1/a$ is the overall size of the diffraction pattern. Thus, the external size of the array controls the size of the individual spike in the diffraction pattern.

Figure 1.18 illustrates how the external shape of the array controls the shape of the individual spikes in the diffraction pattern. In Fig. 1.18a, the element apertures are circular, and the grids are bordered by four different boundaries — circular, rectangular, rectangular tilted at $45°$, and triangular. The corresponding diffraction patterns are shown in Fig. 1.18b.

From Eq. (1.131), the pattern of an individual spike is proportional to $[\text{sinc}(c f_x)$ $\text{sinc}(n f_y)]^2$. As the shape of the outer boundary is rotated, the shape of the individual spike rotates as shown in Figs. 1.18b(2) and 1.18b(3).

Finally, Fig. 1.18b(4) demonstrates that when the outer boundary is an equilateral triangle, the shape of each spike becomes the diffraction pattern of a single triangular aperture.

For large c, Eq. (1.131) can be approximated as

$$\mathcal{F}\{E_0(x_0, y_0)\} = acln\ \text{sinc}(a f_x)[\text{Ш}(b f_x) * \text{sinc}(c f_x)]$$
$$\times \ \text{sinc}(l f_y)[\text{Ш}(m f_y) * \text{sinc}(n f_y)] \tag{1.133}$$

Now, Eq. (1.133) can be separated into two factors of different natures. The factors outside the square brackets are determined solely by the shape of each cell and are called the *element pattern*. The factors inside the square brackets are determined solely

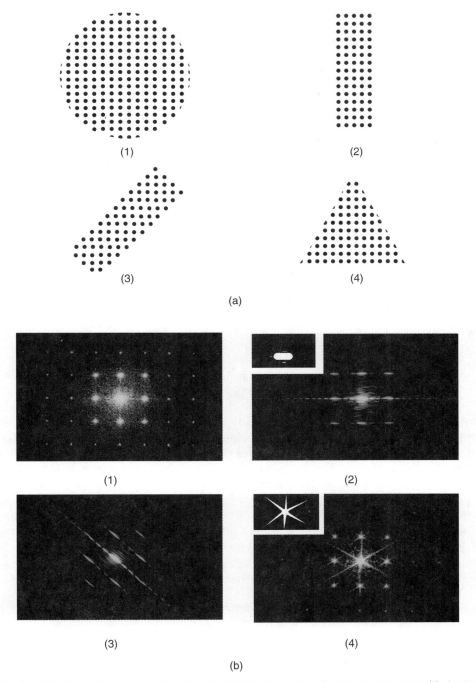

Figure 1.18 The effect of external border shapes on a grid of circular apertures: (1) circular border, (2) rectangular border, (3) tilted rectangular border, and (4) triangular border. (a) Geometry. (b) Diffraction patterns. Each inset is the diffraction pattern of a single aperture with the same shape as the border.

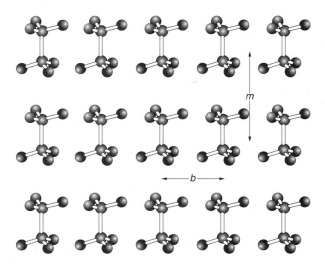

Figure 1.19   Two-dimensional crystal array.

by the spacing and overall dimensions of the lattice, are independent of the shape of each cell, and are called the *array pattern*. The diffraction pattern is the product of these two patterns. The array pattern, which is periodic, provides the lattice dimension. The element pattern, which is the envelope, provides the shape of the cell.

$$(\text{Pattern}) = (\text{Element pattern}) \times (\text{array pattern}) \qquad (1.134)$$

The two-dimensional crystal pattern such as shown in Fig. 1.19 can be obtained by simply replacing $\text{sinc}(a f_x)\, \text{sinc}(l f_y)$ by $G(f_x, f_y) = \mathcal{F}\{g(x_0, y_0)\}$. The amplitude envelope function $G(f_x, f_y)$ contains information about the structure of the unit cell. The structure of the unit cell $g(x_0, y_0)$ can be derived by taking the inverse Fourier transform of the envelope function. This is precisely the principle of X-ray crystallography.

In X-ray crystallography the darkness of each spot of the diffraction pattern in an X-ray photograph is measured by a microdensitometer to obtain the envelope of the intensity distribution $|G(f_x, f_y)|^2$ of the diffraction pattern. This method, however, measures the intensity pattern, and not the amplitude pattern of $G(f_x, f_y)$. The inverse Fourier transform of the intensity pattern gives

$$\mathcal{F}^{-1}|G(f_x, f_y)|^2 = g(x_0, y_0) * g(-x_0, -y_0)$$

which is called the Harker pattern. From the Harker pattern, $g(x_0, y_0)$ is resolved using additional information derived from the physical chemistry of the molecule.

1.5   THE CONVEX LENS AND ITS FUNCTIONS

Geometrical optics is most often applied to find the location and size of the image formed by a lens. Geometrical optics, however, fails to provide information about the wavelength and polarization dependences of the field distribution, and the image resolution for a given lens size. On the other hand, Fourier optics describes the wave nature of optics, corrects these failures, and hence will be explained here.

1.5.1 Phase Distribution After a Plano-Convex Lens

Figure 1.20 shows a thin plano-convex lens made by slicing a small section of a sphere by the plane A–A'. The lens medium is glass and the surrounding medium is air. The phase distribution across the tangent plane B–B' will be calculated. Plane B–B' is parallel to plane A–A'. The incident light is a plane wave whose propagation direction is normal to the plane A–A'. The incident wave is represented as an array of parallel rays, as shown on the left side of Fig. 1.20. The ray passing through the fat lens center suffers the longest phase delay, while that passing through the thin rim undergoes the shortest phase delay. The exact distribution $\phi(x, a)$ will be calculated in the coordinates whose origin coincides with the center of the sphere. A light ray at an arbitrary height x goes through both glass and air to reach plane B–B' from plane A–A'. The total phase delay $\phi(x, a)$ is

$$\phi(x, a) = k[n(\sqrt{a^2 - x^2} - b) + a - \sqrt{a^2 - x^2}] \tag{1.135}$$

where the y component is suppressed. With $(x/a)^2 \ll 1$ and using the binomial expansion, $\phi(x, a)$ is approximated as

$$\phi(x, a) = \phi_0 - k\frac{x^2}{2f_0}$$

where

$$\phi_0 = kn(a - b) \tag{1.136}$$

and

$$f_0 = \frac{a}{n - 1} \tag{1.137}$$

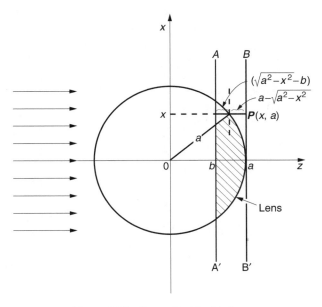

Figure 1.20 Geometry of a thin lens.

Thus, the emergent field from the plane $B-B'$ is

$$E(x) = Ae^{-jkx^2/2f_0} \tag{1.138}$$

The phase ϕ_0 is a constant that causes the same phase delay everywhere. It has no physical significance and is suppressed. In the two-dimensional case of a spherical lens, the emergent field $E(x, y)$ is

$$E(x, y) = Ae^{-jk(x^2+y^2)/2f_0} \tag{1.139}$$

where f_0 is the focal length of the lens, as will be seen shortly. The lens generates a quadratic phase distribution with a negative sign. This negative sign plays an important role in the convex lens.

1.5.2 Collimating Property of a Convex Lens

The collimating property of a lens is one of the simplest Fourier optics examples and is a good starting point for more detailed lens analysis. A delta function source is placed at F in front of a convex lens at a distance f_0 as shown in Fig. 1.21. The field incident on the lens is

$$E(x, y) = \frac{1}{j\lambda f_0} e^{jk[f_0+(x^2+y^2)/2f_0]} \tag{1.140}$$

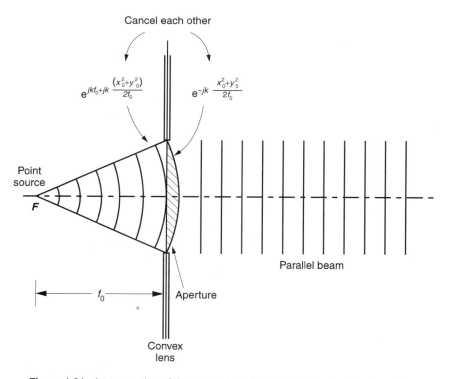

Figure 1.21 Interpretation of the generation of a parallel beam by Fourier optics.

where the coordinates (x, y) are in the plane of the lens. When the field passes through the lens it experiences a phase delay, which has previously been shown in Eq. (1.139) to be $\Delta\phi = -k[(x^2 + y^2)/2f_0]$. Hence, the resultant Fresnel diffraction of this field at a distant $z = z_i$ plane is obtained from the Fresnel–Kirchhoff diffraction formula (Eq. (1.38))

$$E(x_i, y_i) = \frac{1}{j\lambda z_i} e^{jk[z_i + (x_i^2 + y_i^2)/2z_i]}$$

$$\times \mathcal{F}\left\{ \underbrace{\frac{1}{j\lambda f_0} e^{jk[f_0 + (x^2 + y^2)/2f_0]}}_{\text{Input}} \underbrace{e^{-jk(x_2 + y_2)/2f_0}}_{\text{Lens}} \underbrace{e^{jk(x^2 + y^2)/2z_i}}_{\substack{\text{Part of the} \\ \text{point spread} \\ \text{function}}} \right\}_{f_x = x_i/\lambda z_i, f_y = y_i/\lambda z_i}$$

$$(1.141)$$

The divergence of the input field factor is partially cancelled by the lens factor. The rearrangement of Eq. (1.141) results in

$$E(x_i, y_i) = \frac{1}{j\lambda f_0} e^{jkf_0 + jk[(x_i^2 + y_i^2)/2z_i]} \mathcal{F}\left\{ \frac{1}{j\lambda z_i} e^{jk[z_i + (x^2 + y^2)/2z_i]} \right\}_{f_x = x_i/\lambda z_i, f_y = y_i/\lambda z_i} \quad (1.142)$$

The quantity in the curly brackets is the point spread function and its Fourier transform is obtained from Eq. (1.41):

$$E(x_i, y_i) = \frac{1}{j\lambda f_0} e^{jk(f_0 + z_i) + jk[(x_i^2 + y_i^2)/2z_i]} [e^{-j\pi\lambda z_i(f_x^2 + f_y^2)}]_{f_x = x_i/\lambda z_i, f_y = y_i/\lambda z_i} \quad (1.143)$$

Due to cancellation we finally obtain a function independent of x_i and y_i:

$$E(x_i, y_i) = \frac{1}{j\lambda f_0} e^{jk(f_0 + z_i)} \quad (1.144)$$

Hence, the field is a parallel beam and is a plane wave with constant amplitude $1/\lambda f_0$. The amplitude stays constant with distance z_i as shown in Fig. 1.21. This proves that f_0 is indeed the focal length of the lens.

1.5.3 Imaging Property of a Convex Lens

The Gaussian lens formula is derived by geometrical optics as

$$\frac{1}{d_1} + \frac{1}{d_2} = \frac{1}{f_0} \quad (1.145)$$

where f_0 is the focal length of the convex lens, d_1 is the distance from the lens to the object, and d_2 is the distance from the lens to the image. The Gaussian lens formula will be obtained using Fourier optics. The imaging condition is that the light emanating from a point on the object converges back to a point in the image plane.

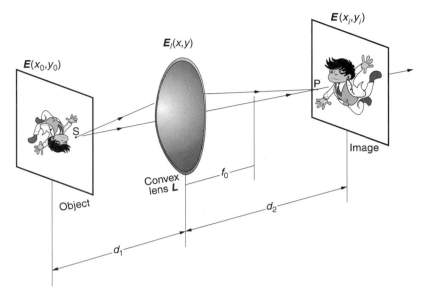

Figure 1.22   Imaging condition of a convex lens.

For simplicity, the object is a delta function source S located on the z axis, as shown in Fig. 1.22. The input field $E_l(x, y)$ to the lens in x, y coordinates is

$$E_l(x, y) = \frac{1}{j\lambda d_1} e^{jk[d_1 + (x^2 + y^2)/2d_1]} \tag{1.146}$$

The Fresnel diffraction from the lens to the image plane at $z_i = d_2$ is, from Eq. (1.38),

$$E(x_i, y_i) = -\frac{e^{jk[d_2 + (x_i^2 + y_i^2)/2d_2]}}{\lambda^2 d_1 d_2}$$

$$\mathcal{F}\left\{ \underbrace{e^{jk[d_1 + (x^2 + y^2)/2d_1]}}_{\text{Input}} \underbrace{e^{-jk(x^2 + y^2)/2f_0}}_{\text{Lens}} \underbrace{e^{jk(x^2 + y^2)/2d_2}}_{\substack{\text{Part of the} \\ \text{point spread} \\ \text{function}}} \right\}_{f_x = x_i/\lambda d_2, \, f_y = y_i/\lambda d_2}$$

$$\tag{1.147}$$

If the observation is made at the particular value of d_2 that satisfies the Gaussian lens formula, Eq. (1.145), then the value inside the curly brackets becomes unity and

$$E(x_i, y_i) = -\frac{1}{\lambda^2 d_1 d_2} e^{jk[d_1 + d_2 + (x_i^2 + y_i^2)/2d_2]} \delta\left(\frac{x_i}{\lambda d_2}\right) \delta\left(\frac{y_i}{\lambda d_2}\right) \tag{1.148}$$

The delta function image P is recovered if the imaging condition is satisfied. Thus, Fourier optics has proved Eq. (1.145).

$E(x_i, y_i)$ will be rewritten in a form more conducive to extracting the physical meaning. Using Eq. (1.109), the last two factors of Eq. (1.148) can be rewritten as

$$\delta\left(\frac{x_i}{\lambda d_2}\right)\delta\left(\frac{y_i}{\lambda d_2}\right) = (\lambda d_1)^2 \delta\left(\frac{d_1}{d_2}x_i\right)\delta\left(\frac{d_1}{d_2}y_i\right) \tag{1.149}$$

Let m be the magnification factor:

$$m = \frac{d_2}{d_1} \tag{1.150}$$

The expression for the image finally becomes

$$E(x_i, y_i) = -\frac{1}{m}e^{jk[d_1+d_2+(x_i^2+y_i^2)/2d_2]}\delta\left(\frac{x_i}{m}\right)\delta\left(\frac{y_i}{m}\right) \tag{1.151}$$

Since the object is a delta function source $\delta(x_0)\delta(y_0)$, its image is also expected to be a delta function and is $\delta(x_i/m)\delta(y_i/m)$. The width of $\delta(x_i/m)$ is wider than $\delta(x_0)$ by m times, and hence the image is magnified by m times. As far as the light intensity is concerned, the amplitude of the light is diluted by $1/m$ times.

The advantage of the Fourier optics approach is that it can be applied to situations where geometric optics is inadequate, such as dealing with the finite size of the lens. For example, the case of the finite-sized square aperture of dimensions $a \times a$ is represented as

$$A(x, y) = \Pi\left(\frac{x}{a}\right)\Pi\left(\frac{y}{a}\right) \tag{1.152}$$

This has the effect of altering the input by this factor. With this insertion into Eq. (1.146), the finite lens equivalent of Eq. (1.148) is

$$E(x_i, y_i) = -\frac{1}{\lambda^2 d_1 d_2}e^{jk[d_1+d_2+(x_i^2+y_i^2)/2d_2]}$$
$$\times [a^2 \operatorname{sinc}(af_x)\operatorname{sinc}(af_y) * \underbrace{\delta(f_x)\delta(f_y)]}_{\text{Image}}|_{f_x=x_i/\lambda d_2, f_y=y_i/\lambda d_2} \tag{1.153}$$

$\underbrace{\phantom{[a^2 \operatorname{sinc}(af_x)\operatorname{sinc}(af_y)}}_{\substack{\text{Due to a finite} \\ \text{square aperture}}}$

The final result is

$$E(x_i, y_i) = \frac{1}{\lambda^2 d_1 d_2}e^{jk[d_1+d_2+(x_i^2+y_i^2)/2d_2]}a^2 \operatorname{sinc}\left(a\frac{x_i}{\lambda d_2}\right)\operatorname{sinc}\left(a\frac{y_i}{\lambda d_2}\right) \tag{1.154}$$

Thus, the image of the delta function object is no longer a delta function but a sinc function whose main lobe size is $(2\lambda d_2/a) \times (2\lambda d_2/a)$. The larger the aperture $a \times a$ is, the smaller the width of the lobe becomes. However, as long as the dimension of the aperture is finite, the lobe is also finite. Hence, the image cannot be the same as the original delta function, even when the lens is designed perfectly and is aberration free. If the only limitation on resolution is due to the finiteness of the aperture causing the diffraction, then the imaging system is said to be *diffraction limited*.

1.5.4 Fourier Transformable Property of a Convex Lens

Besides the properties of converging, diverging, collimating, and forming images, the convex lens also has a Fourier transformable property [1,6,8]. It takes place at the back focal plane of the lens. For instance, when a parallel beam (constant with respect to x_0 and y_0) is incident onto a convex lens, the distribution of light that results on the focal plane behind the lens is a point or delta function. The delta function is the Fourier transform of a constant. In this way, the convex lens can be considered to possess a *Fourier transformable* property.

Now consider a source (an input function or input mask) placed on the back surface of a convex lens, as shown in Fig. 1.23. The screen is located in the near field and hence Fresnel's near-field diffraction equation, Eq. (1.38), is used. The field on the screen is the Fourier transform of the field just behind the lens multiplied by the point spread function.

$$E(x_i, y_i) = \frac{1}{j\lambda z_i} e^{jk[z_i + (x_i^2 + y_i^2)/2z_i]}$$

$$\mathcal{F}\{E(x_0, y_0)e^{-jk(x_0^2 + y_0^2)/2f_0} \cdot e^{jk(x_0^2 + y_0^2)/2z_i}\}_{f_x = x_i/\lambda z_i,\ f_y = y_i/\lambda z_i} \quad (1.155)$$

If the observation is made at a distance $z_i = f_0$, the point spread phase factor is cancelled by the quadratic phase factor of the convex lens and Eq. (1.155)

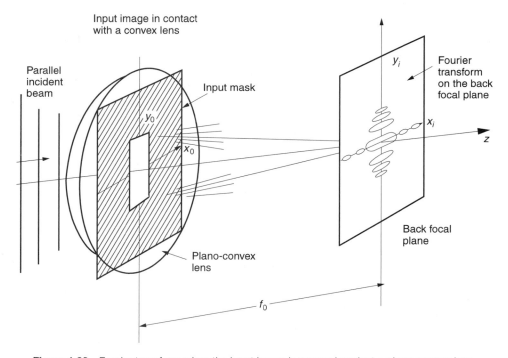

Figure 1.23   Fourier transform when the input image is pressed against a plano-convex lens.

becomes

$$E(x_i, y_i) = \frac{1}{j\lambda f_0} e^{jk[f_0 + (x_i^2 + y_i^2)/2f_0]} \varepsilon \left(\frac{x_i}{\lambda f_0}, \frac{y_i}{\lambda f_0} \right) \qquad (1.156)$$

where

$$\varepsilon(f_x, f_y) = \mathcal{F}\{E(x_0, y_0)\}$$

Thus, the output $E(x_i, y_i)$ is the Fourier transform of the input function. It should be noted, however, that the output is not exactly the Fourier transform but is altered by a quadratic phase factor $\exp[jk(x_i^2 + y_i^2)/2f_0]$. Furthermore, the size of the Fourier transform pattern $\varepsilon(x_i/\lambda f_0, y_i/\lambda f_0)$ depends on the focal length of the lens.

Next, a different problem is investigated. Consider the case when the input function is placed at the focal plane in front of the lens, as arranged in Fig. 1.24. The field incident upon the lens is in the near field. Hence, the Fresnel diffraction pattern of the input function, in the form of convolution with the point spread function as given by Eq. (1.39), is used.

$$E_l(x, y) = E_0(x, y) * f_{f_0}(x, y) \qquad (1.157)$$

Using the previous discovery that the field in the back focal plane of a convex lens is merely the Fourier transform of the input function (Eq. (1.156)), the distribution on

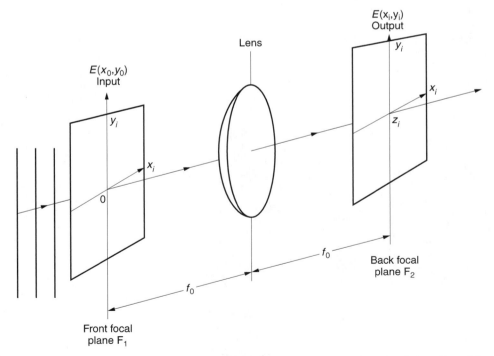

Figure 1.24 Fourier transform by a convex lens when the input is at the front focal plane F_1 and the Fourier transform is on the back focal plane F_2.

the screen in the back focal plane is given by

$$E(x_i, y_i) = \frac{1}{j\lambda f_0} e^{jk[f_0 + (x_i^2 + y_i^2)/2f_0]} \mathcal{F}\{E_0(x, y) * f_{f_0}(x, y)\}_{f_x = x_i/\lambda f_0, \ f_y = y_i/\lambda f_0} \quad (1.158)$$

Using the product rule of Fourier transforms and the Fourier transform of the point spread function from Eq. (1.41), the field is equal to

$$E(x_i, y_i) = \frac{1}{j\lambda f_0} e^{jk[f_0 + (x_i^2 + y_i^2)/2f_0]} [\varepsilon(f_x, f_y) \cdot e^{jkf_0 - j\pi\lambda f_0(f_x^2 + f_y^2)}]_{f_x = x_i/\lambda f_0, \ f_y = y_i/\lambda f_0}$$

$$(1.159)$$

$$E(x_i, y_i) = \frac{e^{j2kf_0}}{j\lambda f_0} \varepsilon\left(\frac{x_i}{\lambda f_0}, \frac{y_i}{\lambda f_0}\right) \quad (1.160)$$

This final result can be compared to the previous findings for the lens system of Fig. 1.23. This formula is very similar except for the absence of the quadratic phase factor.

Thus, the Fourier transform without the quadratic phase factor is obtained when the input is placed in the front focal plane and observed on the back focal plane.

Lastly, consider what happens when the input is placed in an arbitrary converging beam, as shown in Fig. 1.25. In this case, the Fourier transform is obtained in a plane containing the point of convergence. It does not matter whether the converging beam has been made by a single lens or a composite of lenses, as long as a converging spherical wave is the incident input beam.

Before going into detail, the difference in the expressions for diverging and converging rays needs to be understood. Figure 1.26 shows how two phase fronts evolve as time elapses from $t = 1$ to $t = 3$. Both beams are propagating in the positive

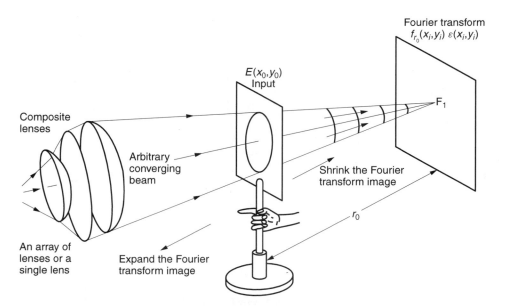

Figure 1.25   Fourier transform by a converging beam.

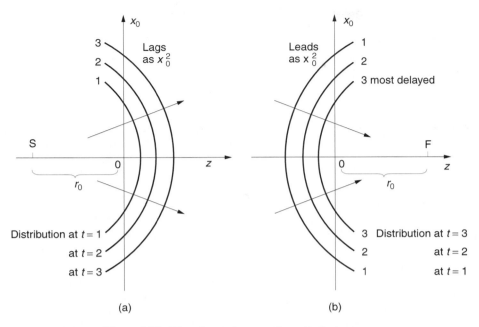

Figure 1.26   Diverging and converging spherical waves.

z direction and have a quadratic phase distribution in the transverse (x_i, y_i) plane. However, the signs of the phase are opposite. The difference in signs can be explained as follows. In Fig. 1.26a, if the phase is observed outward from the origin (increasing $x_0^2 + y_0^2$), the later or the more delayed phase is observed. Therefore, the factor is positive and is expressed as

$$E(x_0, y_0) = E_0 \, e^{jk(x_0^2+y_0^2)/2r_0} \qquad (1.161)$$

where r_0 is the distance between the point source and the screen.

On the other hand, for the converging beam in Fig. 1.26b, the earlier or more leading phase is observed as the point of observation moves away from the center of the lens; thus, the phase factor has to be

$$E(x_0, y_0) = E_0 \, e^{-jk(x_0^2+y_0^2)/2r_0} \qquad (1.162)$$

Now, let us return to the diffraction pattern of the input and consider the case when the input is inserted into a converging beam. From the Fresnel diffraction formula of Eq. (1.38) and the phase factor, the field observed at a distance r_0 from the input is

$$E(x_i, y_i) = \frac{1}{j\lambda z_i} \, e^{jk[r_0+(x_i^2+y_i^2)/2r_0]}$$

$$\mathcal{F}\{e^{-jk(x_0^2+y_0^2)/2r_0}E(x_0, y_0)e^{jk(x_0^2+y_0^2)/2r_0}\}_{f_x=x_i/\lambda r_0,\ f_y=y_i/\lambda r_0} \qquad (1.163)$$

The phase of the converging beam and the point spread function cancel each other, and the final expression becomes

$$E(x_i, y_i) = \frac{e^{jk[r_0+(x_i^2+y_i^2)/2r_0]}}{j\lambda r_0} \varepsilon \left(\frac{x_i}{\lambda r_0}, \frac{y_i}{\lambda r_0} \right) \tag{1.164}$$

The Fourier transform is obtained, but with the inclusion of the quadratic phase factor once again. This time, however, the size of the Fourier transform diffraction pattern can be controlled without changing the focal length of the lens. The size of the image is enlarged as r_0 is increased, until the input is obstructed by the lens.

1.5.5 How Can a Convex Lens Perform the Fourier Transform?

Figure 1.27 gives a pictorial explanation of how a convex lens performs the Fourier transform of the input image. The input is placed in the front focal plane of the convex lens, and the output is observed in the back focal plane. The observation point $P(x_i)$ will always be at $x_i = x_i$ in the back focal plane.

   In case (1), a delta function source is placed at the origin of the input plane. The emergent light from the lens is a parallel beam. The phase at $P(x_i)$ is the same as that at the origin. In case (2), the light source is moved to $x_0 = x_1$. The parallel beam will be tilted and propagates slightly downward. The phase at $P(x_i)$ is leading that at the origin by $-jk(\sin\theta_i)x_i$ radians or approximately $-j(2\pi/\lambda)(x_1/f_0)x_i$ radians. The field at $P(x_i)$ is

$$E(x_i) = E_1 \exp\left(-j\frac{2\pi}{\lambda}\frac{x_i}{f_0}x_1 \right)$$

In case (3), one more source is added at $x_0 = x_2$, and the sum of the contributions of the two sources is

$$E(x_i) = E_1 \exp\left(-j\frac{2\pi}{\lambda}\frac{x_i}{f_0}x_1 \right) + E_2 \exp\left(-j\frac{2\pi}{\lambda}\frac{x_i}{f_0}x_2 \right)$$

In case (4), a distributed source $E(x_0)$ is placed in the input plane. The contribution of the distributed source is expressed by an integral over the source plane.

$$E(x_i) = \int E(x_0) \exp\left(-j\frac{2\pi}{\lambda}\frac{x_i}{f_0}x_0 \right) dx_0$$

Thus, the lens performs the Fourier transform of the input image and the output is

$$E(x_i) = \mathcal{F}\{E(x_0)\}_{f_x=x_i/\lambda f_0}$$

1.5.6 Invariance of the Location of the Input Pattern to the Fourier Transform

Even when the input pattern is moved sideways, the location of the Fourier transform pattern stays in close proximity to the back focal point of the lens. Figure 1.28 shows how the ray paths change as the input object (letter envelope) is moved vertically in the $x-y$ plane and horizontally along the z axis. From the figure, we see that the

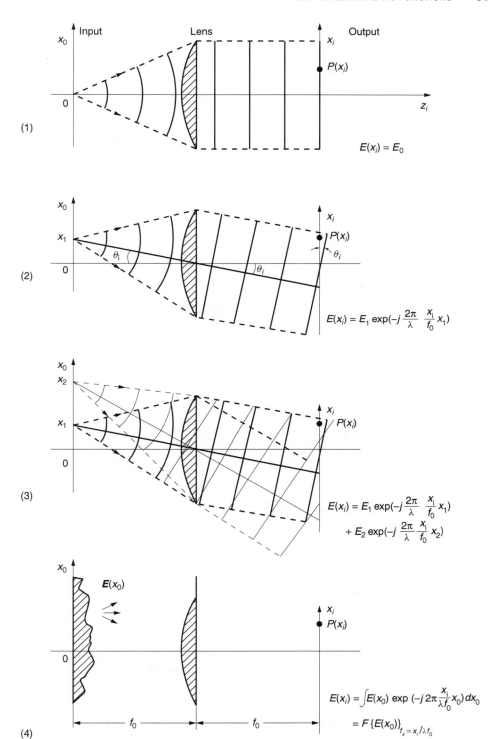

Figure 1.27 How a convex lens performs the Fourier transform.

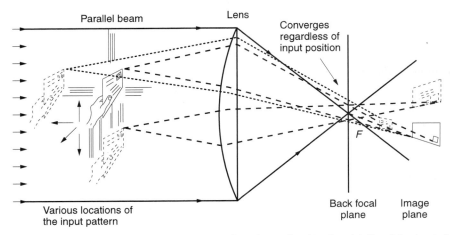

Figure 1.28 Demonstrating that the Fourier transform is confined to the vicinity of the back focal point F regardless of the location of the input image.

incident parallel beam converges to the back focal point and the diffraction pattern of the input remains around the focal point regardless of the location of the input. When the envelope is lowered, the scattered beam starts from a lower location but the scattered field first converges to the center of the back focal plane and then forms an image at a higher location. In fact, the Fourier transform pattern is usually confined within a small range. For instance, with $\lambda = 0.63$ μm, the focal length $f = 10$ cm, and with a maximum spatial frequency of $f_x = 100$ lines/mm, the Fourier transform appears within $x_i = \pm f_x \lambda f = \pm 6.3$ mm at the back focal point. This is one of the most useful features of the Fourier transform nature of a lens. In practical applications, such as mass processing of addresses on letters, this means that stringent positioning requirements of the input image are not necessary.

The reason for this phenomenon mathematically is that the Fourier transform of the shifted input $g(x - a)$ is $e^{-j2\pi f a}G(f)$. The shift causes the phase shift $e^{-j2\pi f a}$. However, the human eye cannot recognize phase shifts and thus the pattern does not show any change when the input is shifted. However, this is only true with a translational shift and not true with a rotational shift.

1.6 SPATIAL FREQUENCY APPROACHES IN FOURIER OPTICS

Another way of calculating the diffraction pattern will be introduced. The given input pattern is first decomposed into its constituent spatial frequencies by Fourier transforming the input, and then the propagated field of each spatial frequency component is calculated. The desired diffraction pattern of the given input pattern is the sum of the propagated patterns of the spatial frequency components [1, 11].

1.6.1 Solution of the Wave Equation by Means of the Fourier Transform

With the geometry shown in Fig. 1.2, the field $\mathbf{E}(x_i, y_i, z, t)$ at $z = z_i$ will be calculated from the spectrum of the input field at $z = 0$. Assuming a sinusoidal time dependence, the field is expressed as

$$\mathbf{E}(x, y, z, t) = \mathbf{E}(x, y, z)e^{-j\omega t} \tag{1.165}$$

Equation (1.165) has to satisfy the wave equation

$$\nabla^2 \mathbf{E} + k^2 \mathbf{E} = 0 \tag{1.166}$$

If Cartesian coordinates are used to express Eq (1.166), the general solutions for E_x, E_y, and E_z are all identical. What makes E_x, E_y, and E_z different is that the boundary conditions depend on the components. Let the solution of Eq. (1.166) be denoted as $E(x, y, z)$. This approach is called the *scalar wave approach*. The scalar wave approach is much simpler than the vector wave approach but is less accurate because it assumes that the same boundary conditions are applicable to both normal and tangential components. Accepting this trade of simplicity for accuracy, the scalar equivalent of Eq. (1.165) is

$$E(x, y, z, t) = E(x, y, z)e^{-j\omega t} \tag{1.167}$$

Thus, the scalar wave equation becomes

$$\frac{\partial^2 E}{\partial x^2} + \frac{\partial^2 E}{\partial y^2} + \frac{\partial^2 E}{\partial z^2} + k^2 E = 0 \tag{1.168}$$

Now, the wave equation (1.168) can be solved using the Fourier transform method [11]. The Fourier transform of E with respect to x in the f_x, f_y domain is

$$\mathcal{F}_x\{E\} = \varepsilon(f_x, y, z) \tag{1.169}$$

The derivative rule of the Fourier transform is

$$\mathcal{F}\left\{\frac{\partial E}{\partial x}\right\} = j2\pi f_x \varepsilon(f_x, y, z) \tag{1.170}$$

The Fourier transform of Eq. (1.168) with respect to x gives

$$(j2\pi f_x)^2 \varepsilon(f_x, y, z) + \frac{\partial^2}{\partial y^2}\varepsilon(f_x, y, z) + \frac{\partial^2}{\partial z^2}\varepsilon(f_x, y, z) + k^2\varepsilon(f_x, y, z) = 0 \tag{1.171}$$

Similarly, the Fourier transform of the above equation with respect to y gives

$$(j2\pi f_x)^2 \varepsilon(f_x, f_y, z) + (j2\pi f_y)^2 \varepsilon(f_x, f_y, z) + \frac{\partial^2}{\partial z^2}\varepsilon(f_x, f_y, z) + k^2\varepsilon(f_x, f_y, z) = 0 \tag{1.172}$$

Hence, the Fourier transform of the scalar wave equation, Eq. (1.168), with respect to both x and y is

$$\left[\frac{\partial^2}{\partial z^2} + (2\pi f_z)^2\right]\varepsilon(f_x, f_y, z) = 0 \tag{1.173}$$

where

$$f_z = \sqrt{f_s^2 - f_x^2 - f_y^2} \tag{1.174}$$

f_s is the spatial frequency in the direction of propagation and $f_s = 1/\lambda$ as defined by Eq. (1.20). Equation (1.173) is a second order partial differential equation with respect to z whose solution is

$$\varepsilon(f_x, f_y, z) = Ae^{j2\pi f_z z} + Be^{-j2\pi f_z z} \tag{1.175}$$

The first term is a forward wave while the second term is a backward wave. The values of A and B are to be found from the boundary conditions of either ε or its derivative. We will restrict ourselves to the simple case where only the forward wave is present, and $B = 0$. By setting $z = 0$ in Eq. (1.175), A is found to be $\varepsilon(f_x, f_y, 0)$. The solution becomes

$$\varepsilon(f_x, f_y, z) = \varepsilon(f_x, f_y, 0)e^{j2\pi f_z z} \tag{1.176}$$

Finally, we solve for the field $E(x_i, y_i, z_i)$ by taking the inverse Fourier transform of Eq. (1.176). This gives

$$E(x_i, y_i, z_i) = \iint_\infty \varepsilon(f_x, f_y, 0)e^{j2\pi\sqrt{f_s^2 - f_x^2 - f_y^2}\cdot z_i} e^{j2\pi f_x x_i + j2\pi f_y y_i} df_x df_y \tag{1.177}$$

where

$$\varepsilon(f_x, f_y, 0) = \iint_\infty E(x_0, y_0, 0)e^{-j2\pi f_x x_0 - j2\pi f_y y_0} dx_0 dy_0 \tag{1.178}$$

The combination of these equations is called the *Rayleigh–Sommerfeld diffraction formula* and provides the field at $z = z_i$ from the Fourier transform of the input field at $z = 0$.

Even though Eqs. (1.177) and (1.178) are simple expressions, it is difficult to obtain a highly accurate representation of $E(x_0, y_0, 0)$ that expresses the field for an aperture of finite size. For example, the field of an aperture illuminated from the back by a plane wave is almost, but not exactly, uniform across the aperture. The aperture contains contributions from the waves scattered by the edges, as well as multiple scatterings between the facing edges. As a result, it is quite difficult to obtain the exact expression of the input field.

Example 1.6   Demonstrate that the Rayleigh–Sommerfeld and Fresnel–Kirchhoff diffraction formulas provide the same answer using the example of the diffraction pattern of a pinhole. The pinhole is located at the origin of the input plane and is illuminated by a plane wave of unit amplitude from behind as shown in Fig. 1.29a.

Solution

1. *Solution by the Rayleigh–Sommerfeld formula.*

   In this method, the Fourier transform ε of the input field is first obtained and then ε is allowed to propagate to the output screen where the inverse Fourier transform is performed to obtain the final result.

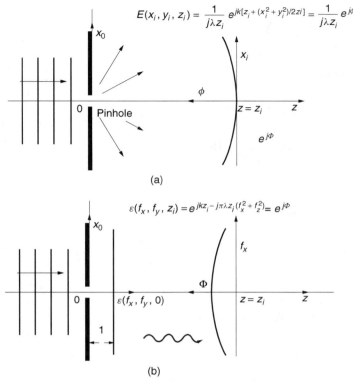

Figure 1.29 Field from a pinhole in space and spatial frequency domains. (a) Space domain. (b) Spatial frequency domain.

The pinhole is represented by

$$E(x_0, y_0) = \delta(x_0)\delta(y_0) \tag{1.179}$$

Inserting Eq. (1.179) into (1.178) gives

$$\varepsilon(f_x, f_y, 0) = 1 \tag{1.180}$$

Since there is no backward wave in the region between the input and output planes, $B = 0$ in Eq. (1.175). Let $\varepsilon(f_x, f_y, 0)$ propagate to the output screen using Eq. (1.176):

$$\varepsilon(f_x, f_y, z_i) = e^{j2\pi f_z z_i} \tag{1.181}$$

If we apply the para-axial approximation, which means that the propagation directions of the component waves are almost along the z axis, then f_s satisfies

$$f_s^2 \gg f_x^2 + f_y^2 \tag{1.182}$$

Recalling that

$$f_s = \frac{1}{\lambda}$$

the binomial expansion of Eq. (1.174) becomes

$$f_z = \frac{1}{\lambda} - \frac{\lambda}{2}(f_x^2 + f_y^2) \tag{1.183}$$

Inserting Eq. (1.183) into (1.181) gives

$$\varepsilon(f_x, f_y, 0)e^{j2\pi f_z z_i} = e^{jkz - j\pi\lambda z_i(f_x^2 + f_z^2)} \tag{1.184}$$

Using the Fourier transform relationship of Eq. (1.44), Eq. (1.184) is inverse Fourier transformed to obtain the final result as

$$E(x_i, y_i, z_i) = \frac{1}{j\lambda z_i}e^{jk[z_i + (x_i^2 + y_i^2)/2z_i]} \tag{1.185}$$

2. *Solution by the Fresnel–Kirchhoff integral.*
Inserting Eq. (1.179) into (1.38) gives

$$E(x_i, y_i, z_i) = \frac{1}{j\lambda z_i}e^{jk[z_i + (x_i^2 + y_i^2)/2z_i]} \iint_{-\infty}^{\infty} \delta(x_0)\delta(y_0)$$
$$\times e^{jk(x_0^2 + y_0^2)/2z_i} \cdot e^{-j2\pi(f_x x_0 + f_y y_0)}dx_0\,dy_0 \tag{1.186}$$

The integral in Eq. (1.186) is unity and Eq. (1.186) is the same as Eq. (1.185).
□

Example 1.7   As shown in Fig. 1.30, light is incident from an optically dense medium with refractive index n into free space. The interface between the medium and free space is in the plane of $z = 0$. The propagation direction of the incident wave is in the x–z plane and the incident angle to the interface is θ_i. Find the field at point (x_i, y_i, z_i) in free space and its propagation direction.

Solution   The wavelength λ_{x_0} in the x_0 direction at the interface is

$$\lambda_{x_0} = \frac{\lambda}{n\sin\theta_i} \tag{1.187}$$

The corresponding spatial frequency from Eq. (1.13) is

$$f_{x_0} = nf_s\sin\theta_i \tag{1.188}$$

Note that B in Eq. (1.175) is zero because there is no wave propagating in the negative z direction in the free-space region. Thus, the field $E(x_0, y_0, 0)$ in the $z = 0$ plane is

$$E(x_0, y_0, 0) = E_0 e^{j2\pi n f_s x_0 \sin\theta_i} \tag{1.189}$$

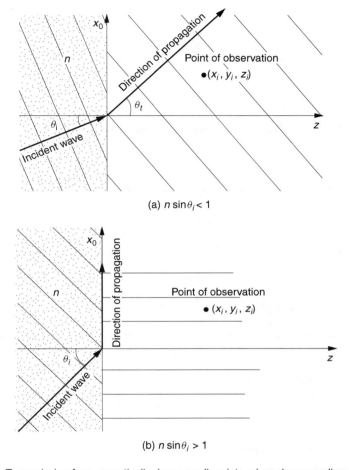

(a) $n \sin \theta_i < 1$

(b) $n \sin \theta_i > 1$

Figure 1.30 Transmission from an optically dense medium into a less dense medium is calculated.

From Eq. (1.189), Eq. (1.178) becomes

$$\varepsilon(f_x, f_y, 0) = E_0 \delta(f_x - n f_s \sin \theta_i) \delta(f_y) \tag{1.190}$$

Thus, the input field contains only one spatial frequency component, $n f_s \sin \theta_i$. Next, Eq. (1.177) will be calculated:

$$E(x_i, y_i, z_i) = \iint E_0 \delta(f_x - n f_s \sin \theta_i) \delta(f_y) e^{j2\pi \sqrt{f_s^2 - f_x^2 - f_y^2} \ z_i} \ e^{j2\pi(f_x x_i + f_y y_i)} \, d f_x \, d f_y \tag{1.191}$$

Applying Eq. (1.102) to (1.191) gives

$$E(x_i, y_i, z_i) = E_0 \exp[j2\pi \sqrt{f_s^2 - n^2 f_s^2 \sin^2 \theta_i} \ z_i + j2\pi n f_s (\sin \theta_i) x_i] \tag{1.192}$$

From Snell's law at the boundary,

$$n \sin \theta_i = \sin \theta_t \tag{1.193}$$

the final result is

$$E(x_i, y_i, z_i) = E_0 \exp[j2\pi f_s (\cos \theta_t) z_i + j2\pi f_s (\sin \theta_t) x_i] \tag{1.194}$$

Equation (1.194) is a plane wave whose propagation unit vector is

$$\hat{\mathbf{e}} = \sin \theta_t \hat{\mathbf{i}} + \cos \theta_t \hat{\mathbf{k}} \tag{1.195}$$

The propagation direction with respect to the normal of the input plane is

$$\tan^{-1} \frac{e_x}{e_z} = \theta_t \tag{1.196}$$

Next, the case of the evanescent wave will be explained. Equation (1.192) can be rewritten as

$$E(x_i, y_i, z_i) = E_0 \exp[-2\pi f_s \sqrt{n^2 \sin^2 \theta_i - 1} \; z_i + j2\pi n f_s (\sin \theta_i) x_i] \tag{1.197}$$

When either θ_i or n is large and the condition $n \sin \theta_i > 1$ is satisfied, the amplitude of the wave decays exponentially in the z direction. This is an example of an evanescent wave. For the evanescent wave, the phase varies in the x_i direction whereas the amplitude of the wave does not, as shown by Eq. (1.197) and Fig. 1.30b. The evanescent wave is a very important subject and will be treated in more detail in the next chapter.   □

1.6.2   Rayleigh–Sommerfeld Integral

This section is devoted to explaining the conceptual differences between the Rayleigh–Sommerfeld and the Fresnel–Kirchhoff diffraction formulas. Mathematically, the two integrals are equivalent, and Appendix A of Volume I presents a general proof of this. Conceptually, there are differences between the two approaches. Understanding these differences is important so that the most appropriate choice is made in solving a given problem.

The Fresnel–Kirchhoff diffraction formula, Eq. (1.28), is repeated here for convenience.

$$E(x_i, y_i, z_i) = \frac{1}{j\lambda} \iint \underbrace{\frac{e^{j2\pi f_s r}}{r}}_{\substack{\text{Spherical} \\ \text{wave}}} \underbrace{E(x_0, y_0) \, dx_0 \, dy_0}_{\substack{\text{Amplitude of the} \\ \text{spherical wave}}} \tag{1.198}$$

$$\text{with} \quad r = \sqrt{(x_i - x_0)^2 + (y_i - y_0)^2 + z_i^2}$$

As mentioned in Section 1.2, $e^{j2\pi f_s r}/r$ represents a spherical wave emanating from the point (x_0, y_0) with $E(x_0, y_0) \, dx_0 \, dy_0$ as its amplitude. The representation of the source function as a collection of spherical wave point sources is indicated by $S_1, S_2, S_3, \ldots, S_n$ in Fig. 1.31a. In order to find the field observed at P, the contributions of all spherical wave sources $S_1, S_2, S_3, \ldots, S_n$ are integrated over the entire shape of the source.

Next, let us examine the Rayleigh–Sommerfeld diffraction formula. Equation (1.177) is rewritten below:

$$E(x_i, y_i, z_i) = \iint \underbrace{e^{j2x\pi f_x x + j2\pi f_y y + j2\pi f_z z}}_{\substack{\text{Plane wave propagating} \\ \text{in the } (f_x \mathbf{i}, f_y \mathbf{j}, f_z \mathbf{k}) \\ \text{direction}}} \underbrace{\varepsilon(f_x, f_y, 0)}_{\text{Amplitude}} d f_x \, d f_y \qquad (1.199)$$

$$\text{with} \quad f_z = \sqrt{f_s^2 - f_x^2 - f_y^2}$$

The first factor of the integral in Eq. (1.199) is a plane wave component propagating in the direction

$$(f_x \mathbf{i}, \ f_y \mathbf{j}, \ f_z \mathbf{k}) \qquad (1.200)$$

The second factor is the amplitude of the plane wave. The integral with respect to f_x and f_y means the integration of the plane wave contributions from all propagating directions. Since $f_s = 1/\lambda$ is given, once f_x and f_y are specified, the value of f_z is accordingly set from the lower equation of Eq. (1.199). There are few noteworthy points with regard to this integral. As shown in Fig. 1.31b, the integral includes not only the plane wave B whose wave normal is aimed at P but also all other wave normals denoted by $A, C, \ldots$. However, a wave normal is not the same thing as a light beam. Any line drawn parallel to the propagation direction of a given wave can be the wave normal of that same wave. For example, the wave represented by B' in Fig. 1.31b is identical with the wave represented by B.

It is also important to include the contributions of plane waves propagating in the negative x_0 direction, such as shown by A in Fig. 1.31b.

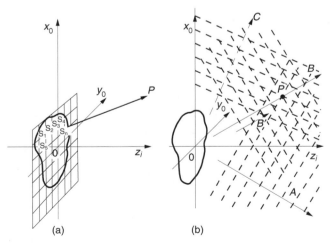

Figure 1.31 Conceptual comparison of the diffraction formulas. (a) Fresnel–Kirchhoff. (b) Rayleigh–Sommerfeld.

According to Eq. (1.199), another significant point to remember is that f_z becomes an imaginary number for

$$f_x^2 + f_y^2 > f_s^2 \tag{1.201}$$

When f_z is an imaginary number, the integrand becomes an evanescent wave whose magnitude decays exponentially with z. The contribution of the evanescent wave is insignificant unless the point of observation is close to the source and z is small.

1.6.3 Identifying the Spatial Frequency Components

The spatial frequency components shown in Fig. 1.31b can be identified by using a convex lens as illustrated in Fig. 1.32. Only two spatial frequency components are considered. Both of them are incident from the left of the convex lens. Because both waves are plane waves, they are focused on the back focal plane. The component with $f_x = f_y = 0$ is incident normal to the lens and is focused at $(0, 0, f)$, where f is the focal length of the convex lens. The component with $f_x \neq 0$, $f_y = 0$ is incident to the lens with an incident angle

$$\theta = \tan^{-1}(f_x/f_z) \tag{1.202}$$

and is focused on the back focal plane at $(x_i, 0, f)$, where

$$x_i = f \tan \theta \doteq f \lambda f_x \tag{1.203}$$

and where the approximation of $f_z = f_s$ was made. The location of the focused light rises along the x_i axis as f_x is increased. The higher the location, the higher the spatial frequency of the converging light. This means that by placing a mask of a predetermined

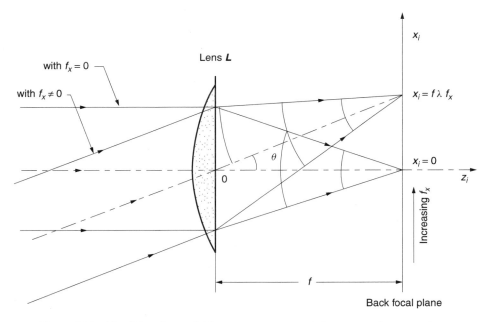

Figure 1.32 Identifying the spatial frequency components by means of a convex lens.

transmittance distribution in the back focal plane of the convex lens, the waves of a particular spatial frequency component can be either selectively transmitted through or blocked by this mask. The following sections elaborate on this technique. (See Section 1.2 for the case without a lens.)

1.7 SPATIAL FILTERS

This section is devoted to a discussion of optical signal processing based on manipulation of spatial frequency components.

1.7.1 Image Processing Filters

Figure 1.33 shows different arrangements for image processing using various types of spatial frequency filters [1,8]. For all the cases, there is a point source that is collimated

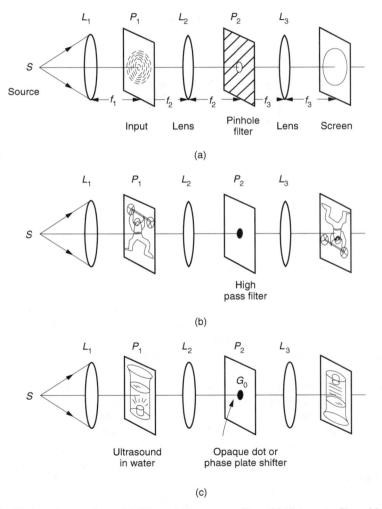

Figure 1.33 Various types of spatial filters. (a) Low-pass filter. (b) High-pass filter. (c) Schlieren camera. (d) Spatial derivative operation. (e) Step and repeat operation.

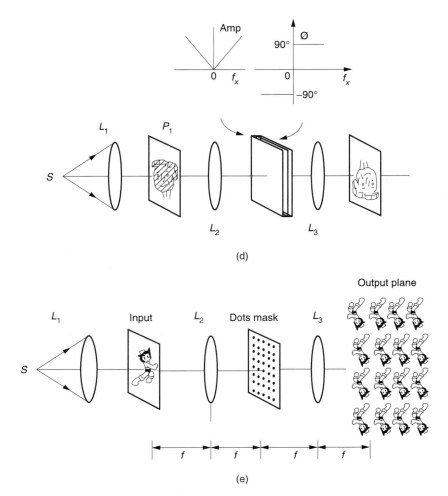

Figure 1.33 (*Continued*)

by lens L_1. The collimated beam illuminates a transparency overlay forming the input image. Lens L_2 Fourier transforms the transparency pattern in its back focal plane. The spatial filter is placed in the back focal plane of L_2 and modifies the pattern. The modified pattern is Fourier transformed again by means of lens L_3, and the processed image is finally projected onto the screen.

Figure 1.33a shows a low-pass spatial filter. In the plane of the spatial filter, the spatial frequency is zero on the z axis and increases linearly with distance from the axis. The higher frequency components, which are diffracted to the area away from the axis, are blocked by the mask. A typical low-pass filter consists of a pinhole and a lens combination, commonly marketed as a "spatial filter."[*] This filter filters out all the spatial frequency components except the zero spatial frequency, resulting in a pure parallel beam.

---

[*] Keep in mind that the true meaning of a spatial filter is broader in scope than the marketplace meaning of spatial filter.

Figure 1.33b shows an example of image processing with a high-pass spatial filter. Here, the lower spatial frequency components are suppressed. In general, sharp edges and fine lines in the image generate higher spatial frequency components, which are then accentuated in the image. Thus, fine point or edge enhancement is achieved by this type of filter.

Figure 1.33c shows an example of a Schlieren camera. This filter is used to view transparent objects. It is normally difficult to view or photograph objects that are transparent although they may have varying indices of refraction within them. Examples include such objects as microbes, turbulent air, or ultrasound patterns in a liquid. The difficulty is that the images of these objects have only phase variations, which neither our eyes nor an ordinary camera can detect.

The Schlieren camera, however, creates an interference pattern between the image and a constant phase reference wave. The constant phase reference wave is generated by placing an opaque dot at the back focal point of lens L_2 on P_2. The generation of the reference wave is explained as follows. The zero spatial frequency is blocked by the opaque dot. Blocking the zero spatial frequency component means there is zero field at the location of the opaque dot, and a zero field is equivalent to the sum of two waves of equal amplitude and opposite phase. Thus, the field at the location of the opaque dot can be expressed as the sum of the original zero spatial frequency wave, and a wave of equal amplitude but opposite phase (reference wave). This reference wave can be thought of as a fictitious point source located at the front focal point of lens L_3 and projecting a constant phase reference field onto the screen, while the input wave is Fourier transformed by L_3, forming an inverted image on the same screen. The superposition of image and reference waves on the screen creates an interference pattern. The contours of the constructive interference are the brightest and those of the destructive intereference are the darkest. The phase variation of the object is converted into an intensity variation by the Schlieren camera.

A quarter-wave phase plate can also be used instead of an opaque dot (Problem 1.10). The effects of a Schlieren camera can also be replicated by placing a knife edge to block the entire lower half domain of the Fourier transform.

Figure 1.34 shows a photograph of a fetal mouse taken by a Schlieren camera. Practically no image is formed if an ordinary microscope is used.

Figure 1.34   Schlieren photograph of a fetal mouse. (Courtesy of Olympus Optical Co., Ltd.)

Figure 1.33d shows an example of the spatial derivative operation. The Fourier transform of the derivative is

$$\mathcal{F}\{g'(x_0)\} = j2\pi f_x \cdot \mathcal{F}\{g\} \tag{1.204}$$

The filter placed at the back focal plane of the Fourier transform lens L_2 has transmittance characteristics of $\tau(f_x) = f_x$ in the spatial frequency domain. The filter, however, has to have this characteristic from $f_x = -\infty$ to $f_x = \infty$ in order for $\tau(f_x) = f_x$ to be realized. The positive f_x region can easily be realized, but the negative f_x region is much more difficult to realize. The phase of the negative f_x region can be reversed by a thin layer of a substance with a higher refractive index. This layer causes a π-radian phase delay. After another Fourier transform through lens L_3, the image of $g'(x_0)$ is obtained on the screen. One of the practical applications of such an operation is in edge enhancement or outlining the input image for easy identification of the shape of the object.

Figure 1.33e shows how to generate the repeated image of the input. A mask of a grid of pinholes with periods a and b in the x and y directions is placed in the back focal plane of lens L_2. The transmittance of such a mask is

$$\frac{1}{ab} \mathrm{III}\left(\frac{x}{a}\right) \mathrm{III}\left(\frac{y}{b}\right) \tag{1.205}$$

Thus, the field distribution after passing through the pinhole mask would be

$$\frac{1}{ab} \frac{e^{jkf}}{j\lambda f} G\left(\frac{x}{\lambda f}, \frac{y}{\lambda f}\right) \mathrm{III}\left(\frac{x}{a}\right) \mathrm{III}\left(\frac{y}{b}\right) \tag{1.206}$$

where $G(f_x, f_y)$ is the Fourier transform of the input $g(x, y)$ function. The effect to the field then, after passing through lens L_3, is an additional Fourier transform producing the field

$$E(x_i, y_i) = -e^{j2kf} g(-x_i, -y_i) * \left\{ \mathrm{III}\left(\frac{ax_i}{\lambda f}\right) \mathrm{III}\left(\frac{by_i}{\lambda f}\right) \right\} \tag{1.207}$$

on the back focal plane of lens L_3. From the above relationship, we see that the resultant image is a grid of repeated input images.

1.7.2   Optical Correlators

Optical correlators determine whether a particular image exists within a given picture. The two most popular types of correlators are the Vander Lugt correlator [1,12,13] and the joint transform correlator [14].

1.7.2.1   *Vander Lugt Correlator*

The Vander Lugt correlator (VLC) was first proposed by Vander Lugt in 1964. The VLC will be explained in two stages: a brief description of the principle in this section, followed by more detailed mathematical expressions in the next section.

Figure 1.35a shows a schematic of the VLC correlator. The input image $h(x_1, y_1)$ is interrogated to determine if it has the same shape as a given reference image $g(x_1, y_1)$.

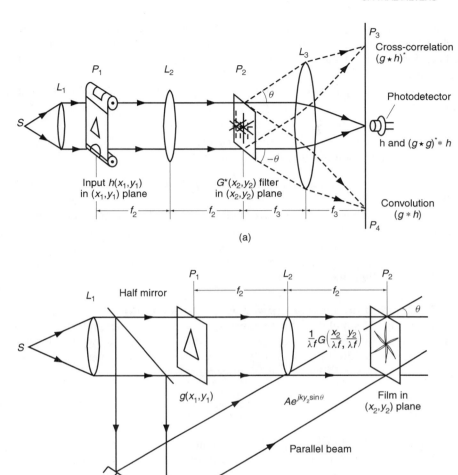

Figure 1.35 Vander Lugt correlator (VLC). (a) Layout. (b) Fabrication of a VLC filter.

First, with the input $h(x_1, y_1) = g(x_1, y_1)$ in the front focal plane of lens L_2, its Fourier transform

$$G(f_x, f_y) = |G(f_x, f_y)|e^{j\phi(f_x, f_y)} \tag{1.208}$$

is projected onto the P_2 plane in the back focal plane of lens L_2. The encoded filter, whose transmission is the complex conjugate of Eq. (1.208),

$$G^*(f_x, f_y) = |G(f_x, f_y)|e^{-j\phi(f_x, f_y)} \tag{1.209}$$

is inserted in the P_2 plane. The light transmitted through the filter is the product of Eqs. (1.208) and (1.209) and

$$E(f_x, f_y) = |G(f_x, f_y)|^2 \tag{1.210}$$

This function $E(f_x, f_y)$ has no spatial variation in phase, and the wavefront of the incident light on lens L_3 is parallel to the lens surface. The incident light converges to the back focal point of lens L_3 as indicated by the solid line where a photodetector is located to measure the light intensity.

If, however, the input is $h(x_1, y_1) \neq g(x_1, y_1)$, then the light projected through the filter is

$$H(f_x, f_y) = |H(f_x, f_y)|e^{j\phi'(f_x, f_y)} \tag{1.211}$$

and the transmitted light through the filter on P_2 becomes

$$E'(f_x, f_y) = |H(f_x, f_y)||G(f_x, f_y)|e^{j\phi'(f_x, f_y) - j\phi(f_x, f_y)} \tag{1.212}$$

The phase distribution is not uniform and not all the light converges to the back focal point of lens L_3. Moreover, the patterns of $|H(f_x, f_y)|$ and $|G(f_x, f_y)|$ may not overlap and the total light power reaching lens L_3 will be less than the case of $h(x_1, y_1) = g(x_1, y_1)$. Hence, the input that best matches the reference image $g(x_1, y_1)$ gives the largest light intensity at the output.

1.7.2.2 Detailed Analysis of the Vander Lugt Correlator

It is not always a simple matter to fabricate a filter with the prescribed complex transmission coefficient. One way, which is similar to fabricating a hologram, is illustrated in Fig. 1.35b. From Eq. (1.160), when the reference image $g(x_1, y_1)$ is put in the front focal plane of lens L_2, its Fourier transform,

$$E(x_2, y_2) = \frac{1}{j\lambda f}G\left(\frac{x_2}{\lambda f}, \frac{y_2}{\lambda f}\right) \tag{1.213}$$

is projected onto the film in the back focal plane. The factor e^{j2kf_0} is supressed. A parallel beam at angle θ to the normal of the film is added at the same time.

The transmittance $t(x_2, y_2)$ of the exposed and then developed film is

$$t(x_2, y_2) = t_0 - \beta \left| Ae^{jky_2 \sin\theta} + \frac{1}{j\lambda f}G\left(\frac{x_2}{\lambda f}, \frac{y_2}{\lambda f}\right) \right|^2 \tag{1.214}$$

where t_0 and β specify the photographic characteristics of the film. Equation (1.214) is rewritten as

$$t(x_2, y_2) = A^2 + \left(\frac{1}{\lambda f}\right)^2 \left| G\left(\frac{x_2}{\lambda f}, \frac{y_2}{\lambda f}\right) \right|^2$$

$$+ \frac{jA}{\lambda f}G^*\left(\frac{x_2}{\lambda f}, \frac{y_2}{\lambda f}\right)e^{jky_2 \sin\theta}$$

$$+ \frac{A}{j\lambda f}G\left(\frac{x_2}{\lambda f}, \frac{y_2}{\lambda f}\right)e^{-jky_2 \sin\theta} \tag{1.215}$$

where t_0 and $-\beta$ were suppressed. The third term of Eq. (1.215) is the most important term. The desired quantity for the filter is $G^*(x_2/\lambda f, y_2/\lambda f)$ but this term has an extra factor of $e^{jky_2 \sin\theta}$. This factor, however, does not harm the operation but just shifts the location of the correlation peak by θ degrees from the center.

Now, the input image $h(x_1, y_1)$ to be interrogated is installed in the front focal plane of lens L_2 in Fig. 1.35a and the filter is installed in the back focal plane of the same lens. The light transmitted through the filter is Fourier transformed by lens L_3 and its field $E(x_3, y_3)$ in the P_3 plane is given by

$$E(x_3, y_3) = \frac{1}{j\lambda f} \mathcal{F}\left\{ \frac{1}{j\lambda f} t(x_2, y_2) H\left(\frac{x_2}{\lambda f}, \frac{y_2}{\lambda f}\right) \right\}_{f_x = x_3/\lambda f, f_y = y_3/\lambda f} \tag{1.216}$$

Equation (1.216) has the same number of terms as Eq. (1.215), and we can designate these terms as $E_1(x_3, y_3)$, $E_2(x_3, y_3)$, $E_3(x_3, y_3)$, and $E_4(x_3, y_3)$.

The first term is

$$E_1(x_3, y_3) = \frac{1}{j\lambda f} \mathcal{F}\left\{ \frac{A^2}{j\lambda f} H\left(\frac{x_2}{\lambda f}, \frac{y_2}{\lambda f}\right) \right\}_{f_x = x_3/\lambda f, f_y = y_3/\lambda f} \tag{1.217}$$

Recall that lens L_3 performs the Fourier transform but not the inverse Fourier transform, and

$$E_1(x_3, y_3) = -A^2 h(-x_3, -y_3) \tag{1.218}$$

The inverted image of the input picture is seen around the origin of the P_3 plane.

The contribution of the second term is

$$E_2(x_3, y_3) = \frac{1}{j\lambda f} \mathcal{F}\left\{ \frac{1}{j(\lambda f)^3} \left| G\left(\frac{x_2}{\lambda f}, \frac{y_2}{\lambda f}\right) \right|^2 H\left(\frac{x_2}{\lambda f}, \frac{y_2}{\lambda f}\right) \right\}_{f_x = x_3/\lambda f, f_y = y_3/\lambda f} \tag{1.219}$$

Changing variables as

$$\frac{x_2}{\lambda f} = \xi, \qquad \frac{y_2}{\lambda f} = \eta$$

Eq. (1.219) becomes

$$E_2(x_3, y_3)$$
$$= -\frac{1}{(\lambda f)^2} \iint_{-\infty}^{\infty} G(\xi, \eta) G^*(\xi, \eta) H(\xi, \eta) \{ e^{-j2\pi\lambda f(f_x\xi + f_y\eta)} d\xi \, d\eta \}_{f_x = x_3/\lambda f, f_y = y_3/\lambda f} \tag{1.220}$$

$$E_2(x_3, y_3)$$
$$= -\frac{1}{(\lambda f)^2} g(-\lambda f f_x, -\lambda f f_y) * g^*(\lambda f f_x, \lambda f f_y) * h(-\lambda f f_x, -\lambda f f_y)_{f_x = x_3/\lambda f, f_y = y_3/\lambda f}$$

Inserting f_x and f_y into the equation and using Rule (5) from the boxed note, $E_2(x_3, y_3)$ is expressed as

$$E_2(x_3, y_3) = -\frac{1}{(\lambda f)^2} [g(x_3, y_3) \star g(x_3, y_3)]^* * h(-x_3, -y_3) \tag{1.221}$$

$E_2(x_3, y_3)$ is the autocorrelation of $g(x_3, y_3)$ convolved with $h(-x_3, -y_3)$. It is spread around the origin and is considered background noise (zero order noise).

The contribution of the third term is the most important one.

$$E_3(x_3, y_3) = \frac{1}{j\lambda f} \mathcal{F}\left\{ \frac{A}{(\lambda f)^2} G^*\left(\frac{x_2}{\lambda f}, \frac{y_2}{\lambda f} \right) H\left(\frac{x_2}{\lambda f}, \frac{y_2}{\lambda f} \right) e^{jky_2 \sin\theta} \right\}_{f_x = x_3/\lambda f, f_y = y_3/\lambda f}$$

(1.222)

Changing variables as

$$\frac{x_2}{\lambda f} = \xi, \qquad \frac{y_2}{\lambda f} = \eta$$

Eq. (1.222) becomes

$$E_3(x_3, y_3)$$
$$= \frac{A}{j\lambda f} \iint_{-\infty}^{\infty} G^*(\xi, \eta) H(\xi, \eta) \{ e^{j2\pi(\sin\theta/\lambda)\lambda f \eta} e^{-j2\pi\lambda f(f_x\xi + f_y\eta)} \, d\xi \, d\eta \}_{f_x = x_3/\lambda f, f_y = y_3/\lambda f}$$

(1.223)

Equation (1.223) can be expressed as the convolution of three factors. Using the rules in the boxed note, Eq. (1.223) is written as

$$E_3(x_3, y_3) = \frac{A}{j\lambda f} g^*(\lambda f f_x, \lambda f f_y) * h(-\lambda f f_x,$$
$$- \lambda f f_y) * \delta\left[\lambda f f_x, \lambda f \left(f_y - \frac{\sin\theta}{\lambda} \right) \right]_{f_x = x_3/\lambda f, f_y = y_3/\lambda f}$$

(1.224)

Inserting f_x and f_y into the equation gives

$$E_3(x_3, y_3) = \frac{A}{j\lambda f} g^*(x_3, y_3) * h(-x_3, -y_3) * \delta(x_3, y_3 - f \sin\theta)$$

(1.225)

Using Rule (5) in the boxed note, Eq. (1.225) is expressed as

$$E_3(x_3, y_3) = \frac{A}{j\lambda f} [g(x_3, y_3) \star h(x_3, y_3)]^* * \delta(x_3, y_3 - f \sin\theta)$$

(1.226)

Equation (1.226) is the expression of the cross-correlation between g and h. When $g(x_1, y_1)$ and $h(x_1, y_1)$ match, the peak value rises at

$$(0, f \sin\theta)$$

(1.227)

in the P_3 plane as indicated by the dotted line.

Finally, the contribution of the fourth term of the Eq. (1.216) is considered.

$$E_4(x_3, y_3) = \frac{1}{j\lambda f} \mathcal{F}\left\{ \frac{-A}{(\lambda f)^2} G\left(\frac{x_2}{\lambda f}, \frac{y_2}{\lambda f} \right) H\left(\frac{x_2}{\lambda f}, \frac{y_2}{\lambda f} \right) e^{-jky_2 \sin\theta} \right\}_{f_x = x_3/\lambda f, f_y = y_3/\lambda f}$$

(1.228)

$E_4(x_3, y_3)$ is quite similar to Eq. (1.221). The differences are the minus sign in the front, the absence of the complex conjugate sign on G, and the minus sign in the exponent

(a) The following are relationships for the Fourier transforms of Fourier transforms:

$$(1) \quad \mathcal{F}\{G(f)\} = \int_{-\infty}^{\infty} G(f)e^{-j2\pi fx} \, df$$

$$= \int_{-\infty}^{\infty} G(f)e^{j2\pi f(-x)} \, df$$

$$= g(-x)$$

$$(2) \quad \mathcal{F}\{G^*(f)\} = \int_{-\infty}^{\infty} G^*(f)e^{-j2\pi fx} \, df$$

$$= \left[\int_{-\infty}^{\infty} G(f)e^{j2\pi fx} \, df \right]^*$$

$$= g^*(x)$$

$$(3) \quad \mathcal{F}\{|G(f)|^2\} = \mathcal{F}\{G(f) \cdot G^*(f)\}$$

$$= g(-x) * g^*(x)$$

(b) The correlation symbol is $\star$ and the correlation operation is defined as

$$(4) \quad g(x) \star h(x) \triangleq \int g(\xi)h^*(\xi - x) \, d\xi$$

$$= \int g(\xi + x)h^*(\xi) \, d\xi$$

The operation $g \star h$ is called the cross-correlation, and the operation $g \star g$ is called the autocorrelation. Thus, the relationship between the convolution and the cross-correlation becomes

$$(5) \quad g(x) * h^*(-x) = \int g(\xi)h^*(\xi - x) \, d\xi$$

$$= g(x) \star h(x)$$

and from Rules (3) and (5)

$$(6) \quad \mathcal{F}\{|G(f)|^2\} = [g(x) \star g(x)]^*$$

of the last factor. Thus, the contribution of the fourth term is obtained directly from Eq. (1.226) as

$$E_4(x_3, y_3) = \frac{jA}{\lambda f} g(-x_3, -y_3) * h(-x_3, -y_3) * \delta(x_3, y_3 + f \sin \theta) \qquad (1.229)$$

The convolution of g and h appears around

$$(0, -f \sin \theta) \qquad (1.230)$$

in the P_3 plane as indicated by the dotted line.

In summary, the cross-correlation term $E_3(x_3, y_3)$ peaks up when $h = g$, indicating a match. The peak appears at $(0, f \sin \theta)$ in the P_3 plane. The value of θ should be chosen large enough to ensure that the $E_3(x_3, y_3)$ peak is well separated from the fields $E_1(x_3, y_3)$ and $E_2(x_3, y_3)$ that are spread around the origin.

Figure 1.36 Fingerprint interrogator. Correlation of fingerprints. *Left*: Reference fingerprint. *Middle*: Sample fingerprints. *Right*: Correlation peak. Only the sample fingerprint that matches the reference fingerprint generates a correlation peak. (Courtesy of A. Bergeron, J. Gauvin, and INO.)

Figure 1.36 shows the results when the VLC is applied in indentifying a specific fingerprint from multiple samples. The similarity to the encoded fingerprint is indicated by the brightness of the cross-correlation of $E_3(x_3, y_3)$.

1.7.2.3 Joint Transform Correlator

The joint transform correlator (JTC) was first proposed by Weaver and Goodman in 1966 [13,14,15]. A schematic diagram of the joint transform correlator is shown in Fig. 1.37. The difference between the VLC and JTC is the arrangement of the input. In a sense, the former is arranged in series and the latter in parallel. With the VLC, the Fourier transform H is projected onto the prefabricated G^* filter to generate HG^*, while with the JTC, the input images are put side by side in plane P_1 to generate G^*H. The JTC is based on the principle of the lateral shift invariance of the Fourier transform mentioned in Section 1.5.6.

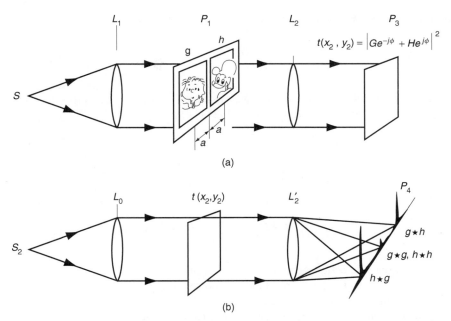

Figure 1.37 Joint transform correlator (JTC). (a) Step 1: Fabrication of the film $t(x_2, y_2)$. (b) Step 2: Display of correlation.

The JTC procedure is a two-step operation. The first step is shown in Fig. 1.37a and is the operation of the square detection of the Fourier transforms of the input images. The second step in Fig. 1.37b is the illumination of the detected Fourier transforms to display the cross-correlation.

In Fig. 1.37a, the reference image is placed a distance a to the left of the center of the P_1 plane, and the input image is placed a distance a to the right of the center. Before considering the general case of an arbitrary input image h, we can gain some useful insight into the JTC by examining the simple case of $g = h$. When $g = h$ in Fig. 1.37a, the input consists of two identical g images located side by side. The Fourier transform of such a pair, which appears in the output plane P_3, is

$$\mathcal{F}\{g(x - a) + g(x + a)\} = G(f)(e^{-j2\pi af} + e^{j2\pi af})$$
$$= 2G(f)\cos 2\pi af$$

Thus, a sinusoidal striation appears in the recorded film pattern. It is this sinusoidal striation that plays a key role in the JTC.

The developed film is now inserted in the front focal plane of lens L_2' in Fig. 1.37b for interrogation. The Fourier transform of the sinusoidal striation generates two peaks in the output plane P_4 (see Problem 1.3), indicating $g = h$.

If, however, $g \neq h$, the sinusoidal striation is absent from the film, and no peaks appear in the output plane P_4, indicating $g \neq h$. A more precise explanation follows.

The total input to the system is

$$g(x, -a, y_1) + h(x_1 + a, y_1) \tag{1.231}$$

The transmittance of the exposed and then developed film is

$$
t(x_2, y_2) = \frac{1}{(\lambda f)^2} \left| G\left(\frac{x_2}{\lambda f}, \frac{y_2}{\lambda f}\right) e^{-j2\pi(a/\lambda f)x_2} + H\left(\frac{x_2}{\lambda f}, \frac{y_2}{\lambda f}\right) e^{j2\pi(a/\lambda f)x_2} \right|^2
$$

(1.232)

where the transmittance of the film was assumed to be linearly proportional to the square of the incident field. Equation (1.232) is expanded as

$$
\begin{aligned}
t(x_2, y_2) = \left(\frac{1}{\lambda f}\right)^2 &\left[\left| G\left(\frac{x_2}{\lambda f}, \frac{y_2}{\lambda f}\right) \right|^2 + \left| H\left(\frac{x_2}{\lambda f}, \frac{y_2}{\lambda f}\right) \right|^2 \right. \\
&+ G^*\left(\frac{x_2}{\lambda f}, \frac{y_2}{\lambda f}\right) H\left(\frac{x_2}{\lambda f}, \frac{y_2}{\lambda f}\right) e^{j4\pi(a/\lambda f)x_2} \\
&\left. + H^*\left(\frac{x_2}{\lambda f}, \frac{y_2}{\lambda f}\right) G\left(\frac{x_2}{\lambda f}, \frac{y_2}{\lambda f}\right) e^{-j4\pi(a/\lambda f)x_2} \right]
\end{aligned}
$$

(1.233)

Thus, the square detection generates a G^*H term.

The second step is the generation of the correlation peaks. As shown in Fig. 1.37b, the film is placed in the front focal plane of lens L_2' and is illuminated by a parallel beam with amplitude A. The field in the back focal plane of L_2' is

$$
E(x_3, y_3) = \frac{A}{j\lambda f} \mathcal{F}\{t(x_2, y_2)\}_{f_x = x_3/\lambda f, f_y = y_3/\lambda f}
$$

(1.234)

Equation (1.234) can be calculated directly by comparing Eq. (1.233) with the VLC results. The first two terms of Eq. (1.233) are compared to Eq. (1.219), and the last two terms with Eq. (1.222). The final result is

$$
\begin{aligned}
E(x_3, y_3) = \frac{A}{j\lambda f} [&g(x_3, y_3) \star g(x_3, y_3) \\
&+ h(x_3, y_3) \star h(x_3, y_3) \\
&+ g(x_3, y_3) \star h(x_3, y_3) * \delta(x_3 - 2a, y_3) \\
&+ h(x_3, y_3) \star g(x_3, y_3) * \delta(x_3 + 2a, y_3)]^*
\end{aligned}
$$

(1.235)

The last two terms of Eq. (1.235) are the cross-correlation terms appearing at $(2a, 0)$ and $(-2a, 0)$. If g and h are pure real, the two peaks are of identical shape.

When $g = h$, both curves are not only identical but their intensities peak up, indicating a match.

The complex conjugate signs appearing in Eqs. (1.221), (1.226), and (1.235) disappear if the (x_3, y_3) coordinates are further transformed as $x_3 \to -x_3$, $y_3 \to -y_3$, namely, rotating the coordinates of (x_3, y_3) by $180°$ in its plane.

1.7.2.4 Comparison Between VLC and JTC

VLC and JTC are compared as follows [16,17]:

1. While VLC needs a prefabricated reference filter, JTC does not.

2. After the reference filter has been made, the VLC can interrogate the input in one step. If the same reference image is used, countless interrogations can be made without changing the filter. The JTC requires a two-step operation. For

every interrogation, a Fourier transform filter has to be made, which then has to be illuminated for the correlation peak.

3. VLC demands a very precise lateral alignment of the filter. Even though the lateral location of the input image h in the P_1 plane is arbitrary in Fig. 1.35a, its Fourier transform H always appears at the same location in the P_2 plane. The location of the filter G^* has to match precisely with this location of H. In fact, it has to match within microns. In short, the location of h is arbitrary but that of G^* has to be very precise. JTC does not demand this precision.

4. The required diameter of the JTC lens is twice that of the VLC lens.

5. A higher signal-to-noise ratio (S/N) is obtainable with VLC because a larger separation from the zero order terms is possible by increasing θ, whereas with JTC, the separation from the zero order term is limited by a and hence by the size of the input lens.

1.7.3 Rotation and Scaling

Even though the JTC is impervious to lateral misalignment, its sensitivity is significantly reduced if the input and reference images are rotated with respect to one another or there is a difference in their sizes.

First, let us consider a rotation countermeasure [18]. This countermeasure is explained by way of example in Fig. 1.38.

The points on the rectangle are replotted in polar coordinates with $\rho(\theta)$ as the vertical axis and θ as the horizontal axis. To obtain the polar graph, the transformation

$$\rho(\theta) = \sqrt{x^2 + y^2}$$
$$\theta = \tan^{-1}\left(\frac{y}{x}\right)$$

(1.236)

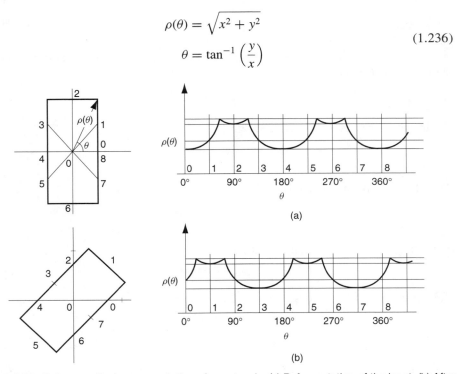

Figure 1.38 Polar coordinate representation of a rectangle. (a) Before rotation of the input. (b) After rotation.

was performed. Figure 1.38a shows the polar graph before the rotation of the rectangle and Fig. 1.38b, after the rotation.

It is clear that the shape of the polar graphs are the same regardless of the rotation of the input image. The only difference is a shift of the curve in the horizontal direction. If the polar coordinate image is used as the input, the property of the lateral shift invariance of the Fourier transform can be utilized.

The shape of the transformed result, however, depends on the choice of the center of rotation. This polar graph method should be used in the Fourier transform domain rather than in the input image domain because the location of the Fourier transformed image is independent of the location of the input in the input plane.

Next, let's look at a countermeasure for differences in scale [19]. Figure 1.39 shows three similar triangles whose heights are in the ratio of 1:5:10. Figure 1.40 shows these same triangles replotted on a logarithmic graph. The triangular shape is distorted but all three are congruent. The locations of the logarithmic images are shifted to the right according to the size of the input image. The logarithmic images can be used as the input for the VLC or JTC processor when the scales of the inputs are different.

Logarithmic scaling for the VLC will be explained using mathematical expressions. The transformation from the (x, y) plane to the (ξ, η) log–log plane is

$$\xi = \log x, \qquad \eta = \log y \tag{1.237}$$

The reference image $g(x, y)$ in the xy plane is transformed to $m(\xi, \eta)$ in the log–log plane.

$$g(x, y) \Rightarrow m(\xi, \eta) \tag{1.238}$$

The image $m(\xi, \eta)$ is used as the input to the VLC in the P_1 plane in Fig. 1.35b in order to fabricate the filter. In the P_2 plane, this quantity is Fourier transformed to $M(u)$. For simplicity, the one-dimensional, rather than the two-dimensional, Fourier transform will be used.

$$M(u) = \int_0^\infty m(\xi)e^{-j2\pi u\xi}\, d\xi \tag{1.239}$$

Taking the derivative of Eq. (1.237) gives

$$\frac{d\xi}{dx} = \frac{1}{x}$$

With the help of this derivative, $M(u)$ is expressed in terms of x as

$$M(u) = \int_0^\infty f(x)e^{-j2\pi u\log x}\frac{dx}{x} \tag{1.240}$$

Let's rewrite the exponential by putting

$$Y = e^{-j2\pi u\log x} \tag{1.241}$$

Taking the log of both sides of Eq. (1.241) gives

$$\log Y = -j2\pi u \log x \log e$$

$$\log Y = \log x^{-j2\pi u\log e}$$

and hence,

$$Y = x^{-j2\pi u\log e} \tag{1.242}$$

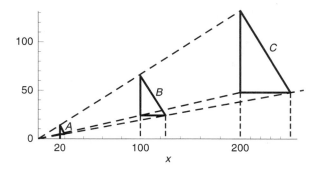

Figure 1.39 Three similar triangles with ratio 1:5:10.

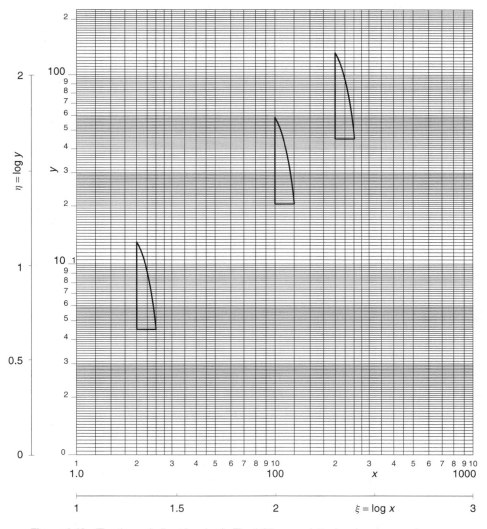

Figure 1.40 The three similar triangles in Fig. 1.39 are replotted on log–log graph paper.

Inserting Eq. (1.242) back into (1.240) gives

$$M(u) = \int_0^\infty f(x) x^{-j2\pi u \log_{10} e - 1} \, dx \tag{1.243}$$

If the same is repeated using a ln–ln graph instead of log–log graph, Eq. (1.243) becomes

$$M(u) = \int_0^\infty f(x) x^{-j2\pi u - 1} \, dx \tag{1.244}$$

where $f(x)$ is defined in $0 \le x < \infty$. This integral is known as the Mellin transform; namely, the Fourier transform of the logarithm of the input function is the Mellin transform of the input function. The VLC filter $M^*(u)$ is made out of $M(u)$ according to the method described in Section 1.7.2.2.

Next, the input image h to be interrogated has to be ln–ln transformed in a similar manner. Let us say the input image h is a times the reference image g, namely,

$$h(x, y) = g\left(\frac{x}{a}, \frac{y}{a}\right) \tag{1.245}$$

The ln–ln transform of Eq. (1.245), $m_a(\xi, \eta)$, is put into the P_1 plane in Fig. 1.35a. The x component of the output in the P_2 plane is

$$M_a(u) = \mathcal{M}\left\{g\left(\frac{x}{a}\right)\right\} \tag{1.246}$$

where $\mathcal{M}\{\ \}$ represents the operation of the Mellin transform.

From Eq. (1.244), $M_a(u)$ is

$$M_a(u) = \int_0^\infty g\left(\frac{x}{a}\right) x^{-j2\pi u - 1} \, dx \tag{1.247}$$

Putting $x/a = X$ gives

$$M_a(u) = a^{-j2\pi u} \int_0^\infty f(X) X^{-j2\pi u - 1} \, dX \tag{1.248}$$

Note the similarity between Eqs. (1.244) and (1.248). In order to rewrite the first factor of Eq. (1.248) as a power of e, let

$$Y = a^{-j2\pi u}$$
$$\ln Y = -j2\pi u \ln a$$
$$Y = e^{-j2\pi u \ln a} \tag{1.249}$$

Putting Eq. (1.249) into (1.248) and comparing with Eq. (1.244) gives the final result:

$$M_a(u) = e^{-j2\pi u \ln a} M(u) \tag{1.250}$$

The enlargement of the input generates an additional phase shift.

$M_a(u)$ is the pattern projected to P_2 in Fig. 1.35a, where the Fourier transform $M^*(u)$ of the reference image has already been placed. Thus, the input to lens L_3 in Fig. 1.35a is

$$M^*(u) M_a(u) = e^{-j2\pi u \ln a} |M(u)|^2 \tag{1.251}$$

where λf was assumed unity for simplicity.

The output from lens L_3 is now proportional to

$$[m(x_3) \star m(x_3)]^* \delta(x_3 + \ln a) \qquad (1.252)$$

Thus, the magnitude of the correlation peak is always

$$[m(x_3) \star m(x_3)]^*$$

regardless of the enlargement factor a. Only the location of the peak shifts in accordance with the enlargement a.

1.7.4 Real-Time Correlation

Correlators have many practical applications [13,15]. They are used as robotic eyes in automatic assembly lines, and as security devices for checking biometric indicators such as fingerprints, facial images, voice, and DNA. For applications such as these, correlators with real-time response are crucial. The bottleneck for the real-time operation of either the VLC or JTC is the recording of G and H by means of photographic film, which acts as a square detector to produce G^*H. The photographic film can be replaced either by a photorefractive (PR) crystal whose index of refraction is changed by light intensity (see Section 5.6) or simply by using a CCD camera with the camera lens removed.

Figure 1.41 shows an example of the real-time operation of the JTC using a photorefractive crystal. The pattern of $|G + H|^2$ is generated by laser light S_1 from the left. This pattern is recorded by the photorefractive crystal in the P_2 plane. The recorded pattern is then read by laser light S_2 from the right. S_1 and S_2 have different wavelengths. The read image is projected into a CCD camera whose camera lens has been removed. As far as the input method is concerned in this particular example, the input scene h has been taken by another CCD camera whose camera lens is intact. The input scene is displayed on the spatial light modulator (SLM) located in the input plane P_1. The SLM is a liquid crystal display panel, which is described in Section 5.10.4.4. One of the advantages of the SLM is its optically flat display surface. The surface

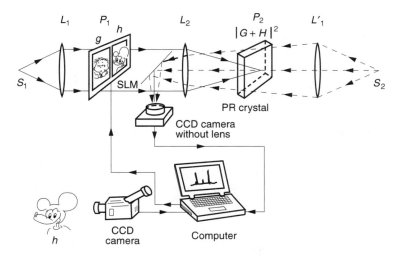

Figure 1.41   Real-time JTC using a photorefractive (PR) crystal.

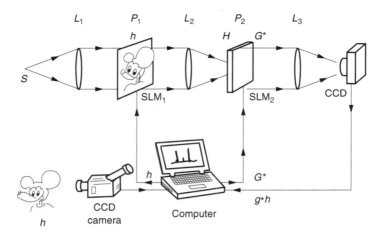

Figure 1.42   Real-time VLC using optics and a computer.

condition of the images is an important factor in the performance of optical signal processors that use coherent light.

Optical signal processing provides high speed and large information capacity while computers provide versatility and reliability. The examples shown in Figs. 1.41 and 1.42 make use of both.

Now let's look at the real-time VLC example shown in Fig. 1.42. The input image, taken by the CCD camera whose lens is intact, is fed into the computer, and is displayed on SLM_1 in the input plane P_1. The input image is then optically Fourier transformed and projected over the computer-generated G^* displayed on SLM_2. The operation of $\mathcal{F}\{G^*H\}$ is again performed optically and is projected onto the CCD camera whose lens has been removed. The signal from the lensless CCD camera is fed back to the computer. The computer processes the results to arrive at a final decision.

Figure 1.43 shows a similar arrangement but with the JTC. The process of obtaining G^*H is quite similar to the previous case. The laser light S, which has been branched off to the bottom of the figure, optically performs the operation of $\mathcal{F}\{G^*H\}$ to provide $g \star h$ on the CCD camera whose lens has been removed.

Figure 1.44 shows a system that relies more heavily on the computer. The scene captured by the lensed CCD camera is displayed on the SLM. The optical system Fourier transforms $(g + h)$ to give $|G + H|^2$ on the lensless CCD camera. The computer takes over the rest of the processing including the operation of $\mathcal{F}\{G^*H\}$ as well as the decision on the result of the interrogation.

1.7.5 Cryptograph

Another special application of the spatial filter is the cryptograph [20,21] (meaning encoding) of an image for security purposes. Optical signal processing needs to provide fast and reliable identification of people and verification of their signatures on a document.

First, the method of encryption will be described referring to Fig. 1.45a. Let the input image to be encrypted be $E_0(x_0, y_0)$. The Fourier transform of $E_0(x_0, y_0)$ is projected onto the back focal plane of L_2. The key card, onto which a white sequence noise pattern $n(x, y)$ is imprinted, is also placed in the back focal plane of L_2. The

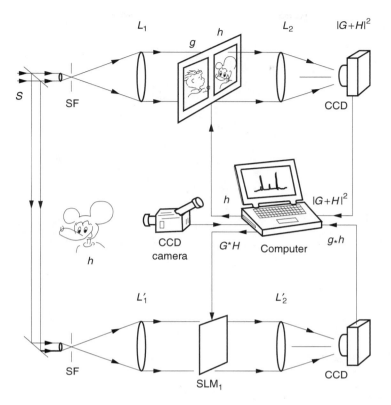

Figure 1.43   Real-time JTC using optics and a computer.

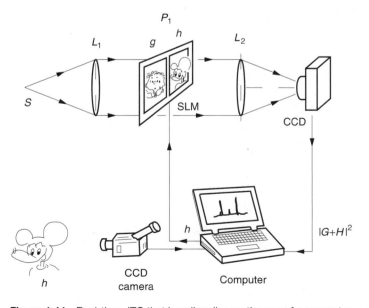

Figure 1.44   Real-time JTC that heavily relies on the use of a computer.

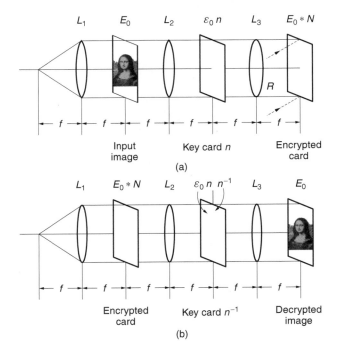

Figure 1.45 Cryptographic spatial filters. (a) Encryption. (b) Decryption.

input to the second Fourier transform lens L_3 becomes

$$\mathcal{F}\{E_0(x, y)\}n(x, y) \tag{1.253}$$

In the back focal plane of L_3, the image given by Eq. (1.253) is further Fourier transformed to

$$E(x_i, y_i) = E_0(-x_i, -y_i) * N(x_i, y_i) \tag{1.254}$$

where $N(x_i, y_i)$ is the Fourier transform of the white noise sequence. The output image represented by Eq. (1.254) does not resemble the original input image $E_0(x_0, y_0)$. The original input image can be recovered only when the key card is used. In order to photographically record $E(x_i, y_i)$ including the phase onto a card, a reference plane wave $R(x_i, y_i)$ (not shown in the figure) is added at the time of recording. Thus, the encrypted card has the intensity pattern $I(x_i, y_i)$:

$$
\begin{aligned}
I(x_i, y_i) &= |R(x_i, y_i) + E_0(-x_i, -y_i) * N(x_i, y_i)|^2 \\
&= |R(x_i, y_i)|^2 + |E_0(-x_i, -y_i) * N(x_i, y_i)|^2 \\
&\quad + R^*(x_i, y_i)[E_0(-x_i, -y_i) * N(x_i, y_i)] \\
&\quad + R(x_i, y_i)[E_0(-x_i, -y_i) * N(x_i, y_i)]^* \tag{1.255}
\end{aligned}
$$

Next, the method of decrypting the original image is explained referring to Fig. 1.45b. The encrypted pattern $I(x_i, y_i)$ is placed in the front focal plane of lens L_2, and all four terms in Eq. (1.255) are Fourier transformed in the back focal plane.

Only the Fourier transform $E_3(x, y)$ of the third term in Eq. (1.255) is of prime concern,

$$E_3(x, y) = R_0 \mathcal{F}\{E_0(-x, -y)\} \cdot n(-x, -y) \qquad (1.256)$$

where for simplicity of expression, the direction of incidence of the plane reference wave was assumed normal to the card surface in order to make $R(x_i, y_i)$ a constant R_0 across the surface of the card. If a key card imprinted with the pattern of $n^{-1}(-x, -y)$ is placed in the back focal plane of L_2, the second factor in Eq. (1.256) is canceled. The input to the Fourier transform lens L_3 becomes $R_0 \mathcal{F}\{E_0(-x, -y)\}$ and in the back focal plane of L_3, the original input image is recovered. Recovery of the original image is only possible for a key card imprinted with $n^{-1}(x, y)$.

A method for fabricating a key card that can be used for both encrypting and decrypting is described as follows. A pseudorandom pattern of 0s and 1s is written onto a half-wavelength thick optical film on a substrate. The light passing through the "0" location experiences no phase shift while that passing through the "1" experiences a π-radian phase shift. The transmission pattern of the key card is then

$$n(x, y) = e^{j\pi b(x, y)} \qquad (1.257)$$

where $b(x, y)$ is the pseudorandom pattern. Note that such a pattern satisfies $n^{-1}(x, y) = n(x, y)$ and can be used for both encrypting and decrypting. The pattern $n^{-1}(-x, -y)$ can be obtained by rotating the card by $180°$ in the plane of the card.

1.8 HOLOGRAPHY

Holography was invented by Dennis Gabor in 1948 when he was trying to improve the quality of electron microscope images. The word *holo* in Greek means complete and *gram* means recording, so that a "hologram" is a complete recording of the wave scattered from an object. Holography uses both phase and amplitude distributions of the scattered light to record the image of the object [1,6,8].

Both conventional photography and holography utilize light-sensitive film as the recording medium. In both cases, the film records light intensity. In conventional photography, the camera's lens generates an image of an object in the film plane, and the film records the image's intensity pattern. In a hologram, the film is directly exposed to the light scattered by an object. By itself, recording the scattered wave's intensity is not sufficient to make a hologram, as phase information would be lost. This shortcoming is overcome by illuminating the holographic film with a reference wave, as well as the scattered object wave. The holographic film records the fringe pattern that results from the interference of the reference wave and the scattered object wave. Fringe contours with high intensity indicate the scattered object wave and the reference wave are in phase; likewise, low intensity contours indicate the waves are out of phase. Note that the phase information has been converted into an intensity pattern. The holographic film thus exposed and developed is the hologram.

In order to see the image from the hologram, a laser beam illuminates the hologram. The laser beam is diffracted by the recorded fringe patterns on the hologram. The diffracted light recreates the image of the original object. An observer looking through the hologram toward the laser will see a view just as if the real, original object were present behind the hologram.

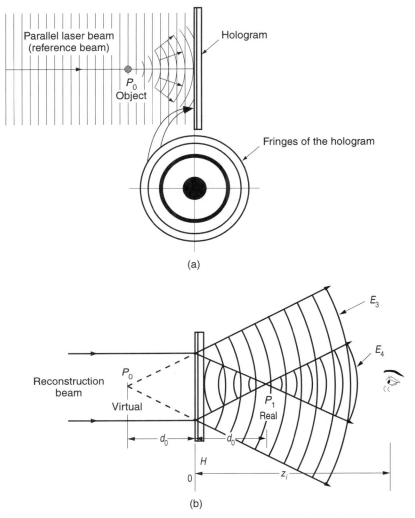

Figure 1.46   Illustration of the Gabor-type hologram. (a) Fabrication of the hologram (point object). (b) Reconstruction of the image.

1.8.1   Gabor-Type Hologram

Figure 1.46a shows an arrangement for fabricating a Gabor-type hologram. A parallel laser beam illuminates both the object and the photographic film. A special feature of the Gabor-type hologram is that only one beam illuminates both the object and the film. In contrast, the Leith–Upatnieks type hologram uses two beams: one for illuminating the object and the other for illuminating the film.

Let's take a closer look at the geometry of the Gabor-type hologram in Fig. 1.46a. For simplicity, the object at P_0 is a point object. The reference beam R is incident along the z axis, which is perpendicular to the film. The object beam is the wave scattered from the object. The film is simultaneously illuminated by both the reference beam and the object beam. The film records the interference pattern between the reference and object beams. For the recording of the interference pattern to be successful, it is necessary to have a mechanically stable setup, a spectrally pure source, and a high-resolution film. After being exposed, the film is developed, and the result is a Gabor-type hologram.

The human eye is not able to decipher the interference pattern of the hologram directly. Instead, the image has to be recreated by a process called reconstruction.

Let's assume that a laser was used to provide the spectrally pure source for creating the hologram, and that a similar laser with the same wavelength is used in the reconstruction process. Figure 1.46b shows the arrangement for reconstructing the image from the hologram. The hologram is illuminated from the back by the reconstruction laser beam. The light scattered from the interference pattern on the hologram forms the image of the original point object.

A more quantitative description is in order. Let's say the film is in the (x, y) plane at a distance d_0 from the object. The field scattered by the point object is a spherical wave centered at the object. The scattered field observed on the film is

$$O(x, y) = \frac{A}{j\lambda} \frac{e^{jkr}}{r} \qquad (1.258)$$

where

$$r = \sqrt{d_0^2 + x^2 + y^2}$$

If d_0 is much larger than the size of the photographic film, Eq. (1.258) can be approximated as in Eq. (1.31)

$$O(x, y) = \frac{A}{j\lambda d_0} e^{jk[d_0 + (x^2 + y^2)/2d_0]} \qquad (1.259)$$

The reference beam is a plane wave propagating in the z direction and is expressed by

$$R = R_0 e^{jkz} \qquad (1.260)$$

where R_0 is a real number.

The photographic film is now exposed to the interference pattern of the object and reference beams. The developed film looks like a Fresnel zone plate and is composed of a set of concentric rings as shown below the hologram in Fig. 1.46a. The expression $t(x, y)$ for the transmission coefficient of the film is

$$t(x, y) = t_0 - \beta |R_0 + O(x, y)|^2$$
$$= t_0 - \beta[R_0^2 + |O(x, y)|^2 + R_0 O(x, y) + R_0 O^*(x, y)] \qquad (1.261)$$

We are now ready to reconstruct the image from the hologram by illuminating it with the reconstruction beam. The Fresnel diffraction pattern forms the reconstructed image. In order to use the Fresnel diffraction formula Eq. (1.38), we need to find the integrand $E(x, y)$, which is the light distribution that has just passed through the hologram. For simplicity, the reconstruction beam is assumed to be the same as the reference beam used at the time of the hologram's fabrication. From Eqs. (1.260) and (1.261), the expression for $E(x, y)$ is

$$E(x, y) = R_0^2 + |O(x, y)|^2 R_0 + R_0^2 O(x, y) + R_0^2 O^*(x, y) \qquad (1.262)$$

where t_0 and β were suppressed. The hologram is placed at $z = 0$ and the diffraction pattern is observed in the (x_i, y_i) plane at $z = z_i$.

The first two terms of Eq. (1.262) do not have much of a spatial variation, which means that this portion of the reconstruction beam passes straight through along the z axis with some attenuation. The Fresnel diffraction pattern $E_3(x_i, y_i)$ associated with the third term of Eq. (1.262) is from Eqs. (1.38), (1.259), and (1.260):

$$E_3(x_i, y_i) = AR_0^2 \frac{e^{jk[z_i + d_0 + (x_i^2 + y_i^2)/2z_i]}}{j\lambda z_i} \frac{1}{j\lambda d_0} \mathcal{F}\{e^{jk(x^2 + y^2)/2d_0} e^{jk(x^2 + y^2)/2z_i}\}_{f_x = x_i/\lambda z_i, f_y = y_i/\lambda z_i}$$

(1.263)

Equation (1.263) can be rewritten as

$$E_3(x_i, y_i) = AR_0^2 \frac{e^{jk[z_i + d_0 + (x_i^2 + y_i^2)/2z_i]}}{j\lambda z_i} \frac{1}{j\lambda d_0} \mathcal{F}\{e^{jk(x^2 + y^2)/2D}\}_{f_x = x_i/\lambda z_i, f_y = y_i/\lambda z_i} \quad (1.264)$$

where

$$\frac{1}{D} = \frac{1}{z_i} + \frac{1}{d_0}$$

Using the Fourier transform relationship Eq. (1.44) gives

$$E_3(x_i, y_i) = AR_0^2 \frac{e^{jk[z_i + d_0 + (x_i^2 + y_i^2)/2z_i]}}{j\lambda z_i \, j\lambda d_0} \cdot j\lambda D\{e^{[-j\pi\lambda D(f_x^2 + f_y^2)]}\}_{f_x = x_i/\lambda z_i, f_y = y_i/\lambda z_i}$$

(1.265)

$$E_3(x_i, y_i) = AR_0^2 \frac{e^{jk(z_i + d_0)}}{j\lambda(z_i + d_0)} \cdot e^{jk(x_i^2 + y_i^2)/2(z_i + d_0)} \quad (1.266)$$

Equation (1.266) is the expression for a diverging spherical wave that would be established if a point source were located at a distance of $z_i + d_0$ from the observer. Referring to Fig. 1.46b, this location is P_0 and is exactly where the point object had been placed. No light rays, however, converge to this point so that the image is a virtual image of the point object.

Next, the diffraction pattern due to the fourth term is obtained in a similar manner and the field observed is

$$E_4(x_i, y_i) = AR_0^2 \frac{e^{jk(z_i - d_0)}}{j\lambda(z_i - d_0)} e^{jk(x_i^2 + y_i^2)/2(z_i - d_0)} \quad (1.267)$$

Equation (1.267) is the expression of a spherical wave that would be established if a point source were located at a distance of $z_i - d_0$ from the observer. Referring to Fig. 1.46b, this location is P_1 and is symmetric to P_0 with respect to the hologram. The spherical wave is convergent first in the region $z_i < d_0$ and actually converges at $z_i = d_0$; then it is divergent again in the region of $z_i > d_0$. If a sheet of paper is placed at $z_i = d_0$, a bright spot is observed. This bright spot is the real image of the point object.

The object was assumed to be a point object, but the analysis can be extended to a more complex object. The complex object can be considered as an ensemble of individual points; thus, the virtual image of the original object is observed at the original object location. This image is called the orthoscopic image.

The real image is in a plane that is symmetric to the object with respect to the hologram. This image is called the pseudoscopic image of the hologram and is explained further in Section 1.8.3. The observer can view this image either by inserting a piece of paper or by positioning his/her eyes at a distance $z_i > d_0$.

A major disadvantage of the Gabor-type hologram is that the wavefronts overlap. With the geometry shown in Fig. 1.46b, the observer sees three wavefronts at the same time; one from the undiffracted portion of the reconstructing beam that passes straight through the hologram, one from the extension of the virtual image P_0, and one from the real image P_1. When the observer tries to focus his/her eyes on P_0, the reconstruction beam and the blurred image of P_1 are in the background and the quality of the reconstructed image of the Gabor-type hologram is not high.

1.8.2 Off-Axis Hologram

The disadvantage of the Gabor-type hologram was overcome by Leith and Upatnieks in 1962. They proposed a scheme for slanting the direction of incidence of the reference beam in order to spatially separate the locations of the reconstructed images. The quality of the images were substantially improved.

Figure 1.47a shows an off-axis reference hologram. The reference beam is incident on the hologram at an angle θ with respect to the normal to the hologram and is expressed by

$$R = R_0 e^{jkx \sin\theta + jkz \cos\theta} \tag{1.268}$$

where R_0 is a real number, again. The same light beam will be used as the reconstruction beam. The change to the slanted reference beam from the straight-on reference beam needs only a minor modification to the previous results obtained with the Gabor-type hologram in Section 1.8.1. The transmission coefficient of the hologram at $z = 0$ is

$$
\begin{aligned}
t(x, y) &= t_0 - \beta |R_0 e^{jkx \sin\theta} + O(x, y)|^2 \\
&= t_0 - \beta [R_0^2 + |O(x, y)|^2 + R_0 e^{-jkx \sin\theta} O(x, y) + R_0 e^{jkx \sin\theta} O^*(x, y)]
\end{aligned}
\tag{1.269}
$$

In the process of reconstructing the holographic image, the hologram is illuminated with a reconstruction beam that is the same as Eq. (1.268). The input pattern $E(x, y)$ to the Fresnel diffraction formula is obtained by the multiplication of $t(x, y)$ and R.

$$E(x, y) = R_0^3 e^{jkx \sin\theta} + |O(x, y)|^2 R_0 e^{jkx \sin\theta} + R_0^2 O(x, y) + R_0^2 e^{j2kx \sin\theta} O^*(x, y) \tag{1.270}$$

The amplitude of the first two terms has practically no spatial variation; these terms are the parallel beams in the θ direction. The third term is identical to the earlier result in Eq. (1.263). This means that tilting the reference beam does not affect the location of the virtual image (as long as the reference beam is used as a reconstruction beam). The virtual image occupies the same location as the original object, which in Fig. 1.47a is along the z axis. The fourth term can be obtained the same way that Eq. (1.267)

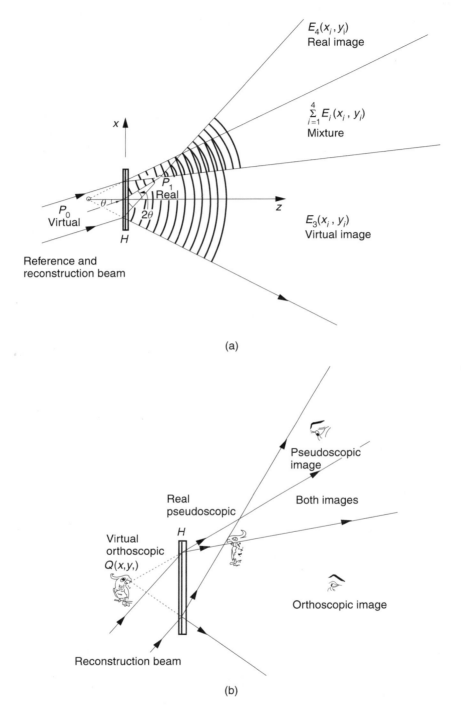

Figure 1.47   Off-axis reference beam hologram. (a) Geometry. (b) Regions for observing pseudo-scopic and orthoscopic images.

was obtained:

$$E_4(x_i, y_i) = AR_0^2 \frac{e^{jk[z_i - d_0 + (x_i^2 + y_i^2)/2z_i]}}{j\lambda z_i} \frac{1}{j\lambda d_0} \mathcal{F}\{e^{jk(x^2 + y^2)/2D'} \cdot e^{j2kx\sin\theta}\}_{f_x = x_i/\lambda z_i, \, f_y = y_i/\lambda z_i}$$

(1.271)

where

$$\frac{1}{D'} = \frac{1}{z_i} - \frac{1}{d_0}$$

(1.272)

The Fourier transform of the second factor in the braces is

$$\mathcal{F}\{e^{j2kx\sin\theta}\} = \delta\left(f_x - \frac{2\sin\theta}{\lambda}\right)$$

(1.273)

Using the convolution relationship and then the delta function property, Eqs. (1.110) and (1.44), the final expression for Eq. (1.271) is

$$E_4(x_i, y_i) = \frac{jAR_0^2}{\lambda(z_i - d_0)}$$

$$\times \exp\left[jk\left(z_1 - d_0 + \frac{(x_i - 2d_0\sin\theta)^2 + y_i^2 + 4d_0(z_i - d_0)\sin^2\theta}{2(z_i - d_0)}\right)\right]$$

(1.274)

Thus, the real image appears at

$$(x_i, y_i, z_i) = (2d_0\sin\theta, 0, d_0)$$

(1.275)

with some aberration for large θ.

For small θ, Fig. 1.47a summarizes the positions of the images. These images are the virtual image at $\theta = 0$, the undiffracted beam at θ, and the real image at 2θ.

Except for the overlap region, the images are clearly separated and can be observed without interference.

1.8.3 Pseudoscopic Image

When the observer views the real image, his/her eyes will see a peculiar image. Let's suppose the object is a bird, as shown in Fig. 1.47b. The object bird was facing the observer when the hologram was made, but the reconstructed real image of the bird is facing away from the observer. The hologram can record only the side of the object that is facing the hologram. The light scattered from the other side never reaches the hologram and cannot be recorded. Because of this, the observer looking at the real reconstructed image sees the tip of the beak further away than the bird's eyes. The observer has the sensation of looking straight through the back surface and seeing the front surface image from the inside out (like looking at the inside of a mask). Such an inside-out image is called a pseudoscopic image.

1.8.4 Volume Hologram

Important applications such as the high-density recording of images and white light or color holograms are based on the volume hologram. As the thickness of the emulsion

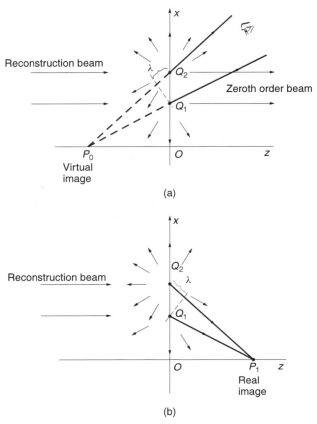

Figure 1.48 Illustration of the reconstruction of the holographic image from a thin emulsion film. (Cross section of the hologram). (a) Reconstruction of the virtual image. (b) Reconstruction of the real image.

of the photographic film is increased, properties that do not exist with thin emulsion holograms begin to surface. Before going into the case of the thick emulsion, the case of the thin emulsion will be reviewed.

The geometry considered is exactly the same as already studied in Fig. 1.46, but Fig. 1.48 is intended to show the finer details of the fringe patterns on the hologram. Points $Q_1, Q_2, \ldots$ (actually rings) form an array of scattering centers. The reconstruction of the image can be treated approximately as the problem of obtaining the scattering pattern from a periodic array of point scatterers. As mentioned earlier in Section 1.4.9, the field pattern can be separated into the element pattern, which is the radiation pattern of the individual element, and the array pattern, which is the pattern determined by the spacing between the elements and the overall dimension of the array. The scattering pattern is the product of these two patterns as indicated by Eq. (1.134).

When the emulsion of the film is thin, the size of the individual Q's are comparable to the light wavelength, and each Q scatters the light in all directions. Hence, the element pattern is omnidirectional when the emulsion is thin.

With the geometry shown in Fig. 1.48a, the path difference between $\overline{P_0Q_1}$ and $\overline{P_0Q_2}$ is exactly one wavelength and the rays scattered from Q_1 and Q_2 enhance each other in the direction of $\overline{P_0Q_2}$. The observer sees this peak of light as the virtual image.

Another peak is observed in the direction of propagation of the reconstruction beam because the light scattered from Q_1 and Q_2 are also in phase in this direction. This peak is associated with the undiffracted or zeroth order beam and is treated as noise.

Furthermore, with the geometry shown in Fig. 1.48b, the path difference between $\overline{Q_1P_1}$ and $\overline{Q_2P_1}$ is exactly one wavelength and the array pattern makes another peak that is associated with the real image. Thus, the array pattern of a thin emulsion hologram has three peaks; namely, in the directions of the virtual image, the zeroth order noise, and the real image. The sharpness of the image is determined by the product of the element and array patterns. In the case of a thin emulsion, the element pattern is omnidirectional and the sharpness of the image is predominantly determined by the array pattern.

Next, the thick emulsion case will be explained. The scattering points $Q_1, Q_2, \ldots$ grow into a set of mirror platelets made out of silver grains as shown in Fig. 1.49a. The mirror platelets are formed by the standing wave pattern created by the interference between the reference and object beams. They are oriented in the plane of the bisector of the angle between the reference and object beams. If the reconstruction beam is the same as the reference beam, the orientation of each mirror platelet is such that the reflected beam is directed toward the extension of the object beams $\overline{P_0Q_1}$ and $\overline{P_0Q_2}$, thus forming the virtual image.

Because of the reflective nature of the platelet mirror surfaces, the element pattern has only one peak, which corresponds to the virtual image at P_0. Light is not reflected toward the real image P_1, and *no pseudoscopic image is observed from the volume hologram.*

As far as the array pattern is concerned, the spacing between the mirror platelets is the same as for the thin emulsion, and the array pattern is also directed toward the extension of the object beam. The diffraction pattern is the product of the element and array patterns, both of which are sharply directed in the same direction. Thus, a very sharp image is reconstructed from a thick emulsion hologram. This type of hologram is often called a volume hologram.

Bragg reflection is the basis for X-ray analysis of atomic layers. Bragg reflection takes place when the directivity established by the path difference between the rays diffracted from adjacent atomic layers matches up with the direction of the specular reflection from the surface of the atomic layer. With the geometry shown in Fig. 1.49c, the Bragg reflection is in the direction given by

$$2d \sin \theta = m\lambda \tag{1.276}$$

where θ is traditionally taken with respect to the surface rather than the normal. In other words, Bragg reflection takes place when the array pattern of the atomic layer matches the element pattern of the atomic layer. So, we can say that the reconstruction of the image from the volume hologram is based on Bragg diffraction.

A volume hologram is capable of storing a large number of images. If the images are each recorded using a different angle of incidence for the reference beam, then the images can selectively be reconstructed by adjusting the angle of the reconstruction beam. The reason we do not see all the images reconstructed simultaneously is

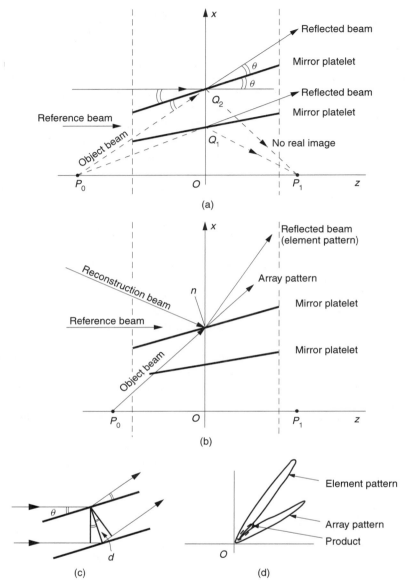

Figure 1.49   The reconstruction of the holographic image from a thick emulsion film. (Cross section of the hologram). (a) Reference and reconstruction beams are identical. (b) Reference and reconstruction beams are not identical. (c) Bragg reflection. (d) Polar diagram of the pattern.

explained as follows. Suppose the reference and reconstruction beams for a given image are not identical, as shown in Fig. 1.49b. With the change in the direction of the reconstruction beam, both the element and the array patterns shift, but they shift differently. Their peaks no longer match, as indicated by the polar diagram in Fig. 1.49c, and the sharp peak disappears from the product of the polar diagram. The image can be reconstructed only when the reconstruction beam is identical to the reference beam. This fact is used for the high-density recording of the volume

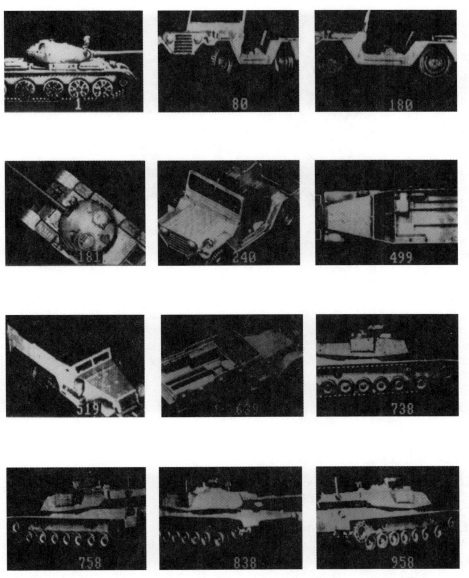

Figure 1.50. Sample reconstructions from 1000 holograms recorded in a lithium niobate crystal. (Courtesy of F. H. Mok [22]).

hologram. Figure 1.50 shows a sample of reconstructions from 1000 holograms recorded in a $2 \times 1.5 \times 1$-cm^3 photorefractive crystal of lithium niobate, Fe:LiNbO$_3$. The direction of the reference beam was stepped at less than $0.01°$ [22,23].

The principle of the thick emulsion hologram is also applied to the white light hologram. The direction of the peak of the array pattern moves with the wavelength of the reconstruction beam but that of the element pattern more or less remains unchanged. Hence, when white light is used for reconstructing the holographic image, the image is reconstructed only by the light spectrum that creates a match between the peaks of the

Figure 1.51   White light hologram. (Courtesy of Dainippon Printing Co.)

element and array patterns. Thus, the reconstruction process is wavelength selective, and the reconstructed image is of a single color. The color is the same as that of the reference beam used to fabricate the hologram. This kind of white light hologram is called either a Lippmann or Denisyuk hologram. An example is shown in Fig. 1.51.

The wavelength selectivity of the thick emulsion hologram is also the basis of the color hologram. As just mentioned, the thick emulsion hologram reconstructs the image in the color used to fabricate the hologram. To produce a color hologram, the thick emulsion hologram is triple exposed by three kinds of lasers whose wavelengths correspond to basic colors such as red, green, and blue. When the hologram is illuminated with white light, images of the three different colors are reconstructed, and a color image is produced.

1.8.5   Applications of Holography

Holography has gained recognition both as an art form and as a measurement tool. A few of the numerous applications are described next [8,24].

1.8.5.1   Three-Dimensional Displays

When a hologram is illuminated, it causes the wavefront that was originally formed by the object to be reconstructed. It is the lens of the observer's eye that forms an image on the retina from the reconstructed wavefront. Depending on the viewing angle of the observer's left and right eyes relative to the hologram, the left and right eyes intercept different portions of the reconstructed wavefront. Figure 1.52a is a photograph of the reconstructed image taken by a camera set at the position of the left eye of the observer and Fig. 1.52b is a photograph of the same image taken with the camera set at the position of the right eye.

Notice the movement of the chess piece in the foreground with respect to the others: Fig. 1.52a is what the left eye sees and Fig. 1.52b is what the right eye sees. This difference of scenes produces the perception of viewing the original object as if it were present in front of the observer in three dimensions. This is called *binocular parallax*. Even with one eye, the side to side movement of the observer's face provides the observer with a differential movement (the foreground appears to move faster than

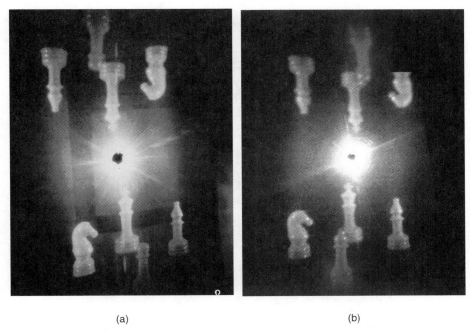

(a)   (b)

Figure 1.52 The reconstructed image of a hologram varies with the viewing angle. The upside down image is the pseudoscopic image. (a) When viewed from the left. (b) When viewed from the right.

the background), and as the observer's face moves hidden portions of the reconstructed image become visible. These movement effects are usually perceived and contribute to the observer's sensation of seeing the image in three dimensions. This is called *movement parallax*. The tendency of the observer's eyes to focus on the object, which is called *accommodation*, also contributes to the three-dimensional perception.

1.8.5.2 Microfiche Recording

Since the hologram does not use a lens in the fabrication process, the reconstructed image has neither limitations on the depth of focus nor aberrations due to the lens.

The scattered wave from the object is spread and recorded over the entire hologram, so that even if a portion of the hologram is missing or damaged, the same image could still be reconstructed. The damage to the hologram results in an overall degradation in quality, but no specific portion of the image is lost. Consequently, the hologram has a high tolerance to mishandling and is ideal for applications such as high-density microfiche recordings.

1.8.5.3 Measurement of Displacement

An ordinary photograph records only the intensity pattern, but the image reconstructed from a hologram has both phase and amplitude information. If the light waves of the reconstructed image are overlayed with those of the original object, an interference pattern is generated. The interference fringes indicate minute distortions of the object between its current state and its prior state when the hologram was recorded. This technique is called *interference holography*. An interference hologram can be fabricated by exposing a hologram twice to the same object, with and without the deformation. The resultant fringe patterns are seen in the reconstructed image.

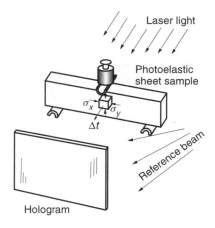

Figure 1.53   Interferometric hologram of a photoelastic sheet under load.

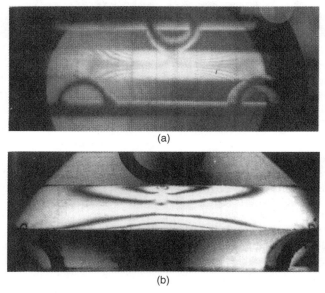

Figure 1.54   A photoelastic sheet under three-point loading. (a) Fringes of the interferometric hologram, which represent $\sigma_x + \sigma_y$. (b) Photoelastic fringes, which represent $\sigma_x - \sigma_y$. (Courtesy of X. Yan, T. Ohsawa, [25] and T. Ozaki.)

An example of combining the interferometric hologram with photoelasticity is presented in Figs. 1.53 and 1.54 [25]. With the geometry shown in Fig. 1.53, an interferometric hologram is made to visualize the pattern of strain established in a photoelastic sheet (e.g., Plexiglas^{TM}). When a photoelastic sheet of uniform thickness is loaded, the thickness becomes nonuniform. When the direction of the illuminating laser beam is normal to the sheet surface, the phase change $\Delta\phi$ associated with the increase in the optical path length is

$$\Delta\phi = kn\,\Delta t \tag{1.277}$$

where Δt is the increase in thickness. The hologram is doubly exposed. The first exposure is made without compression and the second, under compression. Dark fringes will appear in the reconstructed image where

$$\Delta\phi = (2n + 1)\pi \qquad (1.278)$$

due to the destructive interference between the two reconstructed images. This pattern is called the *isopachic fringe pattern*, meaning the locus of points of constant thickness of the sheet [26]. Both of the principal stresses σ_x and σ_y can be attributed to the change in thickness of the sheet. Δt is proportional to the sum of $\sigma_x + \sigma_y$. From the isopachic fringe alone, the contributions of σ_x and σ_y cannot be known separately.

Next, we will describe the procedure for observing the birefringence pattern of the same sample with the same loading. In order to do so, a polariscope (Section 6.7) is built around the sample. A polarizer is placed between the sample and the illuminating laser light, and the hologram is replaced by an analyzer whose transmission axis is perpendicular to that of the polarizer.

When the photoelastic sheet is stressed, the sheet becomes birefringent. That is, if the direction of compression is in the x direction, the index of refraction seen by the light polarized in the x direction is no longer the same as that of the y direction, and the degree of birefringence is proportional to the difference $\sigma_x - \sigma_y$ between the principal stresses. The degree of birefringence can be observed using the polariscope. With the isotropic (uncompressed) sample inserted in between the polarizer and the analyzer, no light gets through. With the birefringent (compressed) sample inserted in between the polarizer and analyzer, light is observed emerging from the analyzer in accordance with the degree of birefringence caused by the stress. Thus, the polariscope displays the pattern of the stress by means of the birefringence in the photoelastic sheet. By combining $\sigma_x + \sigma_y$ from the *isopachic fringes* with $\sigma_x - \sigma_y$ from the *photoelastic fringes*, it is possible to obtain σ_x and σ_y separately.

Figure 1.54a is the isopachic fringe pattern observed by means of the interferometric hologram. Figure 1.54b is the photoelastic fringe pattern of the same sample under the same compression as in Figure 1.54a observed by means of the polariscope.

1.8.5.4 *Measurement of Vibration*

Another application of holography is in recording vibration patterns. The hologram is able to record fringe patterns on the order of microns. While the holographic film is being exposed, both the object and the holographic plate must be very steady, otherwise the fringe patterns cannot be recorded. This fact can be used conversely to record the vibration pattern of an object since the vibrating portions of the object would appear dark in the reconstructed image. Figure 1.55 shows an example of mapping vibration patterns in a loudspeaker [27].

1.8.5.5 *Nonoptical Holographies*

So far, light has been assumed to be the wave source for generating holograms. In fact, other types of waves are capable of forming an interference pattern that can be used for generating holograms. For instance, microwave [8] and acoustic wave [28] holograms are possible. These hologram images can then be viewed optically by photographically reducing the dimensions of the hologram to the ratio of their relative wavelengths. By doing so, one can visualize how microwaves radiate from antennas (radiation patterns) or reflect from objects.

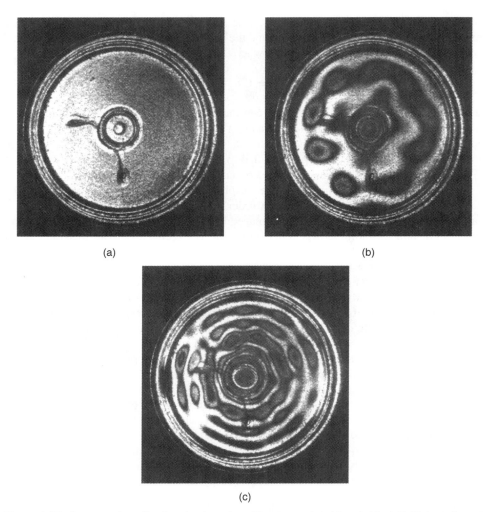

(a)

(b)

(c)

Figure 1.55 Images of a vibrating loudspeaker. (Courtesy of A. Marrakchi, J.-P. Huignard, and J. P. Herriau [27].

Figure 1.56 shows an example of a microwave hologram used to examine the radiation pattern from an antenna [29]. Figure 1.56a shows a photograph of the object, which is a radiating monopole driven at the end of a microwave waveguide and Fig. 1.56b shows the reconstructed image of the plane of the monopole. The hologram reveals that the tip and the driving point are the major sources of radiation, and the function of the antenna wire is simply that of a feed wire for bringing current to the tip of the antenna. Figure 1.56c and 1.56d are the reconstructed images as the point of observation moves away from the antenna.

Figure 1.57a shows the geometry of a side scan sonar. The ship tows the sonar transducer whose beam is swept perpendicular to the direction of the ship's motion. The composite image is constructed as the ship cruises. The transducer is towed instead of being kept aboard the ship in order to decouple the motion of the ship. Figure 1.57b is the recorded image of a sunken ship resting on the sea bottom at a depth of 25 m.

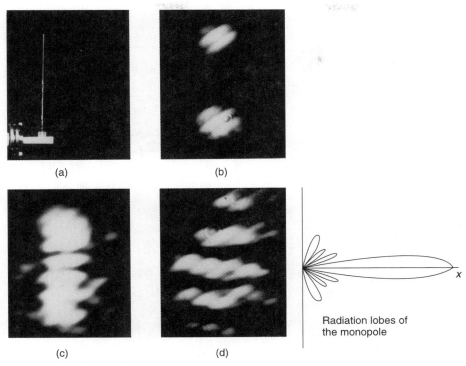

Figure 1.56 Visualization of a radiation pattern from a monopole antenna by means of a microwave hologram. (a) Photograph of the monopole used as the object. (b) At the antenna plane. (c) At a plane in the Fresnel region. (d) At the Fraunhofer region [29].

Figure 1.57c is the image of a bicycle sunken 4.2 m deep in the ocean [30]. The sound wave signal was processed by synthetic aperture [8] holography and even the spokes of the bicycle wheels can almost be resolved.

Another advantage of nonoptical holography is that an object imbedded in an optically opaque medium can be visualized.

A further advantage of the nonoptical hologram over "shadowgrams" such as an X ray is the freedom of choice of the hologram's location with regard to the source and object. With X rays, the film always has to be behind the object, which is not true in holograms. A particularly useful example of this advantage is when nonoptical holograms are used in geological surveys, where placing a film deep beneath the surface is neither efficient nor desirable.

1.8.5.6 Computer-Generated Holograms

Holograms need not be fabricated using actual electromagnetic waves. They can be computer generated from values based on theory [31,32]. For instance, holograms can also be generated using the expected field pattern for a given object that can be calculated using the Fresnel diffraction formula of Eq. (1.38) or (1.39). The interference of the object and reference beams can then be calculated and drawn by computer. This image is then photographically reduced so that the holographic image can be reconstructed by light. Figure 1.58 shows an example of a computer-generated hologram.

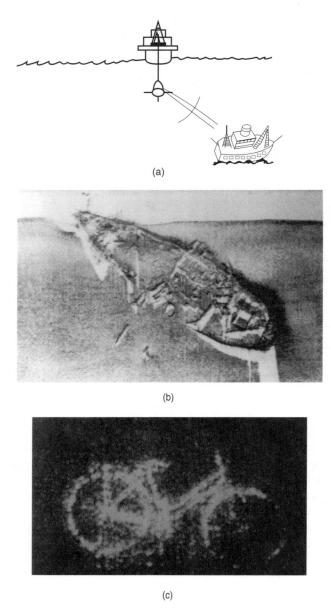

(a)

(b)

(c)

Figure 1.57 High-resolution underwater acoustic images. (a) Geometry of the side scan sonar. (b) *Vineyard Sound Lightship* on the sea bottom off Nantucket, Massachusetts, recorded with a HYDROSCAN. (Courtesy of Klein Associates, Inc.) (c) Synthetic aperture holographic acoustic image at 4.2 m. (Courtesy of K. Mano and K. Nitadori [30].)

Figure 1.59 shows another kind of computer-generated hologram. The image of the hologram is reconstructed by light propagating in the plane of the hologram [33]. As shown in Fig. 1.59a, the reconstruction beam is fed through an optical fiber pigtail. Figure 1.59b is a photograph of a scanning electron microscope image of the computer-generated hologram. The pattern was generated using electron-beam

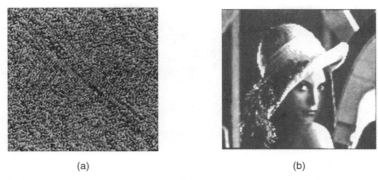

(a)   (b)

Figure 1.58. Computer-generated holography. (a) Hologram. (b) Reconstructed image. (Courtesy of D. Asselin, A. Bergeron, and INO)

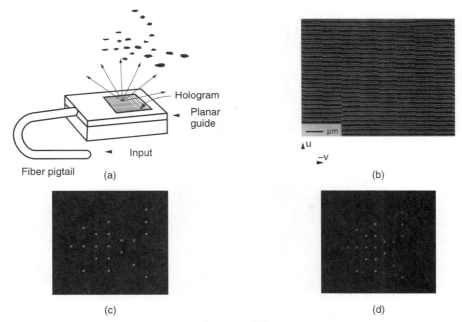

Figure 1.59. Off-plane computer-generated waveguide hologram (OP-CGWH). (a) Geometry. (b) Photograph of scanning electron microscope image of the hologram. (c) Designed pattern. (d) Obtained pattern. (After M. Li et al. [33].)

lithography. Figure 1.59c gives the desired pattern of the reconstructed image. Figure 1.59d shows the reconstructed image from the fabricated computer hologram. A hologram such as this can be used as an optical interconnect, where the output light from an optical fiber has to be connected to multiple terminals through free-space propagation.

1.8.5.7 Holographic Video Display

Figure 1.60 shows a schematic of a holographic video display system [34]. The system is intended to animate computer-generated holograms. The key component of the system is the surface acoustooptic modulator (AOM) on which three of the line fringe patterns are written for three primary color holograms.

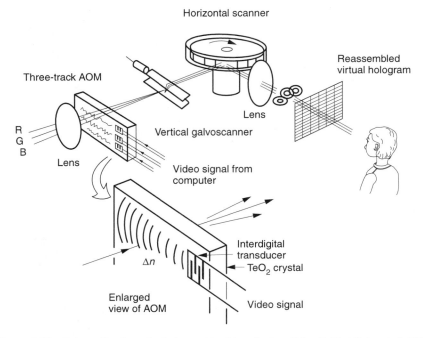

Figure 1.60 Schematic of a color holographic video display. (After P. St.-Hilaire et al. [33].)

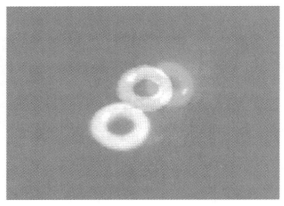

Figure 1.61. A frame of the animated color computer-generated hologram of colorful donuts. (Courtesy of P. St.-Hilaire et al. [33].)

The fringes of the holograms are scanned as a composite of line holograms. The diffracted images from the line holograms are reassembled into a complete image of a computer-generated hologram using horizontal as well as vertical scanners. The AOM is made of a TeO_2 crystal. A pair of interdigital electrodes are deposited on the crystal surface. The interdigital electrodes launch a surface acoustic wave (SAW) due to the piezoelectric effect in accordance with the video signal from the computer. The surface wave spatially modulates the index of refraction due to the acoustooptic effect and writes a one-line hologram on the crystal as it propagates along the crystal (see Example 5.6).

When laser light is illuminated perpendicular to the surface, a diffracted image moving at the speed of the surface acoustic image is generated.

The horizontal scanner is an 18-sided polygonal mirror. It scans in the opposite direction to the movement of the diffracted image to immobilize the image. The horizontal scanner also multiplexes the image of the crystal, creating a virtual crystal that is exactly the same length as one line of the computer-generated hologram.

The galvanometric scanner shifts the horizontal one-line fringe vertically after each horizontal scan completing the reassembly of the computer-generated hologram.

Figure 1.61 shows one of the frames of an animated color computer-generated hologram of colorful donuts. Three primary color lasers were used simultaneously. These are a HeNe laser at $\lambda = 633$ nm for red, a frequency-doubled YAG laser at $\lambda = 532$ nm for green, and a HeCd laser at $\lambda = 442$ nm for blue.

PROBLEMS

1.1 For a plane wave that is propagating (Fig. P1.1) in the direction

$$\theta = 45° \qquad \phi = 45°$$

the light field observed at $P(2, 3, 4) \times 10^{-6}$ m is expressed as

$$E = E_0 e^{j67.32 - j2.44 \times 10^{15} t}$$

(a) Find the wavelength of light in the medium.

(b) Find the index of refraction of the medium.

1.2 The spatial frequencies of an incident wave (Fig. P1.2) were measured along the x and y axes as

$$f_x = 0.6 \ \text{lines/μm}, \qquad f_y = 0.8 \ \text{lines/μm}$$

The wavelength of the received light is $\lambda = 0.84$ μm.

(a) What is the direction of propagation of the incident wave?

(b) What would be the spatial frequency if the measurement were made along a line in the direction $\hat{\mathbf{l}} = 3\hat{\mathbf{i}} + 4\hat{\mathbf{j}}$?

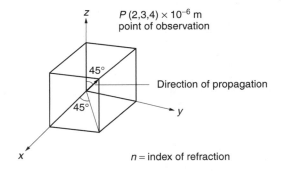

Figure P1.1   Geometry of the propagation.

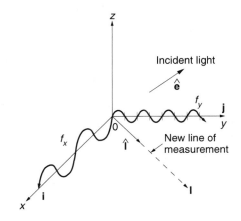

Figure P1.2   Geometry of the incident wave.

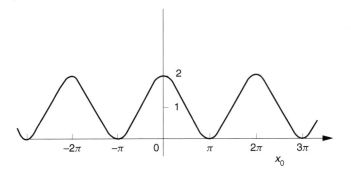

Figure P1.3   Sinusoidal distribution $1 + \cos 2\pi f_{x_0} x_0$.

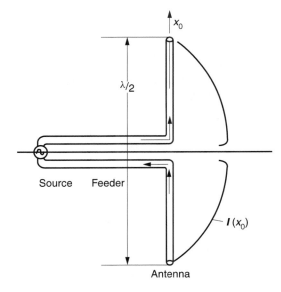

Figure P1.4   Half-wave dipole antenna and its current distribution.

1.3 Find the diffraction pattern from a film whose transmittance is modulated sinusoidally as shown in Fig. P1.3:

$$t(x_0, y_0) = 1 + \cos 2\pi f_{x_o} x_0$$

Assume that the size of the film is infinitely large.

1.4 Derive the radiation pattern of a half-wave dipole using Fourier optics. Assume that the current distribution along the antenna is sinusoidal (Fig. Pl.4).

1.5 Match each of the letter apertures in Fig. P1.5a with its correct diffraction pattern in Fig. P1.5b.

1.6 Figure P1.6a shows apertures of three of the fingerprints that appeared in Fig. 1.36, and an aperture in the shape of a cartoon character. Figure P1.6b shows the photographs of the diffraction patterns. Match the apertures with their respective diffraction patterns.

1.7 The finest possible light spot is to be obtained using a finite size convex lens. The aperture of the lens is square $\Pi(x/a)\Pi(y/a)$ and the focal length is f_0. The incident light is a parallel beam.

(a) What is the smallest spot size obtainable?

(b) What is the required size of a diffraction-limited lens that can resolve 1 μm. The wavelength of the light is $\lambda = 0.555$ μm and $f_0 = 50$ mm. Assume a rectangular lens.

1.8 Explain the principle of operation of a pinhole camera using Fourier optics (Fig. P1.8). Assume the shape of the pinhole is a square.

1.9 An input function $g(x_0, y_0)$ is placed in the front focal plane of lens L_1 with focal length f_1 (Fig. P1.9). A second lens L_2 with focal length f_2 is placed behind lens L_1 at a distance $f_1 + f_2$ from L_1. Find the expression for the field at the back focal plane of lens L_2.

1.10 The optical system shown in Fig. 1.33c is a Schlieren camera made by placing an opaque dot or $\pi/2$-radian phase plate at the back focal point G_0 of lens L_2. Let us say that the object is transparent but with a small variation of the index of refraction so that the input function

$$g(x_0, y_0) = e^{j\phi(x_0, y_0)}$$

can be approximated as

$$g(x_0, y_0) = 1 + j\phi(x_0, y_0)$$

(a) Find the expression for the output intensity distribution $I_a(x_i, y_i)$ with the opaque dot placed at the back focal point G_0 of lens L_2.

(b) Find the output intensity distribution $I_b(x_i, y_i)$ with the $\pi/2$-radian phase plate at the same location as the opaque dot in part (a).

(c) Compare the results of parts (a) and (b).

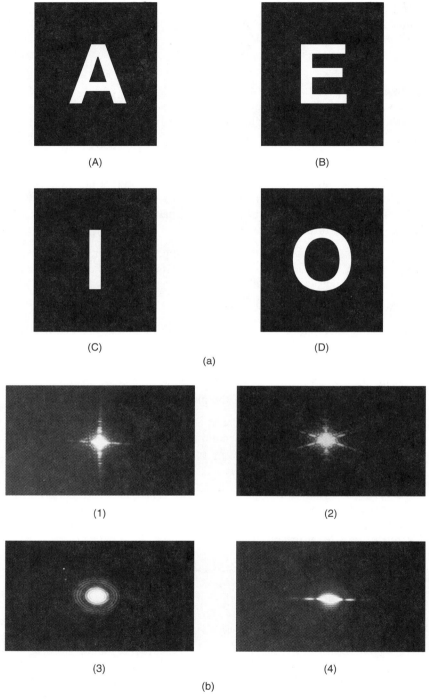

Figure P1.5. Match the aperture with its diffraction pattern. (a) Apertures of letters. (b) Photographs of diffraction patterns.

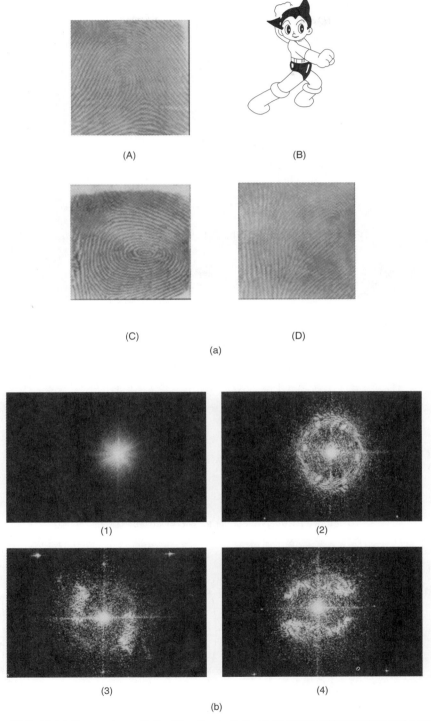

Figure P1.6. Match the aperture with its diffraction pattern. (a) Apertures of fingerprints and a cartoon character. (b) Photographs of diffraction patterns.

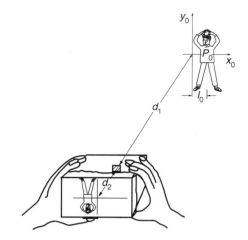

Figure P1.8 Principle of the pinhole camera by Fourier optics.

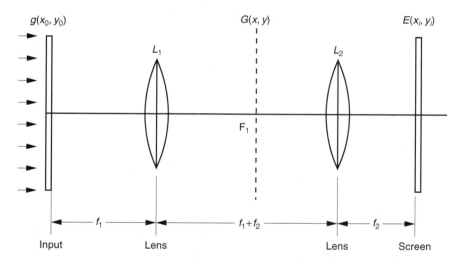

Figure P1.9 Two Fourier transform lenses in series.

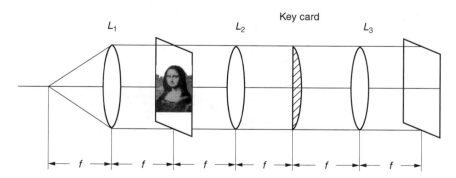

Figure P1.11 A convex lens with focal length f_0 is used as the key card for encryption.

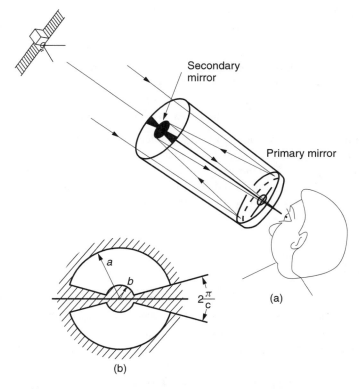

Figure P1.12   Cassegrain reflecting telescope with obstruction. (a) Telescope. (b) Obstruction.

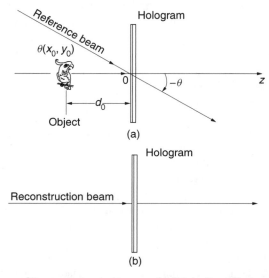

Figure P1.13   Locations of the reconstructed images. (a) Fabrication of hologram. (b) Reconstructing the images.

1.11 With the encryption scheme shown in Fig. P1.11, if a convex lens of focal length f_0 is used as the key card for the encryption, what is the key card for the decryption?

1.12 A Cassegrain telescope has a primary concave mirror and a secondary convex mirror, as shown in Fig. P1.12a. The supports holding the secondary mirror present an obstruction and act like a mask, as shown in Fig. P1.12b. Find the diffraction pattern of the masked aperture shown in Fig. P1.12b.

1.13 A hologram was fabricated with the geometry shown in Fig. P1.13. Assume the thin emulsion case. The object was placed on the z axis, which is perpendicular to the hologram. A plane wave reference beam was incident at angle θ to the normal of the hologram. Find the locations of the reconstructed images when the reconstruction beam is not identical with the reference beam and is incident along the z axis.

REFERENCES

1. J. W. Goodman, *Introduction to Fourier Optics*, McGraw-Hill, New York, 1968.
2. M. V. Klein and T. E. Furtak, *Optics*, 2nd ed., Wiley, New York, 1986.
3. E. Hecht, *Optics*, 3rd ed., Addison-Wesley, Reading, MA, 1998.
4. F. T. S. Yu and X. Yang, *Introduction to Optical Engineering*, Cambridge University Press, Cambridge, 1997.
5. R. D. Guenther, *Modern Optics*, Wiley, New York, 1990.
6. B. E. A. Saleh and M. C. Teich, *Fundamentals of Photonics*, Wiley, New York, 1991.
7. G. R. Fowles, *Introduction to Modern Optics*, 2nd ed., Dover Publications, New York, 1989.
8. K. Iizuka, *Engineering Optics*, 2nd ed., Springer Verlag, Berlin, 1986.
9. C. A. Balanis, *Antenna Theory Analysis and Design*, 2nd ed., Wiley, New York, 1997.
10. C. R. Wylie and L. C. Barrett, *Advanced Engineering Mathematics*, 6th ed., McGraw-Hill, 1995.
11. G. A. Massey, "Microscopy and pattern generation with scanned evanescent waves," *Appl. Opt.* **23**(5), 658–660 (1984).
12. A. Vander Lugt, "Practical considerations for the use of spatial carrier-frequency filters," *Appl. Opt.* **5**(11), 1760–1765 (1966).
13. F. T. S. Yu and Suganda Jutamulia, *Optical Pattern Recognition*, Cambridge University Press, Cambridge, 1998.
14. C. S. Weaver and J. W. Goodman, "A technique for optically convolving two functions," *Appl. Opt.* **5**(7), 1248–1249 (1966).
15. G. Lu, Z. Zhang, S. Wu, and F. T. S. Yu, "Implementation of a non-zero-order joint-transform correlator by use of phase-shifting techniques," *Appl. Opt.* **36**(2), 470–483 (1997).
16. P. Purwosumarto and F. T. S. Yu, "Robustness of joint transform correlator versus Vander Lugt correlator," *Opt. Eng.* **36**(10), 2775–2780 (1997).
17. F. T. S. Yu, Q. W. Song, Y. S. Cheng, and D. A. Gregory, "Comparison of detection efficiencies for Vander Lugt and joint transform correlators," *Appl. Opt.* **29**(2), 225–232 (1990).
18. L. Leclerc, Y. Sheng, and H. H. Arsenault, "Circular harmonic covariance filters for rotation invariant object recognition and discrimination," *Opt. Commun.* **85**, 299–305 (1991).

19. D. Casasent and D. Psaltis, "Position, rotation, and scale invariant optical correlation," *Appl. Opt.* **15**(7), 1795–1799 (1976).

20. B. Javidi, "Optical information processing for encryption and security systems," *Opt. Photonics News*, 29–33 (Mar 1997).

21. B. Javidi and A. Sergent, "Fully phase encoded key and biometrics for security verification," *Opt. Eng.* **36**(3), 935–942 (1997).

22. F. H. Mok, "Angle-multiplexed storage of 5000 holograms in lithium niobate," *Opt. Lett.* **18**(11), 915–917 (1993).

23. G. Barbastathis and D. J. Brady, "Multidimensional tomographic imaging using volume holography," *Proc. IEEE* **87**(12), 2098–2120 (1999).

24. P. Hariharan, *Optical Holography, Principles, Techniques and Applications*, 2nd ed., Cambridge University Press, Cambridge, 1996.

25. X. Yan and T. Ohsawa, "Measurement of the internal local stress distribution of composite materials by means of laser imaging methods," *Composites* **25**(6), 443–450 (1994).

26. J. W. Dally and W. F. Riley, *Experimental Stress Analysis*, 3rd ed., McGraw-Hill, New York, 1991.

27. A. Marrakchi, J.-P. Huignard, and J. P. Herriau, "Application of phase conjugation in $Bi_{12}SiO_{20}$ crystals to mode pattern visualization of diffuse vibrating structures," *Opt. Commun.* **34**(1), 15–18 (1980).

28. B. P. Hildebrand and B. B. Brenden, *An Introduction to Acoustic Holography*, Plenum Press, New York, 1972.

29. K. Iizuka and L. G. Gregoris, "Application of microwave holography in the study of the field form a radiating source," *Applied Physics Letters* **17**(12), 509–512 (1970).

30. K. Mano and K. Nitadori, "An experimental underwater viewing system using acoustical holography," *1977 IEEE Ultrasonics Symposium Proceedings*, Phoenix, Arizona, Oct. 26–28, IEEE Cat. #77CH/264-1SU, pp. 272–277, 1977.

31. A. Bergeron, H. H. Arsenault, J. Gauvin, and D. J. Gingras, "Computer-generated holograms improved by a global iterative coding," *Opt. Eng.* **32**(9), 2216–2226 (1993).

32. H. Yoshikawa and H. Taniguchi, "Computer generated rainbow hologram," *Opt. Rev.* **6**(2), 118–123 (1999).

33. M. Li, J. Bengtsson, M. Hagberg, A. Larsson, and T. Suhara, "Off-plane computer-generated waveguide hologram," *IEEE J. Selected Topics Quantum Electro.* **2**(2), 226–235 (1996).

34. P. St.-Hilaire, S. A. Benton, M. Lucente, and P. M. Hubel, "Color images with the MIT holographic video display," *SPIE Proceedings*, Vol. 1667, *Practical Holography VI*, San Jose, pp. 73–84, 1992.

2

BOUNDARIES, NEAR-FIELD OPTICS, AND NEAR-FIELD IMAGING

When light travels from one medium into another, reflection and refraction usually take place at the boundary. A portion of the light is reflected back into the first medium, and the other portion is transmitted (refracted) into the second medium at the angle set by Snell's law. The laws of reflection and refraction are among the most basic principles of optics.

Optical systems are made up of boundaries. One cannot have a component without boundaries. A clear understanding of the boundary phenomena is most important, particularly in the field of integrated optics, where the boundaries are so close together that interactions become a serious problem. What will be treated here are simple configurations, but an in-depth understanding of these configurations will provide a solid foundation for solving more complex real-life problems, such as the photon tunneling microscope, the scanning near-field optical microscope (SNOM), and the high-density digital video disk (DVD).

2.1 BOUNDARY CONDITIONS

We will consider the simplest case when light is incident from medium 1 onto its border with medium 2, as indicated in Fig. 2.1. Let the indices of refraction of medium 1 and 2 be n_1 and n_2, respectively. On the boundary, there are five important conditions:

1. The frequency of light does not change across the border, unless one of the media happens to be a nonlinear medium. In nonlinear media higher harmonics are generated.

2. The wavelength either expands or contracts, according to the ratio of the indices of refraction:

$$\frac{n_1}{n_2} = \frac{\lambda_2}{\lambda_1} \qquad (2.1)$$

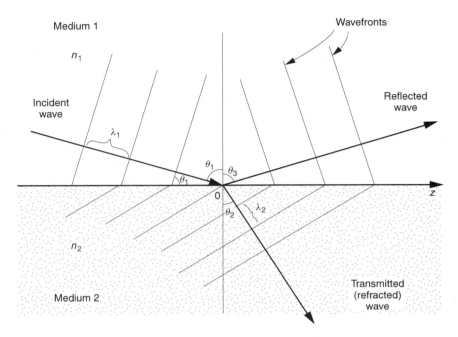

Medium 1

Wavefronts

n_1

Reflected
wave

Incident
wave

λ_1

θ_1

θ_3

θ_1

0

z

θ_2

λ_2

n_2

Transmitted
(refracted)
wave

Medium 2

Figure 2.1 Wavefront and ray directions near the boundary.

The wavelength expands when light goes into a medium with a smaller index
of refraction. A medium with a smaller index of refraction is often called an
optically less dense medium.

3. Speaking in terms of electromagnetic theory, the ratio of light energy, contained
in the form of electric energy, to that in the form of magnetic energy is changed at
the boundary depending on the intrinsic impedance of the medium. The intrinsic
impedance η_1 of medium 1 is

$$\frac{|\mathbf{E}_1|}{|\mathbf{H}_1|} = \frac{|\mathbf{E}_3|}{|\mathbf{H}_3|} = \eta_1$$

$$\eta_1 = \sqrt{\frac{\mu_0\mu_{r1}}{\epsilon_0\epsilon_{r1}}} = \frac{\eta_0}{n_1}$$

(2.2)

where $\mathbf{E}_1$ and $\mathbf{H}_1$ are the electric and magnetic fields of the incident wave, and
$\mathbf{E}_3$ and $\mathbf{H}_3$ are the electric and magnetic fields of the reflected wave. Both waves
are in medium 1. The intrinsic impedance η_0 of the vacuum is

$$\eta_0 = \sqrt{\frac{\mu_0}{\epsilon_0}}$$

(2.3)

and its value in MKS units is $120\pi\,\Omega$. In Eq. (2.2), μ_0 represents the magnetic
permeability of free space, and μ_{r1} is the relative magnetic permeability of
medium 1. With most of the materials treated here, the value of μ_r is unity, and
we will treat it as such unless stated otherwise. In MKS units, $\mu_0 = 1.2566 \times
10^{-6}$ henry/meter. (MKS units will be used throughout the book.) In Eq. (2.3),

the quantities ϵ_0 and ϵ_r are the absolute and relative dielectric constants. The value of the former is $\epsilon_0 = 8.855 \times 10^{-12}$ farad/meter, and that of the latter varies from medium to medium. In optics the index of refraction n is used more often than ϵ_r.

The index of refraction n is defined as the ratio of the wavelength in the medium to that in vacuum. The wavelength λ in a nonmagnetic medium is

$$\lambda = v/f \qquad \text{where } v = 1/\sqrt{\mu_0 \epsilon_0 \epsilon_r}$$

and that in the vacuum is

$$\lambda_0 = c/f \qquad \text{where } c = 1/\sqrt{\mu_0 \epsilon_0}$$

hence,

$$n = \lambda_0/\lambda = \sqrt{\epsilon_r}.$$

With media 1 and 2,

$$\sqrt{\epsilon_{r1}} = n_1$$
$$\sqrt{\epsilon_{r2}} = n_2 \qquad (2.4)$$

Similarly, for medium 2, the intrinsic impedance η_2 is

$$\frac{|\mathbf{E}_2|}{|\mathbf{H}_2|} = \eta_2 \qquad \eta_2 = \sqrt{\frac{\mu_0 \mu_{r2}}{\epsilon_0 \epsilon_{r2}}} = \frac{\eta_0}{n_2} \qquad (2.5)$$

The ratio of the electric to magnetic field has to be changed as soon as the light crosses the boundary.

The manner in which the transition takes place is governed by Maxwell's boundary conditions. These boundary conditions represent the fourth and fifth items in our list of important conditions, and are stated below.

4. The tangential components of $\mathbf{E}$ and $\mathbf{H}$ are continuous across the boundary.
5. The normal components of $\mathbf{D}$ and $\mathbf{B}$ are continuous, where $\mathbf{D}$ and $\mathbf{B}$ are the electric and magnetic flux densities.

The $\mathbf{E}$ field is often related to the voltage V and the $\mathbf{H}$ field to the current I, and the ratio $\mathbf{E}/\mathbf{H}$ is often related to the impedance V/I, even though no tangible medium is present. It is more appropriate to say that the intrinsic impedance is defined by the ratio of $\mathbf{E}$ to $\mathbf{H}$ of a plane wave in that medium.

The angles and amplitudes of the reflected and transmitted light are governed by the above five conditions; three by frequency, wavelength and impedance conditions, and two by Maxwell's continuity conditions. In the following, using these conditions, quantities associated with the boundary are found.

2.2 SNELL'S LAW

The relationship between the angles of incidence, reflection, and transmission (refraction) are found. If the incident light is sinusoidally varying in both time and space, then both the reflected and the transmitted waves have to vary sinusoidally accordingly. To understand this fact, suppose that at a particular instance and at a particular location of the boundary, the oscillation of the incident wave is at its maximum; then both reflected and transmitted waves have to be at their maxima. Otherwise, even though the boundary conditions can be met at a specific instance or at a specific location, they cannot be met throughout time and the entire boundary surface. In other words, in order to meet Maxwell's continuity boundary condition, the wavelengths along the interface surface must have the same temporal and spatial variation,

$$\lambda_{z1} = \lambda_{z2} = \lambda_{z3} \tag{2.6}$$

where λ_{z1}, λ_{z2}, and λ_{z3} are the wavelengths measured in the z direction in the plane of the incident, transmitted, and reflected waves. Instead of Eq. (2.6) in wavelengths, the propagation constant $\beta_{1,2,3} = 2\pi/\lambda_{z1,2,3}$ is often used to express this condition and Eq. (2.6) is rewritten as

$$\beta_1 = \beta_2 = \beta_3 \tag{2.7}$$

Equation (2.7) is known as either β, k, phase, or momentum matching, but all mean the same thing: wavelength matching.

The free-space wavelengths λ_1, λ_2, and λ_3 are determined solely by the refractive indices of the media, and at the moment $n_1 \neq n_2$ and $\lambda_1 \neq \lambda_2$, according to Condition 2. How then is the condition of wavelength matching to be satisfied? The only way to meet the wavelength matching requirement is for the directions of the transmitted and reflected waves to be bent, as shown in Fig. 2.1. By decreasing the incident angle θ_1 from 90° to 0°, the wavelength λ_z along the z direction increases from λ_1 to infinity. Then, the condition for the wavelength matching is

$$\frac{\lambda_1}{\sin \theta_1} = \frac{\lambda_1}{\sin \theta_3} = \frac{\lambda_2}{\sin \theta_2} \tag{2.8}$$

One immediate result from Eq. (2.8) is that the angle of reflection θ_3 is identical to the angle of incidence θ_1:

$$\theta_1 = \theta_3 \tag{2.9}$$

Applying Eq. (2.1) to (2.8) gives

$$n_1 \sin \theta_1 = n_2 \sin \theta_2 \tag{2.10}$$

which is Snell's law. Snell's law is one of the most used laws in optics.

Next, the relative amplitudes of the transmitted and reflected waves are considered.

2.3   TRANSMISSION AND REFLECTION COEFFICIENTS

The electric field transmission coefficient t_E is defined as the ratio of the field $\mathbf{E}_2$ of the transmitted wave to the field $\mathbf{E}_1$ of the incident wave. The electric field reflection coefficient r_E is defined as the ratio of the field $\mathbf{E}_3$ of the reflected wave to the field $\mathbf{E}_1$ of the incident wave. Analogous definitions t_H and r_H hold for the magnetic field $\mathbf{H}$.

The reflection and transmission coefficients depend on the angle of incidence and the directions of polarization. Let us begin with the simplest case of normal incidence and then proceed to the general case of an arbitrary angle of incidence.

2.3.1 Transmission and Reflection Coefficients (at Normal Incidence)

When the direction of propagation is normal to the boundary (see Fig. 2.2), both the **E** and **H** fields are parallel to the interface. As shown in Fig. 2.2, the convention of positive **E** and **H** is chosen with regard to the Poynting vectors $\mathbf{s}_1$, $\mathbf{s}_2$, and $\mathbf{s}_3$, which point down, down, and up, respectively. For a description [1] about the directions of the field, see Example 2.1. The Poynting vector **s** is the time rate of flow of electromagnetic energy per unit area (the power intensity), and $\mathbf{s} = \mathbf{E} \times \mathbf{H}$. Boundary Condition 4 is used, and

$$\mathbf{E}_1 + \mathbf{E}_3 = \mathbf{E}_2 \tag{2.11}$$

$$\mathbf{H}_1 - \mathbf{H}_3 = \mathbf{H}_2 \tag{2.12}$$

From Eqs. (2.2) and (2.5), the above equations are written as

$$\eta_1 \mathbf{H}_1 + \eta_1 \mathbf{H}_3 = \eta_2 \mathbf{H}_2 \tag{2.13}$$

$$\frac{\mathbf{E}_1}{\eta_1} - \frac{\mathbf{E}_3}{\eta_1} = \frac{\mathbf{E}_2}{\eta_2} \tag{2.14}$$

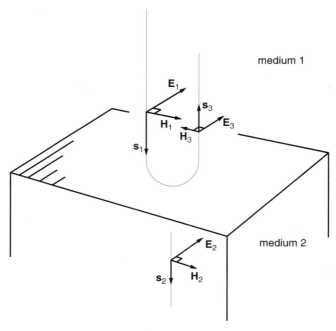

Figure 2.2 Transmission and reflection coefficients at an interface between two media at normal incidence.

From Eqs. (2.11) to (2.14), all of the following expressions for r_E, r_H, t_E, and t_H are derived:

$$r_E = r_H = \frac{E_3}{E_1} = \frac{H_3}{H_1} = \frac{\eta_2 - \eta_1}{\eta_2 + \eta_1} \qquad (2.15)$$

$$t_E = \frac{E_2}{E_1} = \frac{2\eta_2}{\eta_2 + \eta_1} \qquad (2.16)$$

$$t_H = \frac{H_2}{H_1} = \frac{2\eta_1}{\eta_2 + \eta_1} \qquad (2.17)$$

Note that t_E and t_H are not equal.

In optics, the index of refraction n rather than the intrinsic impedance η is used. With the assumption that $\mu_{r_1} = \mu_{r_2}$, Eqs. (2.15) to (2.17) can be rewritten as

$$r_E = r_H = \frac{n_1 - n_2}{n_1 + n_2} \qquad (2.18)$$

$$t_E = \frac{2n_1}{n_1 + n_2} \qquad (2.19)$$

$$t_H = \frac{2n_2}{n_1 + n_2} \qquad (2.20)$$

Example 2.1   Let medium 1 be air and medium 2 be glass. The index of refraction of glass is $n_2 = 1.5$ and of air is $n_1 = 1$. Calculate the coefficients of reflection and transmission associated with the air-glass interface for normal incidence.

Solution   With $n_1 = 1.0$, $n_2 = 1.5$, Eqs. (2.18) to (2.20) become

$$r_E = r_H = -\tfrac{1}{5} \qquad (2.21)$$

$$t_E = \tfrac{4}{5} \qquad (2.22)$$

$$t_H = \tfrac{6}{5} \qquad (2.23)$$

Everything looks all right except Eq. (2.23). Can the transmission coefficient be larger than 1? Let us examine this more closely [2].

Note from Eq. (2.18) when $n_2 > n_1$, the reflection coefficient becomes negative. What does a negative value for the reflection coefficient mean? Reexamine Fig. 2.2 closely. The positive directions of **E** and **H** were chosen such that the positive directions of E_1 and E_3 are the same but those of H_1 and H_3 are opposite; hence, when $r_H = -\tfrac{1}{5}$ it really means H_3 points in the same direction as H_1 and its magnitude is $\tfrac{1}{5}|\mathbf{H}_1|$, as illustrated in Fig. 2.3a. Likewise, since r_E is also negative according to Eq. (2.21), it really means that the direction of $\mathbf{E}_3$ is opposite to $\mathbf{E}_1$ and its magnitude is $\tfrac{1}{5}|\mathbf{E}_1|$, as illustrated in Fig. 2.3a. The resultant **E** field just above the boundary is $|\mathbf{E}_1 + \mathbf{E}_3| = \tfrac{4}{5}|\mathbf{E}_1|$ and that just below the boundary is $|\mathbf{E}_2| = \tfrac{4}{5}|\mathbf{E}_1|$ so that the boundary condition for **E** is satisfied. Similarly, the resultant **H** field just above the interface is $\tfrac{6}{5}|\mathbf{H}_1|$ and that just below is $\tfrac{6}{5}|\mathbf{H}_1|$ and both happily match up with the results of Eqs. (2.22) and

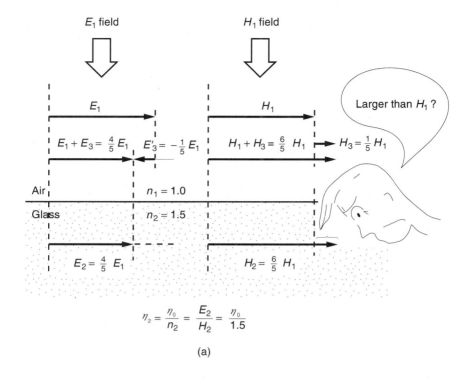

$$\eta_2 = \frac{\eta_0}{n_2} = \frac{E_2}{H_2} = \frac{\eta_0}{1.5}$$

(a)

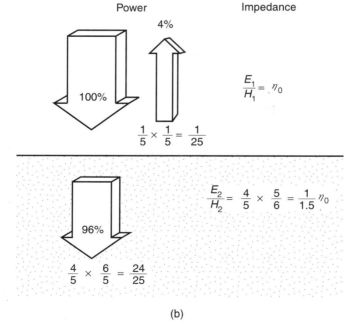

(b)

Figure 2.3   Field and power redistributions and impedance at an air–glass interface: $n_1 = 1, n_2 = 1.5$. (a) Field redistribution. (b) Power redistribution and impedance.

(2.23). The intrinsic impedance in the glass is

$$\eta_2 = \frac{\eta_0}{n_2} = \frac{E_2}{H_2} = \frac{\frac{4}{5}E_1}{\frac{6}{5}H_1} = \frac{\eta_0}{1.5}$$

From this example, the following generalization can be made. When light is incident from a medium of a higher intrinsic impedance (lower index of refraction) onto that of a lower intrinsic impedance (higher index of refraction), the transmitted **E** field decreases, whereas the transmitted **H** field increases as seen from Eqs. (2.16) and (2.17). Whenever $n_2 > n_1$, t_H in Eq. (2.20) is larger than unity!   □

The transmittance T is defined as the power ratio of transmitted light to incident light. The reflectance R is the power ratio of reflected light to incident light. The value of T and R will be calculated for Example 2.1. (Note that transmission and reflection coefficients are field ratios.)

Let P denote light power, **s** denote the Poynting vector, and A denote the area on the interface intercepted by the light beams. For simplicity, A will be taken as a unit area. For the present case, $P = |s|A = |s|$.

The incident light power is therefore

$$|\mathbf{s}_1| = |\mathbf{E}_1 \times \mathbf{H}_1| = \frac{1}{\eta_1}|\mathbf{E}_1|^2 \tag{2.24}$$

The reflected power is

$$|\mathbf{s}_3| = |\mathbf{E}_3 \times \mathbf{H}_3| = \frac{1}{\eta_1}|\mathbf{E}_3|^2 = \frac{1}{\eta_1}|r_E\mathbf{E}_1|^2 \tag{2.25}$$

Hence, the reflectance is

$$R = \frac{|\mathbf{s}_3|}{|\mathbf{s}_1|} = |r_E|^2 = \frac{1}{25} \tag{2.26}$$

The transmitted power is

$$|\mathbf{s}_2| = |\mathbf{E}_2 \times \mathbf{H}_2| = \frac{1}{\eta_2}|\mathbf{E}_2|^2 = \frac{1}{\eta_2}|t_E\mathbf{E}_1|^2 \tag{2.27}$$

Hence, the transmittance T is

$$T = \frac{(1/\eta_2)|t_E\mathbf{E}_1|^2}{(1/\eta_1)|\mathbf{E}_1|^2} = \frac{\eta_1}{\eta_2}|t_E|^2 = \frac{1.5}{1}\left(\frac{4}{5}\right)^2 = \frac{24}{25} \tag{2.28}$$

In terms of the **H** field, one has

$$|\mathbf{s}_1| = \eta_1|\mathbf{H}_1|^2, \quad |\mathbf{s}_2| = \eta_2|t_H\mathbf{H}_1|^2, \quad |\mathbf{s}_3| = \eta_1|r_H\mathbf{H}_1|^2$$

$$R = \frac{|\mathbf{s}_3|}{|\mathbf{s}_1|} = |r_H|^2 = \frac{1}{25} \tag{2.29}$$

$$T = \frac{|\mathbf{s}_2|}{|\mathbf{s}_1|} \frac{\eta_2}{\eta_1} |t_H|^2 = \frac{1}{1.5} \left(\frac{6}{5}\right)^2 = \frac{24}{25} \tag{2.30}$$

which is consistent with the values calculated using the **E** field.

2.3.2   Transmission and Reflection Coefficients (at an Arbitrary Incident Angle)

The values of the transmission and reflection coefficients depend not only on the angle of incidence, but also on the direction of polarization, which is the direction of the **E** field (some books define polarization in terms of **H**).

The propagation direction of the incident light and the normal to the interface define a plane, which is referred to as the plane of incidence. Two polarizations of particular interest are the **E** field polarized in the plane of incidence, and **E** polarized perpendicular to the plane of incidence.* An incident wave with an arbitrary direction of polarization can always be decomposed into these two polarizations.

The in-plane polarization, more commonly called parallel polarization, or p wave, is considered first. In Fig. 2.4a, the $y = 0$ plane is the plane of incidence. The convention of the positive directions of **E** and **H** are taken such that the Poynting vector $\mathbf{s} = \mathbf{E} \times \mathbf{H}$ coincides with the direction of light propagation (see boxed note). The continuity Conditions 4 and 5 are used. In medium 1, components of both incident and reflected waves must be considered. The tangential components of the **E** and **H** fields on both

---

There is more than one combination of **E** and **H** that satisfies this convention. In order to maintain consistency in the derivation of equations, henceforth in this text, we shall impose the following additional constraint on the choice of positive **E** and **H**. For both parallel and perpendicular polarizations, in the limit that the angle of incidence approaches 90°, all **E** vectors shall point in the same direction. (Picture Figs. 2.4a and 2.4b with θ_1 approaching grazing incidence.) This is convenient when dealing with optical guides, where the angles of incidence at boundaries are often large. However, this constraint does lead to a peculiarity at normal incidence. For parallel polarization at normal incidence, incident and reflected **E** vectors point in opposite directions, whereas for perpendicular polarization, incident and reflected **E** vectors point in the same direction, giving $r_\perp = -r_\parallel$ for $\theta_1 = 0$. For example, if you examine Eqs. (2.35) and (2.40) for $\theta_1 = \theta_2 = 0$, you will find $r_\parallel = -r_\perp$. The difference in sign is simply due to the difference in the conventions for positive **E** for the two cases. In Fig. 2.4a, the direction of positive E_{3t} is taken in the opposite direction to that of E_{1t} in Fig. 2.4a, while the positive directions of E_1 and E_3 are taken the same. Whichever convention one chooses, the final result for a given configuration comes out the same, as explained in Example 2.1.

---

* Sometimes these waves are called p waves and s waves, originating from the German terms of parallel wave and senkrecht wave.

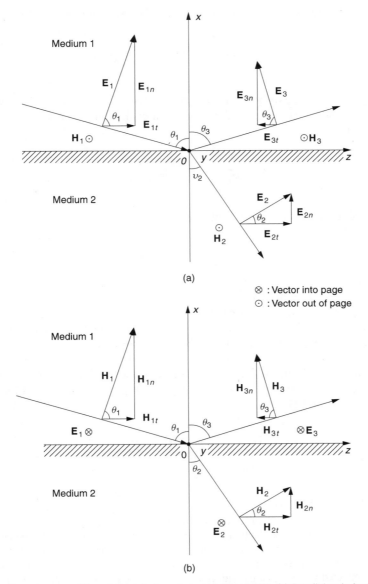

⊗ : Vector into page
⊙ : Vector out of page

Figure 2.4 Boundary condition at the interface between two media. (a) Parallel polarization. (b) Perpendicular polarization.

sides of the boundary are equal; hence,

$$E_1 \cos \theta_1 - E_3 \cos \theta_3 = E_2 \cos \theta_2 \qquad (2.31)$$

$$H_1 + H_3 = H_2 \qquad (2.32)$$

Equation (2.32) can be rewritten further using Eqs. (2.2) and (2.5) as

$$\frac{1}{\eta_1}(E_1 + E_3) = \frac{1}{\eta_2} E_2 \qquad (2.33)$$

From the continuity of the normal component of **D**, and from the equality $\theta_1 = \theta_3$, one obtains

$$\epsilon_{r1}(E_1 + E_3) \sin \theta_1 = \epsilon_{r2} E_2 \sin \theta_2 \tag{2.34}$$

A combination of any two boundary conditions listed above can provide the electric field reflection (see boxed note) and transmission coefficients. The reflection coefficient $r_\parallel$ for the **E** field is

$$r_\parallel = \frac{E_3}{E_1} = \frac{n_2 \cos \theta_1 - n_1 \cos \theta_2}{n_2 \cos \theta_1 + n_1 \cos \theta_2} \tag{2.35}$$

Similarly, the transmission coefficient $t_\parallel$ for the **E** field is

$$t_\parallel = \frac{E_2}{E_1} = \frac{2n_1 \cos \theta_1}{n_2 \cos \theta_1 + n_1 \cos \theta_2} \tag{2.36}$$

The subscript $\parallel$ indicates the case when the **E** field is in the plane of incidence, as shown in Fig. 2.4a. There is a difference between the values $r_\parallel$ and $t_\parallel$ for the **E** field and the same quantities for the **H** field. (See Problem 2.1.)

For the s wave case of light polarized perpendicular to the plane of incidence as shown in Fig. 2.4b, the reflection and transmission coefficients for the **E** field are obtained in a similar manner. Boundary Condition 4 for the continuity of the tangential components of **E** and **H** is used.

$$\mathbf{E}_1 + \mathbf{E}_3 = \mathbf{E}_2 \tag{2.37}$$

$$H_1 \cos \theta_1 - H_3 \cos \theta_3 = H_2 \cos \theta_2 \tag{2.38}$$

Why does the wave reflect when a discontinuity in the index of refraction is encountered? Let's start with the case of reflection from a perfect conductor (mirror). As the electromagnetic wave is incident upon the conductor, a current is induced on the surface of the conductor, just as a current is induced on a metal wire receiving antenna. The induced current sets up a field. The reflected wave is just the induced field. The amount of induced current is such that the induced field is identical to the incident wave but exactly opposite in phase. Thus, the resultant field just above the surface becomes zero. On the other hand, the field just below the surface (inside the conductor) is zero. Thus, the continuity condition is satisfied.

When the second medium is a dielectric medium, the situation is slightly more complicated. In the optical spectrum, when an **E** field impinges on a dielectric medium, the orbits of the electrons in the medium become slightly displaced due to the **E** field, and a dipole moment is induced. The time-varying nature of the incident **E** field causes these dipole moments to vibrate. This vibrating dipole moment establishes the reradiating field. This field radiates both in the direction of the original medium as well as into the second medium. The former is the reflected wave and the latter is the transmitted wave. The relative amplitudes of these two waves are determined such that Eqs. (2.35) and (2.36) are satisfied.

Using Eq. (2.2), and $\theta_1 = \theta_3$, Eq. (2.38) becomes

$$\frac{\cos \theta_1}{\eta_1}(E_1 - E_3) = \frac{E_2}{\eta_2}\cos \theta_2 \tag{2.39}$$

Equations (2.37) and (2.39) are used to derive the expressions for $r_\perp$ and $t_\perp$:

$$r_\perp = \frac{E_3}{E_1} = \frac{n_1 \cos \theta_1 - n_2 \cos \theta_2}{n_1 \cos \theta_1 + n_2 \cos \theta_2} \tag{2.40}$$

$$t_\perp = \frac{E_2}{E_1} = \frac{2n_1 \cos \theta_1}{n_1 \cos \theta_1 + n_2 \cos \theta_2} \tag{2.41}$$

The above four equations for $r_\parallel$, $t_\parallel$, $r_\perp$, and $t_\perp$ are called Fresnel's equations. Note that for $\theta_1 = \theta_2 = 0$, Eqs. (2.40) and (2.41) match Eqs. (2.18) and (2.19), respectively.

In order to calculate Fresnel's equations, it is necessary to know θ_2 using Snell's law, Eq. (2.10). One can also make use of Snell's law to completely eliminate both indices of refraction, n_1 and n_2. With Eq. (2.10), Eq. (2.40) can be reduced to

$$r_\perp = -\frac{\sin(\theta_1 - \theta_2)}{\sin(\theta_1 + \theta_2)} \tag{2.42}$$

Similarly, but with a slightly longer derivation (see Problem 2.2), Eqs. (2.35), (2.36), and (2.41) are reduced to

$$r_\parallel = \frac{\tan(\theta_1 - \theta_2)}{\tan(\theta_1 + \theta_2)} \tag{2.43}$$

$$t_\perp = \frac{2\cos \theta_1 \sin \theta_2}{\sin(\theta_1 + \theta_2)} \tag{2.44}$$

$$t_\parallel = \frac{2\cos \theta_1 \sin \theta_2}{\sin(\theta_1 + \theta_2)\cos(\theta_1 - \theta_2)} \tag{2.45}$$

If Eqs. (2.42) and (2.44) are examined closely, one can verify that

$$1 + r_\perp = t_\perp \tag{2.46}$$

This is a direct consequence of Maxwell's continuity condition, as can be seen by multiplying both sides of Eq. (2.46) by the incident field E_1. The left-hand side of Eq. (2.46) represents the tangential component of the resultant field just above the interface and the right-hand side, the same just below the interface. It should be noted that the same is not true for the relationship between $r_\parallel$ and $t_\parallel$ (see Problem 2.3).

To gain some insight into the actual quantities, let us calculate the values of the transmission and reflection coefficients as a function of incident angle for a particular interface. Figure 2.5 shows curves for the glass ($n_2 = 1.5$) and air ($n_1 = 1$) interface. Figure 2.5 reveals some noteworthy features. All curves turn downward as θ_1 increases. The transmission coefficients $t_\parallel$ and $t_\perp$ both decrease as the incident angle θ_1 increases, and when θ_1 approaches $90°$; that is, when the incident beam becomes nearly parallel to the interface, the amount of transmitted light is almost zero. The decrease is especially significant for $\theta_1 > 70°$.

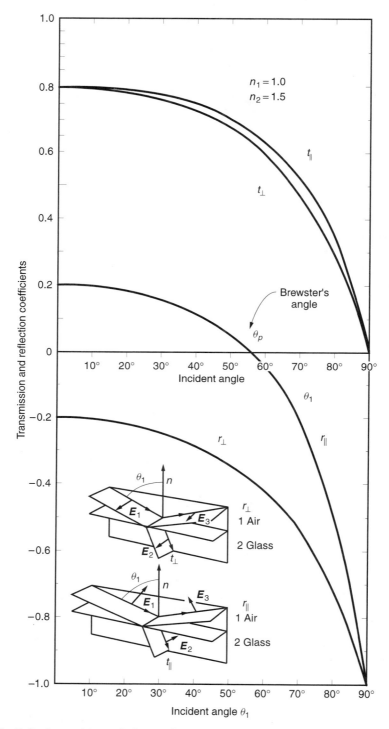

Figure 2.5 Reflection and transmission coefficients as a function of incident angle for air to glass interface ($n_1 = 1.0$, $n_2 = 1.5$).

This information is put to good use in the following practical example. Suppose that a glass lens is used to collect a diverging light source such as the emission from a light-emitting diode (LED). The edge of a double convex lens, as shown in Fig. 2.6a, should be avoided because the angle of incidence is large in this area and transmission is small. The result is an inefficient light collection.

The meniscus lens alleviates this problem, as shown in Fig. 2.6b. But if the meniscus lens is used incorrectly, the situation becomes even worse than with a double convex lens, as indicated in Fig. 2.6c.

Another interesting observation from Fig. 2.5 is that the curve for $r_{\parallel}$ crosses the zero axis. This means that there is no reflected wave at this angle. This angle is called Brewster's angle and is discussed in detail in Section 2.5. At this angle, there is no reflected wave.

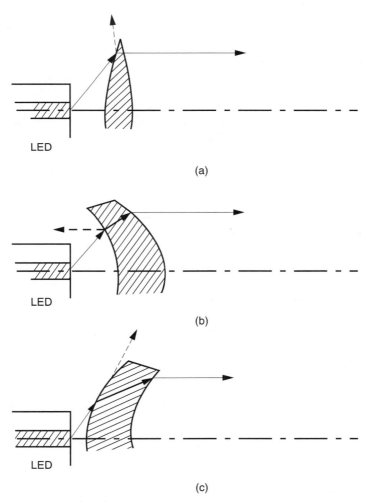

(a)

(b)

(c)

Figure 2.6 Selecting a lens to collect the emission from an LED. (a) Increase in reflection toward the edge of a double convex lens. (b) Minimal increase in reflection toward the edge of a meniscus lens. (c) Increase in reflection toward the edge with a misused meniscus lens.

2.3.3 Impedance Approach to Calculating Transmission and Reflection Coefficients

In the case of normal incidence the formulas for t and r are Eqs. (2.15) to (2.17), which are simpler and easier to remember than the equations for an arbitrary angle of incidence. The simplicity of Eqs. (2.15) to (2.17) is the basis for the impedance approach to calculating transmission and reflection coefficients. One defines an impedance out of certain components of $\mathbf{E}$ and $\mathbf{H}$ chosen such that their Poynting vectors point normal to the interface. From Fig. 2.4b, the impedances referring to the normal direction, which are sometimes called *characteristic wave impedances referred to the x direction* [2], are defined for incident and transmitted waves as

$$\eta_{1x} = \frac{E_1}{H_1 \cos \theta_1} = \frac{\eta_1}{\cos \theta_1} \tag{2.47}$$

$$\eta_{2x} = \frac{E_2}{H_2 \cos \theta_2} = \frac{\eta_2}{\cos \theta_2} \tag{2.48}$$

Replacing η_1 and η_2 in Eqs. (2.15) to (2.17) by η_{1x} and η_{2x} above, and then converting η to n, one can obtain the expressions for an arbitrary angle of incidence given by Eqs. (2.40) and (2.41). Equations (2.35) and (2.36) of the corresponding expressions for the other polarization are also obtained in a similar manner, referring to Fig. 2.4a and noting the sign convention mentioned in the boxed note. The impedance approach is quite powerful in dealing with surface acoustic wave devices, where the index of refraction is modulated in space and time, and the boundary conditions become more complex.

2.4 TRANSMITTANCE AND REFLECTANCE (AT AN ARBITRARY INCIDENT ANGLE)

The previous sections dealt primarily with the transmission and reflection coefficients t and r, which are field ratios. Transmittance and reflectance, which are power ratios, are studied in this section.

Referring to Fig. 2.7, consider the light power passing through a unit area on the interface. According to the law of conservation of energy, the power incident on this unit area has to be identical to the power emergent from the area. If this were not so, the energy would accumulate, resulting in an ever increasing build up of energy on the boundary. In applying conservation of energy, one has to realize two things:

1. Even if the amplitude of $\mathbf{E}$ is the same, the energy of the wave is different if the medium in which the wave is propagating is different. The same is true for $\mathbf{H}$.
2. The power flowing through a unit area on the interface depends on the direction of incidence to the area.

The time rate of flow of the electromagnetic energy per unit area is given by the Poynting vector $\mathbf{s}$, where $\mathbf{s} = \mathbf{E} \times \mathbf{H}$. The magnitude of the Poynting vector, $s = |\mathbf{s}|$, represents the maximum instantaneous value of the electromagnetic power per unit area (s has units of W/m^2). Another quantity often used in optics is the irradiance I.

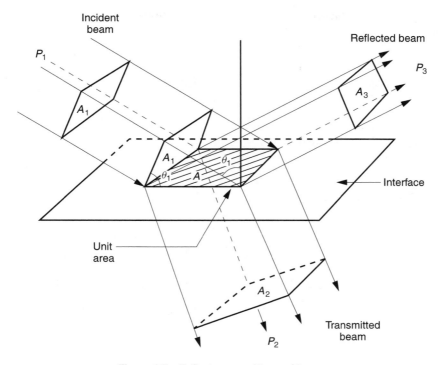

Figure 2.7 Reflectance and transmittance.

The irradiance I is the magnitude of the time average of the Poynting vector, $I = |\langle \mathbf{s} \rangle|$, and it also has units of W/m². Since $\mathbf{E}$ and $\mathbf{H}$ are sinusoidally varying, the irradiance is $I = \frac{1}{2}|\mathbf{E} \times \mathbf{H}|$.

Using Eq. (2.2) or (2.5), the magnitude of the Poynting vector is given by

$$s = |\mathbf{s}| = \frac{1}{\eta}|\mathbf{E}|^2 \tag{2.49}$$

or

$$s = |\mathbf{s}| = \eta|\mathbf{H}|^2 \tag{2.50}$$

Suppose that the same energy flow, s_0, is present in two different media such that $s_0 = (1/\eta_1)|\mathbf{E}_1|^2 = (1/\eta_2)|\mathbf{E}_2|^2$. What this means is that if $\eta_1 > \eta_2$, then $|\mathbf{E}_1^2| > |\mathbf{E}_2|^2$. The medium with the greater impedance has the larger $\mathbf{E}$ field.

As mentioned earlier, the power passing through the unit area in Fig. 2.7 depends on the angle of incidence. Let us use the symbol P to denote maximum instantaneous light power. For the geometry of Fig. 2.7, the incident, transmitted, and reflected light powers are

$$P_1 = s_1 A_1$$
$$P_2 = s_2 A_2 \tag{2.51}$$
$$P_3 = s_3 A_3$$

Also from Fig. 2.7, the areas A_1, A_2, and A_3 are related to the area A on the interface as follows:

$$A_1 = A \cos \theta_1$$
$$A_2 = A \cos \theta_2 \tag{2.52}$$
$$A_3 = A \cos \theta_1$$

As A is assumed to be a unit area, Eqs. (2.49), (2.51), and (2.52) are combined to give the incident, transmitted, and reflected powers:

$$P_1 = \frac{1}{\eta_1} |\mathbf{E}_1|^2 \cos \theta_1$$
$$P_2 = \frac{1}{\eta_2} |\mathbf{E}_2|^2 \cos \theta_2 \tag{2.53}$$
$$P_3 = \frac{1}{\eta_1} |\mathbf{E}_3|^2 \cos \theta_1$$

The transmittance T is defined as the ratio of transmitted power to incident power:

$$T = \frac{P_2}{P_1} = \frac{\eta_1}{\eta_2} \frac{|\mathbf{E}_2|^2}{|\mathbf{E}_1|^2} \frac{\cos \theta_2}{\cos \theta_1} \tag{2.54}$$

When $\mu_{r1} = \mu_{r2}$, $E_2 = E_1 t'_E$

$$T = \frac{n_2 \cos \theta_2}{n_1 \cos \theta_1} |t'_E|^2 \tag{2.55}$$

The value of T depends on the direction of polarization. For parallel ($T_\parallel$) and perpendicular ($T_\perp$) polarizations, $t'_E = t_\parallel$ and $t'_E = t_\perp$, respectively, have to be used in Eq. (2.55).

The reflectance R is the ratio of reflected power to incident power,

$$R = \frac{P_3}{P_1} = \frac{|\mathbf{E}_3|^2}{|\mathbf{E}_1|^2} = |r'_E|^2 \tag{2.56}$$

where R also depends on the polarization, and for parallel ($R_\parallel$) and for perpendicular ($R_\perp$) polarizations, $r'_E = r_\parallel$ and $r'_E = r_\perp$, respectively, have to be used in Eq. (2.56). (For general directions of polarization, see Ref. 1).

It is left as an exercise (see Problem 2.4) to show that

$$R + T = 1 \tag{2.57}$$

for both directions of polarization. In practice, Eq. (2.55) is seldom used to calculate T. Rather, the reflectance is first calculated via the simple relationship $R = |r'_E|^2$ and then the transmittance is found by making use of Eq. (2.57), $T = 1 - R$.

Figure 2.8 shows the plots of transmittance and reflectance at the interface between air and glass ($n_1 = 1.5$) as a function of the incident angle θ_1. Note that light with perpendicular polarization has a narrower range of good transmission. To achieve 90% transmittance, the incident angle θ_1 has to stay smaller than 45° when perpendicular polarization is used, but the incident angle can be extended to 75° if parallel polarization is used.

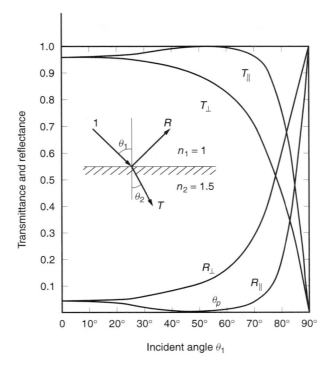

Figure 2.8 Transmittance and reflectance at the air–glass interface.

2.5 BREWSTER'S ANGLE

Of the four curves (for $n_1 = 1.0$, $n_2 = 1.5$) in Fig. 2.5, only one curve r crosses the horizontal axis. The reflection coefficient r becomes zero near $\theta_1 = 56°$, or to be more exact, at $\theta_1 = 56°19'$; and there is no reflected wave.* This angle, labelled as θ_p in Fig. 2.5, is called Brewster's angle. Brewster's angle is used to avoid reflection from the surface. It is important to realize that Brewster's angle exists only when the direction of polarization is parallel to the plane of incidence.

The reflection coefficient r vanishes when the denominator of Eq. (2.43) becomes infinity, namely, when

$$\theta_1 + \theta_2 = 90° \tag{2.58}$$

which is known as Brewster's condition. It is interesting to note that when Brewster's condition Eq. (2.58) is satisfied, the direction in which one would expect to find a reflection if it were to exist (dashed line in Fig. 2.9) coincides with the direction of polarization of the transmitted wave as shown in Fig. 2.9. From a microscopic viewpoint, the reflected wave is generated by the oscillation of electric dipoles in the transmission medium. The oscillating dipole does not radiate in the direction of oscillation.

---

* Substituting $\theta_p = 56°19'$ into Eq. (2.36), one can see that t is not unity at $\theta_1 = \theta_p$, even though r is zero. Why is this so? See Example 2.3.

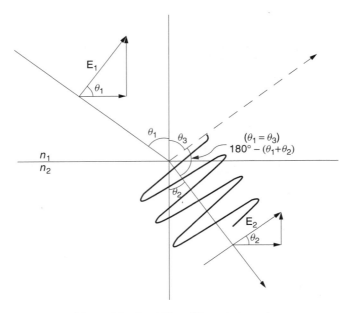

Figure 2.9   Condition of Brewster's angle.

The incident angle for Brewster's condition can be calculated by combining Eq. (2.58) with Snell's law:

$$n_1 \sin \theta_1 = n_2 \sin(90° - \theta_1)$$

$$= n_2 \ \cos \ \theta_1$$

Thus, Brewster's angle θ_p is

$$\theta_p = \tan^{-1}\left(\frac{n_2}{n_1}\right) \tag{2.59}$$

Brewster's condition can be interpreted as being a consequence of impedance matching with reference to the vertical direction. As in Section 2.3.3, the characteristic wave impedances η_{1x} just above the interface and η_{2x} just below the interface are defined for parallel polarization as

$$\eta_{1x} = \frac{E_1}{H_1} \cos \theta_1 = \eta_1 \cos \theta_1 \tag{2.60}$$

$$\eta_{2x} = \frac{E_2}{H_2} \cos \theta_2 = \eta_2 \cos \theta_2 \tag{2.61}$$

where both η_{1x} and η_{2x} refer to the vertical direction. When Brewster's condition, Eq. (2.58), is satisfied, Eqs. (2.60) and (2.61) can be written as

$$\eta_{1x} = \eta_1 \sin \theta_2 \tag{2.62}$$

$$\eta_{2x} = \eta_2 \sin \theta_1 \tag{2.63}$$

Snell's law in impedances can be expressed as

$$\eta_1 \sin \theta_2 = \eta_2 \sin \theta_1 \tag{2.64}$$

With Eqs. (2.62)–(2.64)

$$\eta_{1x} = \eta_{2x} \tag{2.65}$$

Thus, Brewster's angle has led to the condition of impedance matching.

A good example of the use of Brewster's angle is the window of a gas laser tube of the external mirror type, as described in Example 2.2.

Example 2.2   Figure 2.10 shows a diagram of a gas laser tube with external mirrors. The end windows of the tube are tilted at Brewster's angle so that the reflection from the windows will be minimized. What is the Brewster's angle? Is the output of such a laser polarized? If so, in which direction is it polarized? Assume that the index of refraction of the glass window is $n_2 = 1.54$.

Solution   From Eq. (2.59) with $n_1 = 1$, $n_2 = 1.54$, Brewster's angle is $\theta_p = 57°$. Brewster's angle is applicable only to the wave with parallel polarization. The reflection coefficient for the perpendicular polarization $r_\perp$ is calculated for comparison. From Eq. (2.40) or (2.42), $r_\perp = 0.4$. Compared to $r_\parallel = 0$, there is a significant difference. The external mirrors of the laser are usually highly reflective so that a large portion of the light makes thousands of passes back and forth between the two mirrors. For each pass through the laser medium, the light is amplified (this is a characteristic of laser media and represents a situation dealt with in Chapter 14 in Volume II). Because the perpendicular polarization suffers reflection losses at the windows, the losses for this polarization outweigh the gain through the amplifying medium. In comparison, the parallel polarization, which has no reflection losses at the windows, reaps the gain of the amplifying medium. Light exits the laser by transmission through the external mirrors. Even though the mirrors are highly reflective, a small amount is transmitted at each light pass. After the first few passes, the light that exits the laser is predominantly the parallel polarization: that is, the light that exits the laser of the external cavity type

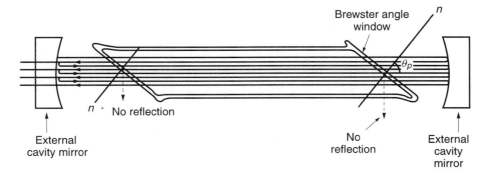

Figure 2.10   Laser cavity of the external mirror type.

is predominately polarized in the plane made by the normal to the window and the
axis of the light beam.   □

Example 2.3   Referring to Fig. 2.5, $r_\parallel = 0$ at Brewster's angle but the corresponding
transmission coefficient is 0.66. Since $r_\parallel = 0$, why isn't $t_\parallel$ unity?

Solution   In applying conservation of energy, the angle dependency of the power flow
as indicated in Fig. 2.8 has to be considered, as well as the difference in **E** due to the
difference in media. These considerations were mentioned in Section 2.4. The law of
conservation of energy does not say that $r_\parallel + t_\parallel = 1$; but the law of conservation of
energy does say that $R_\parallel + T_\parallel = 1$. As a matter of fact, that $r_\parallel + t_\parallel = 1$ is not true is
seen from Eqs. (2.35) and (2.36). Equations (2.55) and (2.56) give

$$R_\parallel = r_\parallel^2 = 0$$

$$T_\parallel = \frac{1.5\cos(33°41')}{\cos 56°19'} \times 0.66^2 \approx 1.0 \qquad\qquad □$$

2.6 TOTAL INTERNAL REFLECTION

Consider the case when the light is incident from an optically dense medium (larger
index of refraction n_1) onto that of an optically less dense medium (smaller index
of refraction n_2) with the geometry shown in Fig. 2.11. If and only if the light is
incident from an optically dense medium onto a less dense medium, is there a chance
that the angle of the emergent light reaches 90° with an incident angle less than

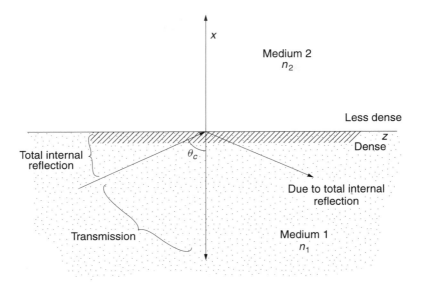

Figure 2.11   Total internal reflection, $n_1 > n_2$, $n = n_2/n_1 < 1$.

Total internal reflection.

90°. As soon as the emergent angle reaches 90°, *total internal reflection** takes place. The angle of total internal reflection follows the law of reflection, which means the angle of reflection is equal to the angle of incidence. The angle θ_c at which total internal reflection starts to take place is called the *critical angle* and is calculated from Eq. (2.10) as

$$\theta_c = \sin^{-1}\left(\frac{n_2}{n_1}\right) \tag{2.66}$$

In the field of fiber optics, total internal reflection at the boundary between the core and cladding layers supports the guided wave in the fiber. The light propagates as it is bounced back and forth.

Because of its durability, total internal reflection is sometimes used as a reflective device in place of a metal-deposited surface mirror, but the disadvantage is the restriction over the choice of incident angle.

2.6.1 Bends in a Guide

Changing the direction of an optical guide is often required. In many cases, a change of direction by a total internal reflection mirror is preferred to gradual bending of the guide. Gradual bending is limited to small bend angles. Typically, the radius of curvature of the bends cannot be less than 10–30 mm because, if any smaller, the incident angle to some portion of the wall of the guide becomes less than the critical angle, as indicated by the dotted line in Fig. 2.12a, and the light leaks.

A sharp bend, however, can be made if a flat mirror is placed at the bend [3,4]. The flat mirror preserves the angle of the zigzag path of the ray in the guide, as shown by the solid line, and the light does not leak. For instance, if the angle of incidence to the flat mirror is 65°, that of the circular bend in this particular geometry is 55°.

---

* The terms total reflection and total internal reflection are interchangeable. Johannes Kepler (1571–1630) experimentally discovered the existence of total internal reflection.

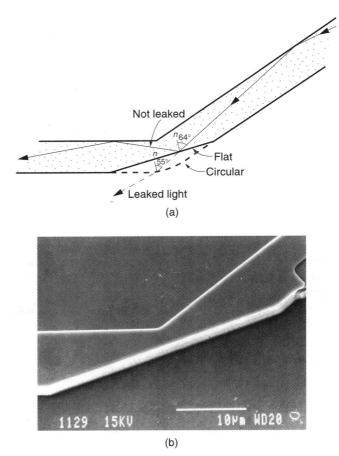

Figure 2.12 Low-loss corner mirrors with 45° deflection angle for integrated optics. (a) Bend by the total internal reflection flat mirror compared with that of a circular bend. (b) Scanning electron microscope picture of the total internal reflection corner mirror. (Photograph courtesy of E. Gini, G. Guekos, and H. Melchior [4].)

Figure 2.12b shows an electron microscope photograph of a GaAs rib guide mirror. The insertion loss[†] of such a bend was 0.3 dB [4].

2.7 WAVE EXPRESSIONS OF LIGHT

Up to this point, boundary phenomena, such as reflection, refraction, total internal reflection, and Brewster's angle, have been described without explicit reference to the wave nature of light. In the next few sections, the wave nature will be emphasized. By using the wave expressions, finer details can be reworked.

[†] Let P_{in} be the power into the device and P_{out} be the power out of the device. The insertion loss of the device is defined as

$$10 \log \frac{P_{in}}{P_{out}}$$

2.7.1 Fields Near the Boundary

The expressions representing the fields near a discrete boundary between media 1 and 2 with the geometry shown in Fig. 2.11 will be described [5].

In medium 1, a plane wave propagating in the direction

$$\mathbf{k} = k_x\hat{\mathbf{i}} + k_y\hat{\mathbf{j}} + k_z\hat{\mathbf{k}} \tag{2.67}$$

is expressed from Eq. (1.22) as

$$\mathbf{E}_1 = |\mathbf{E}_1|e^{j(\mathbf{k}\cdot\mathbf{r}-\omega t)} \tag{2.68}$$

where $\mathbf{r}$ is the position vector

$$\mathbf{r} = x\hat{\mathbf{i}} + y\hat{\mathbf{j}} + z\hat{\mathbf{k}}$$

The incident wave propagating in medium 1 is given by

$$\mathbf{E}_1 = |\mathbf{E}_1|e^{j(k_{1x}x+k_{1y}y+k_{1z}z-\omega t)} \tag{2.69}$$

Similarly, the transmitted wave in medium 2 is

$$\mathbf{E}_2 = |\mathbf{E}_2|e^{j(k_{2x}x+k_{2y}y+k_{2z}z-\omega t)} \tag{2.70}$$

Inserting $\mathbf{E}_1$ and $\mathbf{E}_2$ into the wave equation, $\nabla^2\mathbf{E} + (nk)^2\mathbf{E} = 0$, gives the condition that the sum of the squares of the k components has to equal the square of the propagation constant in that medium; that is,

$$k_{1x}^2 + k_{1y}^2 + k_{1z}^2 = (n_1k)^2 \tag{2.71}$$

and

$$k_{2x}^2 + k_{2y}^2 + k_{2z}^2 = (n_2k)^2 \tag{2.72}$$

Equations (2.71) and (2.72) will be manipulated in order to clarify the allowed and the prohibited regions of propagation.

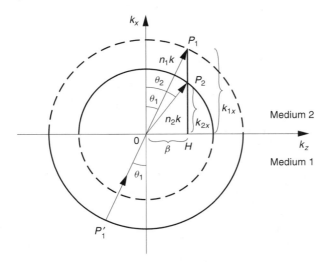

Figure 2.13 Wave vectors propagating in k space.

Equations (2.71) and (2.72) are equations of spheres in k space, with radii $n_1 k$ and $n_2 k$. Figure 2.13 is a two-dimensional representation in k space, with $k_y = 0$. In this figure, propagation takes place only in the x and z directions. The angles that the radial vectors $\overrightarrow{OP_1}$ (translated from $\overrightarrow{P_1'O}$ for convenience) and $\overrightarrow{OP_2}$ make with the k_x axis are the incident and emergent angles, θ_1 and θ_2, respectively. The wavelength matching condition (phase matching condition) on the boundary for the z direction is satisfied by ensuring that the projections of $\overrightarrow{OP_1}$ and $\overrightarrow{OP_2}$ onto the k_z axis are the same, namely,

$$k_{1z} = k_{2z} = \beta \tag{2.73}$$

Equation (2.73) fixes the value of θ_2 for a given θ_1. With the phase matching condition, Eqs. (2.71) and (2.72) become

$$k_{1x}^2 + \beta^2 = (n_1 k)^2$$
$$k_{2x}^2 + \beta^2 = (n_2 k)^2 \tag{2.74}$$

Expressions for E_1 and E_2 will now be found. Referring to Fig. 2.13, one has $k_{2x} = n_2 k \cos \theta_2$, so that the transmitted wave can be expressed as

$$E_2 = |E_2| e^{j n_2 k \cos \theta_2 \cdot x + j \beta z - j \omega t} \tag{2.75}$$

Now, Eq. (2.75) will be rewritten in terms of the incident angle θ_1 and the transmission coefficients $t_{\parallel,\perp}$. Snell's law, Eq. (2.10), gives

$$\cos \theta_2 = \frac{1}{n} \sqrt{n^2 - \sin^2 \theta_1} \tag{2.76}$$

where

$$n = \frac{n_2}{n_1}$$

Henceforth, the symbol n without the suffix is used to denote the ratio of the indices of refraction.

The expressions for E_2 and E_1 are

$$E_2 = t_{\parallel,\perp} |E_1| e^{j n_1 k [\sqrt{n^2 - \sin^2 \theta_1}\, x + (\sin \theta_1) z] - j \omega t} \tag{2.77}$$

and

$$E_1 = |E_1| e^{j n_1 k [(\cos \theta_1) x + (\sin \theta_1) z] - j \omega t} \tag{2.78}$$

Note that Eqs. (2.77) and (2.78) are equivalent if n is unity, as one would expect.

2.8   THE EVANESCENT WAVE

When light is incident from an optically dense medium onto a less dense medium, total internal reflection takes place as soon as the emergent angle θ_2 reaches $90°$. Let us take a look at total internal reflection from the viewpoint of the requirements of the boundary condition on the border. Since the wave is totally reflected back into the denser medium, no field is supposed to be present inside the less dense medium. Does the field abruptly become zero on the boundary? Because of the boundary condition of continuity, the field in the less dense medium cannot abruptly become zero. The presence of the so-called evanescent wave (also called surface wave) solves this problem. The amplitude of the evanescent wave at the boundary is the same as the field on the boundary of the dense medium, but the amplitude of the evanescent wave decays very rapidly as it goes away from the boundary into the less dense medium. Moreover, the evanescent wave moves along the boundary in the z direction at the same speed as the incident wave in order to satisfy the phase matching condition.

Let us study this mysterious evanescent wave quantitatively. Equation (2.77) is the expression for the wave in the transmitted medium. The exponential term in this equation contains a square root. In the present case, $n < 1$, so that the quantity inside the square root becomes negative as θ_1 is increased. This means that, for large enough θ_1, the square root term becomes a pure imaginary number $j\gamma$ and the expression for the transmitted wave becomes

$$E_2 = t_{\parallel,\perp}|E_1|e^{-\gamma x + j(\beta z - \omega t)} \tag{2.79}$$

where

$$\gamma = n_1 k \sqrt{\sin^2 \theta_1 - n^2} \tag{2.80}$$

The wave expressed by Eq. (2.79) is precisely the evanescent wave. Note from Eq. (2.79) that the amplitude of E_2 decays exponentially in the x direction, and there is no sinusoidal phase variation in the x direction. On the other hand, the phase variation in the z direction synchronizes with that of the incident wave. These are the major characteristics of the evanescent wave. The *effective depth h* of the penetration of the evanescent wave, which is defined as the depth where the amplitude decays to $1/e$ of that on the boundary, is $1/\gamma$ from Eq. (2.79). Equation (2.79) is graphically represented in Fig. 2.14. The equiamplitude lines are horizontal. The amplitude decays exponentially with x. The equiphase lines are vertical and they translate in the z direction with the incident wave.

In order to determine the amplitude of the evanescent wave at the boundary, let us first turn our attention to the fields inside the optically dense medium. When total internal reflection occurs, the optically dense medium contains both the incident wave, and the wave that is totally reflected. While their z components of propagation are the same, their x components of propagation are opposite. Because of the opposite directions of propagation, they form a standing wave in the x direction in the optically dense medium. In the less dense medium, only the evanescent wave exists. On the boundary, the amplitude of the standing wave and that of the evanescent wave have to be matched.

In the following sections, the reflection coefficient for the case of total internal reflection will be derived in order to calculate the reflected wave. Then, the reflected wave will be used to find the standing wave in the dense medium, and finally the amplitude of the evanescent wave at the boundary will be found.

2.8.1   Transmission and Reflection Coefficients for Total Internal Reflection

The reflection coefficients [5] for parallel and perpendicular polarizations will be extended to the case of total internal reflection. It will be assumed that Snell's law still holds true even in the case of total internal reflection because after all Snell's law is derived

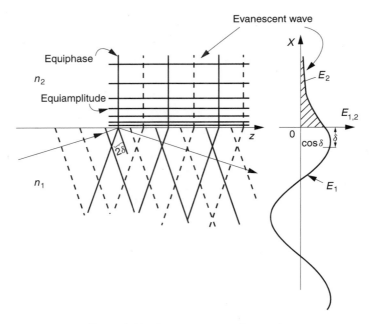

Figure 2.14   Evanescent wave ($n_1 > n_2$).

from the k-matching on the boundary as illustrated in Section 2.2 and has to be satisfied at all times. Writing Eqs. (2.35) and (2.40) in terms of the ratio $n = n_2/n_1$ gives

$$r_\parallel = \frac{\cos\theta_1 - (1/n)\cos\theta_2}{\cos\theta_1 + (1/n)\cos\theta_2} \tag{2.81}$$

$$r_\perp = \frac{\cos\theta_1 - n\cos\theta_2}{\cos\theta_1 + n\cos\theta_2} \tag{2.82}$$

For a given θ_1, the only unknown quantity in Eqs. (2.81) and (2.82) is $\cos\theta_2$. Equation (2.76) gives an expression for $\cos\theta_2$. In the case of total internal reflection, $n < \sin\theta_1$ and Eq. (2.76) is rewritten as

$$\cos\theta_2 = \frac{j}{n}\sqrt{\sin^2\theta_1 - n^2} \tag{2.83}$$

in order to explicitly show that $\cos\theta_2$ is a pure imaginary number. Insertion of Eq. (2.83) into Eqs. (2.81) and (2.82) gives

$$r_\parallel = \frac{\cos\theta_1 - j\sqrt{\sin^2\theta_1 - n^2}/n^2}{\cos\theta_1 + j\sqrt{\sin^2\theta_1 - n^2}/n^2} \tag{2.84}$$

$$r_\perp = \frac{\cos\theta_1 - j\sqrt{\sin^2\theta_1 - n^2}}{\cos\theta_1 + j\sqrt{\sin^2\theta_1 - n^2}} \tag{2.85}$$

Note that, in both the denominator and the numerator, the first term is real while the second is imaginary. Equations (2.84) and (2.85) can be written in simpler form if the polar coordinate representation of a complex number is used. First, since the magnitudes of the denominator and numerator are identical, the absolute value of the reflection coefficient is unity. Next, if the phase angle of the numerator is $-\delta_{\|,\perp}$, and that of the denominator is $\delta_{\|,\perp}$, then

$$r_{\|} = e^{j\phi_{\|}} \qquad \text{where} \quad \phi_{\|} = -2\delta_{\|} \tag{2.86}$$

$$r_{\perp} = e^{j\phi_{\perp}} \qquad \text{where} \quad \phi_{\perp} = -2\delta_{\perp} \tag{2.87}$$

and

$$\tan \delta_{\|} = \frac{\sqrt{\sin^2 \theta_1 - n^2}}{n^2 \cos \theta_1} \tag{2.88}$$

$$\tan \delta_{\perp} = \frac{\sqrt{\sin^2 \theta_1 - n^2}}{\cos \theta_1} \tag{2.89}$$

Thus, the magnitude of the reflection coefficient for the total internal reflection is indeed unity but there is a phase lag on the boundary by $\phi_{\|,\perp}$ radians. The value of $\phi_{\|,\perp}$ is zero at the critical angle and approaches $-\pi$ radians as θ_1 increases. Using a trigonometric relationship with Eqs. (2.88) and (2.89), the difference becomes

$$\tan(\delta_{\|} - \delta_{\perp}) = \frac{\cos \theta_1 \sqrt{\sin^2 \theta_1 - n^2}}{\sin^2 \theta_1} \tag{2.90}$$

The phase shift difference $\phi_{\|} - \phi_{\perp} = -2(\delta_{\|} - \delta_{\perp})$ can be used to make waveplates (Section 6.3). A phase difference of $\pi/2$ radians between the waves of parallel and perpendicular polarizations results in a quarter-waveplate that can convert a linearly polarized wave into a circularly polarized wave. See Example 2.4 for more details. Such a quarter-waveplate is quite useful because it is independent of wavelength. Another good feature of this device is its slow dependence on θ_1, as shown by the dashed line in Fig. 2.15.

Example 2.4   A Fresnel rhomb makes use of the phase shift difference produced by total internal reflection to convert a linearly polarized incident wave into a circularly polarized emergent wave. With the proper choice of incident angle, the Fresnel rhomb is designed to create a 45° phase difference at each of two total internal reflections, in order to achieve the 90° difference between the waves of parallel and perpendicular polarizations. The direction of the polarization of the incident wave is tilted 45° as shown in Fig. 2.16, so that the incident wave is decomposed into equal amplitudes of parallel and perpendicular polarizations.

These equal-amplitude component waves with 90° relative phase difference constitute a circularly polarized wave (mentioned in Chapter 6). Find the angle θ of incidence into the Fresnel rhomb shown in Fig. 2.16. Assume $n = 1/1.5$.

Solution   Since each phase difference $2(\delta_{\|} - \delta_{\perp})$ is $\pi/4$, from Eq. (2.90), we have

$$\tan \frac{\pi}{8} = \frac{\cos \theta_1 \sqrt{\sin^2 \theta_1 - n^2}}{\sin^2 \theta_1}$$

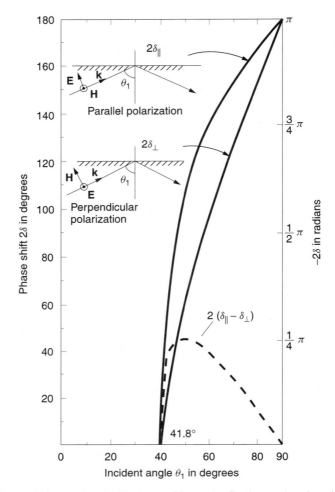

Figure 2.15 Phase shift associated with the total internal reflection at the glass ($n = 1.5$) and air interface.

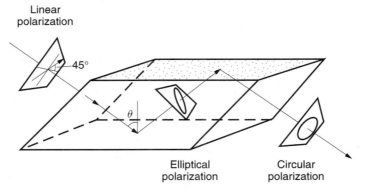

Figure 2.16 Fresnel rhomb.

Solving for $\sin^2 \theta_1$ gives

$$(1 - x)(x - n^2) = A^2 x^2$$

where

$$A = \tan \pi/8$$

$$x = \sin^2 \theta_1$$

This is a quadratic equation,

$$(A^2 + 1)x^2 - (n^2 + 1)x + n^2 = 0$$

with solution

$$x = \frac{(n^2 + 1) \pm \sqrt{(n^2 + 1)^2 - 4(A^2 + 1)n^2}}{2(A^2 + 1)}$$

$$\theta_1 = \sin^{-1}(\sqrt{x})$$

$$\theta_1 = 50.23° \text{ and } 53.26° \qquad \qquad \Box$$

Here, Eqs. (2.88) and (2.89) will be rewritten using the k-diagram mentioned in Section 2.8.3 so that they will be in a form suitable for describing the phase delay at the boundaries of the optical guide in the chapter to come. Let both the numerators and the denominators of Eqs. (2.88) and (2.89) be multiplied by $n_1 k$. Comparing with Eqs. (2.80), the numerators of both equations become γ. From Fig. 2.13, the denominators of Eqs. (2.88) and (2.89) become $n^2 k_{1x}$ and k_{1x}, respectively.
Thus,

$$\tan \delta_{\parallel} = \frac{\gamma}{n^2 k_{1x}} \qquad (2.91)$$

$$\tan \delta_{\perp} = \frac{\gamma}{k_{1x}} \qquad (2.92)$$

In the particle theory of light, a propagation constant represents something proportional to the momentum of a photon, so that $\tan \delta_{\parallel}$ and $\tan \delta_{\perp}$ can then be interpreted as the ratio of momenta in medium 1 and medium 2.

Next, the transmission coefficient of total internal reflection will be considered. Keep in mind that the meaning of the transmission coefficient in the case of the total internal reflection is slightly different from the usual refraction. A simple model is that the evanescent wave is a result of light energy that goes into the less dense medium only for a short distance and then comes back again into the dense medium. The transmission coefficient describes the magnitude on the boundary but just a short distance inside the less dense medium.

An attempt will be made to find out how much light goes into the less dense medium by calculating the transmission coefficient of the total internal reflection at the boundary. Note from Eqs. (2.35) and (2.36) that

$$t_{\parallel} = \frac{1}{n}(1 + r_{\parallel}) \qquad (2.93)$$

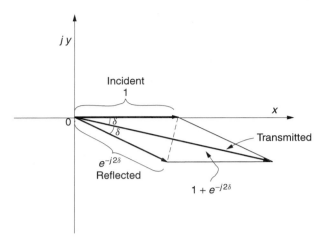

Figure 2.17 Polar coordinates of a complex-valued transmission coefficient.

and from Eqs. (2.40) and (2.41)

$$t_\perp = 1 + r_\perp \tag{2.94}$$

Use of Eqs. (2.86) and (2.87) in Eqs. (2.93) and (2.94) and with the aid of Fig. 2.17 gives

$$t_\parallel = \frac{2}{n} \cos \delta_\parallel e^{-j\delta_\parallel} \tag{2.95}$$

$$t_\perp = 2 \cos \delta_\perp e^{-j\delta_\perp} \tag{2.96}$$

Using Eqs. (2.88) and (2.89), $\cos \delta_\parallel$ and $\cos \delta_\perp$ in Eqs. (2.95) and (2.96) can be further rewritten as

$$t_\parallel = \frac{2n \cos \theta_1}{\sqrt{1 - n^2}\sqrt{1 - (n^2 + 1)\cos^2 \theta_1}} e^{-j\delta_\parallel} \tag{2.97}$$

$$t_\perp = \frac{2 \cos \theta_1}{\sqrt{1 - n^2}} e^{-j\delta_\perp} \tag{2.98}$$

Recall that the momentum perpendicular to the boundary is $kn_1 \cos \theta_1$. With an increase in $\cos \theta_1$, this momentum increases, and also the magnitude of the transmission coefficient into the second medium increases. This is as if the photon in the optically dense medium is being pushed out into the less dense medium by this momentum. The extent to which the evanescent wave protrudes out into the less dense material reaches its maximum at the critical angle. The amount of phase shift, however, reaches zero at the critical angle from Eqs. (2.88) and (2.89). This fact rejects the simple explanation that the phase delay is the time needed for the evanescent wave to go out into the less dense medium and come back into the dense medium. The evanescent wave and phase delay indeed exist, but it cannot safely be said that the phase delay is due to the round-trip time. Another example where the phase delay at the boundary has nothing to do with penetration into the second medium is reflection from a perfect mirror. The reflection from the perfect mirror creates π radians of phase shift but has nothing to do with the round-trip time in the mirror, since light does not go into the mirror at all. The phase shift is needed to match the boundary condition. Goos and Hänchen conducted interesting experiments to clarify these matters as described in the next section.

2.8.2 Goos-Hänchen Shift

Goos and Hänchen investigated the ray path associated with total internal reflection. They wanted to determine whether or not a spatial lateral shift accompanies the phase shift when light undergoes total internal reflection. One possibility is that the light ray is reflected back immediately into the dense medium without a spatial lateral shift but with $\phi_{\parallel,\perp}$-radian phase delay as in the case of reflection from a surface mirror. The other possibility is that the reflected light ray first penetrates into the less dense medium and is laterally shifted along the boundary before going back into the dense medium as well as undergoing the phase delay.

In 1947, Goos and Hänchen [6] looked at the difference between reflection from a silver surface and total internal reflection at a glass–air interface. A clever experiment was performed using a glass prism onto which a narrow strip of silver was deposited along the center. As shown in Fig. 2.18, a wide light beam was incident on one end of the prism at such an incident angle that total internal reflections took place repeatedly inside the prism before the beam exited the prism and was projected onto the screen. These total internal reflections took place only with the portion excluding the silver deposit in the center. In the center, the light is reflected repeatedly by the silver strip as

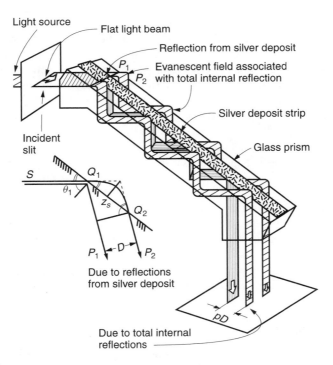

Figure 2.18 Goos and Hänchen experiment.

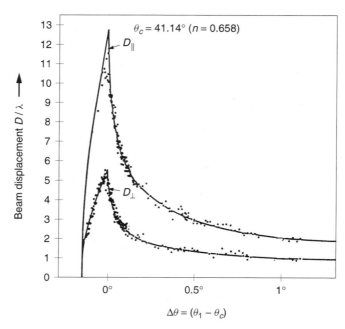

Figure 2.19   The beam displacement of the Goos-Hänchen shift versus the angle difference $(\theta_1 - \theta_c)$ in degrees. (After H. K. V. Lotsch [7] and H. Wolter [8].)

it travels through the prism. On the screen, the translation between the two light beams with different mechanisms of reflection can be compared, as shown in Fig. 2.18.

The explanation of the observed results is indicated by the diagram on the lower left-hand side in Fig. 2.18. While the reflected light from the silver surface takes the direct path SQ_1P_1, the light that has undergone total internal reflection penetrates the optically less dense medium and then comes back into the optically dense medium taking the route $SQ_1Q_2P_2$, thus causing a beam displacement D between the two beams after reflection. If the number of reflections is p, the lateral shift of the output beams on the screen is pD. Goos and Hänchen experimentally verified the shift. Such a translation is known as the Goos–Hänchen shift. The Goos–Hänchen shift is an important quantity to consider when attempting to accurately estimate the penetration depth of the light field outside an optical guide or to fully account for the phase shift associated with an optical guide.

The original Goos–Hänchen experiments were refined to such an extent that the modified ray model agrees quite well with the experimental values, as shown in Fig. 2.19 [7,8]. For these experiments, glass with $n = 1.520$ was used as the optically dense medium, and air as the optically less dense medium. Near the critical angle, where the δ's are at their minimum, the D's are at their maximum. The value of the beam displacement $D_\parallel$ with light polarized parallel to the plane of incidence is over 10 times the wavelength.

2.8.3   Evanescent Field and Its Adjacent Fields

Now we are ready to calculate the fields around the interface. In the dense medium, a standing wave E_{s1} is formed by the incident and reflected waves. Using Eq. (2.86), the expression for the standing wave is given by

$$E_{s1} = E_1[e^{j(k_{1x}x+\beta z)} + e^{j(-2\delta-k_{1x}x+\beta z)}]e^{-j\omega t} \qquad (2.99)$$

and finally

$$E_{s1} = 2E_1\cos(k_{1x}x + \delta)e^{j(\beta z-\omega t-\delta)}, \quad x < 0 \qquad (2.100)$$

where the subscripts $\parallel$ and $\perp$ have been suppressed for convenience. Equation (2.100) is the expression for a standing wave in the x direction and a propagating wave in the z direction. The amplitude of Eq. (2.100) is plotted on the right in Fig. 2.14. The position of maximum intensity is shifted downward by δ radians.

Because of the continuity condition, the amplitude $|E_2|$ of the evanescent wave has to be identical to that of the standing wave E_{s1} evaluated on the boundary; hence, from Eq. (2.79)

$$E_2 = 2E_1e^{-j\delta}(\cos\delta)e^{-\gamma x+j(\beta z-\omega t)}, \quad x > 0 \qquad (2.101)$$

Equations (2.100) and (2.101) together complete the expressions near the boundary.

With a decrease in the incident angle θ_1 from 90° approaching the critical angle θ_c, both δ and γ decrease and $h(=\gamma^{-1})$ increases according to Eqs. (2.80), (2.88), and (2.89). This means that with a decrease in θ_1, the maxima shift toward the interface and the evanescent wave shifts into the optically less dense medium, as illustrated in Fig. 2.20. An analogy can be made with breaking a piece of wood. The sharper the bend, the more the broken ends protrude.

In practice, with the wavelengths 0.85–1.55 μm and $n_1 \simeq 1.55$, $n_2 \simeq 1.54$, which are normally used for the components of fiber optical communication systems, the extent of the evanescent wave is of the order of 1–10 μm. The minimum required diameter of the cladding layer can be estimated (see Chapter 11).

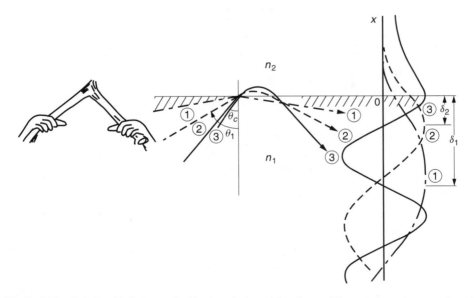

Figure 2.20   Relationship between incident angle θ_1 and the shape of the evanescent waves, $\theta_1 > \theta_c$.

In integrated optics, where the dimensions of the elements are quite small, a thorough understanding of the evanescent wave is very important. A typical example of the use of the evanescent wave is an optical guide coupler, such as shown in Fig. 2.21. An optical coupler is used either for monitoring the signal of the main guide or dividing the power into desired proportions.

When an auxiliary optical guide is put in the evanescent field of the main optical guide, the auxiliary guide is excited by the evanescent field. The amount of excitation can be controlled by either adjusting the spacing between the guides, the indices of refraction of the media surrounding the guides, or the length of the region of interaction. In fact, by adjusting these parameters properly, almost the entire energy of guide 1 can be transferred into guide 2.

Another interesting phenomenon is that if the interaction region is made long enough, the energy that was transferred to the auxiliary guide will itself set up an evanescent field, which in turn excites the main guide and transfers energy back into the main guide. If the guide is made out of an electrooptic material like lithium niobate, $LiNbO_3$, the index of refraction of the guide can be changed by an external electric field (see Chapter 5). By careful design of the parameters of the coupler, an electronically controlled optical switch can be fabricated. The optical switch based on coupling of the evanescent field can achieve nearly 100% switching with a low control voltage, typically less than 10 volts.

The phasefront of the evanescent wave moves at the same speed as that of the bulk wave (wave inside the optically dense medium). It does not leave the surface, and no energy flows out of the interface, and so no loss of energy takes place. However, as

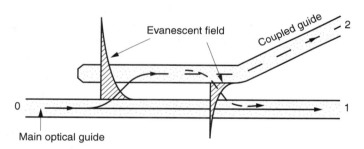

Figure 2.21   Optical guide coupling by the evanescent wave.

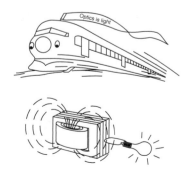

in the example of the optical guide coupler, if there is another guide that can carry the energy away, the evanescent wave provides the means of conveying energy from the main guide to the second guide.

The evanescent wave can be compared to a banner attached to a train. The banner moves with the train but has very little to do with the loss of energy of the train. Should the banner be caught by an external obstacle, it will convey the train's energy into the external obstacle.

Another analogy might be the magnetic flux leaking out of a transformer. The leaked magnetic flux is there all the time but contributes very little to the loss of energy of the main source. But once a loop circuit is placed in the leakage field, the energy of the main source can be transferred into the loop circuit.

2.8.4 *k* Diagrams for the Graphical Solution of the Evanescent Wave

Many of the expressions derived thus far have been expressed in terms of angles. Here the same quantities will be expressed in terms of the k_x vectors. From Fig. 2.13, the expressions for the k_x vectors are $k_{1x} = n_1 k \cos\theta_1$ and $k_{2x} = n_2 k \cos\theta_2$. There is a one-to-one correspondence between k_{1x} and θ_1, and likewise between k_{2x} and θ_2 inside a given medium. It is sometimes easier to use k_{1x} and k_{2x} rather than angles. For example, when analyzing the fields inside an optical guide having a rectangular shape, the rectangular quantities k_{1x} and k_{2x} are much more convenient variables to use than θ_1 or θ_2.

From Eq. (2.74), which was obtained by phase matching, β can be eliminated to give

$$k_{1x}^2 - k_{2x}^2 = (n_1^2 - n_2^2)k^2 \tag{2.102}$$

Equation (2.102) is Snell's law in k-coordinates (see Problem 2.8). Note that when $n_1 > n_2$, the right-hand side of Eq. (2.102) is positive and there is a region where k_{1x} becomes too small (i.e., θ_1 too large) to maintain the left-hand side of Eq. (2.102) positive and there is no real number k_{2x}. Consequently, there is no transmitted light. This is the region of total internal reflection discussed in Section 2.6. The only possible way to keep the left-hand side positive is to make k_{2x} a pure imaginary number $\pm j\gamma$; and in the case of the total internal reflection k_{2x} is imaginary.

Thus, when total internal reflection exists,

$$k_{2x} = \pm j\gamma \qquad k_{1x} = K \tag{2.103}$$

where k_{1x} is more conveniently written without the subscript. Inserting Eq. (2.103) into Eqs. (2.74) and (2.102) gives

$$K^2 + \beta^2 = (n_1 k)^2 \tag{2.104}$$

$$K^2 + \gamma^2 = (n_1^2 - n_2^2)k^2 \tag{2.105}$$

Both of these curves are circles. The graph of Eq. (2.104) is the $K-\beta$ diagram and the radius is $n_1 k$. The graph of Eq. (2.105) is the $K-\gamma$ diagram and the radius is $k\sqrt{n_1^2 - n_2^2}$, as shown in Fig. 2.22. Note that the radius of the $K-\gamma$ diagram is smaller than that of the $K-\beta$ diagram. For a given θ_1, draw in a straight line at θ_1 with respect

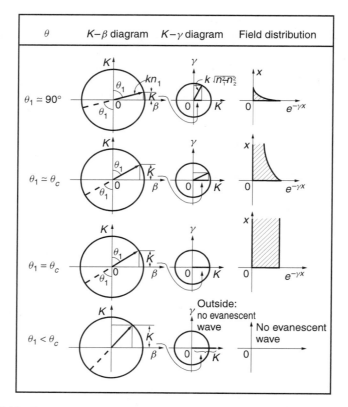

Figure 2.22 K–β and K–γ diagrams for the graphical solution of the evanescent wave.

to the K axis. The K–β diagram provides the values of K and β directly. From the value of K thus obtained, the K–γ diagram gives the value of γ and also determines whether or not total internal reflection takes place. Thus, the two circles combined provide γ for a given θ_1. Figure 2.22 demonstrates how to use the diagram based on what has already been calculated in Section 2.8.3. The diagrams are arranged in descending order of the incident angle θ_1 in the optically dense medium.

In the top row of the diagram, θ_1 is large and is close to $90°$. From the K–β diagram, K is small. For a small value of K, the K–γ diagram provides a large value of γ. A large value of γ means an evanescent wave, which decays very fast with distance away from the boundary. The corresponding situation is labeled ① in Fig. 2.20.

In the diagrams in the second row of Fig. 2.22, θ_1 is reduced to a value just slightly above the critical angle. Compared to the top row diagrams, K increases and hence γ decreases, and the evanescent wave extends further into the optically less dense medium. This corresponds to cases ② and ③ in Fig. 2.20.

The diagrams in the third row explain what happens when $\theta_1 = \theta_c$. The value of K becomes exactly the same as $k\sqrt{n_1^2 - n_2^2}$ and $\gamma = 0$. The field does not decay with x and the energy is lost into the less dense medium.

In the bottom row diagrams, θ_1 is reduced to a value below the critical angle, $\theta_1 < \theta_c$. The value of K now becomes large and falls outside the K–γ diagram. This means that no evanescent wave exists.

2.9 WHAT GENERATES THE EVANESCENT WAVES?

So far, only the evanescent wave that was generated by total internal reflection has been discussed. Is this the only way to generate an evanescent wave? In this section, the evanescent wave will be treated in a more general manner.

2.9.1 Structures for Generating Evanescent Waves

Let us consider a medium with index of refraction n_i. As with Eqs. (2.71) and (2.72), solving the wave equation in this medium requires that

$$k_x^2 + k_y^2 + k_z^2 = (n_i k)^2 \tag{2.106}$$

If $k_x^2 + k_y^2 + k_z^2$ were bigger than $(n_i k)^2$, in order to satisfy the boundary condition, at least one of k_x^2, k_y^2, or k_z^2 has to become negative, and the corresponding k_x, k_y, or k_z becomes an imaginary number. There is an evanescent wave in that direction. For the geometry so far discussed, k_{2z} was made so large (in order to meet the phase match boundary condition in medium 1) that k_{2x} had to be an imaginary number.

A corrugated metal surface is another example of a geometry that supports an evanescent wave [9]. The corrugation is oriented in the z direction, and the teeth in the x direction, as shown in Fig. 2.23a. Assume there is no variation in the y direction, meaning $\lambda_y = \infty$ and $k_y = 0$. A further assumption is made that the electric field is zero on the contour of the corrugation.

In order to match the boundary conditions, a number of component waves have to be summed. We realize immediately that the field distribution of the component

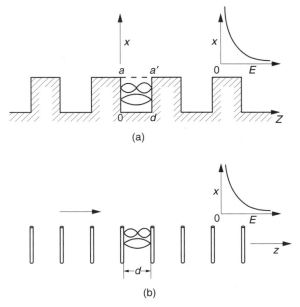

(a)

(b)

Figure 2.23 Structures that support evanescent waves. (a) Corrugated metal surface. (b) Array of metal pins.

waves inside the slot space is proportional to a trigonometric function so as to satisfy the condition of vanishing fields on both sides of the wall. To satisfy the boundary condition, an integral multiple of a half-wavelength $\lambda_{zn}/2$ should fit the wall dimension d:

$$\frac{\lambda_{zn}}{2}n = d \quad \text{or} \quad k_{zn} = \frac{\pi}{d}n \tag{2.107}$$

where n is an integer and the order number of the component waves. Note that k_{zn}, which describes the propagation constant of the nth order component wave in the vicinity of a–a', can be increased by reducing d. From Eq. (2.106), we have

$$k_{xn}^2 = \left(\frac{2\pi}{\lambda}\right)^2 - \left(\frac{\pi}{d}n\right)^2 \tag{2.108}$$

where k_{xn} is the propagation constant of the nth order component wave in the x direction.

The lower order component waves can propagate in both the x and z directions because all values in Eq. (2.106) are real. For the higher order components whose n is larger than a certain value, k_x becomes an imaginary number. In this case, the wave is an evanescent wave in the x direction and a propagating wave in the z direction. As a matter of fact, when d is smaller than one-half of the free-space wavelength, all component waves become evanescent waves.

Another structure that supports evanescent waves is shown in Fig. 2.23b. It is an array of conductors. Again, on the surface of the conductors the field has to be zero. By the superposition of many component waves with various propagation constants, this boundary condition is met. Some of the component waves have to satisfy the condition of Eq. (2.107) and become evanescent waves. These structures are used as evanescent waveguides for both microwaves and light waves.

Example 2.5   A lossy glass is bordered by air in the $x = 0$ plane. The propagation constant in the z direction on the border is $\beta + j\alpha$. Find the expression for the evanescent wave in the air near the boundary. Assume $k_y = 0$.

Solution   Let the lossy glass be medium 1, and the air be medium 2. The propagation constants in air have to satisfy

$$k_{2x}^2 + k_{2z}^2 = k^2 \tag{2.109}$$

while the propagation constant along the boundary in the z direction in the glass is $\beta + j\alpha$. Phase matching along the z direction requires that

$$k_{2z} = \beta + j\alpha \tag{2.110}$$

Inserting Eq. (2.110) into (2.109) gives

$$k_{2x}^2 = k^2 - \beta^2 + \alpha^2 - j2\alpha\beta \tag{2.111}$$

The real part K and imaginary part γ of k_{2x} will be found. Putting

$$A = k^2 - \beta^2 + \alpha^2$$
$$B = 2\alpha\beta \tag{2.112}$$

gives

$$k_{2x}^2 = \sqrt{A^2 + B^2}\, e^{j(-2\phi + 2n\pi)}$$

where $\tan 2\phi = (B/A)$ and n is an integer. Depending on n being even or odd, the value of k_{2x} is

$$k_{2x} = \left\{ \begin{array}{ll} K - j\gamma, & n \text{ is even} \\ -K + j\gamma, & n \text{ is odd} \end{array} \right\} \tag{2.113}$$

where

$$K = \sqrt[4]{A^2 + B^2} \cos\phi$$
$$\gamma = \sqrt[4]{A^2 + B^2} \sin\phi$$

Finally, the expressions for the evanescent field become

$$E_2 = E_{20} e^{\gamma x - \alpha z + j(Kx + \beta z - \omega t)} \tag{2.114}$$

or

$$E_2 = E_{20} e^{-\gamma x - \alpha z + j(-Kx + \beta z - \omega t)} \tag{2.115}$$

Equation (2.114) is not acceptable as a solution because E_2 increases indefinitely with an increase in x. Figure 2.24 illustrates the field distribution expressed by Eq. (2.115). Comparing Fig. 2.24 to the distribution for lossless glass shown in Fig. 2.14, the entire

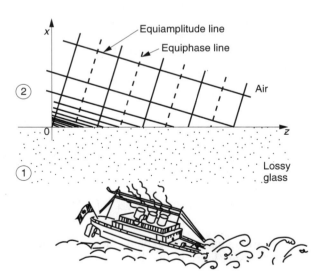

Figure 2.24 Field distribution of a leaky wave.

distribution is tilted and looks as if the wave is sinking into the glass to supply the energy lost in the glass. The angle of the tilt depends on $-K$, and K vanishes when $\alpha = 0$. This kind of wave associated with the surface is called a leaky wave, meaning that it loses energy as it propagates. $\qquad\qquad\qquad\qquad\qquad\qquad\qquad\qquad\qquad\qquad$ □

2.10   DIFFRACTION-UNLIMITED IMAGES OUT OF THE EVANESCENT WAVE

In Chapter 1, the observation regions of the diffraction pattern were classified into two areas: far field and near field. The near field in this classification, however, means that the distance between the aperture and the observation region is still large compared to the light wavelength. In this section, the region of concern is within a few wavelengths of the aperture. It would be more appropriate to call this region the very-near-field region. To avoid confusion, we will use the term "very near field" whenever necessary, but commonly used proper names like SNOM (scanning near-field optical microscope) will be left unchanged.

A lens-type microscope forms a magnified image out of the radiating light wave by means of lenses. The resolution of the microscope, however, is limited to the order of one wavelength of the light.

On the other hand, the evanescent-field-type microscope forms the image out of the evanescent wave generated by the object. An evanescent field to radiating field converter is used to form the image. The resolution of the microscope is 10–20 times better than lens-type microscopes. The resolution of the lens-type microscope is first reviewed in the next section.

2.10.1   Resolution of a Lens-Type Microscope

The minimum detectable variation of a microscope objective lens will be calculated. For the geometry shown in Fig. 2.25, a grating is used as the object. Assume that the grating is illuminated by a parallel laser beam and the light distribution on the surface of the grating is expressed by

$$E(x) = E_0(1 + \cos 2\pi f_g x) \qquad (2.116)$$

where f_g is the spatial frequency of the grating. The spatial frequency is the inverse of the period t; that is,

$$f_g = \frac{1}{t} \qquad (2.117)$$

For the simplicity at the expense of accuracy, the far field approximation will be used. Inserting Eq. (2.116) into (1.36) gives

$$E(\theta) = K \left[\delta \left(\frac{\sin\theta}{\lambda} \right) + \delta \left(\frac{\sin\theta}{\lambda} - f_g \right) + \delta \left(\frac{\sin\theta}{\lambda} + f_g \right) \right] \qquad (2.118)$$

where K absorbs the necessary amplitude and phase factors. Besides the center radiation lobe, Eq. (2.118) gives two side lobes in the directions of

$$\theta = \pm \sin^{-1} \lambda f_g \qquad (2.119)$$

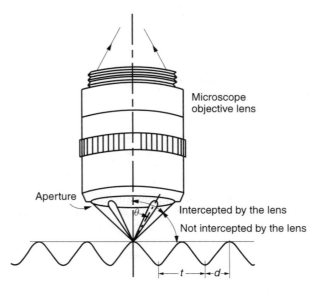

Figure 2.25 Radiating field from the sinusoidal grating into the microscope objective lens.

Note that as the spatial frequency f_g of the grating increases from zero, the direction of radiation of the two side lobes moves from normal to parallel to the grating.

As shown in Fig. 2.25, only the portion of the scattered light that is intercepted by the objective lens contributes to the formation of the image. Then, the maximum spatial frequency f_{gM} of the grating whose radiation lobe can still be intercepted by the aperture of the lens is, from Eq. (2.119),

$$\sin \theta_M = \lambda f_{gM} \tag{2.120}$$

where θ_M is the angle that the aperture subtends with respect to the normal of the grating. The quantity $\sin \theta_M$ is called the numerical aperture (NA) of the objective lens and Eq. (2.120) is rewritten as

$$f_{gM} = \frac{\text{NA}}{\lambda} \tag{2.121}$$

The surface variation d is the distance between the peak and valley of the sinusoidal variation of the grating. The surface variation is expressed as

$$d = \frac{t}{2} \tag{2.122}$$

Then, the minimum detectable variation d_m is, from Eq. (2.117), (2.121), and (2.122),

$$d_m = \frac{\lambda}{2\text{NA}} \tag{2.123}$$

Equation (2.123) is known as Rayleigh's resolution criteria of a diffraction-limited lens (a lens with a perfect shape whose resolution is limited only by the finite size of the lens). The conclusion is that $\sin \theta_M$ cannot be larger than unity and the best lateral (in the x direction) resolution d_m that can be achieved is one half-wavelength. This is the case when the radiating field alone is used for forming the image. Is there any other means of forming a light image and obtaining a resolution better than a half-wavelength of light? The answer is yes, as will be shown next.

Let us turn our attention from the microscope objective to the field scattered out of the grating. Inserting Eq. (2.117) into (2.119) gives the direction of the radiation lobe in terms of the period of the grating:

$$\theta = \sin^{-1}\left(\frac{\lambda}{t}\right) \tag{2.124}$$

What happens if the period of the grating becomes shorter than one wavelength? There is no θ that satisfies Eq. (2.124), meaning there is no radiating field. Still, the boundary conditions have to be satisfied along the contour of teeth of the grating. The evanescent wave coasts the contour of the teeth and appears only in the region within a few wavelengths of light from the surface and decays away exponentially with the distance from the surface. Roughly speaking, information about variations longer than one half-wavelength of light is carried by the radiating portion of the field, while that of variations shorter than a half-wavelength, by the evanescent portion of the field.

The mathematical formulation of the evanescent wave has already been given by Eq. (1.197) in connection with Eq. (1.201) or by Eq. (2.101). Optical microscopes have been devised to collect the evanescent field and convert it into an image. These microscopes have achieved resolutions much shorter than a half-wavelength of light.

2.10.2 Near-Field Optical Microscopes

The microscopes whose operation is based on the evanescent wave may broadly be divided into two categories: the photon tunneling microscope (PTM) and the scanning near-field optical microscope (SNOM) [10,11].

2.10.2.1 *Photon Tunneling Microscope*

Figure 2.26 shows the geometry of the photon tunneling microscope [12]. Input light is incident from the convex side of the plano-convex objective lens, so that the condition of total internal reflection is satisfied on the flat bottom surface of the flexible transducer placed beneath the flat side of the plano-convex objective lens. The evanescent wave is excited in the region of the tunneling gap. The reflected light is fed to the imaging system of a video camera through the other side of the plano-convex objective lens. The photons lost into the sample due to the short tunneling gap are responsible for the reduction in the detected signal, which leads to information about the depth of the tunneling gap.

The depth pattern of the sample is displayed as the brightness distribution in the reconstructed image. The lateral resolution of the photon tunneling microscope is diffraction limited in the same manner as a normal imaging system, but as far as the depth information is concerned, the exponential decay of the evanescent wave is used and the resolution is limited only by the gray scale of the video camera.

Figure 2.27 shows the depth image (gray-scale image) of a single polyethylene crystal and the processed 3D image of the same crystal when measured by a photon tunneling microscope [13].

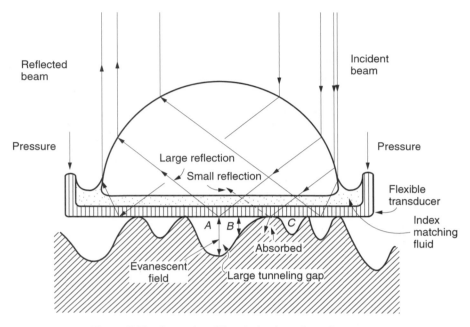

Figure 2.26 Geometry of the photon tunneling microscope.

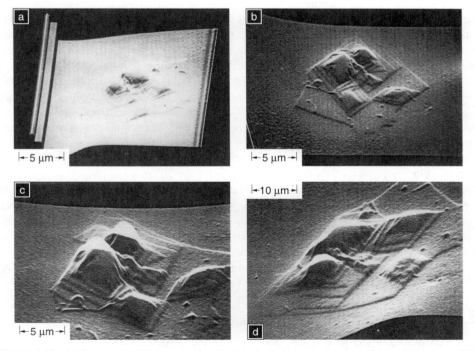

Figure 2.27 (a) Gray-scale PTM image of a single polyethylene crystal. (b,c) Three-dimensional (3D) images showing the topography from different viewpoints. (d) A single large crystal can be imaged with the PTM in contact interference mode. (Courtesy of J. M. Guerra, M. Srinivasarao, and R. S. Stein [13].)

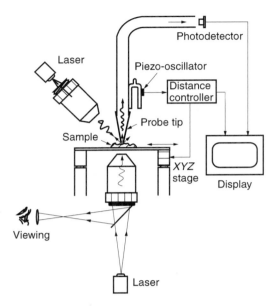

Figure 2.28   Schematics of the scanning near-field optical microscope (SNOM).

2.10.2.2   Scanning Near-Field Optical Microscope (SNOM)

Figure 2.28 shows the geometry of the SNOM [14–16]. Images of subwavelength resolution are obtainable. The surface of the sample can be illuminated from above, underneath, or both by ordinary radiating laser light. The amplitude of the resultant radiating and evanescent fields is collected with a 50–500-nm diameter probe at a distance of 10–60 nm from the sample surface.

The evanescent field decays exponentially with distance away from the surface, and information about the surface variation is lost as soon as the probe is outside the very near field. Placing the tip in the very near field without hitting and possibly damaging the sample is a challenge. There are more than a few techniques for preventing the probe from crashing into the sample. When a voltage is applied between the metalized tip and the sample, an electron tunnel current starts to flow as soon as the two are brought close together. From the tunneling current the clearance is monitored. Another technique is to monitor the change in the frequency of the mechanical vibration that takes place when the probe is placed in proximity to the sample surface. Another method is by the change of the capacitance between the tip and the sample, or by the change of the shear force between the tip and the sample. As a matter of fact, the shear force itself can be used to image the surface [17].

Figure 2.29 is the image of deoxyribonucleic acid, commonly known as DNA, obtained by shear force imaging [18].

2.10.3   Probes to Detect the Evanescent Field

We will now explain how the probe converts the evanescent field into a radiating field [19,20]. The evanescent field that is established on the boundary between the optically dense and less dense media is used as an example. The direction of the polarization of the wave is assumed to be perpendicular to the page or an *s* wave.

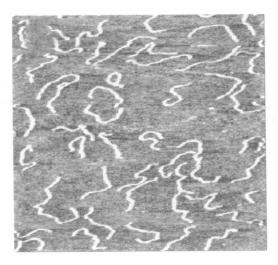

Figure 2.29 Shear-force image of DNA. (Courtesy of M. F. Garcia-Parajo et al. [18].)

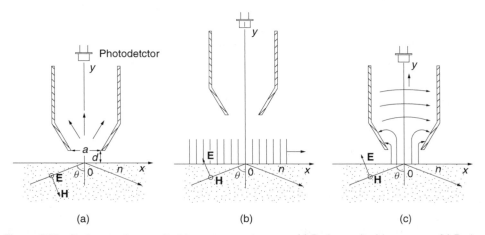

Figure 2.30 Probe aperture excited by evanescent waves. (a) Probe excited by *s* wave. (b) Probe above the *p* wave. (c) Probe excited by the *p* wave.

As shown in Fig. 2.30a, the two-dimensional model is considered. The x axis is taken in the direction of the boundary and the y axis, vertically perpendicular to it. The probe is placed with its aperture parallel to the boundary at a distance d away from the boundary. The evanescent field in which the aperture of the probe is immersed is, from Eq. (1.197),

$$E(d, z) = E_0 \exp[-2\pi f_s \sqrt{n^2 \sin \theta - 1}\, d + j2\pi n f_s (\sin \theta) x] \qquad (2.125)$$

where n is the refractive index of the more dense medium, and the less dense medium is air. The angle θ is the angle of incidence of the light inside the more dense medium to the boundary. The field across the aperture is

$$E_a = \Pi \left(\frac{x}{a}\right) E(x, d) \tag{2.126}$$

where a one-dimensional aperture with width a is assumed. The field reradiated toward the photodetector due to E_a will be found using Eq. (1.177). For this purpose, the Fourier transform of Eq. (2.126) is calculated:

$$\epsilon(f_x, 0) = aA \, \text{sinc} \, [a(f_x - f_0)] \tag{2.127}$$

where

$$A = E_0 \exp(-2\pi f_s \sqrt{n^2 \sin^2 \theta - 1} d)$$

and

$$f_0 = n f_s \sin \theta > f_s$$

where the inequality sign is due to the evanescent wave.

The radiating field component of Eq. (2.126) is distinguished from that of the evanescent field component before the equation is put into Eq. (1.177). The spectrum of Eq. (2.127) is plotted in Fig. 2.31. As explained earlier in Section 2.10.1, whether the component is radiative or evanescent is determined by whether the spatial frequency is smaller or larger, respectively, than $f_s(= 1/\lambda)$. The hatched region in Fig. 2.31 is radiative, and the unhatched region is evanescent. When a photodetector is placed on the y axis at (x, y), the radiating field received by the detector is obtained by inserting Eq. (2.127) into (1.177):

$$E(x, y) = aA \int_{-f_s}^{f_s} \text{sinc}[a(f_x - f_0)] e^{j2\pi\sqrt{f_s^2 - f_x^2}(y-d)} e^{j2\pi f_x x} \, df_x \tag{2.128}$$

The propagation medium between the aperture and the detector is assumed to be free space rather than an optical fiber.

The proper choice of the aperture size of the probe is crucial. Let's first consider what happens if the aperture is too narrow, as shown in Fig. 2.31a. The factor a in front of the integral in Eq. (2.128) becomes small. In addition, the sinc function is spread so wide that the relative portion of the radiative component, which is inside the $-f_s$ to f_s region, is small compared to the entire region. Consequently, the radiating field reaching the detector is weak, and the signal-to-noise ratio of the image deteriorates even though a narrow aperture provides a higher resolution as a microscope. On the other hand, if the aperture is chosen too wide, such as shown in Fig. 2.31b, the sinc function shrinks around $f_x = f_0$. The radiative portion, which is inside $-f_s$ to f_s, is reduced, and the radiating field reaching the detector is again weak. In this case, the spatial resolution of the microscope deteriorates as well.

Next, let's consider the case when the direction of polarization is parallel to the page, or a p wave as shown in Figs. 2.30b and 2.30c. When the aperture stays high above the boundary as shown in Fig. 2.30b, the E field of the evanescent wave is undisturbed. The direction of the E field is parallel to the y axis, and certainly there is no propagation in the y direction because the electromagnetic wave does not propagate in the same

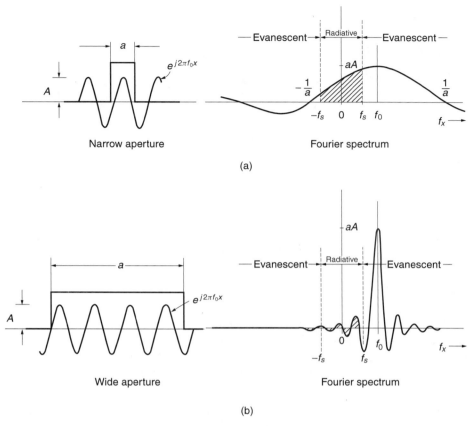

Figure 2.31   Aperture size of the probe and Fourier transform spectrum. (a) A case of narrow aperture. (b) A case of wide aperture.

direction as either the E or H field, but propagates in the direction perpendicular to either the E or H field.

When the aperture is lowered to the boundary, however, the electric lines of force start to curl so that they are terminated perpendicular to the metal surface of the probe in order to satisfy the boundary condition of the E field, as shown in Fig. 2.30c. Now, this curled E field has a component in the x direction and provides the possibility of creating a radiating wave toward the y direction. Note, however, that if the shape of the probe is symmetric, then half of the lines of force are bent toward the positive x direction while the other half are bent toward the negative x direction. One would think that the contributions of the lines of force curled in opposite directions would cancel each other, thereby failing to excite a radiating wave in the y direction. This thinking is not quite correct. Remember that there is a phase lag between the left and right lines of force, and they do not quite cancel each other. The difference contributes to a weak radiating wave toward the y direction. It is safe to say that the probe in this configuration is more sensitive to the s wave than to the p wave.

In short, the probe aperture redistributes the evanescent wave energy into radiative and evanescent field components, and the energy that has been converted into the radiative component reaches the photodetector.

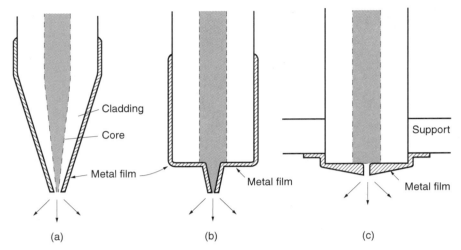

Figure 2.32   Various apertures of SNOM probes at the tip of the optical fiber. (a) Heating and pulling. (b) Selectively etched. (c) Flat tip.

2.10.4   Apertures of the SNOM Probes

The SNOM probes were first fabricated by applying a heating-and-pulling method to an optical fiber. The tip was either bare or coated with such metals as silver (Ag), aluminum (Al), or chromium (Cr). Probes of this type, however, have a long tapered section, such as shown in Fig. 2.32a. The transmission loss is usually high and is in the vicinity of 50 dB for an aperture diameter of 100 nm.

A shorter tip section, such as shown in Fig. 2.32b, is made by chemically etching the tip of the optical fiber using an aqueous mixture of hydrofluoric acid (HF) and ammonium fluoride (NH_4F). The strength of the etchant and the immersion time control the finish of the probe. A smoother etched surface is obtainable by dipping the fiber without removing the acrylate jacket [21]. As the final step, the etched surface is coated with a metal film.

The flat tip shown in Fig. 2.32c almost completely eliminates the tapered section. The metal film is coated directly onto the probe (Fig. 2.32c), the flat end of the optical fiber, and then the film is drilled by a focused ion beam (FIB) [22].

The diameter of the drilled aperture is approximately 50–100 nm. The transmission loss is around 30 dB for an aperture diameter of 100 nm. This is an improvement of 20 dB over the tapered probe.

2.10.5   Modes of Excitation of the SNOM Probes

Five different modes of operation of the SNOM probe are listed in Fig. 2.33. [23]. The collection mode shown in Fig.2.33(a) is the mode that has been dealt with so far. The probe is used to pick up the field scattered from the target.

If the light path of the collection mode is reversed, the result is the illumination mode shown in Fig. 2.33b. In the illumination mode, the optical fiber is driven by a light source and the target is illuminated locally as the probe is scanned, analogous to using a flashlight to illuminate a large object. The advantage of this mode is the minimization of the background light.

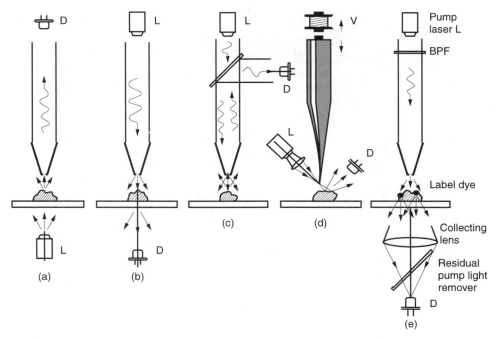

Figure 2.33 Various modes of SNOM operation. (a) Collection. (b) Illumination. (c) Reflection. (d) Apertureless. (e) Fluorescence. L, laser; D, photodetector; V, vibrator; BPF, band-pass filter.

The reflective mode shown in Fig. 2.33c is the combination of the above two modes using just one probe. The probe driven by a laser illuminates the object. The field scattered from the object is collected by the same probe. A half-mirror separates the transmitted and received light. The resultant directivity is the product of the directivities of the collection and illumination modes, and the resultant directivity is narrower than the individual directivity.

The apertureless mode is shown in Fig. 2.33d. The probe is made out of a solid metal wire with a sharp tip. The tip of the probe is mechanically dithered along its axis. The tip is illuminated with laser light, and because of the probe's dithering, the wave scattered from the probe is modulated. The photodetector exclusively detects the modulated signal. The unmodulated background light is eliminated. Because the probe does not contain an aperture, the tip can be made small to minimize the unwanted disturbance of the field due to the presence of the probe.

With all modes of operation, the illuminating light can be either an evanescent or a radiating wave, or even a combination of the two. The illuminating light does not have to be an evanescent wave. However, for the microscope to obtain a subwavelength resolution, the scattered field has to contain the evanescent field.

When a fluorescent substance is illuminated by light of a specific wavelength (pump light), light with wavelengths longer than the pump light is reemitted due to interactions occurring between light and matter. Figure 2.33e is an example of a SNOM probe in the illumination mode that utilizes this fluorescent spectroscopy.

By chemically treating the sample, it is possible to label only one fluorophore per molecule [18]. The area of illumination of the pump light through the SNOM probe is

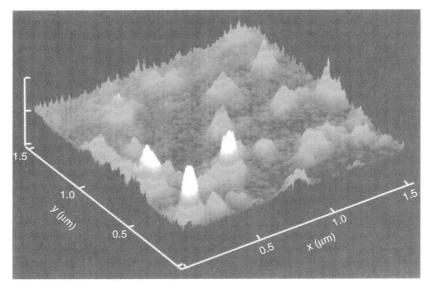

Figure 2.34   Near-field fluorscence image of DNA fragments labelled with rhodamine dye. (Courtesy of M. F. Garcia-Parajo et al. [18].)

restricted to a diameter of approximately 100 nm. Figure 2.34 shows the fluorescent image of a double-stranded DNA fragment labeled at one end with rhodamine dye. Thus, the spectroscopic use of the SNOM makes it possible to monitor the structural changes of a single biomacromolecule.

2.10.6   SNOM Combined with AFM

Figure 2.35 shows a SNOM combined with an atomic force microscope (AFM). In the AFM, a sharp probe tip is attached to the bottom side of the cantilever, and a small mirror is attached on the top side. The orientation of the cantilever is monitored by the change in the direction of the laser beam reflected from the small mirror.

When the tip of the probe is brought closer than several nanometers to the sample surface, an atomic force is experienced between the probe tip and the sample surface in accordance with the distance between them. The cantilever is excited to vibrate by a piezoelectric crystal. The phase and amplitude of the vibration are monitored by the laser beam reflected from the small mirror. The phase and amplitude of vibration vary with the distance of the tip to the sample surface because the atomic force varies. The phase and amplitude are maintained constant as the cantilever is scanned laterally over the sample surface by means of a feedback loop current to another piezoelectric crystal that raises and lowers the sample stage. At the same time, this feedback loop current provides information about the movement of the tip as it coasts the sample surface, and this information is used to form the AFM image.

In Fig. 2.35a, a collection mode SNOM is combined with an AFM. The photodiode is incorporated into the tip of the probe, and the SNOM signal is directly detected by the photodiode. Images obtained by mechanical means, such as by the shear force or atomic force microscopes, are often called topographic images as compared to the images obtained by optical means.

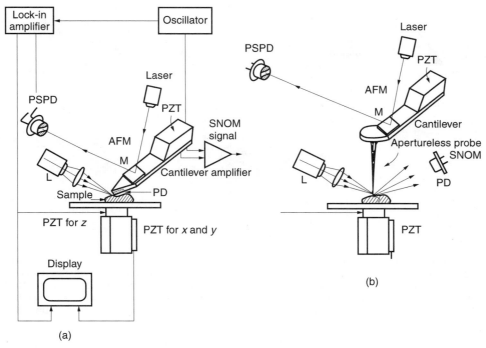

Figure 2.35 Combination of SNOM with the AFM. (a) Collection mode SNOM with AFM. (b) Apertureless SNOM with AFM. PSPD, position-sensitive photodetector; M, mirror; PZT, piezo-electric transducer; PD, photodetector.

In Fig. 2.35b, an apertureless mode SNOM is combined with the AFM. The AFM tip can be used as the apertureless optical probe with minimum modification [24].

Figure 2.36 compares images of the same test sample taken by the AFM and the SNOM in the apertureless mode.

2.10.7 Concluding Remarks

The resolving power of the SNOM is of the order of tens of nanometers, while that of an electron microscope is of the order of a nanometer. Even though the ultimate resolution of the near-field optical microscopes is poorer than that of the scanning electron microscope (SEM), the advantages of the photon tunneling microscope are that it neither requires metallizing the sample nor causes the intrusive effects of electrons to the sample. The very-near-field optical microscope is a valuable tool for revealing faults, especially during the fabrication of optical devices, because the detected signal is directly related to the severity of the faulty function.

The usefulness of the SNOM goes beyond that of being a mere microscope. When combined with optical spectroscopy, the SNOM becomes a unique tool for studying the local interaction of light with matter, even to the extent of allowing the in vivo observation of the movement of a single biomacromolecule.

Another area of importance is the application of the SNOM to the development of a high-density data storage. In the digital video disk (DVD) recorder, a focused beam of laser light is used to read or write the marks and spaces (or "1" and "0" bits) in

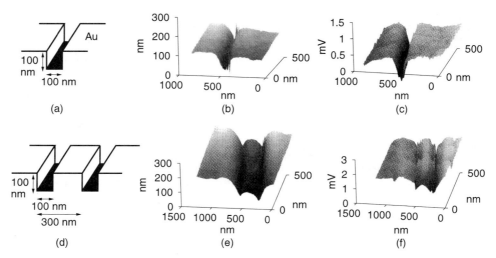

Figure 2.36 Comparison of images taken by AFM and SNOM. (a) Single-groove test sample engraved by X-ray lithography. (b) Topographic image of the single groove determined by AFM. (c) Optical image of the single groove determined by SNOM (amplitude). (d) Double-groove test sample engraved by X-ray lithography. (e) Topographic image of the double groove determined by AFM. (f) Optical image of the double groove determined by SNOM (amplitude). (After R. Bachelot, P. Gleyzes, and A. C. Boccara [24].)

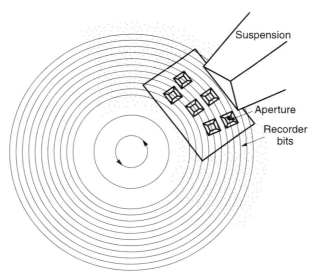

Figure 2.37 Multichannel flat chip used as a scanner over a high-density memory disk. (After M. Kourogi et al. [25].)

memory. Being a focused light beam, the narrowest spacing between the marks and spaces cannot be less than a wavelength of the light. If, however, the light from the SNOM probe is used for writing or reading the memory, the required spacings between codes can be reduced by a factor of 10. The memory area density can be increased by 100 times. The difficulty, however, is how to maintain a 10-nm separation between the probe and the memory disk, which is spinning at a rate equivalent to 1 m/s in linear distance speed. This problem can be alleviated by the multichannel flat chip approach, such as shown in Fig. 2.37. On a flat silicon chip (30 μm $\times$ 150 μm) an array of 80-nm apertures are micromachined, and this chip is slid over the oiled memory disk surface. The flat bottom surface of the chip prevents the crash as well as vibration. The multichannel approach reduces the required speed of spinning the memory disk [25].

PROBLEMS

2.1 Derive the reflection coefficient $r_\parallel$ and transmission coefficient $t_\parallel$ for the **H** field.

2.2 Derive the following equation:

$$r_\parallel = \frac{\tan(\theta_1 - \theta_2)}{\tan(\theta_1 + \theta_2)}$$

2.3 Explain why $1 + r_\parallel \neq t_\parallel$ even though $1 + r_\perp = t_\perp$ is true.

2.4 Prove that $R + T = 1$ for both directions of polarization.

2.5 From Eq. (2.42),

$$r_\perp = -\frac{\sin(\theta_1 - \theta_2)}{\sin(\theta_1 + \theta_2)}$$

Prove that a wave with perpendicular polarization cannot have Brewster's angle.

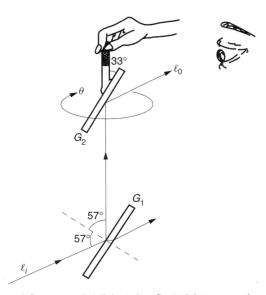

Figure P2.6 When G_1 and G_2 are parallel, light ℓ_0 is reflected; however, when G_2 is rotated 90° from this position, the light ℓ_0 disappears.

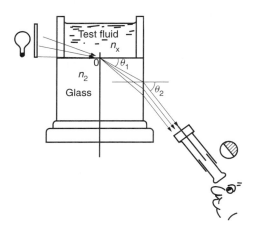

Figure P2.7   Pulfrich's refractometer.

2.6   As shown in the Fig. P2.6, glass plates G_1 and G_2 are initially set parallel to each other. While G_1 remains fixed, G_2 is able to rotate fully. The normals of both glass plates are tilted by 57° from the vertical direction. The light beam l_i is incident at 57° to the normal of the lower glass plate. The reflected light from G_1 is directed vertically upward, reflecting the light l_0. When G_2 is rotated by 90°, however, the exit light disappears. Explain why this exit light disappears.

2.7   Pulfich's refractometer, which measures the index of refraction of fluid, is constructed as shown in Fig. P2.7. As the viewing angle θ_2 of the telescope changes, an angle of θ_2 is created such that half of the field is bright and the other half dark. The surface of the interface between the liquid under test and the base glass is illuminated by diffuse light (light with all angles of incidence). The index of refraction (n_2) of the glass is larger than n_x of the fluid. Find the index of refraction n_x of the fluid when the emergent angle of this half-bright and half-dark condition is θ_2.

2.8   The expression for Snell's law in k coordinates is

$$k_{1x}^2 - k_{2x}^2 = (n_1^2 - n_2^2)k^2$$

By referring to Fig. 2.13, rewrite the k-coordinate expression of Snell's law into the angular expression of Snell's law.

REFERENCES

1. E. Hecht, *Schaum's Outline of Theory and Problems of Optics*, Schaum's Outline Series. McGraw-Hill, New York, 1975.
2. H. H. Skilling, *Fundamentals of Electric Waves*, Wiley, New York, 1948.
3. N. Mizutani and T. Numai, "Analysis of reflectivity for a beveled corner mirror in semiconductor ring lasers," *J. Lightwave Technol.* **19**(2), 222–229 (2001).
4. E. Gini, G. Guekos, and H. Melchior, "Low loss corner mirrors with 45° deflection angle for integrated optics," *Electron. Lett.* **28**(5), 499–501 (1992).

5. G. R. Fowles, *Introduction to Modern Optics*, 2nd ed., Dover Publications, New York, 1989.

6. V. F. Goos and H. Hänchen, "Ein neuer and fundamentaler Versuch zur Totalreflexion," *Ann. Phys.* 6. Folge, Band 1, Heft 7/8, 333–346 (1947).

7. H. K. V. Lotsch, "Beam displacement at total reflection: The Goos–Hänchen effect, II, III, IV" *Optik* **32**, 189–204, 299–319, 553–569 (1971).

8. H. Wolter, "Untersuchungen zur Strahlversetzung bei Totalreflexion des Lichtes mit der Methode der Minimumstrahlkennenzeichnung," *Z. Naturforsch.* **5a**, 143–153 (1950).

9. M. Tsuji, S. Matsumoto, H. Shigesawa, and K. Takiyama, "Guided-wave experiments with dielectric waveguide having finite periodic corrugation," *IEEE Trans. Microwave Theory Tech.* **MTT-31**(4), 337–344 (1983).

10. A. Lewis, "Diffraction unlimited optics," in *Current Trends in Optics*, J. C. Dainty, Ed., Academic Press, London, 1994, Chap. 17.

11. V. P. Tychinsky and C. H. F. Velzel, "Super-resolution in microscopy," in *Current Trends in Optics*, J. C. Dainty, Ed., Academic Press, London, 1994, Chap. 18.

12. J. M. Guerra, "Photon tunneling microscopy applications," *Mat. Res. Soc. Symp. Proc.* **332**, 449–460 (1994).

13. J. M. Guerra, M. Srinivasarao, and R. S. Stein, "Photon tunneling microscopy of polymeric surfaces," *Science* **262**, 1395–1400 (26 Nov. 1993).

14. M. A. Paesler and P. J. Moyer, *Near-Field Optics, Theory, Instrumentation and Applications*, Wiley, New York, 1996.

15. D. W. Pohl, "Some thoughts about scanning probe microscopy, micromechanics and storage," *IBM J. Res. Dev. USA* **39**(6), 701–711 (Nov. 1995).

16. D. W. Pohl and D. Courjon (Eds.), *Near Field Optics*, Kluwer Academic Publishers, Boston, MA, 1993.

17. A. G. Ruiter, J. A. Veerman, K. O. van der Werf, and N. F. van Hulst, "Dynamic behavior of tuning fork shear-force feedback," *Appl. Phys. Lett.* **71**(1), 28–30 (1997).

18. M. F. Garcia-Parajo, J.-A. Veerman, S. J. T. van Noort, B. G. de Grooth, J. Greve, and N. F. van Hulst, "Near field optical microscopy for DNA studies at the single molecular level," *Bioimaging* **6**, 43–53 (1998).

19. J. M. Vigoureux, F. Depasse, and C. Girard, "Superresolution of near-field optical microscopy defined from properties of confined electromagnetic waves," *Appl. Opt.* **31**(16), 3036–3045 (1992).

20. J. M. Vigoureux and D. Courjon, "Detection of nonradiative fields in light of the Heisenberg uncertainty principle and the Rayleigh criterion," *Appl. Opt.* **31**(16), 3170–3177 (1992).

21. P. Lambelet, A. Sayah, M. Pfeffer, C. Philipona, and F. Marquis-Weible, "Chemical etching of fiber tips through the jacket; new process for smoother tips," *5th International Conference on Near Field Optics*, Shirahama, Japan, pp. 218–219, C3, Dec. 6–10, 1998.

22. K. Ito, A. Kikukawa, and S. Hosaka, "A flat probe for a near field optical head on a slider," *5th International Conference on Near Field Optics*, Shirahama, Japan, pp. 480–481, I16, Dec. 6–10, 1998.

23. J.-J. Greffet and R. Carminati, "Image formation in near-field optics," *Prog. Surf. Sci.* **56**(3), 133–237 (1997).

24. R. Bachelot, P. Gleyzes, and A. C. Boccara, "Reflection-mode scanning near-field optical microscopy using an apertureless metallic tip," *Appl. Opt.* **36**(10), 2160–2170 (1997).

25. M. Kourogi, T. Yatsui, S. Ishimura, M. B. Lee, N. Atoda, K. Tsutsui, and M. Ohtsu, "A near-field planar apertured probe array for optical near-field memory," *SPIE* **3467**, 258–267 (1998).

3

FABRY–PÉROT RESONATORS, BEAMS, AND RADIATION PRESSURE

The two topics on Fabry–Pérot resonators and the properties of Gaussian beams complement each other because the Gaussian distribution function describes light beams not only in free space but also inside certain kinds of Fabry–Pérot resonators.

This chapter also covers Bessel beams, which are considered long-range nondiffracting beams, and radiation pressure, which is applied to light tweezers as well as laser cooling.

3.1 FABRY–PÉROT RESONATORS

Fabry–Pérot resonators selectively transmit or reflect a particular wavelength of light and are used for various applications, such as spectroscopy, stabilization of laser oscillation, and interference filters. Along with grating spectroscopy, Michelson interferometry, and Fourier transform spectroscopy, Fabry–Pérot spectroscopy is one of the most important means of analyzing a light spectrum.

Suppose that an individual plate is very opaque: let's say the reflectance R is 99.9%, and almost no light gets through. If two of these plates are placed precisely parallel to each other, something unexpected happens. The plates become transparent with almost nearly 100% transmittance at a particular wavelength, which is determined by the spacing between the reflecting plates, and at all other wavelengths, the plates will become even more reflective than before. The reason for this behavior is due to the large number of multiple reflections of the light inside the resonator. The plates are assumed to be nonabsorbing and are carefully aligned parallel to each other. The light bounces back and forth between the plates a large number of times, with only a small percentage of light being transmitted through the plate at each bounce. The larger the value of R is, the less light gets through the plate per bounce, and the greater the number of bounces. The transmitted light from each pass through the resonator interferes with the transmitted light from other passes. For the special case that the transmitted components from the multiple passes are in phase, the total amount of

Magic opaque screen. Opaque screens become transparent where they overlap.

transmitted light through the resonator becomes nearly 100%. This is known as the *resonance condition*. For the transmitted components to be in phase, the round-trip phase difference must be an even multiple of π radians. Wavelengths that satisfy this condition are called *resonant wavelengths.*

At resonance, the value of the peak transmittance is always nearly 100% regardless of the value of R. Although R does not affect the peak transmittance, the value of R does affect the sharpness of the resonance. The larger the value of R is, the greater the number of multiple reflections, and the more stringent the in-phase condition becomes. This phenomenon has made the Fabry–Pérot resonator quite popular as a device for selecting a particular wavelength of light. Thus, the Fabry–Pérot resonator transmits or reflects very selectively at particular wavelengths of light.

The most important application of the Fabry–Pérot resonator is as a tool for analyzing light spectra. It is also used inside gas as well as semiconductor lasers to host the action of lasing at a specified wavelength or as a component in a circuit for stabilizing the wavelength of the laser oscillation. Moreover, the principle of the Fabry–Pérot resonator is applied to designing such devices as an interference filter that can pick out an extremely narrow spectrum of light or a dichroic mirror whose optimum transmission and reflection take place at two specified wavelengths.

The Fabry–Pérot resonator is sometimes called the Fabry–Pérot etalon or simply etalon. By convention, resonators with a fixed spacing between reflectors are called etalons, and resonators with a variable spacing between reflectors are called interferometers.

3.1.1 Operating Principle of the Fabry–Pérot Resonator

The basic description of the operating principle of the Fabry–Pérot resonator is given in this section, followed by a more comprehensive analysis in the next section.

Figure 3.1a shows a schematic of a Fabry–Pérot resonator. It consists of left and right reflecting plates sandwiching a center medium. For simplicity, each reflecting plate has refractive index n_1, and it is assumed that n_1 is larger than the refractive

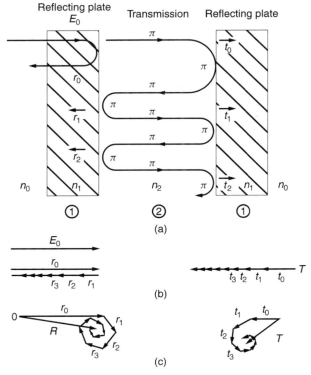

Figure 3.1 Principle of the Fabry–Pérot resonator. (a) Geometry of Fabry–Pérot resonator ($n_1 > n_2$). (b) Phasors at resonance. (c) Phasors off resonance.

index n_2 of the center medium, namely,

$$n_1 > n_2$$

Light with amplitude E_0 is incident from the left to the right and is normal to the reflecting plates. It is further assumed that the spacing between the reflecting plates is adjusted to be an integral multiple of a half-wavelength; namely, the Fabry–Pérot cavity is set at resonance.

The analysis will concentrate on the reflection and transmission at the boundaries between n_1 and n_2. The reflection at the interface between the reflecting plate (n_1) and air (n_0) is not accounted for. First, consider the incident light undergoing reflection at the interface of the left reflecting plate and the center medium. The reflection coefficient r_{12} from Eq. (2.18) in Chapter 2 is

$$r_{12} = \frac{n_1 - n_2}{n_1 + n_2} \tag{3.1}$$

and r_{12} is a positive number. The phase of the reflected light $E_0 r_{12}$ is designated as 0 radians. Let this first reflected light component be called r_0,

$$r_0 = E_0 r_{12} \tag{3.2}$$

The transmission coefficient at the same boundary from Eq. (2.19) is

$$t_{12} = \frac{2n_1}{n_1 + n_2} \tag{3.3}$$

and t_{12} is also a positive number. The light incident into the center medium from the left reflecting plate is $E_0 t_{12}$, and its phase is designated as 0 radians.

The transmission coefficient t_{21} transmitting from the center medium into the right reflecting plate is

$$t_{21} = \frac{2n_2}{n_1 + n_2} \tag{3.4}$$

and t_{21} is a positive number. The light component t_0 emergent from the Fabry–Pérot resonator is

$$t_0 = E_0 t_{12} t_{21} e^{j(2m+1)\pi} \tag{3.5}$$

where m is an integer and the last factor accounts for the phase associated with transmission across the Fabry–Pérot resonator at the resonance condition, which is an odd multiple of π radians. Thus, the phase of the first transmitted component t_0 is π radians.

Next, the light component that is reflected back by the right reflector toward the left reflector is considered. The reflection coefficient r_{21} at the right reflecting plate is

$$r_{21} = \frac{n_2 - n_1}{n_1 + n_2} \tag{3.6}$$

and r_{21} is a negative number. As a complex number, r_{21} can be expressed as

$$r_{21} = |r_{21}| e^{j\pi} \tag{3.7}$$

Thus, there is a phase change of π associated with this reflection.

In simple terms, for each one-way trip through the center medium, the light acquires a phase of $(2m + 1)\pi$ radians, which is equivalent to π radians. There is also a phase change of π radians for each reflection occurring when the light in the center medium hits the reflector boundary, as illustrated in Fig. 3.1a.

The transmitted light t_1 is

$$t_1 = E_0 t_{12} e^{j(2m+1)\pi} r_{21} e^{j(2m+1)\pi} r_{21} e^{j(2m+1)\pi} t_{21}$$

which reduces to

$$t_1 = E_0 t_{12} r_{21}^2 t_{21} e^{j\pi}$$

Components of the transmitted light, $t_0, t_1, t_2 \ldots$, all have the same π-radian phase and add constructively. The transmission through the cavity reaches a maximum at resonance.

On the other hand, the situation of the components $r_0, r_2, r_3, \ldots$ of the light reflected from the cavity is slightly different. The magnitude of the first reflected light component, which has never entered the cavity, is $E_0 r_{12}$ and is much larger than all other reflected components. Table 3.1 summarizes the amplitudes of the various components shown in Fig. 3.1a. The absolute value signs in r_{ij} are suppressed.

Table 3.1 Phase and amplitude of reflected and transmitted components

Reflected Components	Phase	Amplitude	Transmitted Components	Phase	Amplitude
r_0	0	$E_0 r_{12}$	t_0	π	$E_0 t_{12} t_{21}$
r_1	π	$E_0 t_{12} r_{21} t_{21}$	t_1	π	$E_0 t_{12} r_{21}^2 t_{21}$
r_2	π	$E_0 t_{12} r_{21}^3 t_{21}$	t_2	π	$E_0 t_{12} r_{21}^4 t_{21}$
r_n	π	$E_0 t_{12} r_{21}^{2n-1} t_{21}$	t_n	π	$E_0 t_{12} r_{21}^{2n} t_{21}$

The amplitude of the nth component decreases as n increases. The magnitude of the first reflected light r_0 has a phase of 0 radians, while all other smaller components $r_1, r_2, r_3, \ldots$, which eventually come out of the cavity after multiple reflections inside the cavity, all have π-radian phase. The largest component and the accumulation of all the smaller components cancel each other, and the resultant reflected light reduces to a minimum at the resonance of the cavity. It is a common mistake to forget about r_0, and to incorrectly conclude that the reflection also reaches a maximum at resonance due to the accumulation of $r_1, r_2, r_3, \ldots$. The phasors at resonance are shown in Fig. 3.1b.

Finally, the case of off resonance is considered. Off resonance, the phase delay due to the round trip inside the cavity is no longer exactly an integral number of 2π radians. The components of the signal transmitted through the cavity are no longer in phase and the phasors representing them curl up as shown in Fig. 3.1c. The magnitude of the resultant phasors of the transmitted light from the cavity is small.

The phasors of the $r_1, r_2, r_3, \ldots$ light components reflected from inside the cavity also curl up, and the resultant phasor can no longer cancel the large phasor r_0 of the first reflected component that did not enter the cavity. Thus, the reflectance of the cavity reaches a large value when the cavity is not at resonance, as shown by the phasor R on the left side of Fig. 3.1c. In short, the transmitted light decreases off resonance while the reflected light increases.

In the next section, the case when the angle of incidence is not normal to the reflectors is described.

3.1.2 Transmittance and Reflectance of the Fabry–Pérot Resonator with an Arbitrary Angle of Incidence

Referring to the geometry of Fig. 3.2, the general expressions for the transmittance and reflectance of a Fabry–Pérot resonator [1–3] will be derived. These quantities are obtained by summing an array of light beams produced by multiple reflections from a pair of reflectors and reaching point P of the focus of a convex lens. Let the refractive index of the medium of the Fabry–Pérot etalon be n_2, the refractive index of the external medium be n_1, and the spacing between the reflectors be d. There is no reflective film deposited on the etalon, and the reflectors are the discontinuities of the refractive indices between n_1 and n_2. The cases for $n_1 > n_2$ and $n_2 > n_1$ will be treated concurrently. As a matter of fact, the final results are the same for both cases. Also, let the angle of incidence from the normal to the reflector surface be θ_i and the internal angle inside the reflectors be θ. Snell's law gives the relationship between θ and θ_i as

$$n_1 \sin \theta_i = n_2 \sin \theta \tag{3.8}$$

The emergent angle from the Fabry–Pérot resonator is again θ_i.

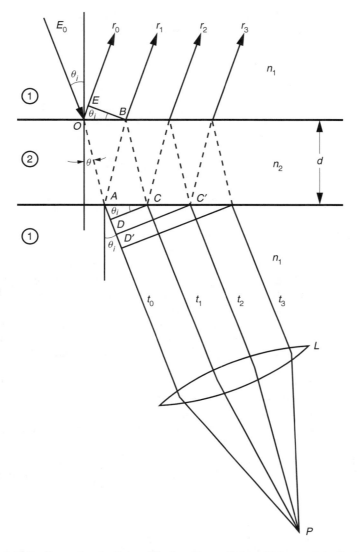

Figure 3.2 Geometry of a Fabry–Pérot cavity consisting of two parallel reflectors.

First, the optical path difference between the first and second transmitted beams reaching point P will be calculated. If line $\overline{CD}$ is perpendicular to both ray t_0 and ray t_1, then the optical path for $\overline{CP}$ is identical to the optical path for $\overline{DP}$. In this case, only the optical path difference between the path from O to D via A, and the path from O to C via reflections at A and B, need be calculated. This optical path difference creates a phase difference ϕ:

$$\phi = k[n_2(\overline{OA} + \overline{AB} + \overline{BC}) - (n_2\overline{OA} + n_1\overline{AD})] + 2\pi \tag{3.9}$$

where k is the free-space propagation constant and 2π accounts for the two phase reversals associated with reflections at A and B if $n_1 > n_2$ is assumed. If $n_2 > n_1$ is assumed, then 2π disappears. Either way, 2π does not influence the subsequent result. Removing the common path and noting that $\overline{AB} = \overline{BC}$, ϕ becomes the optical path difference between $\overline{ABC}$ and $\overline{AD}$

$$\phi = k(2n_2\overline{AB} - n_1\overline{AD}) \tag{3.10}$$

$$\overline{AB} = \frac{d}{\cos\theta}$$

$$\overline{AD} = \overline{AC}\sin\theta_i \tag{3.11}$$

$$\overline{AC} = 2d\tan\theta \tag{3.12}$$

$$\phi = k\left(\frac{2n_2d}{\cos\theta} - 2n_1d\tan\theta\sin\theta_i\right) \tag{3.13}$$

where θ and θ_i are related by Snell's law. Equation (3.8) and ϕ finally become

$$\phi = 2n_2dk\cos\theta \tag{3.14}$$

Next, the array of transmitted beams will be summed to obtain the resultant of the light reaching point P through the Fabry–Pérot resonator. The phases of the multiple reflected beams $t_1, t_2, \ldots, t_n$ are always compared to the phase of the first beam t_0. In order to reach P, optical paths $\overline{C'P}$ and $\overline{D'P}$ are identical and the phase of t_2 at point C' is compared to that at D' on t_0. The media outside and inside the resonator will be denoted by subscripts 1 and 2, respectively. The following notations are used:

E_0 Amplitude of the incident light
t_{12} Amplitude transmission coefficient from medium ① to ②
t_{21} Amplitude transmission coefficient from medium ② to ①
r_{12} Amplitude reflection coefficient when the light is incident from medium ① toward medium ②
r_{21} Amplitude reflection coefficient from medium ② toward medium ①
t_i Amplitude of the transmitted light component after the ith reflection from the top boundary
r_i Amplitude of the reflected light component after the $(i + 1)$th reflection from the top boundary
Δ Phase associated with the optical path length from point O to A
R Reflectance of the reflector, defined as $r_{21}r_{21}$
T Transmittance of the reflector, defined as $t_{12}t_{21}$
ϕ Phase difference due to the optical path difference between $\overline{ABC}$ and $\overline{AD}$ given by Eq. (3.13)

The elements of the transmitted beam are

$$t_0 = E_0t_{12}e^{j\Delta}t_{21}$$
$$t_1 = E_0t_{12}e^{j\Delta}r_{21}r_{21}e^{j\phi}t_{21}$$
$$t_2 = E_0t_{12}e^{j\Delta}r_{21}^2r_{21}^2e^{j2\phi}t_{21} \tag{3.15}$$
$$t_3 = E_0t_{12}e^{j\Delta}r_{21}^3r_{21}^3e^{j3\phi}t_{21}$$
$$t_n = E_0t_{12}e^{j\Delta}r_{21}^{2n}e^{jn\phi}t_{21}$$

The sum E_t of the transmitted light is

$$E_t = E_0 e^{j\Delta} T(1 + Re^{j\phi} + R^2 e^{j2\phi} + R^3 e^{j3\phi} + \cdots) \qquad (3.16)$$

where

$$R = r_{21} r_{21} \qquad (3.17a)$$

$$T = t_{12} t_{21} \qquad (3.17b)$$

$$E_t = E_0 e^{j\Delta} \frac{T}{1 - Re^{j\phi}} \qquad (3.18)$$

If the reflector is lossy and its loss A is defined as

$$T + R + A = 1 \qquad (3.19)$$

then Eq. (3.18) can be written as

$$E_t = E_0 e^{j\Delta} \frac{1 - R - A}{1 - Re^{j\phi}} \qquad (3.20)$$

Similarly, the amplitude of the light reflected from the Fabry–Pérot resonator will now be calculated. The phase of the reflected beam r_i is always referred to that of beam r_0. The reflected light components are

$$r_0 = E_0 r_{12}$$

$$r_1 = E_0 t_{12} r_{21} e^{j\phi} t_{21}$$

$$r_2 = E_0 t_{12} r_{21}^3 e^{j2\phi} t_{21} \qquad (3.21)$$

$$r_3 = E_0 t_{12} r_{21}^5 e^{j3\phi} t_{21}$$

$$r_n = E_0 t_{12} r_{21}^{2n-1} e^{jn\phi} t_{21}$$

where ϕ is the same as before since

$$\overline{OE} = \overline{AD} \qquad (3.22)$$

The total sum E_r of the light reflected toward the source is

$$E_r = E_0 [r_{12} + T r_{21} e^{j\phi} (1 + r_{21}^2 e^{j\phi} + r_{21}^4 e^{j2\phi} + \cdots)]$$

$$= E_0 \left(r_{12} + \frac{T r_{21} e^{j\phi}}{1 - Re^{j\phi}} \right) \qquad (3.23)$$

From Eqs. (3.1) and (3.6) for normal incidence, it is easy to see that

$$r_{12} = -r_{21} \qquad (3.24)$$

which also holds true for either case of $n_1 > n_2$ or $n_2 < n_1$.

Equations (3.17a) and (3.24) give

$$\left.\begin{array}{l} r_{12} = \sqrt{R} \\ r_{21} = -\sqrt{R} \end{array}\right\} \tag{3.25}$$

where r_{12} is a positive number and r_{21} is a negative number for $n_1 > n_2$, and the signs are reversed for $n_1 < n_2$.

Using Eqs. (3.19), (3.23), and (3.25), the expression for E_r is

$$E_r = \pm E_0 \sqrt{R} \frac{1 - (1 - A)e^{j\phi}}{1 - Re^{j\phi}} \tag{3.26}$$

The upper sign is for $n_1 > n_2$, and the lower for $n_1 < n_2$.

Next, the intensity of the transmitted light I_t will be calculated from Eq. (3.20) using the relationship $I_t = E_t E_t^*$:

$$I_t = E_0^2 \frac{(1 - R - A)^2}{(1 - Re^{j\phi})(1 - Re^{-j\phi})} \tag{3.27}$$

$$= I_0 \frac{(1 - R - A)^2}{(1 - R)^2 + 4R \sin^2(\phi/2)} \tag{3.28}$$

Similarly, the intensity of the reflected light is

$$I_r = I_0 R \frac{A^2 + 4(1 - A) \sin^2(\phi/2)}{(1 - R)^2 + 4R \sin^2(\phi/2)} \tag{3.29}$$

Equations (3.28) and (3.29) hold true for either $n_1 > n_2$ or $n_1 < n_2$.

Let's look more closely at the transmitted power. By dividing both denominator and numerator by $(1 - R)^2$, Eq. (3.28) can be rewritten as

$$I_t = I_0 \left(1 - \frac{A}{1 - R}\right)^2 \frac{1}{1 + M \sin^2(\phi/2)} \tag{3.30}$$

where

$$M = \frac{4R}{(1 - R)^2} \tag{3.31}$$

and ϕ is given by Eq. (3.14).

The transmitted power I_t is periodic with respect to ϕ and every time ϕ approaches

$$\phi_m = 2m\pi \tag{3.32}$$

the value of $M \sin^2(\phi/2)$ in Eq. (3.30) becomes zero and I_t reaches a peak value (resonance). When $\phi_m = 2\pi(m + \frac{1}{2})$, I_t reaches a valley (antiresonance). The response curve is shown in Fig. 3.3. The larger the value of M is, the greater the change of

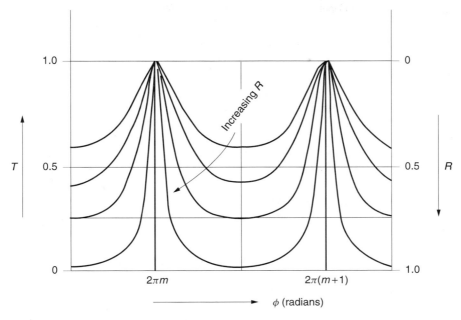

Figure 3.3 Transmittance T and reflectance R of the Fabry–Pérot resonator with respect to ϕ.

$M \sin^2(\phi/2)$ with respect to ϕ, and the sharper the resonance peaks become. For some applications, such as in a monochromator or a laser cavity, the sharper peaks are more desirable, whereas broader peaks are more desirable for other applications. In an optical filter, the bandwidth is manipulated by the value of M.

In monochromator applications, a large value of R is needed, but a sufficient amount of reflection cannot be realized solely by the difference in the indices of refraction n_1 and n_2. A thin metal film is deposited on the inner surfaces of the reflector plates in order to reach the needed value of R. The presence of the metal film can be accounted for with phase $e^{j\gamma}$, loss A, and the new value of R. In the case of the metal film, the phase change of Eq. (3.7) between r_{12} and r_{21} is no longer π radians but has a phase of γ. The energy lost in the metal film is accounted for by A. With these modifications, the above analysis holds true for practical purposes.

Important facts about the Fabry–Pérot resonator are that if $A = 0$, regardless of the value of R, transmission is 100% at resonance, as seen from Eq. (3.30), and moreover, the values of ϕ_m do not depend on the value of R. These two facts are the very reasons why the Fabry–Pérot resonator is so useful as a device for spectroscopy.

The results obtained so far have been derived from a scalar field approach, meaning the direction of the light polarization has not been taken into consideration. As was shown in the previous chapter, the amplitude and phase of the reflection coefficient r depend on the direction of polarization. Fortunately, however, the value of ϕ_m is independent of R and the use of the scalar field approach is justifiable.

On the other hand, if the refractive index n_2 of the medium inside the Fabry–Pérot cavity depends on the direction of light polarization (namely, n_2 is birefringent), then

ϕ_m is polarization dependent, and the vector field approach has to be used [1]. If the medium inside the Fabry–Pérot cavity is a Faraday rotator (see Chapter 5), such as iron garnet, then the multiple reflections taking place inside the Fabry–Pérot resonator enhance the rotation [4]. The measured result becomes dependent on the state of light polarization.

As seen from Eq. (3.14), there are several physical means of sweeping the value of ϕ, such as by changing the (1) spacing d between the reflectors, (2) the internal angle θ, (3) the refractive index of the medium inside the cavity, and (4) the wavelength of the incident light.

Fabry–Pérot devices are best categorized by the means used for sweeping the value of ϕ. These categories, in direct correspondence to the above list of sweeping methods, are (1) the scanning Fabry–Pérot spectrometer, (2) the Fabry–Pérot etalon, (3) the liquid crystal filter, and (4) the laser frequency stabilizer.

Each of the above will be described in more detail in the following sections.

3.2   THE SCANNING FABRY–PÉROT SPECTROMETER

A scanning Fabry–Pérot spectrometer consists of these basic parts: a scanning Fabry–Pérot resonator, a voltage generator, a photodetector, and a display scope. A photograph of a scanning Fabry–Pérot resonator is shown in Fig. 3.4. The resonator has a piezotransducer (PZT), which is driven by a sawtooth voltage generator. As the mirror of the resonator is displaced by the PZT, the transmitted light is detected and displayed on the scope. The block diagram of the scanning Fabry–Pérot spectrometer is shown in Fig. 3.5.

The scanning Fabry–Pérot resonator consists of one stationary and one movable plate. The stationary plate is equipped with a parallelism adjustment assembly onto

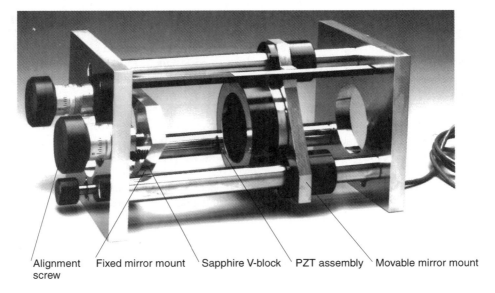

Alignment screw   Fixed mirror mount   Sapphire V-block   PZT assembly   Movable mirror mount

Figure 3.4   Structure of a scanning Fabry–Pérot resonator. (Courtesy of Burleigh Instruments Inc.)

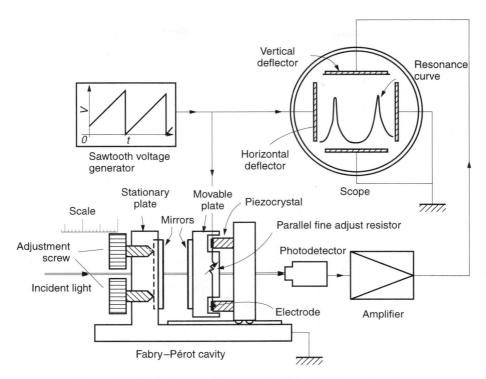

Figure 3.5 Block diagram of the scanning Fabry–Pérot spectrometer.

which one of the two mirrors is fastened. The movable plate can be coarsely set at a distance typically between 50 μm and 15 cm from the stationary plate. A PZT assembly, which holds the other mirror, is installed on the moving plate. Thus, the spacing between the mirrors can be coarsely set between 50 μm to 15 cm and the spacing can be finely scanned for about 2–3 μm by the PZT from the coarsely set position.

The output from the scanning Fabry–Pérot resonator is detected by either a photomultiplier or a photodiode and amplified. The amplified output signal is displayed on an oscilloscope. The output from the sawtooth voltage generator is fed to both the PZT and the horizontal deflection plate of the oscilloscope. Thus, the light output from the scanning Fabry–Pérot device is displayed with respect to displacement of the movable mirror.

The performance of the Fabry–Pérot resonator is characterized quantitatively by two factors: the finesse and the free spectral range. The finesse relates to the sharpness of the resonance, and the free spectral range is approximately equal to the separation between neighboring resonances. This separation can be expressed in terms of wavelength, frequency, or wavenumber (reciprocal of wavelength).

3.2.1 Scanning by the Reflector Spacing

In the case of normal incidence, $\theta = 0°$, into an air-filled Fabry–Pérot cavity with refractive index $n_2 = 1$, the condition for the mth order resonance is, from Eqs. (3.14)

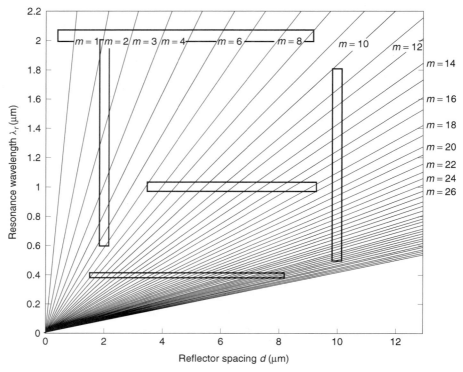

Figure 3.6 Resonance wavelength with respect to the reflector spacing with the mode number m as a parameter: $(\lambda_r/2)m = d$.

and (3.32),

$$\frac{\lambda_r}{2}m = d \tag{3.33}$$

The performance of the Fabry–Pérot resonator is controlled by this equation. Figure 3.6 is a plot of Eq. (3.33) and shows the resonance wavelength λ_r as a function of reflector spacing with the mode number m as a parameter. Equation (3.33) looks quite simple, but there are three variables, and interpretation of Fig. 3.6 needs further explanation. A small section near the origin of Fig. 3.6 has been expanded for detailed illustration in Fig. 3.7, which is described next.

Referring to Fig. 3.6, the intersections between a horizontal line and the resonance curves are equally spaced. For example, if a horizontal line is drawn at $\lambda = 0.4$ μm, the intersections are 0.2 μm apart. If a horizontal line is drawn higher on the graph, say, at 1.0 μm, the intersections are still equally spaced but are now 0.5 μm apart. This is why peaks corresponding to a specific wavelength appear equally spaced in the output display of a scanning Fabry–Pérot interferometer. The intersections between the vertical line and the resonance curves are not equally spaced. The spacing between intersections decreases gradually as the resonance wavelength decreases. In the shorter resonance wavelength regions, the spacings are approximated as uniform for most practical purposes.

3.2.1.1 Fabry–Pérot Resonator with a Fixed Resonator Spacing (Etalon)

Figure 3.7a shows the graph with the reflector spacing fixed at $d = 2$ μm. The resonance takes place at every mode number. The resonance wavelengths for the first three mode numbers are $\lambda_r = 4$ μm for $m = 1$, $\lambda_r = 2$ μm for $m = 2$, and $\lambda_r = 1.33$ μm for $m = 3$. The graph on the left shows the output spectra when light with a continuum spectrum is incident on the resonator. The light field distribution in the resonator for each mode number is shown on the right. The mode of the light spectrum from a Fabry–Pérot cavity-type laser is a good example of this case.

The spacing $\Delta\lambda_r$ between the resonator wavelengths for a large m is

$$\Delta\lambda_r = \frac{2d}{m} - \frac{2d}{m+1} \doteq \frac{2d}{m^2} = \frac{\lambda^2}{2d} \tag{3.34}$$

3.2.1.2 Monochromatic Incident Light with Scanned Reflector Spacing

Figure 3.7b shows the graphs associated with a single spectrum input light with wavelength $\lambda = 2.0$ μm and angle of incidence $\theta_i = 0°$. The reflector spacing is swept from 0 to 3 μm. When the spacing reaches $d = 1$ μm, the resonance wavelength $\lambda_r = 2$ μm of the resonator matches the incident light wavelength. The light is transmitted through the resonator and the first output peak appears. The output is shown in the bottom graph of Fig. 3.7b. The first peak is associated with the $m = 1$ mode. As d is swept further, the $m = 2$ mode resonance wavelength matches that of the incident light and the second peak appears at $d = 2$ μm. The same will be repeated as d is scanned at every $\lambda/2$ or 1 μm.

At the top of Fig. 3.7b, the light field E inside the Fabry–Pérot resonator is drawn for each mode. Every time the cavity length reaches an integral multiple of a half-wavelength, resonance takes place. As a matter of fact, from the interval between the resonance peaks, the wavelength of the incident light can be determined.

3.2.1.3 Free Spectral Range (FSR)

When the incident light has more than one wavelength, especially when the wavelengths are either too far apart or too close to each other, the proper selection of d, (hence m), becomes important.

Let us take the specific case of an incident light spectrum consisting of a main peak at $\lambda = 2.0$ μm. The wavelength of the auxiliary peak $\lambda_a = \lambda + \Delta\lambda$ is larger than the main peak by 20%. The scanning Fabry–Pérot resonator is used to analyze this compound signal. The response of $\lambda = 2.0$ μm alone has already been shown in Fig. 3.7b. While peaks associated with λ appear at an interval of $\lambda/2$ as d is scanned, the peaks associated with $\lambda + \Delta\lambda$ appear at a longer interval of $(\lambda + \Delta\lambda)/2$. Two sets of peaks with different intervals are displayed as d is scanned. The response of both λ and λ_a is shown at the bottom of Fig. 3.7c as the solid and dashed lines, respectively. The optimum region of m (or d) for displaying the spectrum will be sought.

From the bottom graph in Fig. 3.7c, the separation of the peak of λ_a from that of λ becomes larger and larger as m (or d) is increased, and the resolution between the two spectral lines increases with m (or d). There is, however, a limit on the value of m (or d). As the value of m exceeds $m = 5$ or $d = 5$ μm, the shift of the peak of λ_a from that of λ becomes equal to or longer than the regular interval of 1 μm of the peaks of λ. The peak of λ_a, which belongs to $m = 5$, overlaps the peak of λ, which belongs to $m = 6$, and the peaks of λ and λ_a start to intermingle. The mode pattern at $m = 7$ restarts that of $m = 1$ and goes back to the case of the lowest resolution.

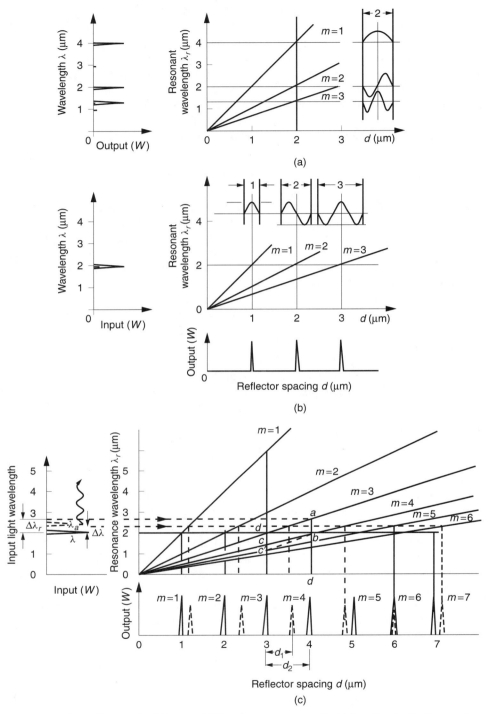

Figure 3.7   Expanded graph of Fig. 3.6. (a) Continuum spectrum light into a fixed reflector spacing. (b) Monochromatic light into scanned reflector spacing. (c) Illustration of how to select the reflector spacing.

In conclusion, the best choice of m is $4 \leq m \leq 5$, not only from the viewpoint of the highest resolution but also avoiding ambiguity.

Next, the question is asked in reverse. What is the maximum spread of the spectrum that can be determined by a Fabry–Pérot resonator without ambiguity for a given wavelength, resonator spacing, and scan range?

Let us say $d = 3$ μm and $\lambda = 2$ μm are given. This example calculates the maximum deviation $\Delta \lambda$ of the wavelength from λ that can be determined by the Fabry–Pérot resonator.

The order of mode m of the output peak appearing at $d = 3$ μm is, from Eq. (3.33),

$$m = \frac{2d}{\lambda} = \frac{2(3)}{2} = 3$$

The output peak of the auxiliary wavelength, $\lambda + \Delta \lambda$, shifts from that of the third order mode of λ toward the fourth order mode of λ as $\Delta \lambda$ is increased. The value $\Delta \lambda$ for which the auxiliary output peak reaches that of the fourth order mode of λ is

$$m \frac{(\lambda + \Delta \lambda)}{2} = (m + 1) \frac{\lambda}{2} \tag{3.35}$$

With $m = 3$, the answer is

$$\Delta \lambda = 0.67 \ \mu m$$

The maximum width of the spectrum that can be determined without ambiguity is called the *free spectral* range (FSR). The FSR of the Fabry–Pérot resonator whose reflector spacing is 3 μm is 0.67 μm at the wavelength $\lambda = 2.0$ μm.

From Eqs. (3.33) and (3.35), the value of the FSR for the general case with spacing d, around the wavelength λ, is

$$\Delta \lambda_{FSR} = \frac{\lambda^2}{2d} \tag{3.36}$$

The FSR decreases with an increase in the reflector spacing.

Next, it will be shown that the free spectral range $\Delta \lambda_{FSR}$ is approximately equal to the spacing between the adjacent resonance wavelengths of a resonator with a fixed reflector spacing d and operated at λ. Referring to Fig. 3.7c, as $\Delta \lambda$ is increased from zero to $\Delta \lambda_r$, the intersection of the horizontal dotted line with $\overline{da}$ moves from d to a. Line $\overline{ab}$ represents the free spectral range $\Delta \lambda_{FSR}$; and line $\overline{dc}$ represents the wavelength spacing between the $m = 3$ and $m = 4$ resonant wavelengths for a fixed value of $d = 3$ μm. From the parallelogram $abcd$, $\overline{ab}$ can be approximated as $\overline{cd}$. Thus, the free spectral range is almost equal to the spacing between the resonant wavelengths for a fixed value of d. This approximation becomes better as m gets larger.

$$\Delta \lambda_{FSR} \doteqdot \Delta \lambda_r \tag{3.37}$$

It is sometimes useful to express the free spectral range in terms of the light carrier frequency. In terms of frequency, the free spectral range is

$$\Delta \nu_{FSR} = \frac{c}{\lambda} - \frac{c}{\lambda + \Delta \lambda_{FSR}} \doteqdot \frac{c}{\lambda^2} \Delta \lambda_{FSR} \tag{3.38}$$

With Eq. (3.36), Eq. (3.38) becomes

$$\Delta \nu_{FSR} = \frac{c}{2d} \tag{3.39}$$

Thus, it should be noted that $\Delta \nu_{FSR}$ is independent of the light frequency. The appropriate selection of the spacing of the mirrors, which determines the value of the FSR, is important for operating with the highest resolution, yet without ambiguity.

Example 3.1 Find the spacing d and the mode number of a scanning Fabry–Pérot spectrum analyzer that can display the spectrum of a superluminescent laser diode (SLD). The center wavelength of the SLD is 1.540 μm and the width of the spectrum is ±10 nm from the center wavelength.

Solution The free spectral range in wavelength has to be 20 nm and, from Eq. (3.36),

$$d = \frac{(1.54)^2}{2(0.02)} = 59.3 \text{ μm}$$

The mode number of operation is

$$m = \frac{2d}{\lambda} = \frac{2(59.3)}{1.54} = 77 \qquad \square$$

From the output display of a Fabry–Pérot resonator, such as shown at the bottom of Fig. 3.7c or in Fig. 3.8 of the next example, the input spectrum is to be determined. In the output display, the peak associated with λ_a is located at $(d_1/d_2)\lambda_{FSR}$ from one of the main peaks associated with λ, as shown at the bottom of Fig. 3.7c:

$$(\lambda_a - \lambda) = \frac{d_1}{d_2}\Delta \lambda_{FSR} \tag{3.40}$$

where $\Delta \lambda_{FSR}$ is the free spectral range given by Eq. (3.36). In terms of frequency, the equivalent of Eq. (3.40) is

$$(\nu - \nu_a) = \frac{d_1}{d_2}\Delta \nu_{FSR} \tag{3.41}$$

Example 3.2 shows the calculation for a specific case.

Example 3.2 Figure 3.8 shows the display of a scanning Fabry–Pérot resonator when a helium–neon laser beam is phase modulated by an MNA (2-methyl-4-nitroaniline) crystal. The wavelength is $\lambda = 0.6328$ μm and the mirror spacing is swept around $d = 600$ μm.

(a) What is the scan length of the PZT from point a to point b in the display in Fig. 3.8?
(b) From Fig. 3.8, find the frequency of the phase modulation of the He–Ne laser light due to the MNA crystal. The phase modulation creates two side lobes centered at the carrier frequency.

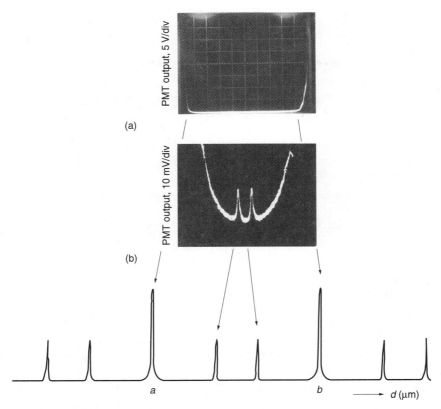

Figure 3.8 Display of a scanning Fabry–Pérot resonator with $d = 600$ μm and $\lambda = 0.6328$ μm. (After C. Wah, K. Iizuka, and A. P. Freundorfer [15].) (a) Output on the oscilloscope at a vertical scale of 5 V/div; only the carrier is visible. (b) Output on the oscilloscope at a vertical scale of 10 mV/div; the first order sidebands appear.

Solution

(a) From the resonance condition

$$\frac{\lambda_r}{2}m = d_m$$

the scan length between adjacent modes is $\lambda/2$:

$$\frac{\lambda_r}{2} = \frac{0.6328}{2} = 316 \text{ nm}$$

(b) The solution is obtained by referring to Fig. 3.7b and Eq. (3.36). The FSR in wavelength can be converted into that in frequency using Eq. (3.39):

$$\Delta \nu_{\text{FSR}} = \frac{c}{2d}$$

$$\Delta \nu_{\text{FSR}} = \frac{3 \times 10^{14}}{2(600)} = 250 \text{ GHz}$$

From the display, the ratio of the distance to the side lobe to that of the adjacent main peak is measured to be 0.38; hence, the frequency of the phase modulation is $250 \times 0.38 = 95$ GHz.   □

Example 3.3   As shown in Part I of Fig. 3.9, the amplitude of the input spectrum is of a triangular shape with respect to the wavelength λ. This spectrum is observed by a Fabry–Perot resonator.

(a) The spacing d between the reflectors is set so that the free spectral range fits the spectral width of the input light. Draw the amplitude of the spectrum when the spacing d is swept by $\lambda/2$.

(b) Draw the display when the spacing of the reflector is expanded to $2d$, and d is swept by $\lambda/2$.

(c) Draw the display when the spacing of the reflector is reduced to $d/2$ and d is swept by $\lambda/2$.

Solution   The mode lines that are inside the $\Delta\lambda \times (\lambda/2)$ windows are responsible for the output. The answers are shown in Part II of Fig. 3.9.

(a) When $\Delta\lambda = \Delta\lambda_{FSR}$, the appropriate output display is obtained.

(b) When $\Delta\lambda > \Delta\lambda_{FSR}$, more than one mode line are inside the $\Delta\lambda \times (\lambda/2)$ window. The output displays overlap and the display becomes ambiguous.

(c) When $\Delta\lambda < \Delta\lambda_{FSR}$, even though the display is correct, the power of resolution is not the highest.

The lesson to be learned from this example is that when the spectral range of the input light is not known, it is good practice to start the measurement with a small value of d (or a large value of $\Delta\lambda_{FSR}$) and gradually increase d for improved resolution until overlapping output starts to display.   □

3.2.2   Scanning by the Angle of Incidence

As seen from Eq. (3.14), as θ is increased, ϕ is decreased, and the Fabry–Pérot resonator can be swept by sweeping θ_i. Figure 3.10 shows an arrangement for displaying the resonance rings without any moving mechanism. A diffuser is placed between the source and the Fabry–Pérot resonator. First, the case of monochromatic incident light will be considered. A diffuser creates wavelets of various incident angles to the Fabry–Pérot resonator. Among the incident wavelets, only those whose incident angles match the resonance angles $\theta_m, \theta_{m-1}, \theta_{m-2}, \ldots, \theta_{m-i}$ that satisfy Eqs. (3.14) and (3.32) can transmit through the cavity. All other incident angles are reflected back by the Fabry–Pérot cavity toward the source.

The resonance condition is not limited to just the wavelets propagating in the x–z plane; the same condition is satisfied in any plane containing the z axis. The emergent light forms a series of concentric fringe rings on the screen. From the pattern of the fringe rings, the spectrum of the source light is analyzed.

Next, the case of multiple spectra is considered. Light of each wavelength forms its own set of fringe rings, and without the prism, the rings will be superimposed, making it difficult to decode the fringe rings. A dispersive prism, which is placed between the Fabry–Pérot resonator and the output screen in Fig. 3.10, disperses the locations of

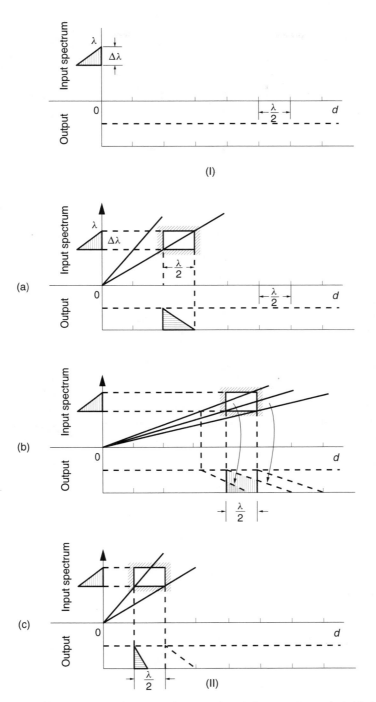

Figure 3.9 Input spectrum (I) and output displays (II) with different values of d. (a) $d = \lambda^2/2\Delta\lambda$. (b) $d = \lambda^2/\Delta\lambda$. (c) $d = \lambda^2/4\Delta\lambda$.

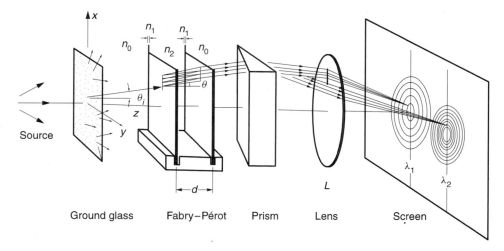

Figure 3.10   Fabry–Pérot spectroscopy using a prism to separate the fringe rings.

the fringe rings belonging to the different wavelengths, so as to avoid the overlapping of the fringe rings.

Example 3.4   Referring to Fig. 3.10, find the radii of the first three maxima of the fringe rings. The wavelength of the diffused source is 0.6328 µm. The spacing d between the reflectors of the Fabry–Pérot resonator is 30.13 µm. The refractive index n_2 of the medium in the Fabry–Pérot resonator is 1.05, and the refractive index n_0 of the medium outside the resonator is 1.00. The focal length f of the lens L is 50 mm.

Solution   Combining Eqs. (3.14) and (3.32) gives

$$2n_2d \cos \theta = m\lambda$$

With the given parameters, the order of the fringe ring m at $\theta = 0$ is

$$2(1.05)(30.13) = m(0.6328)$$

$$m = 100$$

The angle θ for $m = 99$ is

$$2(1.05)(30.13) \cos \theta = (99)(0.6328)$$

$$\cos \theta = \frac{99}{100}$$

$$\theta = 8.11°$$

In the parallel-plate case, the general Snell's law of

$$n_i \sin \theta_i = \text{constant}$$

holds true. The angle θ_i in air can immediately be related to the angle θ in the center medium without dealing with transmission through the plate with index of refraction n_1, namely,

$$n_2 \sin \theta = n_0 \sin \theta_i$$

$$\theta_i = 8.52° \quad \text{for } m = 99$$

Similarly, θ_i for the next two orders is

$$\theta_i = 12.06° \quad \text{for } m = 98$$

$$\theta_i = 14.79° \quad \text{for } m = 97$$

The incident wave to the lens in the x direction is

$$E = E_0 e^{jkx \sin \theta_i}$$

The pattern on the back focal plane is the Fourier transform of the input to the lens (see Eq. (1.160)):

$$E_i(x_i) = F\{E\}_{f_x = \frac{x_i}{\lambda f}}$$

The pattern is cylindrically symmetric, and $x_i = r_i$.

$$E(r_i, f) = E_0 \delta \left[\frac{1}{\lambda} \left(\frac{r_i}{f} - \sin \theta_i \right) \right]$$

$$r_i = f \sin \theta_i$$

Thus, the radii for $m = 99, 98, 97$ are

$$7.41 \text{ mm} \quad \text{for } m = 99$$

$$10.4 \text{ mm} \quad \text{for } m = 98$$

$$12.8 \text{ mm} \quad \text{for } m = 97 \qquad \qquad \square$$

An angle adjustment scheme using a parallel beam incident to an etalon is shown in the fiber ring laser of Fig. 3.11. The etalon transmits light only at certain wavelengths, which are determined by the angles of incidence specified by Eqs. (3.14) and (3.32). The light of a particular wavelength is fed back to the erbium-doped fiber amplifier (see Chapter 13) and the wavelength of the fiber ring laser can be tuned by rotating the etalon [5].

3.2.3 Scanning by the Index of Refraction

Extreme fine tuning of the Fabry–Pérot resonator without disturbing the reflector alignment can be achieved by controlling the refractive index of the medium inside the Fabry–Pérot resonator. For instance, by increasing the pressure of 100% nitrogen

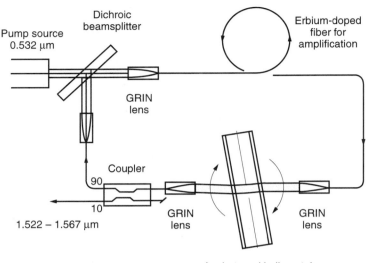

Figure 3.11 A tunable fiber ring laser.

gas by 1 atmosphere, the refractive index changes by 0.0003, and fine control of the refractive index is possible by controlling the pressure of the nitrogen gas.

Another material used is a liquid crystal whose index of refraction can be changed by an external electric field. The electrooptic effect of the nematic-type liquid crystal is large and the control voltage is small, typically 1–5 volts. The refractive index can be controlled by as much as 20% by means of these control voltages.

Figure 3.12 shows a schematic of a liquid crystal fiber filter. A nematic-type liquid crystal is sandwiched between indium tin oxide (ITO) transparent electrodes

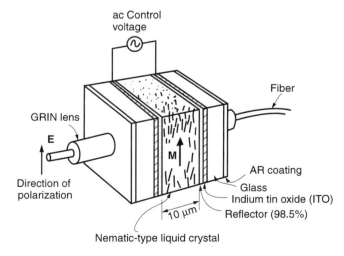

Figure 3.12 Electronically tunable liquid crystal Fabry–Pérot resonator.

deposited on a glass plate. The reflectors of the Fabry–Pérot cavity are deposited over each surface of the ITO electrode. The outer surfaces of the glass plates are antireflection (AR) coated. The whole liquid crystal compartment is connected to GRIN lenses (graded index lenses whose refractive indices are quadratic with the radius and function like thin lenses) [6–11].

Such a tunable filter is useful, for instance, for tuning in to a particular wavelength in a wavelength division multiplexing (WDM) fiber-optic communication system.

Another example of refractive index scanning is found in an electronically tunable semiconductor laser. When the electrons are injected into the guiding layer of a semiconductor laser, its index of refraction is reduced due to the plasma effect (see Section 14.4.3.2) of the electrons. It is decreased as much as 0.35%. A numerical calculation is given in Example 3.5 and the general description can be found in Section 14.10.

Example 3.5   Consider a Fabry–Pérot resonator cavity filled with a semiconductor material with refractive index n, as shown in Fig. 3.13. The spacing between the reflectors is 63.4 μm and the free spectral range around the wavelength $\lambda_m = 1.5517$ μm is $\Delta\lambda_{FSR} = 5.42$ nm.

(a) Find the value n of the refractive index of the semiconductor material.

(b) Find the value m of the order of the longitudinal mode for $\lambda_m = 1.5517$ μm.

(c) With the above arrangement, if an electric current is injected into the semiconductor, the refractive index n of the semiconductor will be decreased due to the plasma effect of the injected current. In this manner, an electronically tunable Fabry–Pérot device can be fabricated. Estimate the refractive index change needed to sweep the wavelength of the resonance peaks by 5.42 nm around $\lambda = 1.5517$ μm by this scheme. Assume the same mode is retained before and after the injection of the current.

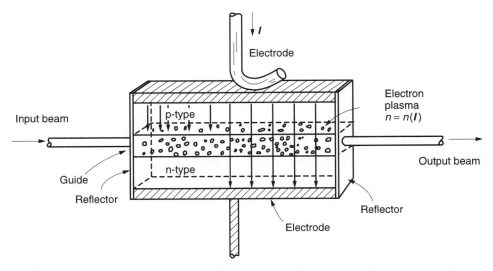

Figure 3.13   Electronically tunable semiconductor Fabry–Pérot cavity.

Solution

(a) From Eq. (3.36) including the index of refraction n, the free spectral range is

$$\Delta \lambda_{\text{FSR}} = \frac{\lambda^2}{2nd}$$

$$n = \frac{\lambda^2}{2\Delta\lambda_{\text{FSR}}d}$$

$$= \frac{(1.5517)^2}{2(5.42 \times 10^{-3})(63.4)}$$

$$= 3.5$$

(b) To find the value of m, we use

$$\frac{\lambda_m}{2}m = nd$$

$$m = \frac{2nd}{\lambda_m} = \frac{2(3.5)(63.4)}{1.5517} = 286$$

(c) Let λ_m be the wavelength of the resonance before injection of the current and let λ'_m be the wavelength after the injection of current.

$$\frac{\lambda_m}{2}m = nd$$

$$\frac{\lambda'_m}{2}m = n'd$$

Thus,

$$n - n' = \left(\frac{\lambda_m - \lambda'_m}{2d}\right)m$$

$$= \frac{(5.42 \times 10^{-3})(286)}{2(63.4)}$$

$$= 0.0122$$

In percentage, the change is

$$\frac{n - n'}{n} = 0.35\% \qquad \square$$

3.2.4   Scanning by the Frequency of Incident Light

By scanning the frequency of the incident light, a series of equally spaced resonance peaks are observed in the output of the Fabry–Pérot resonator at frequencies of

$$f_m = m\frac{c}{2nd}$$

This technique has many applications, such as controlling the channel spacing of wavelength division multiplexing (WDM) in a fiber-optic communication system.

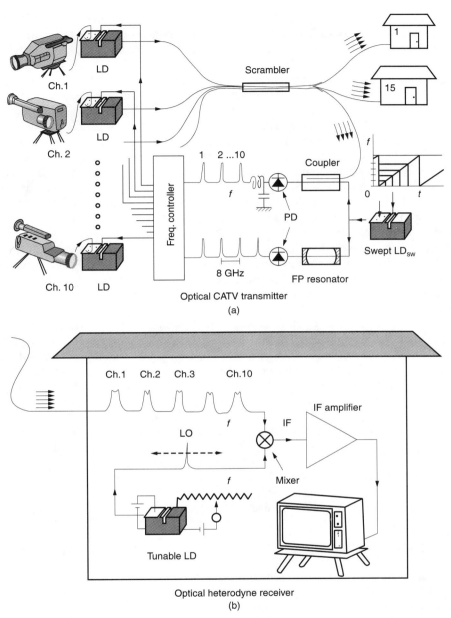

Figure 3.14 Use of a Fabry–Pérot resonator for controlling the channel spacing of the WDM of a fiber-optic communication system. (a) Channel space locking by the reference pulse method. (b) Optical heterodyne receiver at home.

Figure 3.14a shows a block diagram of such a WDM system used in the transmitting station of a fiber-optic cable television system. There are 10 separate tunable laser diodes, which have been tuned to achieve a frequency spacing of 8 GHz between carrier frequencies. Each laser diode is modulated by its own television camera so that 10 TV channels are sent concurrently in one fiber.

The principle of operation is as follows. The outputs from the laser diodes are all fed to the scrambler so that each output fiber from the scrambler transmits the same signal, which contains 10 different TV channels, to each subscriber at home. Subscribers can select any of the 10 channels for viewing at home.

In such a system, it is important to tightly maintain the carrier frequency spacing of 8 GHz so as to avoid crosstalk between the channels. This example demonstrates the reference pulse method [12], which employs a Fabry–Pérot resonator to lock the channel spacing. The reference pulse scheme consists of a swept laser diode LD_{sw}, a coupler, and a Fabry–Pérot (FP) resonator, as shown in the lower right portion of Fig. 3.14a. The carrier frequency of LD_{sw} is linearly swept with respect to time over 80 GHz. One-half of the output from the swept LD_{sw} is fed to the coupler that leads to a photodiode mixer followed by a low-pass filter. The output from one of the scrambler ports is fed to the same coupler. Whenever the frequency of the swept LD_{sw} matches with that of any one of the channels, an electrical pulse appears at the output of the low-pass filter. At each sweep of LD_{sw}, a chain of 10 pulses appears.

On the other hand, the other half of the output from LD_{sw} is fed directly to the Fabry–Pérot resonator. Whenever the swept frequency matches the resonance frequencies of the resonator, the light will reach the photodiode and an electronic pulse appears at the output of the photodiode. The spacing between the resonance frequencies of the Fabry–Pérot resonator is set at 8 GHz. If every pair of pulses from the scrambled signal and from the Fabry–Pérot resonator is synchronized, the frequencies of all channels are properly spaced. The frequency controller compares each pair of pulses. If the pulses of the pair are not synchronized, an error signal is issued to that particular channel to correct the carrier frequency of the laser diode. With this method, the relative frequency fluctuation can be maintained within ±25 MHz, which is 1.25×10^{-7} of the carrier frequency of the transmitter LD. It may be added that this scheme is much more economical than combining 10 individually frequency-stabilized laser diodes.

Lastly, a word about the scheme of the receiver at the subscriber's home will be added. In Fig. 3.14b, an optical heterodyne receiver (see Chapter 13) is used. The optical heterodyne is essentially a converter from an optical frequency to an intermediate frequency (IF), which is normally in the radiofrequency range. This conversion is achieved by mixing with a local oscillator (LO) light. A particular channel can be selected by tuning the frequency of the tunable local oscillator laser diode.

3.3 RESOLVING POWER OF THE FABRY–PÉROT RESONATOR

This section looks more closely at the problem of resolving two closely spaced spectra. Although the mode lines in Fig. 3.6 are all drawn without spectral width, in reality, the output from the Fabry–Pérot resonator does have finite width. The achievable resolution of the scanning Fabry–Pérot resonator as a spectrum analyzer depends on the lineshape of the transmitted light.

The resolving power R will be calculated. The expression for the transmitted light I_t with respect to the frequency ν is given by Eqs. (3.14) and (3.30) as

$$I_t = I_0 \frac{1}{1 + M \sin^2[(2\pi\nu/c)d]} \tag{3.42}$$

where $A = 0$, $n_2 = 1$, and $\theta = 0$. The lineshape of I_t depends on two factors: one is M, and the other is $2d/c$. The value of M is associated with the reflectance R of the reflector through Eq. (3.31), and $2d/c$ is exactly the inverse of the FSR in frequency, as given by Eq. (3.39).

Figure 3.15 shows a graph of I_t as a function of ϕ. Let $\phi_{\pm 1/2}$ be the value of ϕ near the mth order mode for which the intensity drops to half its peak value,

$$I_t(\phi_{\pm 1/2}) = \tfrac{1}{2} I_0 \tag{3.43}$$

Let $\Delta\phi$ be the full width at half maximum (FWHM) in ϕ. From Eq. (3.32), $\phi_{\pm 1/2}$ is expressed as

$$\phi_{\pm 1/2} = 2m\pi \pm \frac{\Delta\phi}{2} \tag{3.44}$$

When the denominator of Eq. (3.30) becomes 2, Eq. (3.43) is satisfied. This condition, in combination with Eq. (3.44), means $\Delta\phi$ must satisfy

$$1 = M \sin^2 \left(\frac{2m\pi \pm \Delta\phi/2}{2} \right)$$

$$= M \left(\sin m\pi \cos \frac{\Delta\phi}{4} \pm \cos m\pi \sin \frac{\Delta\phi}{4} \right)^2$$

$$= M \sin^2 \frac{\Delta\phi}{4}$$

For $\sin(\Delta\phi/4) \ll 1$, $\Delta\phi$ is approximated as

$$\Delta\phi \doteqdot \frac{4}{\sqrt{M}} \tag{3.45}$$

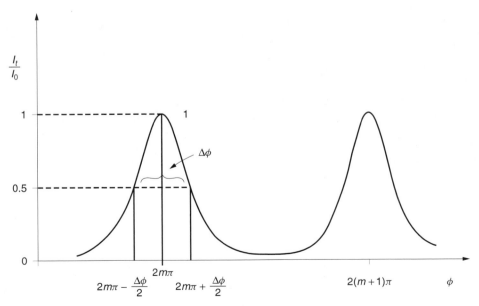

Figure 3.15 Resonance curve of the Fabry–Pérot resonator with ϕ. $\Delta\phi$ is the full width at half maximum (FWHM).

Using Eq. (3.14) with $n_2 = 1$ and $\theta = 0$, ϕ is rewritten in terms of the frequency ν as

$$\phi = \frac{4\pi d}{c}\nu \tag{3.46}$$

Inserting Eq. (3.45) into (3.46), the final result is obtained as

$$\Delta\nu_{1/2} = \frac{\Delta\nu_{\text{FSR}}}{F} \tag{3.47}$$

where

$$F = \frac{\pi}{2}\sqrt{M} \tag{3.48}$$

F is called the *finesse* and is further rewritten with the help of Eq. (3.31) as

$$F = \pi\frac{\sqrt{R}}{1-R} \tag{3.49}$$

F is sometimes called the reflection finesse F_r to distinguish it from other types of finesse, as mentioned below. The finesse increases with an increase in R. With an increase in F, the FWHM $\Delta\nu_{1/2}$ becomes finer. By rewriting Eq. (3.47) using Eqs. (3.33) and (3.39), the value of the resolving power R_s, which is defined as

$$R_s = \frac{\nu}{\Delta\nu_{1/2}} \quad \text{or} \quad \frac{\lambda}{\Delta\lambda_{1/2}} \tag{3.50}$$

is obtained as

$$R_s = mF \tag{3.51}$$

Thus, the resolving power R_s of the Fabry–Pérot resonator is the product of $m(= 2d/\lambda)$ and F, where m can be thought of as the reflector spacing measured in terms of half-wavelengths, and the finesse F increases with the reflectance R of the reflector. The resolving power R_s is known as the quality value Q of the cavity in the field of electrical engineering. In practice, m is typically of the order of 10^3, F is of the order of 10^2, and R_s is of the order of 10^5. Thus, the scanning Fabry–Pérot resonator has the capability of determining the wavelength of infrared light with a resolution better than 0.01 nm.

The analysis thus far has been idealized in that factors such as flatness, parallelism, or finiteness of the reflectors have been ignored. All of these factors decrease the value of the finesse. Among the neglected factors, the most important factor is the flatness. The effective finesse F_t is approximated as

$$\frac{1}{F_t^2} = \frac{1}{F_r^2} + \frac{1}{F_d^2} \tag{3.52}$$

where F_r is the reflection finesse and F_d is the flatness finesse. These finesses are defined as

$$F_r = \pi\frac{\sqrt{R}}{1-R} \tag{3.53}$$

$$F_d = \frac{S}{2} \tag{3.54}$$

where the flatness S of the surface of the reflector is defined as the inverse of the depth of deviation d_s in terms of the wavelength λ.

$$S = \frac{1}{d_s/\lambda} \qquad (3.55)$$

Example 3.6   The tunable fiber Fabry–Pérot filter [13] shown in Fig. 3.16 is used in an application that requires a FSR of 30 nm in wavelength and a finesse F_t of 100 at a wavelength of $\lambda = 1.55$ μm. Design the spacing between the fiber ends and the coated reflectivity. The flatness finesse is $F_d = 400$.

Solution

$$\Delta\lambda_{FSR} = \frac{\lambda^2}{2nd}$$

With $n_2 = 1$, the spacing is

$$d = \frac{\lambda^2}{2\Delta\lambda_{FSR}} = \frac{1.55^2}{2(0.03)} = 40 \text{ μm}$$

From Eq. (3.52), the effective finesse is

$$\frac{1}{F_t^2} = \frac{1}{F_r^2} + \frac{1}{F_d^2}$$

Solving for F_r gives

$$F_r^2 = \frac{F_d^2 F_t^2}{F_d^2 - F_t^2}$$

With $F_d = 400$ and $F_t = 100$, the value of F_r is

$$F_r = \sqrt{\frac{(400)^2(100)^2}{400^2 - 100^2}} = 103.3$$

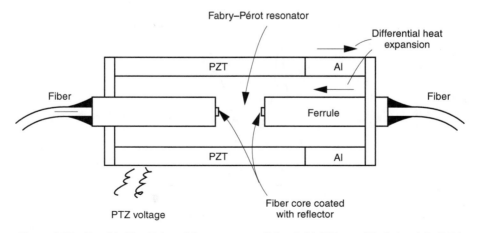

Figure 3.16   Tunable fiber Fabry–Pérot resonator. (After C. M. Miller and F. J. Janniello [13].)

From Eq. (3.53),

$$F_r = \frac{\pi\sqrt{R}}{1-R}$$

and solving for $\sqrt{R}$ gives

$$\sqrt{R} = \frac{-\pi \pm \sqrt{\pi^2 + 4F_r^2}}{2F_r}$$

$$= \frac{-\pi \pm \sqrt{\pi^2 + 4(103.3^2)}}{2(103.3)}$$

$$\sqrt{R} = 0.985$$

$$R = 0.970 \qquad\qquad \square$$

Example 3.7   A scanning Fabry–Pérot resonator is used to obtain the output shown in Fig. 3.17, which is the spectrum from an erbium-doped fiber amplifier (EDFA). Determine the Fabry–Pérot spectrometer's value of finesse and the spacing d of the reflectors required to produce such a display. Assume a free spectral range of 110 nm in order to cover the entire tail of the spontaneous emission spectrum.

Solution   The spacing d of the reflectors is found from Eq. (3.36):

$$d = \frac{\lambda^2}{2\Delta\lambda_{\text{FSR}}} = \frac{1.55^2}{2(0.11)} = 10.92 \ \mu\text{m}$$

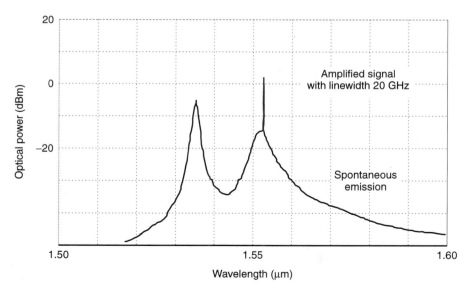

Figure 3.17   Spectrum of the output from an optical EDFA.

The mode number from Eq. (3.33) is

$$m = \frac{2d}{\lambda} = \frac{2(10.92)}{1.55} = 14$$

The linewidth of the amplified signal is 20 GHz, and this determines the required resolving power of the Fabry–Pérot resonator.

The needed resolving power is

$$R_s = \frac{v}{\Delta v}$$

$$v = \frac{c}{\lambda} = \frac{3 \times 10^{14}}{1.55} = 1.94 \times 10^{14} \text{ Hz}$$

$$R_s = \frac{1.94 \times 10^{14}}{20 \times 10^9} = 0.97 \times 10^4$$

From Eq. (3.51), the finesse is

$$F = \frac{R_s}{m} = \frac{0.97 \times 10^4}{14} = 693 \qquad \square$$

Example 3.8 Figure 3.18 shows a schematic of an erbium-doped fiber amplifier (EDFA). Both the signal light with wavelength $\lambda_s = 1.554 \pm 0.020$ μm and the pump light with wavelength $\lambda_p = 1.480$ μm have to be present in the erbium-doped core of the fiber, and the pump light has to be removed from the amplified light after the amplification. The Fabry–Pérot dichroic filter whose transmission specification is shown in Fig. 3.19 is to be designed. It transmits most of λ_s and is highly reflective at λ_p. In all the calculations, assume $n_2 = 1$, $\theta = 45°$, and $A = 0$.

(a) Find the reflector spacing of the Fabry–Pérot resonator.

(b) What is the reflectance R of the reflector if the full width at half maximum (FWHM) of the signal light is 40 nm?

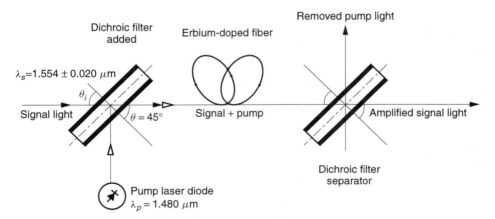

Figure 3.18 Dichroic filter used for an erbium-doped fiber amplifier (EDFA). Passband $\lambda_s = 1.554 \pm 0.020$ μm and stopband $\lambda_p = 1.480$ μm.

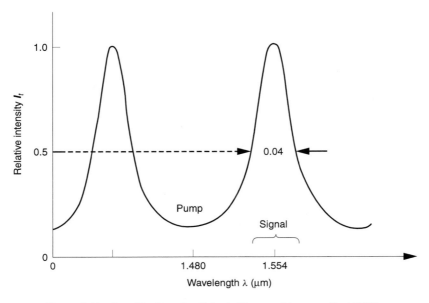

Figure 3.19   Specification of a dichroic filter used for an optical EDFA.

(c) What is the reflectance I_r/I_0 of the dichroic filter at the wavelength of the pump light?

(d) What happens if two of the same filters are staggered? Assume that there are no multiple reflections between the filters.

Solution

(a) From Eqs. (3.14) and (3.32)

$$d \cos \theta = m \frac{\lambda}{2}$$

At the passband, this condition is

$$d \cos \theta = m \frac{\lambda_s}{2}$$

and at the stop band,

$$d \cos \theta = (m + 0.5) \frac{\lambda_p}{2}$$

With $\lambda_s = 1.554$ μm, $\lambda_p = 1.48$ μm, and $\theta = 45°$, the value of m is

$$m = 10.0$$

and the spacing is

$$d = \frac{m \lambda_s}{2 \cos \theta}$$

$$= \frac{(10)(1.554)}{2/\sqrt{2}} = 10.918 \text{ μm}$$

(b) From Eqs. (3.50) and (3.51), the resolving power is

$$R_s = \frac{\lambda_s}{\Delta\lambda_{1/2}} = mF$$

$$\frac{1.554}{0.04} = 10F$$

$$F = 3.89.$$

From Eq. (3.49),

$$F = \frac{\pi\sqrt{R}}{1 - R}$$

$$R = 0.455$$

(c) From Eq. (3.29) with $A = 0$ and $\sin(\phi/2) = 1$, the reflected intensity is

$$\frac{I_r}{I_0} = \frac{4R}{(1 - R)^2 + 4R}$$

$$= \frac{4(0.455)}{(1 - 0.455)^2 + 4(0.455)} = 0.86$$

or

$$\frac{I_t}{I_0} = 0.14$$

(d) The maximum transmittance remains unity. The bandwidth of the passband becomes narrower. The minimum transmittance at the pump light becomes $(0.14)^2 = 0.02$. $\qquad\square$

3.4 PRACTICAL ASPECTS OF OPERATING THE FABRY–PÉROT INTERFEROMETER

This section describes practical techniques for operating the Fabry–Pérot interferometer in the laboratory.

3.4.1 Methods for Parallel Alignment of the Reflectors

The perpendicularity of mirror M_1 with respect to the incident beam can be examined by placing a card with a pinhole as shown in Fig. 3.20. If the laser beam reflected from M_1 goes through the pinhole, M_1 has been adjusted properly.

Mirror M_2 can be adjusted by projecting the transmitted beam onto a distant screen. When the mirror M_2 is aligned, the projected beam converges into a single spot.

Next, Figure 3.21 illustrates a method for fine adjusting a large-diameter reflector. Diffused visible incoherent light is used as the light source. A human eye probes the parallelism of the reflectors. The angles of the emergent light are discrete, as defined by

$$\cos\theta_m = m\frac{\lambda}{2d} \qquad (3.56)$$

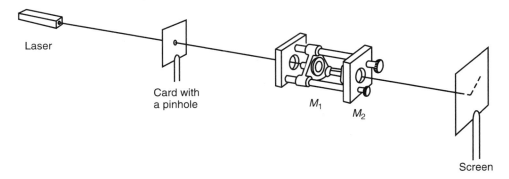

Figure 3.20 A method for aligning the Fabry–Pérot cavity.

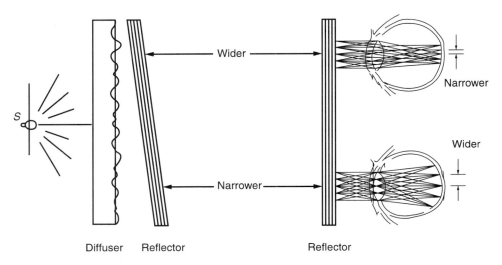

Figure 3.21 A method for the fine adjustment of a pair of large-diameter Fabry–Pérot reflectors.

which is derived from Eqs. (3.14) and (3.32). The discreteness of the angles is also explained in Fig. 3.10 and Example 3.4. The human eye's lens performs the same function as in Fig. 3.10. When the eye is focused on a far away object, an interference fringe is formed on the retina of the eye.

According to Eq. (3.56), regions with a wider reflector spacing project the interference fringe pattern with a narrower period. On the other hand, regions with a narrower reflector spacing, project patterns with a wider period. Parallelism of the reflectors is accomplished by scanning the eye across the reflector and adjusting the tilt of the reflector so that the period of the fringes remains constant.

Next, a method for aligning gas laser mirrors [14] is explained. With gas laser mirrors, not only the mirror parallelism, but also the perpendicularity of the gas tube axis to the mirror surface has to be achieved.

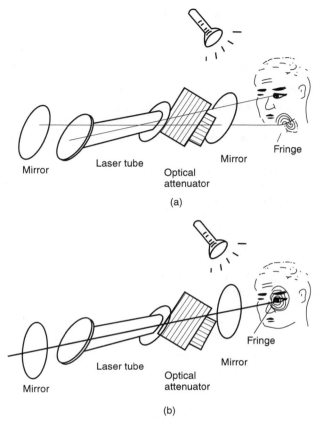

Figure 3.22 A method for aligning laser mirrors. (a) The beam is perpendicular to both mirrors but not collinear to the tube axis. (b) The beam is perpendicular to both mirrors and collinear to the tube axis.

As shown in Fig. 3.22, an optical attenuator is inserted to control the light intensity and protect the eyes from inadvertent laser action. Let us assume that the parallelism of the mirrors has already been established using the previous method, and that interference fringes are already observable in the emergent light.

Figure 3.22a shows a state where the mirrors are parallel but the mirror surface is not perpendicular to the axis of the laser tube. The eye is positioned so that the glow inside the laser tube can be seen cylindrically symmetric. According to Eq. (3.56), the radii of the interference rings are very small for a long laser tube. If the mirror surface is not perpendicular to the laser tube, the interference fringe pattern will be projected onto the observer's face away from the eye. The parallelism of the mirrors is readjusted without moving the position of the eye so that the center of the concentric interference fringe comes to the center of the eyeball. The observer judges whether or not the interference fringe comes to the eyeball by looking at the image of his/her own eye reflected from the end mirror of the laser tube. If the image of the eye is too dark, then the eye is illuminated by a lamp. Figure 3.22b shows the geometry of the well-adjusted mirrors and laser tube.

3.4.2 Method for Determining the Spacing Between the Reflectors

The true spacing between the mirrors can be difficult to measure if the mirrors are extremely close together or if the mirror surfaces are not readily accessible because of the support structure. When the mirrors have to be set within a few microns of each other, utmost care has to be exercised because the mirrors are easily damaged when they touch each other.

Two methods are suggested here. In one method, the same incident beam is measured with two different angles, and in the other method, the same incident beam is measured with two different mirror spacings avoiding the necessity of the absolute measurement of d. The former is recommended when the mirrors have to be set close to each other and is explained in this section. The latter is explained in the next section. Measurement by the different angles technique has been divided into five steps as follows:

1. Using visible light, the parallelism of the mirrors is first adjusted by the method shown in Fig. 3.21. The visible laser beam and the Fabry–Pérot resonator, whose mirror spacing is to be measured, are placed about 1 meter apart, as shown in Fig. 3.23.

2. The entire body of the Fabry–Pérot resonator is tilted such that the back reflected beam comes about one beam diameter away from the laser aperture. This location of the spot is designated as point P_1.

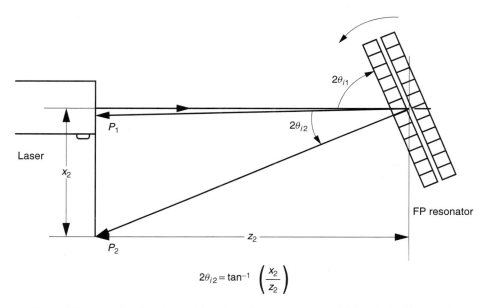

$$2\theta_{i2} = \tan^{-1}\left(\frac{x_2}{z_2}\right)$$

Figure 3.23 A method for determining the reflector spacing by shifting the incident angle.

3. The spacing of the mirrors is adjusted so that the intensity of the spot at P_1 changes to a dark spot. The mirror spacing at the mth resonance condition is, from Eq. (3.14),

$$\frac{4\pi d \cos \theta_1}{\lambda} = 2\pi m \tag{3.57}$$

where θ_1 is the internal angle of the laser beam in the Fabry–Pérot resonator mirror and $n_2 = 1$ is assumed. (θ_1 can be approximated by the external incident angle θ_{i1}.)

4. The entire body of the Fabry–Pérot resonator is then tilted further. Not only the location of the spot shifts, but also the intensity of the spot changes, first increasing to a maximum, then decreasing and becoming dark again. The location of such a dark spot is noted as P_2, which tells us the internal angle θ_2 of the next higher order resonance mode and

$$\frac{4\pi d \cos \theta_2}{\lambda} = 2\pi (m + 1) \tag{3.58}$$

5. From these two angles, d can be calculated as

$$d = \frac{\lambda}{2(\cos \theta_2 - \cos \theta_1)} \tag{3.59}$$

3.4.3 Spectral Measurements Without Absolute Measurement of d

In this section, another method for measuring the spectrum by a scanning Fabry–Pérot resonator without a knowledge of the absolute value of the spacing of the reflectors will be explained [15].

Measurements are repeated with the same input light but with two different reflector spacings, which are translated by a known amount.

Let us assume that the incident light consists of two spectra, a main peak at λ and an auxiliary peak at $\lambda_a = \lambda + \Delta\lambda$. As explained earlier using Fig. 3.7c, the output display of the scanning Fabry–Pérot resonator would be something like the one shown in the encircled areas in Fig. 3.24. Figure 3.24a corresponds to a reflector spacing of $d = d_1$ while the spacing in Fig. 3.24b is $d = d_2$. These two displays have sufficient information to determine the spectrum, without absolute measurement of the reflector spacing.

The reflector spacing d is electronically scanned using a PZT drive assembly in the vicinity of $d = d_1$ in Fig. 3.24a and $d = d_2$ in Fig. 3.24b. Referring to Fig. 3.24a, as d is scanned, the output peaks appear in the following sequence. The (m_1)th order mode of λ appears first, at $d = d_1$, then that of λ_a appears at $d = d_1 + \Delta d_1$, and finally that of the $(m_1 + 1)$th order mode of λ appears at $d = d_1 + \lambda/2$. These three peaks are encircled at the bottom of the figure. The expressions describing the locations of these three peaks are, from Eq. (3.33),

$$d_1 = \frac{\lambda}{2} m_1 \tag{3.60}$$

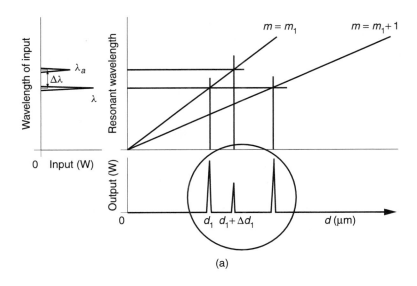

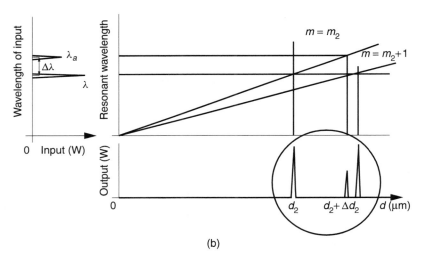

Figure 3.24 Measuring spectrum without absolute measurement of d. Operation of the Fabry–Pérot resonator (a) near $d = d_1$ and (b) near $d = d_2$.

$$d_1 + \Delta d_1 = \frac{(\lambda + \Delta\lambda)}{2} m_1 \tag{3.61}$$

$$d_1 + \frac{\lambda}{2} = \frac{\lambda}{2}(m_1 + 1) \tag{3.62}$$

Having recorded the spectra at $d = d_1$, the reflector is now shifted manually using the adjusting screw on the stationary plate (shown at the left in Fig. 3.5) to a new location $d = d_2$. The new output display resulting from this operation is illustrated in Fig. 3.24b and the expressions describing the new locations of the three peaks are

$$d_2 = \frac{\lambda}{2} m_2 \tag{3.63}$$

$$d_2 + \Delta d_2 = \frac{(\lambda + \Delta\lambda)}{2} m_2 \tag{3.64}$$

$$d_2 + \frac{\lambda}{2} = \frac{\lambda}{2}(m_2 + 1) \tag{3.65}$$

The combination of the two sets of measurements provides the final result. From Eqs. (3.60) and (3.61), Δd_1 is

$$\Delta d_1 = m_1 \frac{\Delta\lambda}{2} \tag{3.66}$$

and from Eqs. (3.63) and (3.64),

$$\Delta d_2 = m_2 \frac{\Delta\lambda}{2} \tag{3.67}$$

Equations (3.66) and (3.67) give

$$\Delta d_2 - \Delta d_1 = \frac{\Delta\lambda}{2}(m_2 - m_1) \tag{3.68}$$

From Eqs. (3.60) and (3.63), the difference $d_2 - d_1$ is

$$d_2 - d_1 = \frac{\lambda}{2}(m_2 - m_1) \tag{3.69}$$

From Eqs. (3.68) and (3.69), the final result is

$$\frac{\Delta\lambda}{\lambda} = \frac{\Delta d_2 - \Delta d_1}{d_2 - d_1} \tag{3.70}$$

The denominator on the right-hand side of Eq. (3.70) is the manual shift by the adjustment screw on the stationary plate, which can be read from the vernier scale, and the numerator is the amount of electronic sweep by the PZT drive assembly.

The wavelength λ of the main spectrum is obtained from the period $d = \lambda/2$ of repetition of the pattern of the main peaks.

3.5 THE GAUSSIAN BEAM AS A SOLUTION OF THE WAVE EQUATION

In geometrical optics, a straight line represents the propagation direction of the wavefront. If the lightwave is indeed a plane wave, both the **E** and **H** fields have to extend to infinity, or if the lightwave is a spherical wave, its wavefront diverges in all directions. These mathematical expressions, although adequate approximations in many situations, are not physically realizable. The Gaussian beam is a more realistic approximation from the viewpoints of both wave and ray optics. The energy of the Gaussian beam is confined within the vicinity of a straight line.

The wavefront of the Gaussian beam is unique in that it behaves like a plane wave in the vicinity of the beam waist, but it gradually converts into a spherical wave as the distance from the waist increases. The beam energy, however, is always confined within a finite divergence angle. The nature of Gaussian beam propagation in free space, as well as inside a cavity, will be explored. The treatment of the propagation in the cavity is especially important because the field inside a Fabry–Pérot resonator with spherically curved reflectors generally has a Gaussian distribution.

3.5.1 Fundamental Mode

Let the expression for a Gaussian beam be [16]

$$u = \psi(x, y, z)e^{jkz} \tag{3.71}$$

The amplitude function $\psi(x, y, z)$ can be a complex function. The Gaussian beam has to satisfy the wave equation:

$$\nabla^2 u + k^2 u = 0 \tag{3.72}$$

By inserting Eq. (3.71) into (3.72), $\psi(x, y, z)$ has to satisfy

$$\frac{\partial^2 \psi}{\partial x^2} + \frac{\partial^2 \psi}{\partial y^2} + j2k\frac{\partial \psi}{\partial z} = 0 \tag{3.73}$$

where the variation of the amplitude function ψ with respect to z was assumed slow, and

$$\frac{\partial^2 \psi}{\partial z^2} = 0 \tag{3.74}$$

ψ can be obtained by solving Eq. (3.73). The function

$$\psi = Ae^{j[P(z)+k(x^2+y^2)/2q(z)]} \tag{3.75}$$

is tried as a solution of Eq. (3.73). The factor A is assumed constant. This solution is called the fundamental mode. The solutions where A is a function of x and y are called higher order modes and are presented in Section 3.7. The quantity $q(z)$ is a complex number referred to as the *q parameter*. The q parameter controls both the phase and the amplitude distribution in the transversal plane, and it plays an important role in describing the type of field. Insertion of Eq. (3.75) into (3.73) gives

$$2k\left(\frac{j}{q} - \frac{\partial P}{\partial z}\right) + \left(\frac{k}{q}\right)^2\left(\frac{\partial q}{\partial z} - 1\right)(x^2 + y^2) = 0 \tag{3.76}$$

The condition that Eq. (3.76) has to be satisfied at any point (x, y, z) gives

$$\frac{\partial q}{\partial z} = 1 \tag{3.77}$$

$$\frac{j}{q} = \frac{\partial P}{\partial z} \tag{3.78}$$

From Eq. (3.77), q satisfies

$$q(z) = z + c \tag{3.79}$$

where c is a constant.

Let

$$q = q_0 \quad \text{at } z = 0$$

Equation (3.79) is written as

$$q(z) = z + q_0 \tag{3.80}$$

Before going further, the nature of the q parameter will be examined. The constant q_0 is expressed explicitly in terms of real and imaginary parts by the real numbers s and t as

$$q_0 = s + jt \tag{3.81}$$

$$q(z) = s + z + jt \tag{3.82}$$

which leads to

$$\frac{1}{q(z)} = \frac{1}{R(z)} - j\frac{1}{Q(z)} \tag{3.83}$$

where

$$\frac{1}{R(z)} = \frac{s+z}{(s+z)^2 + t^2} \tag{3.84}$$

$$\frac{1}{Q(z)} = \frac{t}{(s+z)^2 + t^2} \tag{3.85}$$

ψ is obtained by inserting Eq. (3.83) into (3.75), and then u is obtained from Eq. (3.71) as

$$u = Ae^{\underbrace{jP(z)}_{\substack{\text{Correction} \\ \text{factor}}} + \underbrace{jkz + jk(x^2+y^2)/2R(z)}_{\substack{\text{Parabolic phase} \\ \text{front}}} + \underbrace{k(x^2+y^2)/2Q(z)}_{\text{Amplitude}}} \tag{3.86}$$

3.5.2 Properties of the q Parameter

A single complex number $1/q(z)$ specifies both the phase and amplitude distribution of the Gaussian beam. The real part specifies the phase and the imaginary part specifies the amplitude. It is just like $\gamma = \beta + j\alpha$ in the case of a plane wave (see Example 2.5). Quantities obtainable from $q(z)$ will be summarized.

3.5.2.1 Beam Waist

If the radius of the beam $W(z)$ is defined as the distance $r = \sqrt{x^2 + y^2}$ from the beam axis where the light amplitude u decays to e^{-1} of that on the axis, then from Eq. (3.86)

$$k\frac{W^2(z)}{2Q(z)} = -1 \tag{3.87}$$

and with Eq. (3.85),

$$W^2(z) = -2\frac{(s+z)^2 + t^2}{kt} \tag{3.88}$$

From Eq. (3.88), the minimum radius exists at the location where $z = -s$. Thus, the waist W_0 is

$$W_0^2 = -2\frac{t}{k} \tag{3.89}$$

In conclusion, from Eqs. (3.82) and (3.89), the q parameter is related to the waist as

$$\text{Im}\, q(-s) = t = -\frac{k}{2}W_0^2 \tag{3.90}$$

Thus, the imaginary part of $q(z)$ at $z = -s$ provides the size of the waist.

3.5.2.2 Location of the Waist

As obtained above, the location of the waist is where $z = -s$. This is the location where the real part of $q(z)$ given by Eq. (3.82) is zero; that is, where $q(z)$ is a pure imaginary number.

$$\text{Re}\, q(z) = 0$$
$$q(z) = jt \tag{3.91}$$

3.5.2.3 Radius of Curvature of the Wavefront

The radius of curvative $R(z)$ is, from Eq. (3.84),

$$R(z) = (s+z)\left[1 + \left(\frac{t}{s+z}\right)^2\right] \tag{3.92}$$

First, at the location of the waist where $z = -s$, Eq. (3.92) becomes $R(z) = \infty$. Hence, at the location of the waist, the wavefront is more like that of a plane wave and there is no variation in phase in a plane perpendicular to the direction of propagation. For a large $|s+z|$, $R(z)$ is approximated as

$$R(z) = s + z$$

and the phase factor in Eq. (3.86) represents a parabolic or, approximately speaking, a spherical wavefront with a correction factor $P(z)$. Note that $R(z)$ changes its sign at the waist where $s + z = 0$ and

$$
\begin{aligned}
R(z) > 0 \quad &\text{for } s + z > 0 \\
R(z) < 0 \quad &\text{for } s + z < 0
\end{aligned}
\tag{3.93}
$$

In the region of $s + z > 0$, the wavefront is diverging, while in the region of $s + z < 0$, the wavefront is converging. At far distances, $R(z) \to \infty$ and the wavefront starts to resemble that of a plane wave again.

3.5.3 With the Origin at the Waist

If the origin of the z axis is taken at the location of the waist, the expressions become simpler. With all of the following analyses, the origin will be taken at the waist unless otherwise specified.

3.5.3.1 Focal Parameters

If the origin is shifted to the location of the waist, then the new coordinate z' is

$$
z' = z + s
$$

The prime in z will be suppressed. The q parameter in the new coordinates is, from Eqs. (3.82) and (3.90),

$$
q(z) = z - j\frac{k}{2}W_0^2
\tag{3.94}
$$

From Eqs. (3.88) and (3.89), the beam radius or beamwidth at z is

$$
W^2(z) = W_0^2 \left[1 + \left(\frac{2z}{kW_0^2} \right)^2 \right]
\tag{3.95}
$$

From Eqs. (3.90) and (3.92), the radius of curvature at z is

$$
R(z) = z \left[1 + \left(\frac{kW_0^2}{2z} \right)^2 \right]
\tag{3.96}
$$

From Eq. (3.95), the distance $z = z_0$ where the cross-sectional area of the beam becomes twice as much as that of the waist is

$$
z_0 = \frac{k}{2}W_0^2
\tag{3.97}
$$

Such a distance z_0 is called the *focal beam parameter* or the *Rayleigh range*. It is the distance where the cross-sectional area inflates to twice that of the waist and indicates how fast the beam diverges from the waist. The shorter z_0 is, the faster the divergence is; or in other words, the narrower waist diverges faster.

Inserting Eq. (3.97) into Eqs. (3.95) and (3.96) gives

$$W^2(z) = W_0^2 \left[1 + \left(\frac{z}{z_0} \right)^2 \right] \tag{3.98}$$

$$R(z) = z \left[1 + \left(\frac{z_0}{z} \right)^2 \right] \tag{3.99}$$

From Eqs. (3.98) and (3.99), the relationship between $W^2(z)$ and $R(z)$ will be obtained. Equation (3.98) can be rewritten by multiplying by $(z_0/z)^2/(z_0/z)^2$ as

$$W^2(z) = W_0^2 \left[\frac{1 + (z_0/z)^2}{(z_0/z)^2} \right]$$

Inserting Eqs. (3.97) and (3.99) into this equation gives

$$W^2(z) = \frac{4R(z)}{k^2 W_0^2} z \tag{3.100}$$

Equation (3.100) provides the value of $W^2(z)$ from the given values of $R(z)$ and z, or vice versa.

3.5.3.2 Correction Factor

Next, the value of P will be calculated. Inserting Eq. (3.94) into (3.78) gives

$$\frac{\partial P}{\partial z} = \frac{-1}{(k/2)W_0^2(1 + j2z/kW_0^2)}$$

The solution of this differential equation is

$$P = j \ln \left(1 + j \frac{2z}{kW_0^2} \right)$$

The argument in the logarithm is expressed in polar form as

$$P = j \ln \sqrt{1 + \left(\frac{2z}{kW_0^2} \right)^2} - \phi \tag{3.101}$$

where

$$\phi = \tan^{-1} \frac{2z}{kW_0^2} \tag{3.102}$$

With the help of Eq. (3.95), Eq. (3.101) is rewritten as

$$e^{jP} = \frac{W_0}{W(z)} e^{-j\phi} \tag{3.103}$$

3.5.4 Gaussian Beam Expressions

Various expressions and parameters associated with the Gaussian beam will be summarized to facilitate the use of the formulas.

3.5.4.1 Amplitude Distribution

Using Eqs. (3.87) and (3.103), the expression for the amplitude of the Gaussian beam, Eq. (3.86), is rewritten as

$$u(z) = \underbrace{A}_{\text{Constant}} \underbrace{\frac{W_0}{W(z)}}_{\substack{\text{Contraction} \\ \text{ratio}}} \underbrace{e^{-j\phi + jk(z + r^2/2R(z))}}_{\text{Spherical wavefront}} \underbrace{- \, r^2/W^2(z)}_{\substack{\text{Transverse} \\ \text{amplitude} \\ \text{distribution}}} \tag{3.104}$$

where

$$r^2 = x^2 + y^2$$

$$z_0 = \frac{k}{2} W_0^2 \tag{3.97}$$

$$W^2(z) = W_0^2 \left[1 + \left(\frac{z}{z_0} \right)^2 \right] \tag{3.98}$$

$$R(z) = z \left[1 + \left(\frac{z_0}{z} \right)^2 \right] \tag{3.99}$$

$$\phi(z) = \tan^{-1} \left(\frac{2z}{kW_0^2} \right) \tag{3.102}$$

The field distribution is illustrated in Fig. 3.25a.

The first factor in Eq. (3.104) is the amplitude and the second factor is the change in amplitude due to the change in the radius of the beam. The third factor is the phasefront, which approaches that of a plane wave at the waist as well as in the far field. For intermediate distances, the phasefront approaches a spherical wavefront with radius of curvature $R(z)$, but with a correction factor of ϕ. The last factor represents a bell-shaped transverse field distribution.

From Eq. (3.104), we see that only three parameters are needed to specify the Gaussian beam: the size of the beam waist, the distance from the beam waist, and the wavelength of the light.

3.5.4.2 Intensity Distribution

The intensity distribution $I(r, z)$ is given by uu^* and from Eq. (3.104) it is

$$I(r, z) = I_0 \left(\frac{W_0}{W(z)} \right)^2 e^{-2r^2/W^2(z)} \tag{3.105}$$

As shown in Fig. 3.25b, the transverse distribution at a particular value of z is Gaussian. The longitudinal distribution $I(0, z)$ along the beam axis has its maximum at the

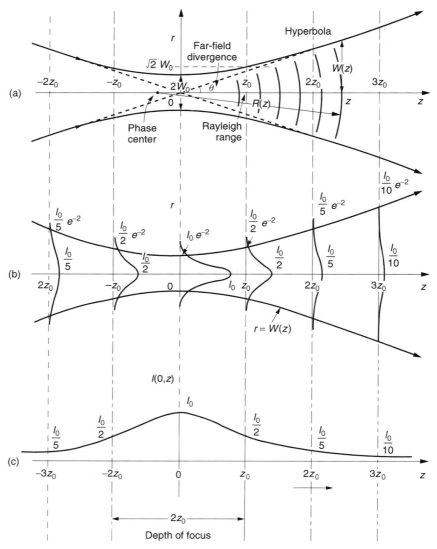

Figure 3.25 Gaussian beam. (a) Amplitude distribution. (b) Cross-sectional intensity distribution and beam radius. (c) Intensity distribution along the beam axis.

waist and decays monotonically as $[W_0/W(z)]^2$, as shown in Fig. 3.25c. The intensity distribution of the Gaussian beam is symmetric with respect to the waist. As soon as the distribution of $I(r, z)$ on one side of the waist is known, that on the other side is obtained by the mirror image.

3.5.4.3 Angle of the Far-Field Divergence
Equation (3.98) can be rewritten as

$$\left(\frac{W(z)}{W_0}\right)^2 - \frac{z^2}{z_0^2} = 1$$

Thus, the beam radius of the Gaussian beam is a hyperbolic function of z. Referring to Fig. 3.25a, the beam expansion at a far distance is

$$\lim_{z \to \infty} \tan \theta = \lim_{z \to \infty} \left(\frac{W(z)}{z} \right)$$

and from Eqs. (3.97) and (3.98),

$$\lim_{z \to \infty} \tan \theta = \frac{2}{kW_0} \tag{3.106}$$

The spread is inversely proportional to the size of the waist W_0. This is similar to the spread of the far-field diffraction pattern of a plane wave through an aperture with radius W_0 (see Section 1.2).

3.5.4.4 Depth of Focus

The region of the waist bordered by $-z_0 < z < z_0$ is called the depth of focus D_0 and, from Eq. (3.97),

$$D_0 = 2z_0 = kW_0^2 \tag{3.107}$$

A smaller spot size means a shorter depth of focus unless the wavelength of the light is shortened.

Example 3.9 The beamwidth and the radius of curvature of a Gaussian beam with wavelength λ were measured at a certain location z. They were R_1 and W_1, respectively.

(a) Find the q parameter.
(b) Find the beamwidth W_2 and the radius of curvature R_2 at a new location a distance d away from the measured location.

Solution The unknowns in this problem are the distance z from the point of observation to the location of the waist and the size W_0 of the waist. These unknowns will be found first.

(a) From Eqs. (3.98) and (3.99)

$$z = \frac{R_1}{1 + (z_0/z)^2}$$

$$W_0^2 = \frac{W_1^2}{1 + (z/z_0)^2}$$

Inserting z from Eq. (3.100) and z_0 from Eq. (3.97) into the denominators of these two equations gives

$$z = \frac{R_1}{1 + (2R_1/kW_1^2)^2}$$

$$W_0^2 = \frac{W_1^2}{1 + (kW_1^2/2R_1)^2}$$

The q parameter is obtained by inserting the above two equations into Eq. (3.94) as

$$q(z) = z - j\frac{k}{2}W_0^2$$

(b) From Eq. (3.79), the q parameter at distance d away is obtained just by adding d in the real part, and therefore $q(z + d)$ is

$$q(z + d) = z + d - j\frac{k}{2}W_0^2$$

The desired $R(z + d)$ and $W(z + d)$ will be found from $q(z + d)$:

$$\frac{1}{q(z + d)} = \frac{1}{R(z + d)} - j\frac{1}{Q(z + d)} \tag{3.108a}$$

$$\frac{1}{R(z + d)} = \frac{z + d}{(z + d)^2 + z_0^2} \tag{3.108b}$$

$$\frac{1}{Q(z + d)} = \frac{-z_0}{(z + d)^2 + z_0^2} \tag{3.108c}$$

where Eqs. (3.89) and (3.97) were used.

From Eq. (3.108b), $R(z + d)$ is obtained as

$$R(z + d) = (z + d)\left[1 + \left(\frac{z_0}{z + d}\right)^2\right]$$

From Eqs. (3.87) and (3.97) and Eq. (3.108c), $W(z + d)$ is obtained as

$$W^2(z + d) = W_0^2\left[1 + \left(\frac{z + d}{z_0}\right)^2\right]$$

It should be noted that the solutions for $R(z + d)$ and $W(z + d)$ are obtained by simply replacing z with $z + d$, once the values of z and W_0 are found in part (a).   □

3.6   TRANSFORMATION OF A GAUSSIAN BEAM BY A LENS

A Gaussian beam propagating in free space cannot have more than one waist (just like a human being). However, if a lens is introduced into the beam, then it is possible to form a new beam waist whose location and size are different from the original. Thus, convex and concave lenses and mirrors are valuable tools for transforming Gaussian beams into desired dimensions. The transformation of the Gaussian beam by a lens is the subject of this section.

3.6.1 Transformation of the q Parameter by a Lens

A thin convex lens with focal length f is placed a distance d_0 to the right of the waist of the incident Gaussian beam as shown in Fig. 3.26. The quantities associated with the incident beam will be denoted by the subscript 0, and those of the emergent beam from the lens, by the subscript 1. The radius and phasefront of the emergent light at an arbitrary distance from the lens, as well as the location and size of the new waist, will be calculated.

The lens not only converges the phasefront but also changes the radius of the beam. In other words, the lens changes both $R(z)$ and $W(z)$, and it is best to deal with the basic parameter q. Not only the lens but the propagation itself changes the q parameter. The process of the change in q will be followed. Let the q parameter at the incident waist be

$$q = q_0 \tag{3.109}$$

At the waist, q_0 is a purely imaginary number and its value is $-j(k/2)W_0^2$ from Eqs. (3.90) and (3.91); but at other points, q is a complex number. After propagating a distance d_0 to the front surface of the lens, from Eq. (3.80), q becomes

$$q_0' = q_0 + d_0 \tag{3.110}$$

Inside the lens, the beam acquires a quadratic phase distribution specified by Eq. (1.139) in Chapter 1, and from Eqs. (1.139) and (3.75) the field immediately after the lens is

$$u = A e^{j[P(d_0) + k(x^2 + y^2)/2q_0' - jk(x^2 + y^2)/2f]} \tag{3.111}$$

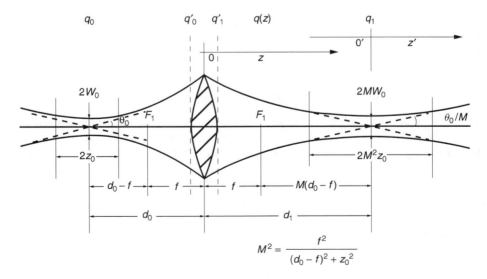

$$M^2 = \frac{f^2}{(d_0 - f)^2 + z_0^2}$$

Figure 3.26 Transformation of Gaussian beam by a lens.

Thus, the q parameter just behind the lens is effectively

$$\frac{1}{q_1'} = \frac{1}{q_0'} - \frac{1}{f} \tag{3.112}$$

After further propagation by a distance z, the q parameter is

$$q(z) = q_1' + z \tag{3.113}$$

where the origin of the z coordinate of the emergent beam is taken at the location of the lens (rather than at the waist).

The rule for calculating the q parameter is that the effect of the propagation is just the addition of the distance, as illustrated in Eqs. (3.110) and (3.113), while the effect of the lens is the addition of the inverse of q and the inverse of the focal length of the lens, as illustrated in Eq. (3.112).

Working backward from Eq. (3.113) to (3.110), the q parameter is expressed in terms of d_0, z, and f as

$$q(z) = \frac{d_0(f - z) + fz + q_0(f - z)}{(f - d_0) - q_0} \tag{3.114}$$

Insert

$$q_0 = -j\frac{k}{2}W_0^2 = -jz_0$$

into Eq. (3.114) to obtain

$$q(z) = \frac{(z - f)[(d_0 - f)^2 + z_0^2] - f^2(d_0 - f) - jf^2z_0}{(d_0 - f)^2 + z_0^2} \tag{3.115}$$

From the real and imaginary parts of $q(z)$, the size and location of the waist will be found.

3.6.2 Size of the Waist of the Emergent Beam

The size of the waist of the emergent beam will be found. The analysis makes use of the fact that the q parameter becomes a pure imaginary number at the waist. As Eq. (3.89) was obtained, the size of the waist W_1 is found from the imaginary part of Eq. (3.115) at $z = d_1$:

$$W_1^2 = -\frac{2}{k}\,\mathrm{Im}\,q(d_1)$$

Inserting Eq. (3.97) into the above equation, the ratio M of the new waist to the old waist is

$$M = \left(\frac{W_1}{W_0}\right) = \frac{f}{\sqrt{(d_0 - f)^2 + (z_0)^2}} \tag{3.116}$$

The curve for M is plotted in Fig. 3.27a with z_0/f as a parameter.

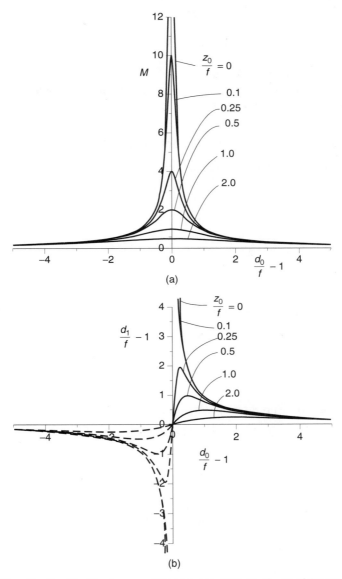

Figure 3.27 Magnification M of the beam waist and the location ($d_1/f - 1$) of the emergent waist as a function of the location ($d_0/f - 1$) of the input waist. (a) Magnification of the waist. (b) Location of the emergent waist.

3.6.3 Location of the Waist of the Emergent Beam

From the condition in Eq. (3.91), the location d_1 of the waist is found from the location where the real part vanishes as

$$d_1 = f + \frac{f^2(d_0 - f)}{(d_0 - f)^2 + z_0^2} \tag{3.117}$$

Combining Eq. (3.116) with (3.117) gives

$$d_1 - f = M^2(d_0 - f) \tag{3.118}$$

Graphs of $(d_1/f - 1)$ as a function of $(d_0/f - 1)$ are plotted in Fig. 3.27b, again with z_0/f as a parameter. In the negative $(d_0/f - 1)$ region, as shown by the dotted line, there is a virtual image of the waist.

3.6.4   Rayleigh Range of the Emergent Beam

The Rayleigh range z_1 of the emergent beam is given by the size of the waist W_1 of the emergent beam and

$$z_1 = \frac{k}{2} W_1^2$$

From Eqs. (3.97) and (3.116), the Rayleigh range is

$$z_1 = M^2 z_0 \tag{3.119}$$

The depth of focus D_1 of the emergent beam, from Eqs. (3.107) and (3.119), is therefore

$$D_1 = M^2 D_0 \tag{3.120}$$

Both z_0 and D_0 increase by a factor of M^2.

3.6.5   Angle of the Far-Field Divergence of the Emergent Beam

The angle of the far-field divergence is obtained from

$$\lim_{z \to \infty} \tan \theta_1 = \lim_{z \to \infty} \left(\frac{W_1(z)}{z - d_1} \right)$$

For a small angle θ_1, the following approximation holds:

$$\theta_1 \doteq \frac{2}{kW_1} \quad \text{rad}$$

and, hence,

$$\theta_1 = \frac{\theta_0}{M} \quad \text{rad} \tag{3.121}$$

3.6.6   Comparison with Ray Optics

In regions far away from the focal depth, the image formed by a Gaussian beam becomes closer to that formed by ray optics. As $(d_0 - f)$ becomes much larger than z_0, the expression for the location of the Gaussian beam waist, Eq. (3.117), approaches

$$f^2 = (d_0 - f)(d_1 - f) \tag{3.122}$$

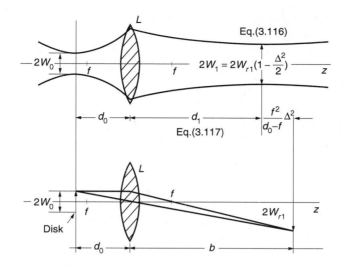

Figure 3.28   Gaussian beam output waist $2W_1$ and ray optics image $2W_{r1}$.

which is identical to Gauss's equation of imagining:

$$\frac{1}{f} = \frac{1}{d_0} + \frac{1}{d_1}.$$

As for the magnification, Gauss's expression for the magnification is

$$\beta = \frac{f}{d_0 - f} \tag{3.123}$$

and Eq. (3.116) also approaches Eq. (3.123) as

$$\Delta = \frac{z_0}{d_0 - f}$$

becomes smaller than unity.

More quantitatively, the case when the input waist is replaced by a circular disk object is calculated using Gauss's ray imaging formula and is compared in Fig. 3.28. The distance d_1 of the output waist of the Gaussian beam is shorter than b of the output disk image by

$$b - d_1 = \frac{f^2}{d_0 - f} \Delta^2$$

The dimension W_1 of the waist given by Eq. (3.116) is smaller than that of the disk approximately by a factor of $(1 - \Delta^2/2)$.

3.6.7   Summary of the Equations of the Transformation by a Lens

Equations describing the transformation are tabulated in Table 3.2. The associated figures are shown in Figs. 3.26 and 3.27.

Table 3.2   Gaussian beam transformation by a lens

Parameter	Input Beam	Output Beam	Equation Number
Beam waist	W_0	$W_{10} = MW_0$	(3.116)
Waist location	$d_0 - f$	$d_1 - f = M^2(d_0 - f)$	(3.118)
Rayleigh range	$z_0 = \dfrac{k}{2}W_0^2$	$z_1 = M^2 z_0 = \dfrac{k}{2}W_{10}^2$	(3.119)
Depth of focus	$D_0 = 2z_0$	$D_1 = M^2 D_0$	(3.120)
Angle of far-field divergence	$\theta_0 = \tan^{-1}\left(\dfrac{2}{kW_0}\right)$	$\theta_1 = \dfrac{\theta_0}{M}$	(3.121)
Magnification	1	$M^2 = \dfrac{f^2}{(d_0 - f)^2 + z_0^2}$	(3.116)
		$= \dfrac{d_1 - f}{d_0 - f}$	(3.118)
		$= \dfrac{\beta^2}{1 + \Delta^2}$	

where

$$\beta = \frac{f}{d_0 - f} \qquad (3.123)$$

$$\Delta = \frac{z_0}{d_0 - f}$$

If we shift the origin of the z coordinate from the lens to the emergent beam waist and name the new coordinate z', Eqs. (3.98) and (3.99) for the beam radius and the radius of curvature for the input beam are converted into those for the emergent beam and expressed as

$$W_1^2(z') = W_{10}^2\left[1 + \left(\frac{z'}{z_1}\right)^2\right]$$
$$R_1(z') = z'\left[1 + \left(\frac{z_1}{z'}\right)^2\right] \qquad (3.124)$$

where W_{10} is the waist of the emergent beam.

3.6.8   Beam Propagation Factor m^2

Focused laser beams with small spot sizes are utilized for such applications as reading and writing a digital video disk (see Section 2.10.7), laser printer heads, and drilling holes in a stainless steel sheet. Figure 3.29 shows a configuration for reducing the waist (spot size) of a laser beam [17]. A large-diameter Gaussian beam waist is incident onto the surface of a lens as shown in Fig. 3.29. The wavefront is parallel to the lens surface and the light intensity along the lens surface decays as the edge of the lens is approached. This bell-shaped distribution contributes to the apodizing effect (see Section 1.4.2) and no harmful side lobes appear in the focused light spot. The location

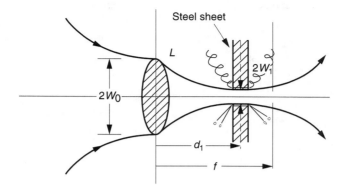

Figure 3.29   Gaussian beam drill spot by a lens. (Input waist is on the surface of the lens).

d_1 of the emergent waist is obtained from Eq. (3.117) with $d_0 = 0$.

$$d_1 = f \frac{1}{1 + (f/z_0)^2} \qquad (3.125)$$

The waist appears at a location shorter than the focal length of the lens and the size of the waist is, from Eq. (3.116),

$$\left(\frac{W_1}{W_0}\right)^2 = \frac{1}{1 + (z_0/f)^2}$$

When a short focal length lens with $z_0 \gg f$ is used with Eq. (3.97), the output spot size diameter $2W_1$ approximately becomes

$$2W_1 = \frac{4\lambda f}{\pi D} \qquad (3.126)$$

where the input waist $2W_0$ is chosen as wide as the lens diameter D. Thus, a smaller laser beam spot is obtained if a shorter wavelength, or a shorter focal length, or a larger diameter lens is selected.

If a lens with a diameter of $D = 4f/\pi$ is used, a spot size of one wavelength should be obtainable. In practice, however, the input laser beam is not a perfect Gaussian beam. This imperfection has been shown to increase the spot size, and in actual situations, the spot size becomes m times as large as $2W_1$, and the Rayleigh range becomes

$$z_0' = m^2 z_0 \qquad (3.127)$$

The factor m^2 is called the m^2 *beam propagation factor* and is used for expressing the quality of the incident laser beam. As a matter of fact, when the incident light is a quasi-Gaussian beam, approximate values of $W_1(z')$ and $R_1(z')$ in Eq. (3.124) are calculated by simply replacing z_1 by $m^2 z_1$ [18].

Example 3.10   Write out a sequence of equations (just equations) to transfer the q parameter through a convex lens followed by a concave lens whose focal lengths are f_1 and f_2 and with spacing among the source, lenses, and observation point as shown in Fig. 3.30.

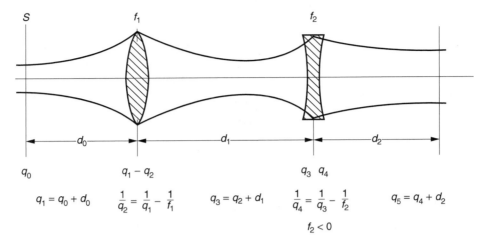

Figure 3.30   Transformation of q parameters; f_1 is a positive quantity, and f_2 is a negative quantity.

Solution   Starting from point S, the sequence of equations for the q parameter is

$$q = q_0$$

$$q_1 = q_0 + d_0$$

$$\frac{1}{q_2} = \frac{1}{q_1} - \frac{1}{f_1}$$

$$q_3 = q_2 + d_1$$

$$\frac{1}{q_4} = \frac{1}{q_3} - \frac{1}{f_2}, \quad \text{where } f_2 \text{ is a negative quantity}$$

$$q_5 = q_4 + d_2 \qquad\qquad\qquad\qquad\qquad\qquad \square$$

Example 3.11   A Gaussian beam with the following parameters is incident onto a convex lens. Find the location and size of the waist of the emergent beam.

$$f = 10 \text{ cm}$$

$$d_0 = 20 \text{ cm}$$

$$W_0 = 0.1 \text{ mm}$$

$$\lambda = 0.63 \text{ μm}$$

Solution   From Eq. (3.117), the location of the emergent beam waist is calculated as

$$d_1 - f = \frac{100^2(200 - 100)}{(200 - 100)^2 + (\pi/0.63 \times 10^{-3})^2(0.1)^4}$$

$$= 80 \text{ mm}$$

$$d_1 = 18 \text{ cm}$$

Note that the contribution of $[(k/2)W_0^2]^2$ is not negligible and, but for this, $d_1 = 20$ cm. From Eqs. (3.116) and (3.118), the emergent beam waist is

$$W_1 = 0.1\sqrt{\frac{8}{10}} = 0.09 \text{ mm}$$ □

3.7 HERMITE GAUSSIAN BEAM (HIGHER ORDER MODES)

So far, the value of A in the solution

$$\psi = Ae^{j[P+k(x^2+y^2)/2q)]} \tag{3.75}$$

was assumed a constant, but this is not the only solution. A more general solution ψ will now be obtained by assuming A is a function of x, y, and z [19]. The derivation of the solution is slightly lengthy, but many interesting tricks are used to find the solution, and the derivation is well worth following. Let

$$\Psi = g\left(\frac{x}{W}\right) h\left(\frac{y}{W}\right) \psi(x, y, z)\, e^{j\Gamma(z)} \tag{3.128}$$

where $\psi(x, y, z)$ is the solution that has already been obtained with A equal to a constant, given by Eq. (3.75). A was replaced by $g(x/W)h(y/W)e^{j\Gamma(z)}$, where $W(z)$ is presented here without the argument z. The functions $g(x/W)$ and $h(y/W)$ that enable Eq. (3.128) to satisfy

$$\frac{\partial^2 \Psi}{\partial x^2} + \frac{\partial^2 \Psi}{\partial y^2} + j2k\frac{\partial \Psi}{\partial z} = 0$$

will be found.

The partial derivatives of Ψ are

$$\frac{\partial^2 \Psi}{\partial x^2} = \left(\frac{1}{W^2}g''h\psi + 2j\frac{kx}{Wq}g'h\psi + gh\frac{\partial^2 \psi}{\partial x^2}\right)e^{j\Gamma} \tag{3.129}$$

$$\frac{\partial^2 \Psi}{\partial y^2} = \left(\frac{1}{W^2}gh''\psi + 2j\frac{ky}{Wq}gh'\psi + gh\frac{\partial^2 \psi}{\partial y^2}\right)e^{j\Gamma} \tag{3.130}$$

$$j2k\frac{\partial \Psi}{\partial z} = \left[j2kx\frac{\partial}{\partial z}\left(\frac{1}{W}\right)g'h\psi + j2ky\frac{\partial}{\partial z}\left(\frac{1}{W}\right)gh'\psi\right.$$
$$\left. - 2gh\psi k\frac{\partial \Gamma}{\partial z} + j2kgh\frac{\partial \psi}{\partial z}\right]e^{j\Gamma} \tag{3.131}$$

where

$$g' = \frac{\partial g}{\partial(x/W)}, \qquad h' = \frac{\partial h}{\partial(y/W)}$$

The sum of the last terms in Eqs. (3.129), (3.130), and (3.131) are zero because ψ itself satisfies Eq. (3.73). The total sum divided by Ψ/W^2 gives

$$\frac{g''}{g} + \frac{h''}{h} + j2\left(x\frac{g'}{g} + y\frac{h'}{h}\right)\left[\frac{kW}{q} + kW^2\frac{\partial}{\partial z}\left(\frac{1}{W}\right)\right] - 2kW^2\frac{\partial\Gamma}{\partial z} = 0 \qquad (3.132)$$

The factor in the third term can be simplified by noting that

$$\frac{\partial}{\partial z}\left(\frac{1}{W}\right) = -\frac{1}{W^2}\frac{\partial W}{\partial z} \qquad (3.133)$$

From Eq. (3.95), $\partial W/\partial z$ is

$$\frac{\partial W}{\partial z} = \frac{1}{W}\frac{4z}{(kW_0)^2} \qquad (3.134)$$

With Eqs. (3.100), (3.133), and (3.134), the last term of the last bracket of Eq. (3.132) becomes

$$kW^2\frac{\partial}{\partial z}\left(\frac{1}{W}\right) = -\frac{kW}{R(z)} \qquad (3.135)$$

With Eqs. (3.83), (3.87), and (3.135), the sum inside the last bracket of Eq. (3.132) becomes a single number $j(2/W)$. Thus, Eq. (3.132) finally becomes

$$\underbrace{\frac{g''}{g} - 4\frac{x}{W}\frac{g'}{g}}_{\text{Function of } x} + \underbrace{\frac{h''}{h} - 4\frac{y}{W}\frac{h'}{h}}_{\text{Function of } y} - \underbrace{2kW^2\frac{\partial\Gamma}{\partial z}}_{\text{Function of } z} = 0 \qquad (3.136)$$

The differential equation, Eq. (3.136), is solved by the method of separation of variables.

The function of x in Eq. (3.136) is

$$\frac{1}{g}\frac{d^2g(\bar{x})}{d\bar{x}^2} - 4\bar{x}\frac{1}{g}\frac{dg(\bar{x})}{d\bar{x}} \qquad (3.137)$$

where

$$\bar{x} = \frac{x}{W}$$

The differential equation of a Hermite polynomial of order n is

$$\frac{\partial^2 H_n}{\partial x^2} - 2x\frac{\partial H_n}{\partial x} + 2nH_n(x) = 0 \qquad (3.138)$$

where n is a positive integer.

Table 3.3 The Hermite polynomials of the nth order

n	$H_n(x)$
0	1
1	$2x$
2	$4x^2 - 2$
3	$8x^3 - 12x$

The expressions of $H_n(x)$ are shown in Table 3.3.

We see that $g(\bar{x})$ almost satisfies the differential equation of the Hermite polynomial except that the coefficient is 4 instead of 2 in the first order derivative. Inserting the change of variables

$$s = \sqrt{2}\,\bar{x}$$

$$\frac{dg}{d\bar{x}} = \sqrt{2}\,\frac{dg}{ds}$$

$$\frac{d^2g}{d\bar{x}^2} = 2\frac{d^2g}{ds^2}$$

into Eq. (3.137) gives

$$\frac{2}{g(s)}\left(\frac{d^2g(s)}{ds^2} - 2s\frac{dg(s)}{ds}\right)$$

The coefficient of the first derivative is 2 and fits for the differential equation of the Hermite polynomial. A similar change of variables is made for $\bar{y} = y/W$.

$$\bar{y} = \frac{y}{W}$$

$$t = \sqrt{2}\,\bar{y}$$

Equation (2.136) now becomes

$$\underbrace{\frac{1}{g}\left(\frac{d^2g}{ds^2} - 2s\frac{dg}{ds}\right)}_{-2n} + \underbrace{\frac{1}{h}\left(\frac{d^2h}{dt^2} - 2t\frac{dh}{dt}\right)}_{-2m} \underbrace{-kW^2\frac{\partial\Gamma}{\partial z}}_{2(m+n)} = 0 \qquad (3.139)$$

The first two terms are functions of s only, the second two terms are functions of t only, and the last term is a function of z only. For Eq. (3.139) to be satisfied everywhere in space, the functions of x, y, and z have to be independently constant: $-2n$, $-2m$, and $2(n+m)$.

$$\frac{d^2g(s)}{ds^2} - 2s\frac{dg(s)}{ds} + 2ng(s) = 0 \qquad (3.140a)$$

$$\frac{d^2h(t)}{dt^2} - 2t\frac{dh(t)}{dt} + 2mh(t) = 0 \tag{3.140b}$$

$$kW^2\frac{\partial \Gamma}{\partial z} = -2(m+n) \tag{3.140c}$$

Hence, from Eq. (3.138),

$$g(s) = H_n\left(\sqrt{2}\frac{x}{W}\right)$$

$$h(t) = H_m\left(\sqrt{2}\frac{y}{W}\right)$$

The value of Γ is found by inserting Eq. (3.95) into Eq. (3.140c):

$$\frac{\partial \Gamma}{\partial z} = -\frac{2(n+m)}{kW_0^2\left[1 + (2z/kW_0^2)^2\right]} \tag{3.141}$$

and

$$\Gamma = -(n+m)\tan^{-1}\left(\frac{2z}{kW_0^2}\right) \tag{3.142}$$

Finally, the expression for a higher order Gaussian beam, which is known as the Hermite Gaussian beam, is

$$\Psi = \frac{W_0}{W}H_m\left(\sqrt{2}\frac{x}{W}\right)H_n\left(\sqrt{2}\frac{y}{W}\right)$$

$$\exp\left\{j\left[\frac{k}{2q}(x^2+y^2) - (m+n+1)\tan^{-1}\left(\frac{2z}{kW_0^2}\right)\right]\right\} \tag{3.143}$$

Figure 3.31 shows measured mode patterns inside a cavity with concave mirrors. The TEM_{00} mode in Fig. 3.31 corresponds to the case when $m = n = 0$, where $H_0(\sqrt{2}x/W) = H_0(\sqrt{2}y/W) = 1$ and Eq. (3.143) reduces exactly to the fundamental mode given by either Eq. (3.75) or (3.104). The cross-section field distribution has its maximum on the z axis and has a Gaussian distribution, such as shown in Fig. 3.25.

The TEM_{11} mode uses the first order Hermite polynomials in both the x and y directions. Using the values in Table 3.3, the front factor of Eq. (3.143) becomes

$$\left(\frac{W_0}{W}\right)\left(2\sqrt{2}\frac{x}{W}\right)\left(2\sqrt{2}\frac{y}{W}\right) = \frac{8W_0}{W^3}xy$$

Thus, the TEM_{11} mode is zero at the origin.

As a matter of fact, the mode number corresponds to the number of zero fields in the transverse plane. Different modes have different field distributions. The operation of the "light tweezers" mentioned later in this chapter critically depends on the field distribution; hence, the proper mode configuration is essential to the operation of the light tweezers.

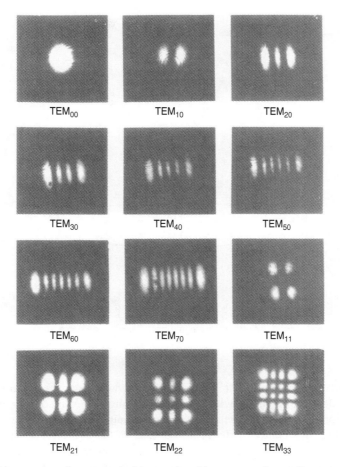

TEM$_{00}$ TEM$_{10}$ TEM$_{20}$

TEM$_{30}$ TEM$_{40}$ TEM$_{50}$

TEM$_{60}$ TEM$_{70}$ TEM$_{11}$

TEM$_{21}$ TEM$_{22}$ TEM$_{33}$

Figure 3.31 Measured mode patterns inside a cavity with concave mirrors. (Courtesy of H. Kogelnik and T. Li [16].)

These different modes have different phase velocities and possess different cavity resonance frequencies. In order to properly interpret the output from a Fabry–Pérot resonator, an understanding of the modes is important.

3.8 THE GAUSSIAN BEAM IN A SPHERICAL MIRROR CAVITY

Since around 1960, spherical mirror cavities have been widely used as Fabry–Pérot interferometers for analyzing the spectrum of light. The sharp frequency dependence of the transmitted light from the cavity is used to find the frequency spectrum of the incident light.

Most of the cavities of gas lasers are made of two concave mirrors facing each other. The actual laser cavities have a gain medium inside the cavity, and the mirrors are not 100% reflective so that some of the laser energy escapes the laser cavity. The characteristics of the laser cavity, however, can be well approximated by those in an

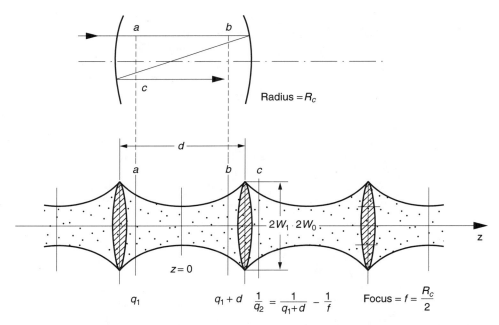

Figure 3.32   Equivalence of the multiple path in a cavity and the path in a lens array with focal length $f = R_c/2$.

ideal air-filled cavity with 100% reflecting, infinitely large concave mirrors. Figure 3.32 shows the structure of a cavity with two identical mirrors. At resonance, the phases of the beam before and after the round trip have to match. This phase match has to take place not only at the center of the beam but also in the entire plane transverse to the laser beam. Such a condition can be realized if the q parameters of the beam before and after the round trip are identical. The resonance condition will be expressed in terms of the radius of curvature and spacing between the mirrors. The repeated reflections by a spheroidal mirror with radius of curvature R_c is equivalent to transmission through an infinite array of lenses with focal length

$$f = \frac{R_c}{2} \tag{3.144}$$

In the present configuration, identical end mirrors are used and the condition after one-half round trip is sufficient.

Referring to Fig. 3.32, let the q parameter immediately after the lens be q_1 and the q parameter immediately after the next lens be q_2. Then the q parameters q_1 and q_2 satisfy

$$\frac{1}{q_2} = \frac{1}{q_1 + d} - \frac{1}{f} \tag{3.145}$$

The resonance condition is

$$q_1 = q_2 = q \tag{3.146}$$

With Eq. (3.146), Eq. (3.145) becomes

$$\frac{df + q(q + d)}{fq(q + d)} = 0 \tag{3.147}$$

Setting the numerator in Eq. (3.147) equal to zero, and dividing by fdq^2, the result is

$$\frac{1}{q^2} + \frac{1}{qf} + \frac{1}{fd} = 0 \tag{3.148}$$

Thus, the required q parameter is

$$\frac{1}{q} = \frac{-1}{2f} \pm j\sqrt{\frac{1}{fd} - \frac{1}{4f^2}} \tag{3.149}$$

The $+j$ term in Eq. (3.149) is not physically acceptable, as the field intensity would grow indefinitely as $(x^2 + y^2)$ increases. The real and imaginary parts of $1/q$ represent $1/R$ and $1/Q$ as in Eq. (3.83). At resonance, from the real part of Eq. (3.149), the radius of curvature of the beam just past the lens or just in front of the mirror is $2f$. Using Eq. (3.144),

$$R = -R_c \tag{3.150}$$

The imaginary part of Eq. (3.149) provides the size of the beam W_1 just in front of the mirror and is, from Eq. (3.87),

$$W_1^2 = \frac{2R_c}{k\sqrt{2R_c/d - 1}} \tag{3.151}$$

Next, let us find the beam waist. Equation (3.100) is the relationship developed taking the origin at the waist, as shown in Fig. 3.32. Assuming that the radius of the beam immediately after the lens is the same as that immediately before the lens, W_1 is expressed as

$$W\left(\frac{d}{2}\right) = W_1$$

The radius of curvature immediately before the lens is the same as that immediately after the lens except for the sign reversal and, from Eq. (3.150),

$$R\left(\frac{d}{2}\right) = R_c$$

Thus, Eq. (3.100) with $z = d/2$ becomes

$$W_1^2 = \frac{4R_c}{k^2 W_0^2} \cdot \frac{d}{2} \tag{3.152}$$

Inserting Eq. (3.151) into (3.152) gives

$$W_0^2 = \frac{d}{k}\sqrt{\frac{2R_c}{d} - 1} \tag{3.153}$$

Let us interpret the results obtained. Equation (3.150) states that, at resonance, the radius of curvature of the wave matches the radius of curvature of the end reflector mirror. In other words, the contour of the constant phase matches the surface of the concave mirror. According to Eq. (3.151), for W_1^2 to be real, there is a range restriction on d for a given R_c:

$$0 < d < 2R_c \tag{3.154}$$

Figure 3.33 shows the field inside a resonant cavity as the spacing d is varied for a fixed radius of curvature R_c. As the spacing is increased, the beam radius W on the surface of the end mirrors expands, whereas the size of the waist in the middle shrinks. As soon as d exceeds $2R_c$, the beam is no longer contained inside the cavity. In this case the cavity becomes unstable, and the resonance no longer exists, as shown in Fig. 3.33e.

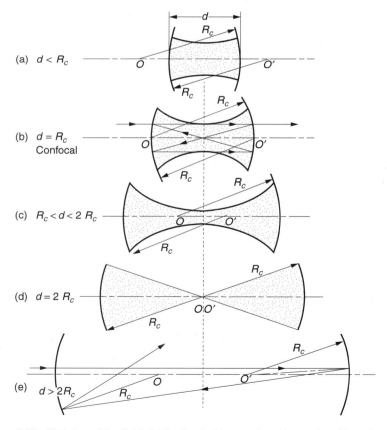

Figure 3.33 Variation of the field distribution inside a cavity with spacing d for a fixed R_c.

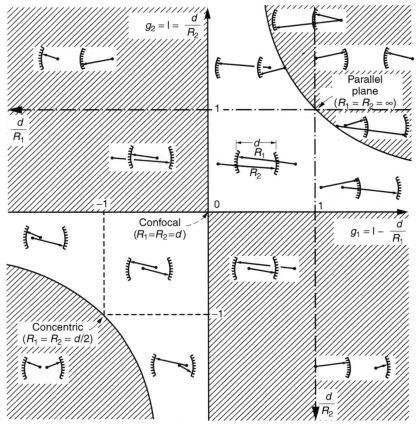

Figure 3.34 Stability diagram. Unstable resonator systems lie in shaded regions. (After H. Kogelnik and T. Li [16].)

The range of existence of the resonance for various combinations of the radii of the end mirrors and the spacing between the mirrors has been studied by Kogelnik and Li [16]. This range is illustrated in Fig. 3.34. The conclusion of the Kogelnik–Li study gives the range of the stable resonance as

$$0 < g_1 g_2 < 1 \qquad (3.155)$$

where

$$
\begin{aligned}
g_1 &= 1 - \frac{d}{R_1} \\
g_2 &= 1 - \frac{d}{R_2}
\end{aligned}
\qquad (3.156)
$$

and where R_1 and R_2 are the radii of curvature of the end mirrors. In the present case of

$$R = R_1 = R_2 = R_c$$

the cavity becomes unstable when

$$g_1 g_2 > 1 \tag{3.157}$$

which is equivalent to $d > 2R_c$, as given in Fig. 3.33e.

3.9 RESONANCE FREQUENCIES OF THE CAVITY

Excitation of the cavity by nothing but the fundamental mode or longitudinal mode $(m + n = 0)$ can be achieved only when the incident beam is of the fundamental mode and when the mirrors are on the same sphere and perfectly aligned, or the radius of curvature $R_c \rightarrow \infty$ as in the case of planar mirrors. It is easier to achieve this condition when the radius of the beam is narrower. In general, higher order modes $(m + n \neq 0)$ are expected to be present.

Let us first find the resonance frequency of the fundamental mode in a cavity with identical end mirrors. Resonance takes place when the phases of the beam before and after a round trip are in phase.

For a cavity with identical end mirrors, the waist is located in the middle of the cavity. The phase distribution in the cavity is readily pictured if the origin of Fig. 3.25a is set in the center of the cavity. The phase of the field at the right end mirror is identical with that in Fig. 3.25a with $z = d/2$, where d is the spacing between the mirrors. The phase correction factor ϕ in Eq. (3.104) or in Eq. (3.143) with $m + n = 0$ can be found from Eq. (3.102) with $z = d/2$.

$$\phi = \tan^{-1} \frac{d}{kW_0^2} \tag{3.158}$$

The value of W_0^2 in Eq. (3.158) can readily be obtained from Eq. (3.153), and ϕ becomes

$$\phi = \tan^{-1} \frac{1}{\sqrt{2R_c/d - 1}} \tag{3.159}$$

For every distance of $d/2$, an additional delay of $-\phi$ exists. The phase delay for one round trip $2d$ has to be an integral multiple of 2π. The round-trip phase delay is

$$2d \frac{2\pi f_p}{v} - 4 \tan^{-1} \frac{1}{\sqrt{2R_c/d - 1}} = 2\pi p \tag{3.160}$$

where v is the velocity of light, p is an integer, and the resonance frequency f_p is called the resonance frequency of the pth longitudinal mode. Solving for f_p, Eq. (3.160) becomes

$$f_p = \frac{v}{2d} \left(p + \frac{2}{\pi} \tan^{-1} \frac{1}{\sqrt{2R_c/d - 1}} \right) \tag{3.161}$$

The frequency spacing Δf between the pth and $(p + 1)$th longitudinal mode is

$$\Delta f_0 = \frac{v}{2d} \tag{3.162}$$

The resonance frequencies are equally spaced, as shown in Fig. 3.35a, and are the same as those of a planar mirror resonator.

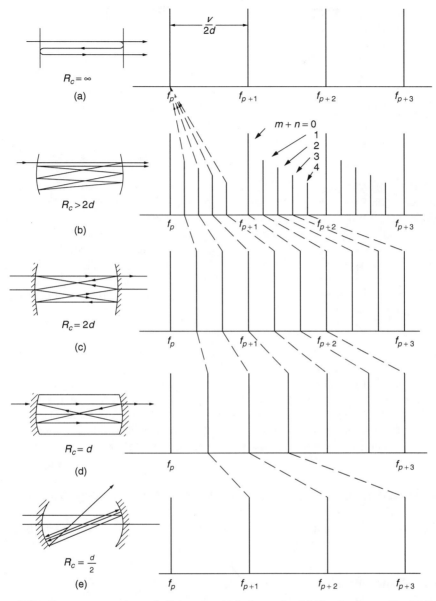

Figure 3.35 Frequency spectrum of a Fabry–perot interferometer. (a) Planar mirror cavity. (b) General spherical mirror cavity. (c) Focal cavity. (Foci are on the confronting mirrors.) (d) Confocal cavity. (Foci of the two mirrors coincide.) (e) Spherical cavity. (Mirrors are on the same sphere.)

Next, the resonance frequencies of the higher order modes ($m + n \neq 0$) or transverse modes will be obtained. The additional phase delay Γ has to be included. From Eqs. (3.142) with $z = d/2$ and (3.153), the phase delay Γ is

$$\Gamma = -(m + n)\tan^{-1}\frac{1}{\sqrt{2R_c/d - 1}} \qquad (3.163)$$

which is similar to Eq. (3.159) except for $(m + n)$. The total phase correction factor at every $d/2$ is $\phi + \Gamma$, and the equation of the resonance frequency $f_{p,m,n}$ becomes

$$f_{p,m,n} = \frac{v}{2d} \left(p + \frac{2}{\pi}(m + n + 1)\tan^{-1} \frac{1}{\sqrt{2R_c/d - 1}} \right) \qquad (3.164)$$

The spacing between $f_{p,0,0}$ and $f_{p,1,0}$ is

$$\Delta f_1 = \frac{v}{\pi d} \tan^{-1} \frac{1}{\sqrt{2R_c/d - 1}} \qquad (3.165)$$

The spacing between $f_{p,1,0}$ and $f_{p,1,1}$ is also the same as Eq. (3.165). As long as the difference $\Delta(m + n)$ is unity, the spacing is given by Eq. (3.165) and if $\Delta(m + n) = 2$ the departure from f_p becomes twice Δf_1. The spectrum with $f_{p,m,n}$ is shown in Fig. 3.35b. There is a cluster of higher order mode resonances at the higher frequency side of each fundamental mode. These higher order modes are disturbing when the cavity resonator is used as a Fabry–Pérot interferometer. As seen from Eq. (3.165), the spacing Δf_1 can be widened if a smaller value of R_c is chosen for a fixed d, as indicated by the dotted lines in Fig. 3.35. In fact, when $R_c = d$ (confocal cavity), Δf_1 in Eq. (3.165) becomes $v/4d$ and

$$\Delta f_1 = \frac{\Delta f_0}{2} \qquad (3.166)$$

which means that the higher order mode shows up midway between adjacent resonant frequencies of the fundamental mode. When $R_c = d/2$, the higher order mode exactly overlaps the fundamental mode; however, as shown in Fig. 3.34, the focal cavity is on the edge of the stability condition and is not recommended. Of the cavities shown in Fig. 3.35, the confocal cavity of Fig. 3.35d is generally the most practical.

3.10   PRACTICAL ASPECTS OF THE FABRY–PÉROT INTERFEROMETER

On the basis of the conclusions reached in the previous section, the practical aspects of the Fabry–Pérot interferometer will be described.

3.10.1   Plane Mirror Cavity

When the mirror is perfectly flat, $R_c = \infty$ and Γ in Eq. (3.163) is zero, which means there is no concern about generating higher order modes. However, when flat mirrors are used for end mirrors, only a parallel beam can be excited in the cavity. In using such an interferometer, it is important to assure the parallelism of the mirrors as well as the perpendicular incidence of the light, otherwise the beam eventually wanders beyond the edges of the mirrors and the sharpness of the resonance (finesse) is poor. For instance, if the reflectivity of the mirror is 99.9%, the beam will bounce back and forth on average 1000 times before leaving the cavity. Unless the mirrors are perfectly parallel, this large number of bounces cannot successfully be completed. If the angle of incidence is other than the normal, the spacing between the mirrors is effectively changed. This angular dependence is also used to analyze the spectrum but only with poor finesse.

The spacing between the resonant frequencies (free spectral range, FSR) is increased as the spacing between the mirrors is reduced, as indicated by Eq. (3.39). The derivative of Δf_0 with respect to d becomes large for small d, which means the interferometer is more susceptible to mechanical vibrations and temperature fluctuations. The FSR should be selected according to the width of the spectrum of the light under test (see Section 3.2.1.3).

3.10.2 General Spherical Mirror Cavity

Only with a narrow beam and a perfect mirror arrangement can the zeroth order beam be excited, otherwise the higher order modes are excited, as indicated in Fig. 3.35b. If higher order modes are present, the interpretation of the spectrum is more difficult.

One way to interpret the generation of the higher order modes or transverse modes is as follows. A beam of finite size can be thought of as a bundle of narrow beams. The narrow beam at the center represents the longitudinal mode, and it takes a direct path across the cavity. For this direct path, the distance for the resonance condition is one round trip (across the cavity and back). As illustrated in Fig. 3.36, some of the narrow beams take a zigzag path around the cavity such that it takes several traversals of the cavity before the beam actually rejoins itself. Upon rejoining, the resonance condition requires that the rejoined beam be in phase with the original beam. For the zigzag path, the rejoining distance is greater than the round-trip distance of the longitudinal mode. If the zigzag path is equivalent to N round trips of the longitudinal mode, then the spectral separation of the higher order mode is $\Delta f_0/N$ because it is equivalent to a cavity N times as long.

Next, the higher order mode patterns are considered. As a matter of fact, the mode patterns shown earlier in Fig. 3.31 are nothing but standing wave patterns in the transverse plane generated by off-axis propagating wave components. For instance, the rays shown in Fig. 3.36c are represented by two component waves whose propagation vectors are $\mathbf{k}_1$ and $\mathbf{k}_2$.

The transverse component of $\mathbf{k}_1$ is pointing up while that of $\mathbf{k}_2$ is pointing down. These two components pointing in opposite directions set up a standing wave pattern in the transverse plane.

3.10.3 Focal Cavity

When the length of the cavity becomes the focal length of the mirror,

$$d = \frac{R_c}{2} \tag{3.167}$$

the cavity is called a focal cavity. For the focal cavity, Eq. (3.165) is $\Delta f_1 = v/6d$ and Δf_1 is one-third of Δf_0. This can be verified by tracing the beam reflected by concave mirrors of the cavity. As shown in Fig. 3.36c, after three round trips, the beam lines up with the original direction and the separation between the higher order spectra becomes one-third that of the fundamental modes. Every third spectra matches up with the fundamental, and there are two additional spectra between the fundamental modes. The tolerance of the incident angle is not as large as for the confocal cavity, as described next. The focal resonator is also on the borderline of stability.

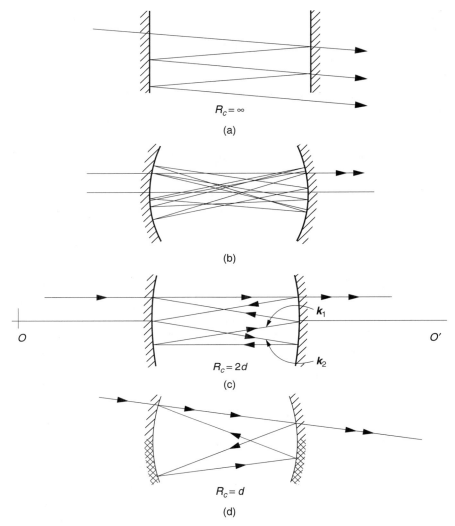

Figure 3.36 Various types of spherical mirrors in a Fabry–perot interferometer. (a) Planar mirror. (b) General. (c) Focal. (d) Confocal.

3.10.4 Confocal Cavity

When the radius of curvature of the spherical mirrors is identical to the spacing between the mirrors,

$$R_c = d \tag{3.168}$$

then such a cavity is called a confocal cavity. For the confocal cavity, Eq. (3.165) is $\Delta f_1 = v/4d$ and Δf_1 is one-half of Δf_0. The spacing of the resonances associated with the transverse mode becomes exactly one-half of that of the longitudinal modes and the ambiguity due to the transverse mode disappears. This can be explained by tracing the path of light. As shown in Fig. 3.36d, after two round trips, the path of the emergent beam lines up with the direction of the incident beam. This is as if the resonance took place in a cavity of twice the length, which means half the spectral separation of the fundamental modes. From the figure, the angle of incidence for the confocal cavity is far less critical than for the plane mirror cavity. The large tolerance on the angle of incidence is a definite advantage for the confocal resonator. A variation on this interferometer is to use end mirrors whose reflectivity on the bottom half of the mirror is much larger than the top half of the mirror. This allows easier entrance but higher reflection in the cavity. Such a Fabry–Pérot interferometer has a higher value of finesse.

3.11   BESSEL BEAMS

The Bessel beam propagates in free space with minimum spread in the transverse direction over distances of more than several meters. Because of this unusual property, the beam has been dubbed a diffraction-free beam.

3.11.1   Features of the Bessel Beam

Special features of the Bessel beam are the following:

1. The diameter of the intensity distribution in the transverse plane remains constant over distances of more than several meters.
2. The intensity decays abruptly at a distance $z = z_{max}$.
3. The propagation constant along the beam is less than that of free space, and the value is adjustable.
4. With an adjustment of the condition of excitation, the nth order beams $J_n(\alpha r)$ can also be excited.

The constant-diameter feature is illustrated in Figs. 3.37–3.39. Figure 3.37 shows the intensity distributions of the Bessel beam in four transverse planes located at $z = 0$, $z = 2$ m, $z = 4$ m, and $z = 5.5$ m, where z is the distance along the beam from the input aperture [20]. The diameter of the main beam remains unchanged at 200 μm over the entire distance of 5.5 m. Figure 3.38 is a photograph of the intensity distribution of a Bessel beam in the plane at $z = 9.6$ m [21] taken with the aid of a projector lens (see Fig. 3.42). The diameter of the main beam is 200 μm. In Fig. 3.39, the transverse field distributions of a Gaussian beam (dotted line) and Bessel beam (solid line) are compared in the following planes: (a) $z = 0$, (b) $z = 10$ cm, and (c) $z = 1$ m. The Gaussian beam displays diffraction spread as well as a rapid decrease in intensity. In order for the Gaussian beam intensity to be visible on the graph, the Gaussian curves (b) and (c) have been magnified 30 times and 2000 times, respectively. In contrast, there is no discernable change in the Bessel beam [22].

The abrupt decay of the Bessel beam at $z = z_{max}$ is shown in Fig. 3.40. The field intensity of the main lobe of the Bessel beam is plotted with respect to the distance z of propagation. Again, for the sake of comparison, the Bessel beam is represented by the solid line and the Gaussian beam, by the dotted line. While the Gaussian beam displays a rapid decay, the Bessel beam maintains its intensity up to $z = z_{max} = 1$ m, oscillating around a certain value with distance.

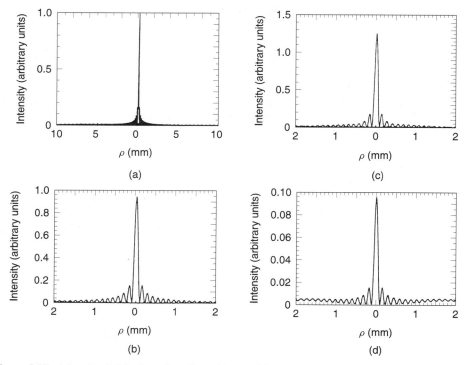

Figure 3.37 Intensity distributions for a Bessel beam: (a) $z = 0$ m, (b) $z = 2$ m, (c) $z = 4$ m, and (d) $z = 5.5$ m. (After J. Durnin [20].)

Figure 3.38 Photograph of the Bessel beam at 9.6 m from the aperture using the setup shown in Fig. 3.42c. (Courtesy of R. M. Herman and T. A. Wiggins [21].)

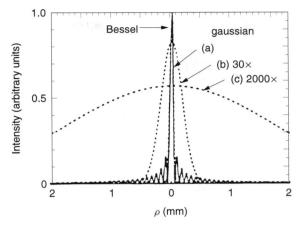

Figure 3.39 The transverse spread of the Bessel beam (solid line) is compared to that of a Gaussian beam (dotted line). The spot sizes of both beams are the same at $z = 0$. The transverse spread of the Gaussian beam is shown at (a) $z = 0$, (b) $z = 10$ cm with $30\times$ magnification in intensity, and (c) $z = 100$ cm with $2000 \times$ magnification in intensity. The spot size of the Bessel beam remains the same. (After J. Durnin, J. J. Miceli Jr., and J. H. Eberly [22].)

3.11.2 Practical Applications of the Bessel Beam

A number of applications have been considered for incorporating the special features of the Bessel beam.

3.11.2.1 Precision Optical Measurement
Since the amount of spread during the propagation can be made much smaller than that of the Gaussian beam, the Bessel beam is useful for precision optical measurements [23].

Not only $J_0(\alpha r)$ but also $J_n(\alpha r)$ can be excited. The $J_1(\alpha r)$ beam, which has its null on the beam axis, may be advantageous for a finer definition in the precision alignment.

3.11.2.2 Power Transport
Due to the sharp fall-off of the transmission power at a predetermined distance, the Bessel beam might be useful as a means of power transport.

3.11.2.3 Nonlinear Optics
Bessel beams have been used for nonlinear optics experiments in long liquid cells of carbon disulfide, CS_2, or acetone, $(CH_3)_2CO$ [24–26]. The beam provides not only a long interaction length but also confined high-intensity light, both of which are essential in nonlinear optics experiments.

An additional advantage of the Bessel beam is easy adjustability of the propagation constant, which is often needed for optimizing the nonlinear interaction [26].

3.11.3 One-Dimensional Model

The one-dimensional model in Figs. 3.41a and 3.14b will be used to explain how the Bessel beam is generated. The z axis is taken in the beam direction and the x and

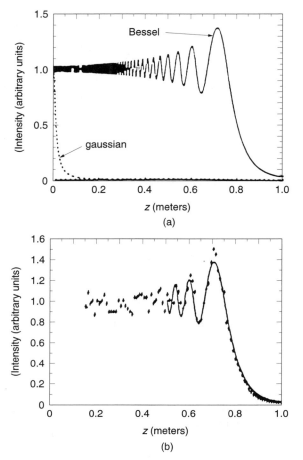

Figure 3.40 (a) Calculated on-axis intensity of the Bessel beam (solid line) and Gaussian beam (dotted line) with respect to the propagation distance z. (b) The corresponding measured results for the Bessel beam. (After J. Durnin, J. J. Miceli, Jr., and J. H. Eberly [22].)

y axes are in the plane transverse to the beam direction. Two delta function sources $\delta(x - a)$ and $\delta(x + a)$ are placed in the front focal plane of a convex lens L, at $x = a$ and $x = -a$. Since the sources are in the front focal plane, the light emerging from the lens is in the form of two parallel beams B_1 and B_2 with angles of inclination $\theta = \pm\tan^{-1}(a/f)$, where f is the focal length of the lens. The resultant field $E(x, z)$ is expressed as

$$E(x, z) = A(e^{j(\beta z + \alpha x)} + e^{j(\beta z - \alpha x)})$$

$$= 2A \cos \alpha x e^{j\beta z} \tag{3.169}$$

where

$$\beta = k \cos \theta \tag{3.170}$$

$$\alpha = k \sin \theta \tag{3.171}$$

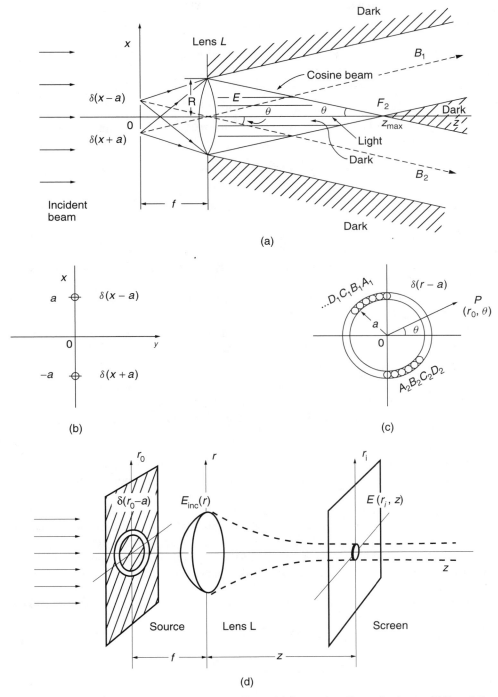

Figure 3.41 Geometries of cosine and Bessel beams. (a) Generation of a cosine beam. (b) Two delta function sources. (c) Bessel beam source $\delta(r - a)$. (d) Generation of the Bessel beam.

As far as the field distribution in the x–z plane is concerned, the resultant field creates a cosinusoidal standing wave pattern in the x direction and a traveling wave in the z direction.

Let this wave be called the cosine beam. The spacing Λ between the standing wave peaks is

$$\Lambda = \frac{\lambda}{2\sin\theta} \tag{3.172}$$

This one-dimensional model (cosine beam) displays almost all features of the Bessel beam mentioned above.

1. The wave propagates while maintaining the same cosine transverse pattern $2A\cos\alpha x$.

2. The beam abruptly ends at

$$z_{\max} = R\tan^{-1}\theta = R\frac{f}{a} \tag{3.173}$$

   where the crossover of the two beams ends. Since the radius R of the lens is finite, $z_{\max}$ is always finite.

3. The propagation constant along the beam (z axis) is less than that of free space and can be adjusted by means of θ.

4. By placing a π-radian phase shifter in front of one of the delta function sources, the cosine beam can be converted into a sine beam.

Next, in order to generate the Bessel beam, the two delta function sources are replaced by a delta function ring source, $\delta(r-a)$. For now, the ring source is divided into paired sections, A_1, A_2; B_1, B_2; C_1, C_2; D_1, D_2; and so on, as shown in Fig. 3.41c. The points $A_2, B_2, C_2, D_2, \ldots$ are located diagonally opposite to $A_1, B_1, C_1, D_1, \ldots$, respectively.

Sections A_1 and A_2 generate exactly the same pattern as the cosine beam mentioned above. Sections B_1 and B_2 do the same but in a plane tilted from the x–z plane. The contributions of all the sections of the ring source are superimposed. Superposition, however, has to be performed taking the phase into consideration.

The distances to all subsections are the same as long as the point of observation is on the z axis, and in this case, the contributions of the subsections are all added in phase. As soon as the point of observation moves away from the z axis, phase mismatch occurs among the contributions of the subsections of the ring source. Thus, the field intensity reaches its maximum on the z axis and decays in a rippling fashion with distance away from the z axis.

Approximately, the field distribution in the transverse plane is a Bessel function and the beam is expressed as

$$E(z, r) = Ae^{j(\beta z - \omega t)}J_0(\alpha r) \tag{3.174}$$

where α is given by Eq. (3.171).

3.11.4   Mathematical Expressions for the Bessel Beam

Mathematical expressions for the Bessel beam will be developed using the geometry with the delta function ring source in Fig. 3.41d [27].

First, an expression is sought for the input field $E_{inc}(r)$ to the convex lens. This expression will be used later for calculating the Bessel beam field distribution. $E_{inc}(r)$ is calculated using the Fresnel approximation of the Fresnel–Kirchhoff diffraction formula, Eq. (1.38). The Fourier transform Eq. (1.66) in cylindrical coordinates is used.

$$E_{inc}(r) = \frac{1}{j\lambda f} e^{jk(f+r^2/2f)} 2\pi \int_0^\infty \delta(r_0 - a) e^{jk(r_0^2/2f)} J_0(2\pi\rho r_0) r_0 \, dr_0 \Big|_{\rho=r/f\lambda} \tag{3.175}$$

where r_0 and r are radial coordinates in the planes of the ring source and the lens L, respectively.

The term $e^{jk(r_0^2/2f)}$ is a part of the point spread function in cylindrical coordinates and corresponds to Eq. (1.40) of rectangular coordinates. The result of the integration from Eq. (1.102) with $dx\,dy = r\,dr\,d\theta$ is

$$E_{inc}(r) = Ae^{jk(r^2/2f)} J_0 \left(k\frac{a}{f}r \right) \tag{3.176}$$

where

$$A = \frac{k}{jf} e^{jk(f+a^2/2f)} \tag{3.177}$$

Thus, $E_{inc}(r)$ on the front surface of the convex lens is the zero order Bessel function of the first kind combined with a quadratic phase factor $e^{jk(r^2/2f)}$. This quadratic phase factor is removed by passing through the convex lens whose transmittance is $e^{-jk(r^2/2f)}$ as given by Eq. (1.139). The field on the output surface from the convex lens is $AJ_0(k(a/f)r)$.

The Fresnel approximation of the field $E(r_i, z)$ emergent from the convex lens is obtained again by the Fresnel–Kirchhoff integral as

$$E(r_i, z) = A' \int_0^R J_0 \left(k\frac{a}{f}r \right) e^{jk(r^2/2z)} J_0(2\pi\rho r) r \, dr \Big|_{\rho=r_i/\lambda z} \tag{3.178}$$

$$A' = A\frac{k}{jz} e^{jk(z+r_i^2/2z)} \tag{3.179}$$

where r_i is the radial coordinate of the point of observation at a distance z away from the lens L.

The expression for the on-axis field distribution ($r_i = \rho = 0$) along the z axis is much simpler than Eq. (3.178).

By putting

$$r_i = \rho = 0$$

$$q = \frac{k}{j2z}$$

$$\alpha = k\frac{a}{f} \tag{3.180}$$

$$x = \sqrt{q}\,r$$

Eq. (3.178) is simplified to

$$E(0, z) = \frac{A'}{2q} \int_0^{\sqrt{qR}} 2xJ_0\left(\frac{\alpha}{\sqrt{q}}x\right) e^{-x^2}\, dx \tag{3.181}$$

The Bessel integral formula is available [28] for Eq. (3.181).

$$\int_0^a 2xJ_0(\gamma x)e^{-x^2}\, dx = \begin{cases} e^{-\gamma^2/4} - \displaystyle\sum_{n=0}^{\infty} \left(\frac{-\gamma}{2a}\right)^n J_n(\gamma a)e^{-a^2} & \text{for } \left|\dfrac{2a}{\gamma}\right| < 1 \\[2em] \displaystyle\sum_{n=0}^{\infty} \left(\frac{2a}{\gamma}\right)^n J_n(\gamma a)e^{-a^2} & \text{for } \left|\dfrac{2a}{\gamma}\right| > 1 \end{cases} \tag{3.182}$$

By comparing Eq. (3.181) with Eq. (3.182) and by putting

$$\begin{aligned} a &= \sqrt{q}R \\ \gamma &= \frac{\alpha}{\sqrt{q}} \end{aligned} \tag{3.183}$$

the expression for $E(0, z)$ becomes

$$E(0, z) = \frac{A'}{2q} \begin{cases} e^{-\alpha^2/4q} - \displaystyle\sum_{n=0}^{\infty} \left(\frac{-\alpha}{2qR}\right)^n J_n(\alpha R)e^{-qR^2} & \text{for } \left|\dfrac{\alpha}{2qR}\right| > 1 \\[2em] \displaystyle\sum_{n=0}^{\infty} \left(\frac{2qR}{\alpha}\right)^n J_n(\alpha R)e^{-qR^2} & \text{for } \left|\dfrac{\alpha}{2qR}\right| < 1 \end{cases} \tag{3.184}$$

Various factors appearing in Eq. (3.184) will be rewritten.
From Eqs. (3.179) and (3.180) and $r_i = 0$, the factor $A'/2q$ becomes

$$\frac{A'}{2q} = Ae^{jkz} \tag{3.185}$$

From Eq. (3.180), the factor $\alpha/2qR$ is rewritten as

$$\frac{\alpha}{2qR} = j\frac{za}{Rf}$$

From Eq. (3.173), this factor becomes

$$\frac{\alpha}{2qR} = \frac{jz}{z_{\max}} \tag{3.186}$$

Thus, inserting Eqs. (3.185) and (3.186) into Eq. (3.184) gives

$$
E(0, z) = Ae^{jkz}
\begin{cases}
e^{-j(\alpha^2/2k)z} - \displaystyle\sum_{n=0}^{\infty} \left(\frac{jz}{z_{\max}}\right)^n J_n(\alpha R)e^{jk(R^2/2z)} & \text{for } z > z_{\max} \\
\displaystyle\sum_{n=0}^{\infty} \left(\frac{z_{\max}}{jz}\right)^n J_n(\alpha R)e^{jk(R^2/2z)} & \text{for } z < z_{\max}
\end{cases}
\tag{3.187}
$$

Calculation of Eq. (3.187) produces curves such as the one shown in Fig. 3.40a.

The calculation of the integral of Eq. (3.178) for the off-axis field distribution is more involved and can be found in Ref. 27. Calculated results with various combinations of physical parameters are shown in the same reference.

3.11.5 Methods of Generating Bessel Beams

In order to explain the principle governing the generation of the Bessel beam, a delta function ring source was used in Fig. 3.41 but such a source is inefficient and impractical. Figure 3.42 summarizes more practical methods. The method shown in Fig. 3.42a employs a conical lens [22] whose thickness is linearly reduced as the rim of the lens is approached.

The conical lens generates two parallel beams crossing each other. The pattern shown in Fig. 3.42a corresponds to the field pattern in the $x-y$ plane, but the field patterns in any plane that includes the optical axis are the same as shown in the figure. The conical lens generates the Bessel beam. The region of focus is a line from point F_1 to F_2 on the optical axis. This type of conical lens forms a line image rather than point image and it is sometimes called an axicon, meaning the axis image [29].

Despite the simplicity and high power output, the arrangement with an axicon alone has the disadvantage that the region of the Bessel beam starts immediately behind the axicon. Usually, a certain distance is required from the source to the beginning of the measuring device, as, for example, when the Bessel beam is used for triangulation [23].

The convex projector lens L in Figs. 3.42a relocates the region of the Bessel beam. The region is transformed from the input region F_1-F_2 to the output region $F_1' - F_2'$. The role of the projector lens is to form the output image F_n' from the input point F_n.

In Fig. 3.42b, positive and negative axicons are combined to transfer the Bessel beam further away from the axicon. The input axicon is a negative axicon and diverges the beam, while the second axicon is a positive axicon and converges the beam. As shown in Fig. 3.42b, there is a hatched dark region in the center of the output axicon so that the two beams meet at a smaller angle and transfer the Bessel beam further away from the output axicon. This combination of a positive with a negative axicon is called a teleaxicon [23].

In Fig. 3.42c, the conical lenses in Fig. 3.42b are replaced by ordinary spherical lenses [21]. The hatched dark region in the center is created by an opaque disk. The resulting beam pattern is quite similar to that from Fig. 3.42b.

In order for the Bessel beam to be created, the region has to be illuminated by nothing but the tilted beams B_1 and B_2. If the opaque disk is not installed in the center, a third beam propagating parallel to the z axis will be present, in addition to B_1 and B_2, and the Bessel beam will be disturbed.

A holographic approach to generating a Bessel beam is illustrated in Fig. 3.42d [30]. The Bessel beam is generated by simply illuminating a computer-generated hologram by a parallel laser beam.

Instead of computer-generated holograms, an actual hologram fabricated directly from a Bessel beam [31] can be used. Figure 3.42e shows an arrangement for the fabrication of such a hologram. The laser beam is split into object and reference waves by means of a half-mirror HM. Each wave is spatially filtered and expanded by a combination of two lenses and a pinhole: L_1, P_1, L_2 for the object wave and L_3, P_2, L_4 for the reference wave.

The object wave is converted into a Bessel beam by means of the delta function annular slit placed in the front focal plane of lens L_5. The photographic plate H is placed in the region $F_1 - F_2$ of the Bessel beam. The off-axis reference wave is added to the photographic plate. After the exposure, the plate is developed. Bleaching after development increases the diffraction efficiency of the hologram.

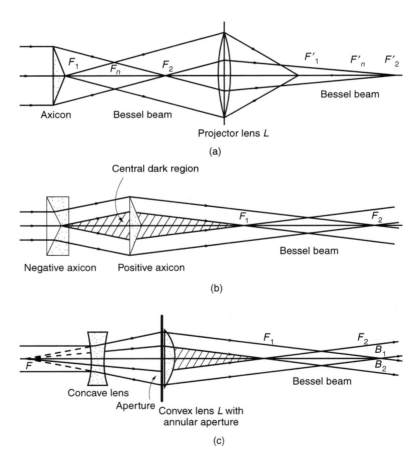

Figure 3.42 Various methods for generating Bessel beams. (a) By means of an axicon and a projector lens. (b) By means of a teleaxicon. (c) By means of a spherical convex lens with an annular iris. (d) By means of a holographic plate. (e) Fabrication of a Bessel beam hologram. (f) By means of β-beam reflector telescope. (After T. Aruga et al. [32].)

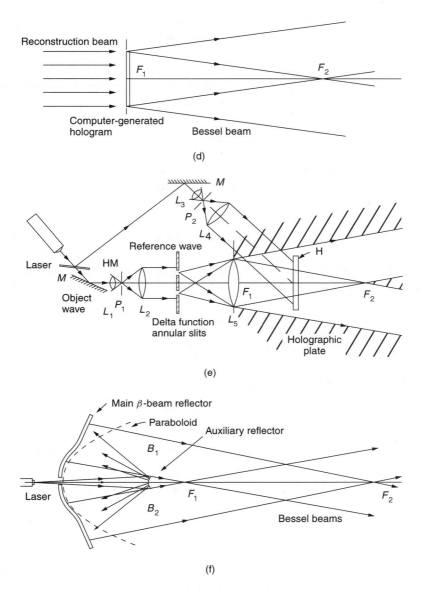

Figure 3.42 (*Continued*)

The Bessel beam is immediately generated by illuminating this hologram with the reconstruction beam that is the reference beam used for fabrication. The advantage of the holographic Bessel beam is that neither an annular slit nor a convex lens is needed to reconstruct the Bessel beam.

Figure 3.42f shows a reflector-type arrangement [32] for generating a Bessel beam. The surface of the main or auxiliary reflector of the telescope is deformed from the paraboloid so that tilted beams B_1 and B_2 are generated. The reflector type is advantageous over the lens or holographic type when a large-diameter device is needed for higher sensitivity.

Example 3.12   For the Bessel beam geometry shown in Fig. 3.43, derive expressions for (a) the range F_1–F_2 and (b) the diameter of the main lobe using Eq. (3.174).

Solution   Figure 3.43 shows a cross section of the x–z plane of the axicon. In order to find the propagation direction θ of beam B_1, the phase ϕ of a plane wave starting from point S to point P is calculated using wave optics as

$$\phi = knx\tan\Delta + k(z - x\tan\Delta) \tag{3.188}$$

where n is the refractive index of the axicon. The expression for the constant phase ϕ_0 is obtained by setting $\phi = \phi_0$. The vector $\mathbf{s}$ of the propagation constant is the gradient of the constant phase line $\nabla\phi_0$, and

$$\mathbf{s} = \hat{\mathbf{i}}[k(n-1)\tan\Delta] + \hat{\mathbf{k}}k \tag{3.189}$$

where $\hat{\mathbf{i}}$ and $\hat{\mathbf{k}}$ are unit vectors in the x and z directions, respectively. Thus, the direction of propagation θ of beam B_1 is

$$\tan\theta = (n-1)\tan\Delta \tag{3.190}$$

(a) From the geometry, the focal range $\overline{F_1F_2}$ is

$$\overline{F_1F_2} \doteq \frac{R}{(n-1)\tan\Delta} \tag{3.191}$$

where R is the radius of the axicon.

(b) α in Eq. (3.180) is the x component of the propagation constant and is, from Eq. (3.189),

$$\alpha = k(N-1)\tan\Delta \tag{3.192}$$

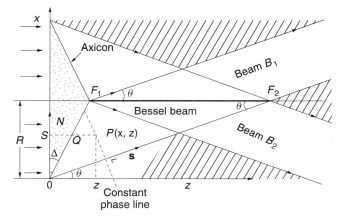

Figure 3.43   Geometry for generating a Bessel beam by means of an axicon lens.

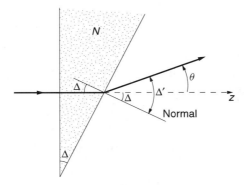

Figure 3.44 Expanded view of the axicon.

$J_0(\alpha r)$ has its first zero at $\alpha r = 2.4$ and the diameter d of the main lobe is

$$d = \frac{2.4\lambda}{\pi(N-1)\tan\Delta} \tag{3.193}$$

It should be pointed out that the same answer is obtained using geometrical optics. An enlarged section is shown in Fig. 3.44. Snell's law gives

$$n\sin\Delta = \sin\Delta'$$

$$\theta = \Delta' - \Delta$$

When $\Delta \ll 1$, θ is approximated as

$$\theta \doteq (n-1)\Delta \tag{3.194}$$

which matches with the approximation of Eq. (3.190) when $\Delta \ll 1$.   □

3.12 MANIPULATION WITH LIGHT BEAMS

In 1619 Johannes Kepler (1571–1630) introduced the concept of radiation pressure to explain why comets' tails are always trailing away from the sun. More recently, the radiation pressure [33,34] of a laser beam has been utilized for trapping micron-sized dielectric spheres. These trapping devices are more commonly known as "*optical tweezers*." Radiation pressure also has been used for quieting down the random motion of atoms and molecules, a technique known as laser cooling. Even though both optical tweezers and laser cooling use the force generated by the changes in the momuntum of the laser beam, the mechanisms of the interaction between the object and the laser beam are different.

3.12.1 Radiation Pressure of Laser Light

Each photon has quantum energy of

$$E = h\nu \tag{3.195}$$

and momentum of

$$p = \frac{h}{\lambda} \qquad (3.196)$$

where h is the Planck's constant and is $h = 6.6 \times 10^{-34}$ J·s. Thus, the total number of photons in light with energy W is

$$N = \frac{W}{h\nu} \qquad (3.197)$$

The momentum $P_m = Np$ is, from Eqs. (3.196) and (3.197),

$$P_m = \frac{W}{h\nu} \cdot \frac{h}{\lambda} = \frac{n_1 W}{c} \qquad (3.198)$$

where n_1 is the index of refraction of the medium.

The force is the time derivative of the momentum, and the force in a medium of index of refraction n_1 is

$$F = n_1 P/c \qquad (3.199)$$

where P is the incident light power in watts. The magnitude of the force is not large, but for a particle of a small mass like an atom or molecule, it is significant. The acceleration of a 1-μm-diameter sphere by the radiation pressure force [33,34] calculated from Eq. (3.199) is over 10^5 times larger than that of gravitation [35].

Next, the force generated by a laser beam transmitted through a sphere will be calculated, referring to Fig. 3.45.

It is assumed that not only the surface reflection is negligible, but also the sphere is perfectly transparent and no heat effect is involved. The refractive index of the dielectric sphere is assumed larger than that of the surrounding medium.

Let vector **a** represent the momentum of the incident light and **a′**, that of the emergent light. The momentum c_a transferred to the sphere is obtained from the law of conservation of momentum:

$$\mathbf{a} = \mathbf{a}' + \mathbf{c}_a \qquad (3.200)$$

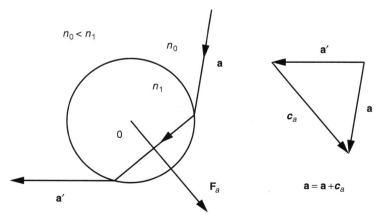

Figure 3.45   The direction of the force on the sphere is known from the conservation of momentum of the laser beam.

The difference c_a between **a** and **a'** provides the direction of the force $\mathbf{F}_a$ acting on the sphere by the transmitted laser beam. $\mathbf{F}_a$ is parallel to c_a.

3.12.2 Optical Tweezers

A TEM$_{00}$ mode Gaussian beam is incident onto a large NA value convex lens. The dielectric sphere to be manipulated is placed near the focus f of the lens. Figure 3.46a shows the case when the sphere is displaced above the focus of the lens. (The lens

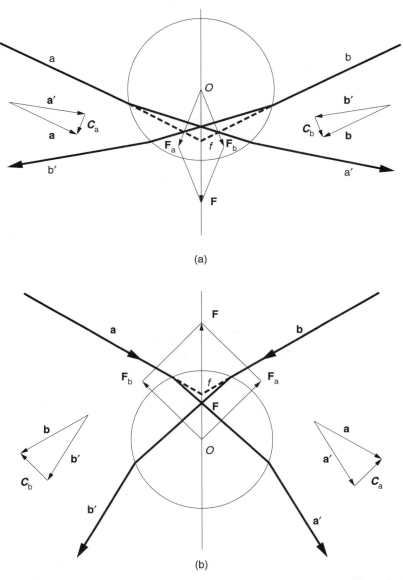

(a)

(b)

Figure 3.46 Restoring forces from the axial displacements of a dielectric sphere: $n_0 = 1.0, n_1 = 1.5$. (a) Displacement above the trap focus f. (b) Displacement below the trap focus f. (After A. Ashkin [35].)

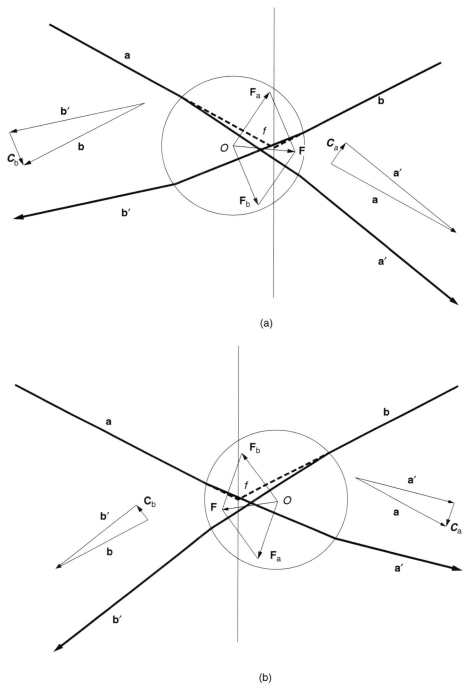

(a)

(b)

Figure 3.47 Restoring forces from the transverse displacement of a dielectric sphere: (a) Displacement to the left of the trap focus f. (b) Displacement to the right of the trap focus f. (After A. Ashkin [35].)

is above the sphere and is not shown.) Lines a and b are representative component beams from the converging lens. From the difference between the vectors $\mathbf{a}$ and $\mathbf{a}'$, $\mathbf{F}_a$ of beam a is found, and similarly from vectors $\mathbf{b}$ and $\mathbf{b}'$, $\mathbf{F}_b$ of beam b is found. The direction of the resultant vector

$$\mathbf{F} = \mathbf{F}_a + \mathbf{F}_b \qquad (3.201)$$

is the direction of the force. The direction of $\mathbf{F}$ is downward or, in other words, in the direction of restoring the axial displacement.

In the same way, Fig. 3.46b shows the case when the sphere is displaced below the focus f. The resultant force F points up and restores the displacement.

The same is repeated in Figs. 3.47a and 3.47b to show the restoring force for the transversally displaced sphere. Figures 3.46 and 3.47 confirm stable operation as an *optical tweezer*.

Next, the TEM_{00} and TEM_{11} Gaussian modes are compared to see which makes a better optical tweezer. The beam components with a larger apex angle generate a larger trapping force.

The TEM_{11} Gaussian mode beam is more efficient than the TEM_{00} Gaussian mode for trapping the sphere, because as shown in Fig. 3.31 the TEM_{11} mode has a null intensity in the center and the light intensity is more concentrated at the larger apex angles where a larger trapping force is generated.

The optical tweezers provide a high degree of control over the dynamics of small particles and play an important role in physical and biological sciences, especially in manipulating living cells. Figure 3.48 shows the sequence of a DNA molecule being pulled through a polymer solution with the use of an optical tweezer. The DNA molecule is seen to relax along a path defined by its contour [36].

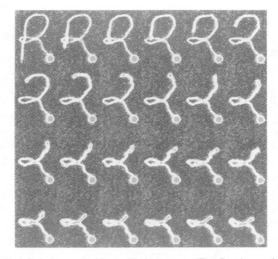

Figure 3.48 DNA molecule being pulled by optical tweezers. The first image in the upper left shows a stained DNA molecule pulled into the shape of the letter R with the optical tweezers. The remaining images (read left to right, top to bottom) show the relaxation of the DNA molecule along its contour. (Courtesy of Steven Chu [36].)

3.13   LASER COOLING OF ATOMS

The random movement of an atom can be slowed down by means of the radiation pressure of a laser beam tuned to an atomic resonance. Such a technique is known as *laser cooling* [37,38]. Because of the minuteness of the target, the laser beam generally passes through the atoms without being disturbed, and no momentum of the laser beam is transferred to the atom. The situation, however, becomes quite different when the frequency of the illuminating laser beam is tuned to the resonance frequency ν_0 of an atomic transition. The transition energy is expressed in terms of the resonance frequency as

$$h\nu_0 = E_1 - E_0 \tag{3.202}$$

where E_0 is the energy level of the ground state of the atom and E_1 is that of the next higher energy level. At the resonant frequency ν_0, the scattering cross section markedly increases.* The focused laser beam is almost entirely absorbed by the atoms in its path and the laser beam momentum is efficiently transferred to the atoms.

Each time an atom in the ground state absorbs a photon, it makes a transition to the excited state and at the same time it is pushed by the radiation pressure of the photon. The change in the velocity of the motion of the atom due to one push is

$$v_c = \frac{h}{\lambda m} \tag{3.203}$$

where m is the mass of the pushed atom and h/λ is the momentum of the photon.

Once the atom makes a transition to the excited state, the atom no longer interacts with the laser beam until the atom returns to the ground state. The atom returns to the ground state by reradiating a photon with energy $h\nu_0$ by spontaneous emission. The direction of the reradiation is statistically random and the net contribution of the recoils is zero. The atom that returned to the ground state is ready to absorb another photon, and the process continues.

For a typical atom, the spontaneous emission lifetime is 10^{-8} second, which means the repetition of the absorption–emission process takes place 10^8 times a second.

Another important aspect of the mechanism of laser cooling is the differential absorption of the photon energy associated with the Doppler shift. The laser frequency ν_l is set lower than the resonance frequency ν_0 of the atom with an offset frequency v/λ, where the v is the velocity of the atom. When the atom is moving toward the laser source, the atom sees a frequency higher than the laser frequency due to the Doppler shift, and with this offset frequency, the atom sees exactly the resonance frequency. The interaction is high. On the other hand, when the atom is moving away from the laser source, the atom sees a frequency lower than the laser frequency due to the Doppler shift. This atom sees a frequency further away from the resonance frequency, and the interaction is low. The differential absorption generates a net force resisting the movement of the atom.

As the atom is cooled down the velocity reduces and the frequency of the laser source has to be raised accordingly (frequency chirp) in order to maintain optimum interaction

---

* At the resonance frequency, the scattering cross section reaches a value of approximately λ^2. Off resonance, the value of the scattering cross section is almost zero.

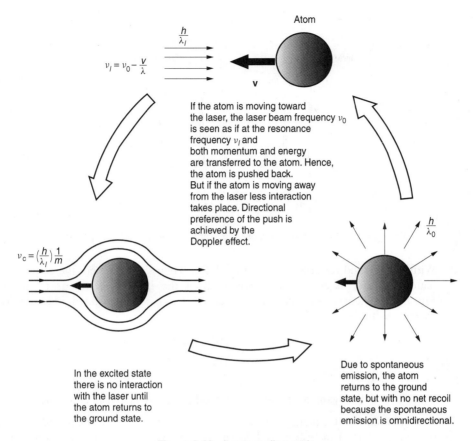

Atom

$$\frac{h}{\lambda_l}$$

$$\nu_l = \nu_0 - \frac{v}{\lambda}$$

v

If the atom is moving toward
the laser, the laser beam frequency ν_0
is seen as if at the resonance
frequency ν_l and
both momentum and energy
are transferred to the atom. Hence,
the atom is pushed back.
But if the atom is moving away
from the laser less interaction
takes place. Directional
preference of the push is
achieved by the
Doppler effect.

$$\frac{h}{\lambda_0}$$

$$\nu_c = \left(\frac{h}{\lambda_l}\right)\frac{1}{m}$$

In the excited state
there is no interaction
with the laser until
the atom returns to
the ground state.

Due to spontaneous
emission, the atom
returns to the ground
state, but with no net recoil
because the spontaneous
emission is omnidirectional.

Figure 3.49   Laser cooling cycle.

with the laser beam. This method is called Doppler cooling in order to distinguish it from another method of cooling called polarization gradient cooling [36]. For the atoms whose velocities are reduced by this process, laser cooling quiets down the motion of the atoms to almost complete rest. Figure 3.49 summarizes the laser cooling cycle.

It should be pointed out that the quantum lost from an atom due to spontaneous emission is $h\nu_0$ while that of the laser source is $h\nu_l$. Since ν_0 is larger than ν_l, the lost energy is larger than the supplied energy to the atom. The difference accounts for the cooling of the atom.

Laser cooling has made a significant impact on high resolution spectroscopy and frequency standards like atomic clocks [39].

PROBLEMS

3.1 What happens to the pattern on the screen in Fig. 3.10, when the lens L and prism are removed? Assume that the incident light is monochromatic. Draw the patterns in the $x-z$ plane when the screen is very near, moderately far, and very far from the Fabry–Pérot resonator.

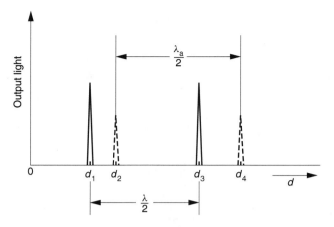

Figure P3.2   Measured peaks using a scanning Fabry–perot resonator.

3.2   With the method described in Section 3.4.3, when d is scanned, the same peak pattern repeats at every half-wavelength of the incident light, as shown in Fig. P3.2. Theoretically speaking, λ and λ_a can be determined individually, by measuring

$$\lambda = 2(d_3 - d_1)$$

$$\lambda_a = 2(d_4 - d_2)$$

What is the disadvantage of such a method as far as $\Delta\lambda/\lambda$ is concerned?

3.3   The modulation frequency of a He–Ne laser beam was measured by the method explained in Section 3.4.3. The He–Ne wavelength is $\lambda = 0.6328$ μm. Determine the frequency of modulation from the following measured quantities:

$$\Delta d_1 = 0.0829 \text{ μm}$$

$$\Delta d_2 = 0.1441 \text{ μm}$$

$$d_1 - d_2 = 300 \text{ μm}$$

3.4   Figure P3.4 shows the display of a scanning Fabry–Pérot resonator when phase modulated light is incident onto the resonator. The mirror spacing is 420 μm.

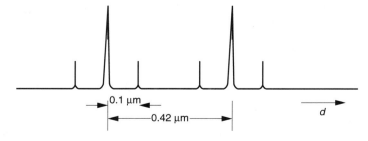

Figure P3.4   The display of the scanning Fabry–perot resonator used to find λ.

(a) What is the carrier wavelength λ?

(b) What is the mode number m of the resonance?

(c) What is the modulation frequency $\Delta\nu$?

3.5 The fringe rings of a Fabry–Pérot spectroscope are shown in Fig. P3.5. Using the key dimensions in the figure, find the wavelength λ of the incident wave.

3.6 It is believed that the gravitational wave causes ground strain when it arrives from extragalactic sources. A 300-m long Fabry–Pérot cavity such as shown in Fig. P3.6 was built [40] to detect the ground strain. What is the resolution

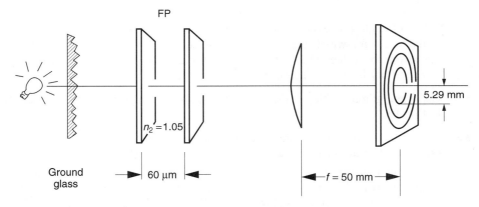

Figure P3.5 Fringe rings of the Fabry–perot spectroscope used to find λ.

Figure P3.6 The 300-m Fabry–perot cavity of the TAMA gravitational wave detector. (Courtesy of A. Araya et al. [40].)

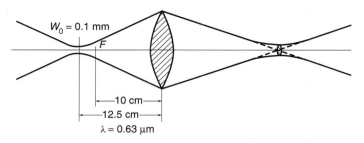

Figure P3.7   Geometry of a Gaussian beam.

$\Delta L/L$ of such a detection system, where L is the length of the cavity and ΔL is the strain for which the output intensity of the cavity drops to one-half of the resonance value. The wavelength of the laser is $\lambda = 1064$ nm. The length of the cavity is 300 m. The finesse of the cavity is $F = 516$.

3.7   Using the graphs in Fig. 3.27, obtain M, d_1, W_1, z_1, and θ_1 for the case shown in Fig. P3.7.

3.8   A free-space optical communication link is to be established between two satellites. The distance between the satellites is 3.6×10^3 km

    (a) A beam waist of 1 m on the receiver satellite is desired. A 10-m focal length lens is installed on the transmitter satellite, and the light wavelength is 0.63 µm, as shown in Fig. P3.8a. What are the size and location of the waist of the launching beam?

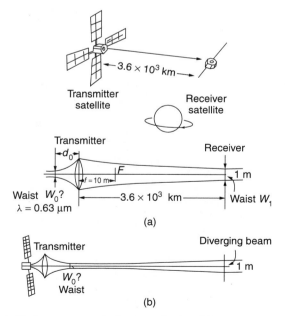

(a)

(b)

Figure P3.8   Inter satellite free-space optical communication. (a) Waist of the beam is on the receiver side. (b) Waist of the beam is on the transmitter side.

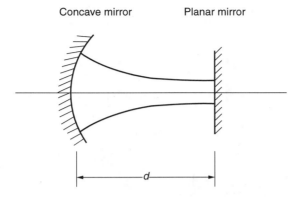

Concave mirror   Planar mirror

Figure P3.9 Combination of a concave mirror and a planar mirror.

(b) Assume this time that the beam waist is located at the surface of the transmitting satellite, as shown in Fig. P3.8b. The Gaussian beam expands to a radius of 1 m at the surface of the receiving satellite. Estimate the size of the beam waist launched from the transmitting satellite.

3.9 A cavity is constructed with a concave mirror with radius of curvature R_c and a plane mirror spaced by a distance d, as shown in Fig. P3.9. Find the radii of the beam on the end mirrors.

3.10 A combination of an annular slit and a convex lens, as shown in Fig. 3.41d, is used to generate a Bessel beam with the following characteristics. The size of the main lobe, which is defined in terms of the dimension ρ from the center to the first zero of the main lobe, is

$$\rho = 60 \ \mu m$$

The maximum beam distance is

$$z_{\max} = 1 \ m$$

The radius of the annular slit is 2.5 mm. The wavelength is 0.63 μm. Find the focal length and radius of the convex lens.

3.11 In 1873 Sir William Crooks (1832–1919) proposed that a radiometer, such as the one shown in Fig. P3.11, could measure radiation pressure. Which way should it rotate if it is indeed measuring the radiation pressure?

3.12 An experiment is conducted to demonstrate radiation pressure by showing how much a reflective sphere is pushed by a laser beam (see Fig. P3.12) [35]. The radius of the sphere is equal to the wavelength of the laser beam and its density is 1 g/cm^3.

   An argon ion laser with an output light power of 1 watt, and with $\lambda = 0.5145$ μm is used. Assume that only 7% of the available force is used for pushing the reflective sphere.

Figure P3.11   Is it a radiometer or a radiation pressure gauge?

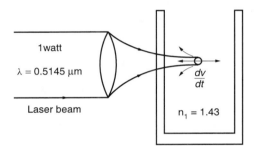

Figure P3.12   Radiation pressure experiment.

(a) How much is the radiation pressure force acting on the sphere?

(b) What is the acceleration that this sphere acquires due to the pushing force of the laser beam?

(c) Compare the acceleration by radiation pressure with that by gravity.

3.13   A convex lens is used to focus a parallel light beam. What is the direction of the radiation pressure acting on the convex lens?

REFERENCES

1. G. Hernandez, *Fabry–Pérot Interferometers*, Cambridge University Press, Cambridge, 1986.

2. J. M. Vaughan, *The Fabry–Pérot Interferometer, History, Theory, Practice and Applications*, Adam Hilger, Bristol, 1989.

3. A. Yariv, *Optical Electronics*, Saunders, Philadelphia, 1991.

4. M. Mansuripur, "Fabry–Pérot etalons in polarized light," *Opt. and Photonics News* 39–44 (Mar. 1997).

5. C. Y. Chen, M. M. Choy, M. J. Andrejco, M. A. Saifi, and C. Lin, "A widely tunable erbium-doped fiber laser pumped at 532 nm," *IEEE Photonics Technol. Lett.* **2**(1), 18–20 (1990).

6. J. R. Andrews, "Low voltage wavelength tuning of an external cavity diode laser using a nematic liquid crystal-containing birefringent filter," *IEEE Photonics Technol. Lett.* **2**(5), 334–336 (1990).

7. J. S. Patel and Y. Silverberg, "Liquid crystal and grating-based multiple wavelength cross-connect switch," *IEEE Photonics Technol. Lett.* **7**(5), 514–516 (1995).

8. J. S. Patel and M. W. Maeda, "Tunable polarization diversity liquid-crystal wavelength filter," *IEEE Photonics Technol. Lett.* **3**(8), 739–740 (1991).

9. P. J. Collings and J. S. Pantel, *Handbook of Liquid Crystal Research*, Oxford University Press, 1997.

10. J. S. Patel and S. D. Lee, "Electronically tunable and polarization insensitive Fabry–Pérot etalon with a liquid-crystal film," *Appl. Phys. Lett.* **58**(22), 2491–2493 (1991).

11. J. S. Patel, "Polarization insensitive tunable liquid-crystal etalon filter," *Appl. Phys. Lett.* **59**(11), 1314–1316 (1991).

12. S. Yamazaki, M. Shibutani, N. Shimosaka, S. Murata, T. Ono, M. Kitamura, K. Emura, and M. Shikada, "A coherent optical FDM CATV distribution system," *J. Light Technol.* **8**(3), 396–405 (1990).

13. C. M. Miller and F. J. Janniello, "Passively temperature-compensated fibre Fabry–Pérot filter and its application in wavelength division multiple access computer network," *Electron. Lett.* **26**(25), 2122–2123 (1990).

14. K. Matsudaira, *Fundamentals and Experiments on Lasers* (in Japanese), Kyoritsu Publishing, Tokyo, 1973.

15. C. K. L. Wah, K. Iizuka, and A. P. Freundorfer, "94 GHz phase modulation of light by organic MNA (2 methyl 4 nitroaniline) crystal," *Appl. Phys. Lett.* **63**(23), 3110–3112 (1993).

16. H. Kogelnik and T. Li, "Laser beams and resonators," *Appl. Opt.* **5**(10), 1550–1567 (1966).

17. B. E. A. Saleh and M. C. Teich, *Fundamentals of Photonics*, Wiley, New York, 1991.

18. T. F. Johnston, Jr., "Beam propagation (M^2) measurement made as easy as it gets: the four-cuts method," *Appl. Opt.* **37**(21), 4840–4850 (1998).

19. D. Marcuse, *Light Transmission Optics*, 2nd ed., Van Nostrand Reinhold, New York, 1982.

20. J. Durnin, "Exact solutions for nondiffracting beams. I. The scalar theory," *J. Opt. Soc. Am. A* **4**(4), 651–654 (1987).

21. R. M. Herman and T. A. Wiggins, "Production and uses of diffractionless beams," *J. Opt. Soc. Am. A* **8**(6), 932–942 (1991).

22. J. Durnin, J. J. Miceli, Jr., and J. H. Eberly, "Diffraction-free beams," *Phys. Rev. Lett.* **58**(15), 1499–1501 (1987).

23. G. Bickel, G. Häusler, and M. Maul, "Triangulation with expanded range of depth," *Opt. Eng.* **24**(6), 975–977 (1985).

24. S. Sogomonian, S. Klewitz, and S. Herminghaus, "Self-reconstruction of a Bessel beam in a nonlinear medium," *Opt. Commun.* **139**, 313–319 (1997).

25. L. Niggl, T. Lanzl, and M. Maier, "Properties of Bessel beams generated by periodic gratings of circular symmetry," *J. Opt. Soc. Am. A* **14**(1), 27–33 (1997).

26. T. Wulle and S. Herminghaus, "Nonlinear optics of Bessel beams," *Phys. Rev. Lett.* **70**(10), 1401–1404 (1993).

27. P. L. Overfelt and C. S. Kenney, "Comparison of the propagation characteristics of Bessel, Bessel–Gauss and Gaussian beams diffracted by a circular aperture," *J. Opt. Soc. Am. A* **8**(5), 732–745 (1991).

28. A. S. Chai and H. J. Wertz, "The digital computation of the far-field radiation pattern of a truncated Gaussian aperture distribution," *IEEE Trans. Antennas Propag.* **AP13**, 994–995 (1965).

29. J. H. McLeod, "The axicon: a new type of optical element," *J. Opt. Soc. Am.* **44**(8), 592–597 (1954).

30. A. Vasara, J. Turunen, and A. T. Friberg, "Realization of general nondiffracting beams with computer-generated holograms," *J. Opt. Soc. Am. A* **6**(11), 1748–1754 (1989).

31. A. J. Cox and D. C. Dibble, "Holographic reproduction of a diffraction-free beam," *Appl. Opt.* **30**(11), 1330–1332 (1991).

32. T. Aruga, S. W. Li, S. Yoshikado, M. Takabe, and R. Li, "Nondiffracting narrow light beam with small atmospheric turbulance-influenced propagation," *Appl. Opt.* **38**(15), 3152–3156 (1999).

33. A. Ashkin, "Forces of a single-beam gradient laser trap on a dielectric sphere in the ray optics regime," *Biophys. J.* **61**, 569–582 (1992).

34. A. Ashkin, "Optical trapping and manipulation of neutral particles using lasers," *Proc. Natl. Acad. Sci. USA* **94**, 4853–4860 (1997).

35. A. Ashkin, "Acceleration and trapping of particles by radiation pressure," *Phys. Rev. Lett.*, **24**(4), 156–159 (1970).

36. S. Chu, "The manipulation of neutral particles," *Rev. Mod. Phys.* **70**(3), 685–706 (1998).

37. M. Watanabe, "The Nobel Prize in Physics 1997 for development of methods to cool and trap atoms with laser light," *Rev. Laser Eng.* **26**(2), 195–198 (1998).

38. W. D. Phillips, "Laser cooling and trapping of neutral atoms," *Rev. Mod. Phys.* **70**(3), 721–741 (1998).

39. C. N. Cohen-Tannoudji, "Manipulating atoms with photons," *Rev. Mod. Phys.* **70**(3), 707–719 (1998).

40. A. Araya, S. Telada, K. Tochikubo, S. Taniguchi, R. Takahashi, K. Kawabe, D. Tatsumi, T. Yamazaki, S. Kawamura, S. Miyoki, S. Moriwaki, M. Musha, S. Nagano, M. Fujimoto, K. Horikoshi, N. Mio, Y. Naito, A. Takamori, and K. Yamamoto, "Absolute length determination of a long-baseline Fabry–Pérot cavity by means of resonating modulation sidebands," *Appl. Opt.* **38**(13), 2848–2856 (1999).

4

PROPAGATION OF LIGHT IN ANISOTROPIC CRYSTALS

Electrooptic materials change their indices of refraction when an external field is applied. These materials have many applications and are used in integrated optic devices such as phase or amplitude modulators, optical switches, optical couplers, optical deflectors, and harmonic frequency generators. They are extremely important materials in integrated optics. Both electrooptic and acoustooptic media, however, are generally anisotropic and the optical properties in one direction are different from those in other directions.

The manner of light propagation in anisotropic media is more complicated than propagation in isotropic media. A good understanding of how light propagates in electrooptic materials is essential in order to avoid undesirable effects or to take advantage of special properties to build more sophisticated devices.

With the right approach, the analysis is not all that complicated. A good starting point is to allow for the fact that only two types of waves exist in the medium. The first type follows the usual laws of propagation and does not create new problems. However, the second wave does not necessarily propagate in the expected direction perpendicular to the wavefront, and herein lies the challenge. Added to this challenge is the fact that the index of refraction varies with the direction of propagation. Fortunately, the two types of waves are easily identifiable by the direction of polarization. Moreover, the two waves can be treated separately, one at a time, and the two results can be added to obtain the final result.

This chapter and the next are twin chapters. They can almost be considered as one chapter. Chapter 4 is devoted to the manner of propagation in a crystal for a given set of anisotropy parameters, while Chapter 5 looks at what kind of external control field is needed to change the anisotropy parameters so as to achieve the desired manner of light propagation.

Chapter 4 starts with a brief explanation of the meaning of polarization in a crystal and the qualitative difference between propagation in isotropic and anisotropic media. Then, differences are derived quantitatively using the wave equation. There are two popular methods of obtaining a graphical solution to the wave equation—the wavevector method and the indicatrix method. Each of these is explained. Using both graphical and analytical methods, the laws of refraction across the boundary between two anisotropic media are examined. As a practical application, in Chapter 6 various optical elements are described that manipulate the state of polarization of the light. These optical elements are used to construct devices such as polarizers, quarter-waveplates, and half-waveplates.

4.1 POLARIZATION IN CRYSTALS

Solids can broadly be classified as either crystalline or amorphous. Crystalline solids are characterized by a regular periodic sequence of the building block molecules. Amorphous solids such as glass or plastic have an irregular molecular arrangement. Optical, electrical, or mechanical properties of the crystalline solids are generally direction sensitive because of the anisotropy.

Figure 4.1a is a two-dimensional model of a crystalline solid. Coulomb's forces acting between the nucleus and the electrons, or among the electrons themselves, are stably balanced within the constraint of overall electrical neutrality and minimum energy. When an external electric field is applied to the crystal by placing the crystal between two capacitor plates, the entire electron pattern is translated toward the positive electrode, as indicated in Fig. 4.1b. The translation of the electron cloud creates a positive excess charge layer on the top surface of the crystal, and a negative charge layer at the bottom surface. These surface charges establish an additional electric field in the crystal. Such a phenomenon is called polarization of the crystal due to the external field [1].

The switch of the battery in Fig. 4.1b is first turned on so as to charge up the capacitor plates with free electrons. The switch remains open so that the surface density of the free electrons on the plate remains fixed. Then, the crystal is inserted. The fields before and after the insertion of the crystal are compared to determine the degree of polarization of the crystal.

As shown in Fig. 4.1c, the direction of the field established by the polarization is opposite to that of the field due to the original charges on the electrode, resulting in a

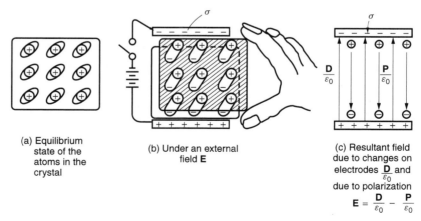

(a) Equilibrium
state of the
atoms in the
crystal

(b) Under an external
field **E**

(c) Resultant field
due to changes on
electrodes $\dfrac{\mathbf{D}}{\varepsilon_0}$ and
due to polarization

$$\mathbf{E} = \frac{\mathbf{D}}{\varepsilon_0} - \frac{\mathbf{P}}{\varepsilon_0}$$

Figure 4.1 Change in the electric field **E** in the capacitor. First, the air-filled capacitor is charged by the battery and then the switch is left open during the experiment. The **E** field is measured before and after the insertion of the crystal.

reduction of the electric field between the capacitor plates. The field strength is reduced to ϵ_r^{-1} of the original field.

If the same crystal is rotated by 90° in the plane of the page, ϵ_r will be different for anisotropic crystalline structures. Chemists use polarization information to study molecular structure.

There are basically three different mechanisms for inducing polarization: (1) atomic polarization, (2) orientation polarization, and (3) space-charge polarization. These are illustrated in Fig. 4.2.

A typical example of atomic polarization is the polarization associated with a hydrogen atom shown in Fig. 4.2a. Without the external electric field, the nucleus is surrounded by a spherically symmetric electron cloud and the "center of gravity" of the negative charges coincides with that of the positive charges and no dipole moment exists. When, however, an external electric field is applied, both the nucleus and the electron cloud shift away from each other along the direction of the external electric field and the centers of gravity of the positive and negative charges no longer coincide. A dipole moment is created, resulting in the polarization.

Some molecules posses a permanent dipole moment, which is already present even before the external electric field is applied. A good example of this kind of molecule is the water molecule. The atoms of the water molecule are arranged in a triangular shape, as shown in Fig. 4.2b. The "center of gravity" of the positive charges of the three nuclei

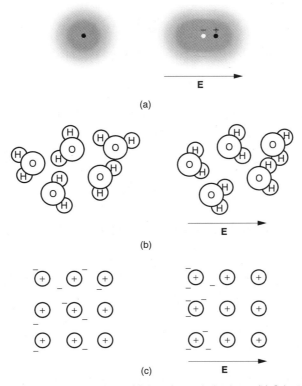

Figure 4.2 Mechanisms of polarization. (a) Atomic polarization. (b) Orientation polarization. (c) Space-charge polarization.

would be somewhere inside the triangle. The location of the center of the three electron clouds (two belonging to the hydrogen atoms and one belonging to the oxygen atom) will be at the same location as the center of gravity of the nuclei, provided the electron clouds are perfectly spherically symmetric. However, in reality, this symmetry is broken by the binding process. When a molecular bond is formed, electrons are displaced toward the stronger binding atom. In the water molecule, oxygen is the stronger binding atom, so that electrons are displaced toward the oxygen atom. In a water molecule, the centers of gravity of the positive and negative charges no longer coincide, and a dipole moment exists even without an external electric field. Ironically, this permanent dipole moment is $10^3 - 10^4$ times larger than that of the atomic dipole moment of a hydrogen atom under an external electric field of 3×10^4 V/cm, which is the maximum external electric field that can be applied without causing electrical arcing [2].

Both atomic and orientation polarizations are produced by the displacement or reorientation of the bound charges. The space-charge polarization shown in Fig. 4.2(c), however, is produced by the buildup of traveling charge carriers within a specific volume or on the surface at an interface.

Note in Fig. 4.1 that the direction of the electric field $\mathbf{P}/\epsilon_0$ due to the polarization is opposite to that of $\mathbf{D}/\epsilon_0$. The resultant field $\mathbf{E}$ including the contribution from the original charge on the electrodes is

$$\mathbf{E} = \frac{\mathbf{D}}{\epsilon_0} - \frac{\mathbf{P}}{\epsilon_0} \tag{4.1}$$

Hence,

$$\mathbf{D} = \epsilon_0 \mathbf{E} + \mathbf{P} \tag{4.2}$$

In the linear range of polarization and in an isotropic medium, $\mathbf{P}$ is proportional to $\mathbf{E}$ and

$$\mathbf{P} = \epsilon_0 \chi \mathbf{E} \tag{4.3}$$

where χ is the electric susceptibility. Inserting Eq. (4.3) into (4.2) gives

$$\mathbf{D} = \epsilon_0 n^2 \mathbf{E} \tag{4.4}$$

where

$$\epsilon_r = n^2 = (1 + \chi) \tag{4.5}*$$

4.2 SUSCEPTIBILITY OF AN ANISOTROPIC CRYSTAL

The susceptibility χ of an anisotropic crystal varies according to the direction. For an anisotropic crystal, χ in Eq. (4.3) has to be replaced by a general form of susceptibility tensor defined as

$$\begin{bmatrix} P_x \\ P_y \\ P_z \end{bmatrix} = \epsilon_0 \begin{bmatrix} \chi_{11} & \chi_{12} & \chi_{13} \\ \chi_{21} & \chi_{22} & \chi_{23} \\ \chi_{31} & \chi_{32} & \chi_{33} \end{bmatrix} \begin{bmatrix} E_x \\ E_y \\ E_z \end{bmatrix} \tag{4.6}$$

---

* ϵ_r of water is known to be 81. Does this mean $n = 9$? It is important that ϵ_r and n are measured at the same frequency. $\epsilon_r = 81$, however, is the value measured at 114 MHz and $n = 9$ is not applicable at optical frequencies.

The values of the tensor elements depend on the choice of coordinates with respect to the crystal axis. In general, it is possible to choose the coordinates so that $\overset{\leftrightarrow}{\chi}$ becomes a diagonal matrix [2–4].

If the new coordinates are chosen in the directions parallel to the eigenvectors, the susceptibility tensor converts into a much more manageable form. This form is

$$\overset{\leftrightarrow}{\chi} = \begin{bmatrix} \chi'_{11} & 0 & 0 \\ 0 & \chi'_{22} & 0 \\ 0 & 0 & \chi'_{33} \end{bmatrix} \tag{4.7}$$

where χ'_{ii} are called the principal susceptibilities. The coordinate axes for which the susceptibility tensor is reduced to diagonal form are called the principal axes. The corresponding index of refraction tensor becomes

$$\mathbf{D} = \epsilon_0 \begin{bmatrix} n_\alpha^2 & 0 & 0 \\ 0 & n_\beta^2 & 0 \\ 0 & 0 & n_\gamma^2 \end{bmatrix} \mathbf{E} \tag{4.8}$$

where n_α, n_β, and n_γ are called the principal refractive indices. Crystals with two identical principal indices are called uniaxial crystals, and those with three different principal indices are called biaxial crystals.

Analysis with biaxial crystals is significantly more complicated than with uniaxial crystals. Since most of the optically transparent crystals that are used for electrooptic devices are uniaxial, this chapter will concentrate on uniaxial crystals.

With a uniaxial crystal it is always possible to find an axis of rotational symmetry and such an axis is called the *optic axis* of the crystal. Sometimes it is casually called the *crystal axis* or simply the *c axis*. Uniaxial crystals have only one optic axis (taken as the z axis in this chapter), while biaxial crystals have two optic axes.

With the uniaxial crystal,

$$n_\alpha = n_\beta = n_o$$
$$n_\gamma = n_e \tag{4.9}$$

Equation (4.8) becomes

$$\begin{bmatrix} D_x \\ D_y \\ D_z \end{bmatrix} = \epsilon_0 \begin{bmatrix} n_o^2 & 0 & 0 \\ 0 & n_o^2 & 0 \\ 0 & 0 & n_e^2 \end{bmatrix} \begin{bmatrix} E_x \\ E_y \\ E_z \end{bmatrix} \tag{4.10}$$

Solving for E_x, E_y, and E_z leads to

$$\epsilon_0 \begin{bmatrix} E_x \\ E_y \\ E_z \end{bmatrix} = \begin{bmatrix} \dfrac{1}{n_o^2} & 0 & 0 \\ 0 & \dfrac{1}{n_o^2} & 0 \\ 0 & 0 & \dfrac{1}{n_e^2} \end{bmatrix} \begin{bmatrix} D_x \\ D_y \\ D_z \end{bmatrix} \tag{4.11}$$

If $n_e > n_o$, the birefringence is called *positive birefringence*, and if $n_e < n_o$, it is called *negative birefringence*.

4.3   THE WAVE EQUATION IN AN ANISOTROPIC MEDIUM

The propagation of a lightwave in an anisotropic medium will be analyzed [5,6]. Maxwell's equations for isotropic and anisotropic media are quite similar. The only difference is that the dielectric constant is a tensor $\overset{\leftrightarrow}{\epsilon}_r$ in the anisotropic case, whereas it is a scalar in the isotropic case. Maxwell's equations are

$$\nabla \times \mathbf{H} = -j\omega \mathbf{D} \tag{4.12}$$

$$\nabla \times \mathbf{E} = j\omega\mu \mathbf{H} \tag{4.13}$$

where

$$\mathbf{D} = \epsilon_0 \overset{\leftrightarrow}{\epsilon}_r \mathbf{E} \tag{4.14}$$

$$\nabla \cdot \mathbf{D} = \rho \tag{4.15}$$

and where ρ is the free-charge density. The curl operation in Eq. (4.13) with the use of Eq. (4.12) immediately leads to

$$\nabla \times \nabla \times \mathbf{E} = \omega^2 \mu \mathbf{D} \tag{4.16}$$

Equation (4.16) is essentially a wave equation and is one of the most basic expressions governing propagation in any kind of media. It will be used often in this chapter to deal with anisotropy. Sometimes, a simple recast of Eq. (4.16) provides an answer.

In this entire chapter, the analysis is restricted to plane waves. The plane wave implies that the sole temporal and spatial dependency in $\mathbf{E}$ and $\mathbf{H}$ is a factor expressed by

$$e^{-j\omega t + j\mathbf{k}\cdot\mathbf{r}} \tag{4.17}$$

where

$$\mathbf{k} = k_x\hat{\mathbf{i}} + k_y\hat{\mathbf{j}} + k_z\hat{\mathbf{k}} \tag{4.18}$$

is the propagation vector.

With the assumption of Eq. (4.17), $(\partial/\partial x, \partial/\partial y, \partial/\partial z)$ in the curl operation becomes (jk_x, jk_y, jk_z) and the curl operation is simply

$$\nabla \times \mathbf{E} = j\mathbf{k} \times \mathbf{E} \tag{4.19}$$

Then, Eqs. (4.12) and (4.13) become

$$\mathbf{k} \times \mathbf{H} = -\omega \mathbf{D} \tag{4.20}$$

$$\mathbf{k} \times \mathbf{E} = \omega\mu \mathbf{H} \tag{4.21}$$

Equation (4.20) specifies that $\mathbf{D}$ is perpendicular to $\mathbf{H}$, and Eq. (4.21) specifies that $\mathbf{E}$ is also perpendicular to $\mathbf{H}$. Thus, both $\mathbf{E}$ and $\mathbf{D}$ have to be perpendicular to $\mathbf{H}$. As shown in Fig. 4.3, both $\mathbf{E}$ and $\mathbf{D}$ are in the same plane but this does not necessarily

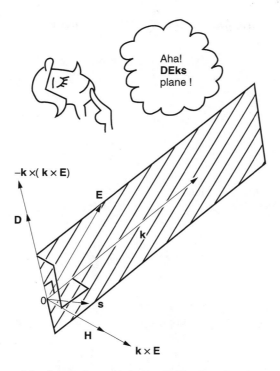

Figure 4.3   Orientations of **k**, **E**, **k** × **E**, **H**, −**k** × (**k** × **E**), and **D**.

mean that **E** and **D** are parallel. Equation (4.21) also specifies that **k** is perpendicular to **H**; thus, **E**, **D**, and **k** are all in the same plane. Equations (4.20) and (4.21) lead to

$$-\mathbf{k} \times (\mathbf{k} \times \mathbf{E}) = \omega^2 \mu \mathbf{D} \tag{4.22}$$

In short, we can make the following observations: (1) **D**, **E**, and **k** are in the same plane, which is perpendicular to **H** and (2) **D** is perpendicular to **k**.

Multiplying both sides of Eq. (4.22) by ϵ_0 and using Eq. (4.11) gives

$$-\mathbf{k} \times \left[\mathbf{k} \times \left(\frac{\overleftrightarrow{1}}{n^2} \right) \mathbf{D} \right] = k_0^2 \mathbf{D} \tag{4.23}$$

Equation (4.23) is called the *generalized wave equation*.

4.4   SOLVING THE GENERALIZED WAVE EQUATION IN UNIAXIAL CRYSTALS

It will be shown that, in an anisotropic medium, only two waves are allowed to propagate. These waves are the *ordinary wave*, or *o-wave*, and the *extraordinary wave*, or *e-wave*. The existence of these two waves, and only these two waves, is the most important fact about propagation in anisotropic media.

4.4.1   Graphical Derivation of the Condition of Propagation in a Uniaxial Crystal

If the optic axis of the unaxial crystal is taken along the z direction, the medium is cylindrically symmetric with respect to the z axis. In this case, it is always possible to choose coordinates such that $k_y = 0$,

$$\mathbf{k} = \begin{bmatrix} k_x \\ 0 \\ k_z \end{bmatrix} \tag{4.24}$$

meaning that $\mathbf{k}$ is in the x–z plane.

$\mathbf{D}$, which is in the plane perpendicular to $\mathbf{k}$, is represented as

$$\mathbf{D} = \begin{bmatrix} D_x \\ D_y \\ D_z \end{bmatrix} \tag{4.25}$$

Only certain directions of the vector $\mathbf{D}$ within this plane will satisfy Eq. (4.23). A short proof follows. Let the vector $\mathbf{T}$ represent the left-hand side of Eq. (4.23).

$$\mathbf{T} = -\mathbf{k} \times \left[\mathbf{k} \times \left(\frac{\overset{\leftrightarrow}{1}}{n^2} \right) \mathbf{D} \right] \tag{4.26}$$

In order for $\mathbf{D}$ to satisfy the wave equation, $\mathbf{T}$ must point in the same direction as $\mathbf{D}$, which is the right-hand side of Eq. (4.23).

Referring to Fig. 4.4a, let us start with an arbitrary vector $\mathbf{D}$ except that it is in a plane perpendicular to $\mathbf{k}$ satisfying Eq. (4.20). The first operation of Eq. (4.26) is that

$$\mathbf{D}' = \left(\frac{\overset{\leftrightarrow}{1}}{n^2} \right) \mathbf{D} \tag{4.27}$$

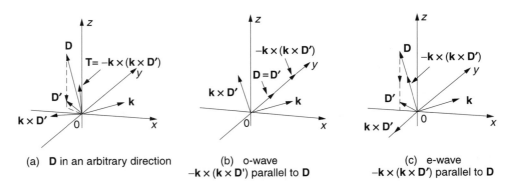

(a) **D** in an arbitrary direction

(b) o-wave
$-\mathbf{k} \times (\mathbf{k} \times \mathbf{D}')$ parallel to **D**

(c) e-wave
$-\mathbf{k} \times (\mathbf{k} \times \mathbf{D}')$ parallel to **D**

Figure 4.4 Vector diagram for explaining that only o-waves and e-waves exist in an anisotropic medium.

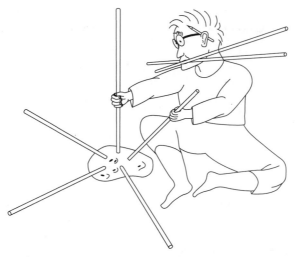

By poking stick-like objects, such as chop sticks, knitting needles, or skewers, into a potato you can make your own study kit.

where $\mathbf{D}'$ is a quantity proportional to $\mathbf{E}$ from Eq. (4.11). With the hypothetical example of

$$\left(\frac{\overset{\leftrightarrow}{1}}{n^2}\right) = \frac{1}{n_o^2}\begin{bmatrix} 1 & 0 & 0 \\ 0 & 1 & 0 \\ 0 & 0 & \frac{1}{10} \end{bmatrix}$$

$\mathbf{D}'$ is expressed as

$$\mathbf{D}' = \frac{1}{n_o^2}\begin{bmatrix} 1 & 0 & 0 \\ 0 & 1 & 0 \\ 0 & 0 & \frac{1}{10} \end{bmatrix}\begin{bmatrix} D_x \\ D_y \\ D_z \end{bmatrix} = \frac{1}{n_o^2}\begin{bmatrix} D_x \\ D_y \\ \frac{1}{10}D_z \end{bmatrix} \tag{4.28}$$

Note that D_z in this example has been reduced to one-tenth of its original value.

Next, vector $\mathbf{k} \times \mathbf{D}'$ is in a plane perpendicular to $\mathbf{k}$. Finally, vector $\mathbf{T}$ in Eq. (4.26) is drawn. $\mathbf{T}$ is also in a plane perpendicular to $\mathbf{k}$. Thus, $\mathbf{D}$, $\mathbf{k} \times \mathbf{D}'$, and $\mathbf{T}$ are all in the same plane perpendicular to $\mathbf{k}$. In this plane, even though $\mathbf{T}$ is perpendicular to $\mathbf{k} \times \mathbf{D}'$, $\mathbf{D}$ is not necessarily perpendicular to $\mathbf{k} \times \mathbf{D}'$; thus, $\mathbf{T}$ and $\mathbf{D}$ are not necessarily in the same direction, and Eq. (4.23) is not satisfied in general. There are, however, two particular directions of $\mathbf{D}$ for which $\mathbf{T}$ and $\mathbf{D}$ point in the same direction to satisfy Eq. (4.23). The waves corresponding to these two particular directions of $\mathbf{D}$ are the ordinary wave (o-wave) and the extraordinary wave (e-wave). In the o-wave, $\mathbf{D}$ is polarized in the direction perpendicular to both $\mathbf{k}$ and the optic axis, namely, in the y direction and in the e-wave, $\mathbf{D}$ is polarized in the plane defined by $\mathbf{k}$ and the optic axis z, namely, in the x–z plane as shown in Figs. 4.4b and 4.4c, respectively.

We will now further explain how the o- and e-waves satisfy Eq. (4.23). First, with the o-wave, the operations of Eq. (4.26) are performed as shown in Fig. 4.4b. Since **D** is polarized in the y direction, Eq. (4.25) becomes

$$\mathbf{D} = \begin{bmatrix} 0 \\ D_y \\ 0 \end{bmatrix} \tag{4.29}$$

The operation of

$$\mathbf{D}' = \left(\frac{\overset{\leftrightarrow}{1}}{n^2} \right) \mathbf{D} \tag{4.30}$$

does not change the direction of **D**, and both **D** and **D**' point in the same direction. Next, the vector $\mathbf{k} \times \mathbf{D}'$ is in the x–z plane, and both **k** and $\mathbf{k} \times \mathbf{D}'$ are in the same plane. Finally, the vector $\mathbf{T} = -\mathbf{k} \times (\mathbf{k} \times \mathbf{D}')$ points in the y direction, which is the direction of **D** and Eq. (4.23) is satisfied.

Next, with the e-wave, the operations of Eq. (4.26) are performed as shown in Fig. 4.4c. **D** is polarized in the x–z plane and is also perpendicular to **k**. Eq. (4.25) for the e-wave is

$$\mathbf{D} = \begin{bmatrix} D_x \\ 0 \\ D_z \end{bmatrix} \tag{4.31}$$

The first operation

$$\mathbf{D}' = \left(\frac{\overset{\leftrightarrow}{1}}{n^2} \right) \mathbf{D} \tag{4.32}$$

brings **D** to **D**'. Using the hypothetical example in Eq. (4.28), **D**' becomes

$$\mathbf{D}' = \frac{1}{n_o^2} \begin{bmatrix} D_x \\ 0 \\ \frac{1}{10} D_z \end{bmatrix} \tag{4.33}$$

Even though the z component of **D**' is reduced to one-tenth of D_z, **D**' is still in the x–z plane. Next, the vector $\mathbf{k} \times \mathbf{D}'$ points in the $-y$ direction. What is important here is that $\mathbf{k} \times \mathbf{D}'$ always points in the $-y$ direction regardless of the value of D_z'.

The final cross product with **k** brings the vector

$$\mathbf{T} = -\mathbf{k} \times (\mathbf{k} \times \mathbf{D}')$$

into the x–z plane and also is perpendicular to **k**, which is the original direction of **D**. Thus, the directions of **T** and **D** match and the wave equation Eq. (4.23) is satisfied.

Thus, in conclusion, we have learned that there are two types of waves in an anisotropic crystal: o- and e-waves. The directions of polarization of **D** for both o- and e-waves are in the plane perpendicular to **k**, but that of the o-wave is the y direction and that of the e-wave is in the plane containing **k** and the optic axis. The directions of polarization of **D** for these waves are orthogonal to each other. No other directions of polarization are allowed. How then can light incident with an arbitrary direction of polarization be analyzed? The answer is to decompose the incident wave into two waves, each with the allowed direction of polarization. Propagation of each component wave is treated separately and summed at the exit of the crystal to obtain the expression of the transmitted light.

4.4.2   Analytical Representation of the Conditions of Propagation in a Uniaxial Crystal

In the previous section, a vector diagram argument was presented to show that only e-waves and o-waves propagate in a uniaxial crystal. Here, an analytical representation of each operation will be made. Let us start with an arbitrary $\mathbf{D}$ given by

$$\mathbf{D} = \begin{bmatrix} D_x \\ D_y \\ D_z \end{bmatrix} \tag{4.34}$$

Since $\mathbf{D}$ and $\mathbf{k}$ are perpendicular to each other from Eq. (4.20),

$$\mathbf{k} \cdot \mathbf{D} = 0 \tag{4.35}$$

Inserting Eqs. (4.24) and (4.34) into Eq. (4.35) gives

$$D_z = -\frac{k_x}{k_z} D_x \tag{4.36}$$

Let $(\overset{\leftrightarrow}{1}/n^2)$ be

$$\left(\frac{\overset{\leftrightarrow}{1}}{n^2}\right) = \frac{1}{n_o^2} \begin{bmatrix} 1 & 0 & 0 \\ 0 & 1 & 0 \\ 0 & 0 & \dfrac{n_o^2}{n_e^2} \end{bmatrix} \tag{4.37}$$

Combining Eqs. (4.34), (4.36), and (4.37) gives

$$\left(\frac{\overset{\leftrightarrow}{1}}{n^2}\right)\mathbf{D} = \frac{1}{n_o^2} \begin{bmatrix} D_x \\ D_y \\ -\dfrac{n_o^2}{n_e^2}\dfrac{k_x}{k_z} D_x \end{bmatrix} \tag{4.38}$$

The value inside the square bracket of Eq. (4.26) is

$$\mathbf{k} \times \left[\left(\frac{\overset{\leftrightarrow}{1}}{n^2}\right)\mathbf{D}\right] = \frac{1}{n_o^2} \begin{bmatrix} \hat{\mathbf{i}} & \hat{\mathbf{j}} & \hat{\mathbf{k}} \\ k_x & 0 & k_z \\ D_x & D_y & -\dfrac{n_o^2}{n_e^2}\dfrac{k_x}{k_z} D_x \end{bmatrix}$$

$$= \frac{1}{n_o^2} \begin{bmatrix} -k_z D_y \\ \left(k_z + \dfrac{n_o^2}{n_e^2}\dfrac{k_x^2}{k_z}\right) D_x \\ k_x D_y \end{bmatrix} \tag{4.39}$$

Another cross product with $\mathbf{k}$ is performed to find $\mathbf{T}$ of Eq. (4.26):

$$\mathbf{T} = -\mathbf{k} \times \mathbf{k} \times \left[\left(\overset{\leftrightarrow}{\frac{1}{n^2}} \right) \mathbf{D} \right] = \frac{1}{n_o^2} \begin{bmatrix} \left(k_z^2 + \dfrac{n_o^2 k_x^2}{n_e^2} \right) D_x \\ k^2 D_y \\ -\dfrac{k_x}{k_z} \left(k_z^2 + \dfrac{n_o^2 k_x^2}{n_e^2} \right) D_x \end{bmatrix} \tag{4.40}$$

where

$$k^2 = k_x^2 + k_z^2 \tag{4.41}$$

The conditions for $\mathbf{T}$ to become parallel to $\mathbf{D}$ are found by setting the cross product of the two to zero:

$$\mathbf{D} \times \mathbf{T} = 0 \tag{4.42}$$

Inserting Eqs. (4.34), (4.36), and (4.40) into Eq. (4.42) gives

$$\mathbf{D} \times \mathbf{T} = \frac{1}{n_o^2} \begin{bmatrix} \dfrac{k_x}{k_z} \left[k^2 - \left(k_z^2 + k_x^2 \dfrac{n_o^2}{n_e^2} \right) \right] D_x D_y \\ 0 \\ \left[k^2 - \left(k_z^2 + k_x^2 \dfrac{n_o^2}{n_e^2} \right) \right] D_x D_y \end{bmatrix} \tag{4.43}$$

Thus, for $\mathbf{D}$ and $\mathbf{T}$ to be parallel requires that each component of Eq. (4.43) is zero. These conditions are categorized as follows.

Case 1: $n_o^2/n_e^2 = 1$

This condition means that the medium is isotropic. From Eq. (4.41), the value in the inner square bracket inside Eq. (4.43) becomes zero; thus D_x and D_y can be arbitrary. Note, however, that D_z has to satisfy Eq. (4.36) so that $\mathbf{D}$ remains perpendicular to $\mathbf{k}$.

Case 2: $n_o^2/n_e^2 \neq 1$

The medium is anisotropic. This situation can be split into three subcases.

(a) *o-Wave*

$$D_x = 0$$
$$D_y \neq 0 \tag{4.44}$$
$$D_z = -\frac{k_x}{k_z} D_x = 0$$

This condition means $\mathbf{D}$ is parallel to the y direction and is perpendicular to the plane of the optic axis (x–z plane) and $\mathbf{k}$. Thus, this case fits the description of the o-wave.

(b) *e-Wave*

$$D_x \neq 0$$

$$D_y = 0 \tag{4.45}$$

$$D_z = -\frac{k_x}{k_z} D_x \neq 0$$

This means that **D** is in the plane made by the optic axis of the crystal and **k** and yet is perpendicular to **k**. This is precisely the description of the e-wave.

(c) *Propagation along the z axis.* The factors in the inner square bracket of the first and third row in Eq. (4.43) can be rewritten using Eq. (4.41) as

$$k_x^2 \left(1 - \frac{n_o^2}{n_e^2} \right) = 0 \tag{4.46}$$

which means

$$k_x = 0$$

Since k_y is already zero, only k_z is nonzero, meaning that propagation is along the z axis. D_x and D_y can be arbitrary but D_z has to be zero because k_x in Eq. (4.36) is zero.

4.4.3 Wavenormal and Ray Direction

The direction of the energy flow of light, which is called the ray direction, is expressed by the Poynting vector

$$\mathbf{s} = \mathbf{E} \times \mathbf{H} \tag{4.47}$$

The ray direction **s** is perpendicular to both **E** and **H**, and **s** is also included in the **DEks** plane in Fig. 4.3. The vector **s** shows the direction of energy flow and is the direction that your eyes have to be positioned at, if you want to see the light. It is, however, **k** but not **s** that follows Snell's law.

Now, referring to Fig. 4.5, **H** is taken along the y axis, hence **s** is in the x–z plane. The relationship between the directions of **s** and **k** will be calculated. Let

$$\mathbf{s} = s_x \hat{\mathbf{i}} + s_z \hat{\mathbf{k}} \tag{4.48}$$

and

$$\tan \phi = \frac{s_x}{s_z} \tag{4.49}$$

where ϕ is the angle that **s** makes with the z axis (or the optic axis). The vector **s** is perpendicular to **E** from Eq. (4.47) and $\mathbf{E} \cdot \mathbf{s} = 0$; hence,

$$E_x s_x + E_z s_z = 0 \tag{4.50}$$

With Eq. (4.50), Eq. (4.49) becomes

$$\frac{E_z}{E_x} = -\tan \phi \tag{4.51}$$

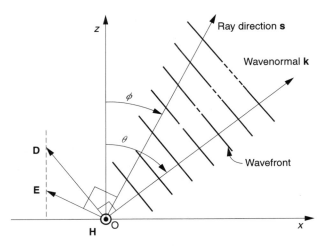

Figure 4.5 The wavefront, wave normal, and ray direction of the e-waves. (The y axis is into the page.)

Similarly,

$$\mathbf{k} = k_x\hat{\mathbf{i}} + k_z\hat{\mathbf{k}}$$

and

$$\tan\theta = \frac{k_x}{k_z} \tag{4.52}$$

Since $\mathbf{k}$ and $\mathbf{D}$ are perpendicular, $\mathbf{k}\cdot\mathbf{D} = 0$:

$$k_xD_x + k_zD_z = 0 \tag{4.53}$$

From Eqs. (4.52) and (4.53), we have

$$\frac{D_z}{D_x} = -\tan\theta \tag{4.54}$$

From Eq. (4.10), the e-wave components are related as

$$\frac{D_z}{D_x} = \left(\frac{n_e}{n_o}\right)^2 \frac{E_z}{E_x} \tag{4.55}$$

Equations (4.51), (4.54), and (4.55) give

$$\tan\theta = \left(\frac{n_e}{n_o}\right)^2 \tan\phi \tag{4.56}$$

The results of the e-wave calculations are shown in Fig. 4.5. Vector $\mathbf{s}$, which indicates the direction of the flow of light energy, does not coincide with $\mathbf{k}$, which indicates the direction of the normal to the wavefront. Also, the direction of polarization of $\mathbf{E}$ is not on the surface of the wavefront.

An analogy can be made with a sheet of cardboard paper blown into the air. The direction of the flight is not necessarily normal to the surface of the cardboard.

Another analogy is that of the skier. When a downhill skier climbs up a steep hill, the skis are set parallel to the contour line of the hill to prevent slips and the skier climbs off the fall line to decrease the effective slope. The parallel lines of the ski tracks resemble the wavefront, the fall line of the hill resembles the wavenormal **k**, and the movement of the skier resembles the ray path **s**.

Now, the double image seen through a calcite crystal (Iceland spar) can be explained. As shown in Fig. 4.6, a spotlight P is placed on the left surface of a calcite crystal

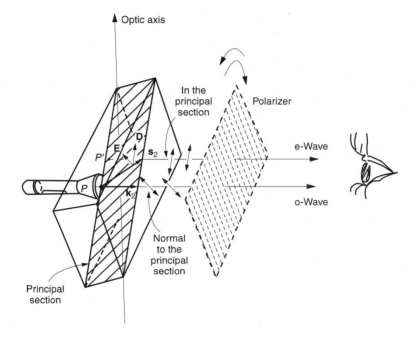

Figure 4.6   Birefringent image made by calcite crystal.

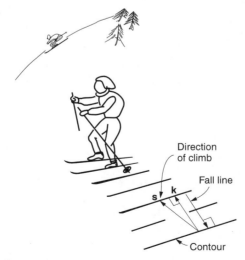

When a skier climbs a steep hill, the skiis are set parallel to the contour lines of the hill to avoid sliding, and the path of the climb is set off the fall line to decrease the effective slope. The direction of the skier is **s**.

and is observed from the right. The hatched plane, which contains the optic axis and also is perpendicular to the pair of cleaved surfaces is called a *principal section*. The incident light is polarized in an arbitrary direction. The energy of the o-wave polarized perpendicular to the principal section goes straight through following $\mathbf{k}_2$ while the energy of the e-wave polarized in the principal section follows the direction of $\mathbf{s}_2$. To the observer, one spot appears as if it were two spots. If the e- and o-wave images pass through a polarizer, then spots P and P' can be selectively seen by rotating the polarizer. The quantitative treatment of the same configuration will be described in Example 4.3.

Example 4.1   Draw the allowed directions of $\mathbf{D}$, $\mathbf{E}$, $\mathbf{s}$, and $\mathbf{H}$ for o- and e-waves for the following three cases: (a) $\mathbf{k}$ is along the x axis, (b) $\mathbf{k}$ is at an arbitrary angle θ with respect to the z axis but in the x–z plane, and (c) $\mathbf{k}$ is along the z axis. The crystal axis is along the z axis.

Solution   The answers are summarized in Fig. 4.7. The direction of $\mathbf{H}$ is always perpendicular to the **DEks** plane.   □

Example 4.2   As shown in Fig. 4.8, the unit vector $\hat{\mathbf{s}}$ of the ray direction $\mathbf{s}$ in an anisotropic crystal is $(1/2, 0, \sqrt{3}/2)$. The optic axis of the crystal is along the z axis. The tensor refractive index is

$$|n^2| = \begin{bmatrix} 2 & 0 & 0 \\ 0 & 2 & 0 \\ 0 & 0 & 3 \end{bmatrix}$$

Find the directions (unit vectors) $\hat{\mathbf{e}}$, $\hat{\mathbf{d}}$ and $\hat{\boldsymbol{k}}$, respectively, for $\mathbf{E}$, $\mathbf{D}$, and $\mathbf{k}$. Find those for both the o- and e-waves. (Note that in order to distinguish the unit vector for the z direction from the unit vector for the wavenormal, script $\hat{\boldsymbol{k}}$ is used for the unit vector for the wavenormal.)

Solution   For the o-wave, the unit polarization vectors $\hat{\mathbf{e}}$ and $\hat{\mathbf{d}}$ are $(0, 1, 0)$ or along the y axis, and $\hat{\boldsymbol{k}} = \hat{\mathbf{s}} = (1/2, 0, \sqrt{3}/2)$.

For the e-wave, the fact that $\mathbf{E}$ and $\mathbf{s}$ are perpendicular gives

$$\mathbf{s} \cdot \mathbf{E} = 0$$

$$\frac{1}{2}E_x + \frac{\sqrt{3}}{2}E_z = 0$$

$$\frac{E_x}{E_z} = -\sqrt{3}$$

$$\tan^{-1} -\sqrt{3} = -60°$$

The unit vector $\hat{\mathbf{e}}$ is $(-\sqrt{3}\hat{\mathbf{i}}, 0\hat{\mathbf{j}}, \hat{\mathbf{k}})/\sqrt{1+3}$:

$$\hat{\mathbf{e}} = \frac{\mathbf{E}}{|\mathbf{E}|} = \frac{1}{2}(-\sqrt{3}\,\hat{\mathbf{i}}, 0\hat{\mathbf{j}}, \hat{\mathbf{k}})$$

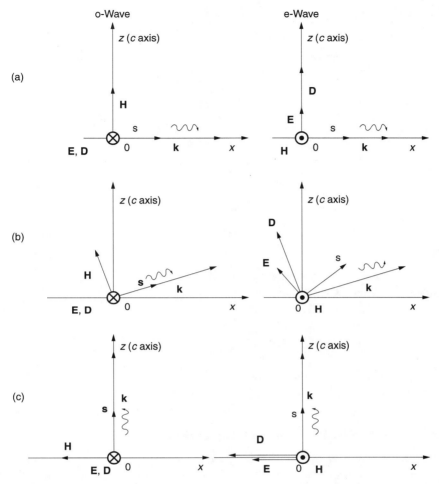

Figure 4.7 **D**, **E**, **H**, and **s** for various **k** of the o- and e-waves. The crystal axis is along the z axis. (The y axis is into the page.)

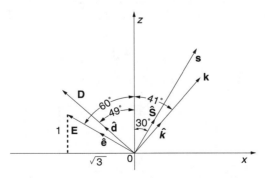

Figure 4.8 Given **s**, find the directions of **E** and **D** of the e-wave in an anisotropic crystal with $|\overset{\leftrightarrow}{n^2}| = \begin{bmatrix} 2 & 0 & 0 \\ 0 & 2 & 0 \\ 0 & 0 & 3 \end{bmatrix}$. (The y axis is into the page.)

The tensor operation is performed as follows:

$$\mathbf{D} = \epsilon_0 (\overset{\leftrightarrow}{n}^2) \hat{e} E$$

$$\mathbf{D} = \epsilon_0 \frac{E}{2} \begin{bmatrix} 2 & 0 & 0 \\ 0 & 2 & 0 \\ 0 & 0 & 3 \end{bmatrix} \begin{bmatrix} -\sqrt{3} \\ 0 \\ 1 \end{bmatrix}$$

$$= \epsilon_0 E \begin{bmatrix} -\sqrt{3} \\ 0 \\ 3/2 \end{bmatrix}$$

$$\hat{\mathbf{d}} = \frac{\mathbf{D}}{|\mathbf{D}|} = \frac{1}{\sqrt{4 \times 3 + 9}} (-2\sqrt{3}\hat{\mathbf{i}}, 0\hat{\mathbf{j}}, 3\hat{\mathbf{k}})$$

$$= \frac{1}{\sqrt{21}} (-2\sqrt{3}\hat{\mathbf{i}}, 0\hat{\mathbf{j}}, 3\hat{\mathbf{k}})$$

The angle that $\hat{\mathbf{d}}$ makes with the optical axis is 49°.
Next, $\hat{\mathbf{k}}$ is found. $\mathbf{k}$ is perpendicular to $\mathbf{D}$.

$$-2\sqrt{3}k_x + 3k_z = 0$$

$$\theta = \tan^{-1}\left(\frac{k_x}{k_z}\right) = \tan^{-1}\left(\frac{3}{2\sqrt{3}}\right) = 41°$$

$$\hat{\mathbf{k}} = \frac{\mathbf{k}}{|\mathbf{k}|} = \frac{1}{\sqrt{21}} (3\hat{\mathbf{i}}, 0, 2\sqrt{3}\hat{\mathbf{k}}) \qquad\qquad \square$$

4.4.4   Derivation of the Effective Index of Refraction

The emphasis has been placed on the direction of the propagation constant $\mathbf{k}$, and the magnitude k has not yet been obtained. The value of k depends on the direction of $\mathbf{D}$. When $\mathbf{D}$ is pointed along the optic axis, the dielectric constant is $\epsilon_0 n_e^2$ and the propagation constant is $n_e k_0$; on the other hand, when $\mathbf{D}$ is in the direction perpendicular to the optic axis, the propagation constant is $n_o k_0$. The magnitude of the wavenormal $\mathbf{k}$ from an arbitrary direction of $\mathbf{D}$ will be obtained here. Using Eq. (4.26), Eq. (4.23) can be rewritten as

$$\mathbf{T} = k_0^2 \mathbf{D} \qquad\qquad (4.57)$$

Equation (4.40) is rewritten in diagonal matrix form as

$$\mathbf{T} = \begin{bmatrix} \dfrac{k_z^2}{n_o^2} + \dfrac{k_x^2}{n_e^2} & 0 & 0 \\ 0 & \dfrac{k^2}{n_o^2} & 0 \\ 0 & 0 & \dfrac{k_z^2}{n_o^2} + \dfrac{k_x^2}{n_e^2} \end{bmatrix} \begin{bmatrix} D_x \\ D_y \\ D_z \end{bmatrix} \qquad (4.58)$$

where Eq. (4.36) was used to convert D_x into D_z.

Inserting Eq. (4.58) into (4.57) gives

$$
\begin{bmatrix}
\dfrac{k_z^2}{n_o^2} + \dfrac{k_x^2}{n_e^2} - k_0^2 & 0 & 0 \\[2ex]
0 & \dfrac{k^2}{n_o^2} - k_0^2 & 0 \\[2ex]
0 & 0 & \dfrac{k_z^2}{n_o^2} + \dfrac{k_x^2}{n_e^2} - k_0^2
\end{bmatrix}
\begin{bmatrix}
D_x \\[1ex] D_y \\[1ex] D_z
\end{bmatrix} = 0
\tag{4.59}
$$

Since the right-hand side is zero, the determinant has to vanish in order for D_x, D_y, and D_z to have nonzero solutions:

$$
\left(\frac{k_z^2}{n_o^2} + \frac{k_x^2}{n_e^2} - k_0^2 \right)^2 \left(\frac{k^2}{n_o^2} - k_0^2 \right) = 0
\tag{4.60}
$$

Equation (4.60) is called the characteristic equation of k, which can be separated into

$$
\frac{k_z^2}{n_o^2} + \frac{k_x^2}{n_e^2} = k_0^2
\tag{4.61}
$$

which is the equation of an ellipse in the k_x-k_z plane and

$$
\frac{k^2}{n_o^2} = k_0^2
\tag{4.62}
$$

which is the equation of a circle. Equation (4.59) is rewritten as

$$
\left(\frac{k_z^2}{n_o^2} + \frac{k_x^2}{n_e^2} - k_0^2 \right) D_x = 0
\tag{4.63}
$$

$$
\left(\frac{k^2}{n_o^2} - k_0^2 \right) D_y = 0
\tag{4.64}
$$

$$
\left(\frac{k_z^2}{n_o^2} + \frac{k_x^2}{n_e^2} - k_0^2 \right) D_z = 0
\tag{4.65}
$$

If Eq. (4.61) is the condition, Eqs. (4.63) and (4.65) mean that D_x and D_z can be nonzero, and this is precisely the condition of the e-wave. Similarly, if Eq. (4.62) is the condition, then D_y can be nonzero, and this is the condition for the o-wave.

With the e-wave, the magnitude k of the propagation constant depends on the direction of **k**. **k** is inclined at an angle θ with respect to the z axis, and

$$
k_x = k \sin \theta, \qquad k_z = k \cos \theta
\tag{4.66}
$$

Inserting Eq. (4.66) into Eq. (4.61) gives the value of k for a given direction of propagation.

$$
k = \pm n_{\text{eff}}(\theta) k_0
\tag{4.67}
$$

where

$$n_{\text{eff}}(\theta) = \frac{1}{\sqrt{\left(\dfrac{\cos\theta}{n_o}\right)^2 + \left(\dfrac{\sin\theta}{n_e}\right)^2}} \tag{4.68}$$

For the e-wave, $n_{\text{eff}}(\theta)$ is the effective index of refraction when the angle of incidence is θ.

Similarly, for the o-wave from Eq. (4.62), the magnitude of the propagation constant is

$$k = \pm n_o k_0 \tag{4.69}$$

and k is independent of the angle of incidence.

4.5   GRAPHICAL METHODS

Methods that provide **D**, **E**, **k**, **H**, and **s** graphically, thereby alleviating some of the calculation, will be presented in this section. Two kinds of graphical methods will be explained: the wavevector and indicatrix methods. The wavevector method makes use of the **k**-space concept and emphasis is placed on propagation of the wavefront. The indicatrix method is based on the space of the indices of refraction and emphasis is on the optical properties of the medium.

4.5.1   Wavevector Method

The wavevector method [7] combines the ellipses calculated in Section 4.4.4 with a method for obtaining **k** and **s**.

For simplicity, the uniaxial crystal case is considered. The characteristic equations (4.61) and (4.62) are represented in k_x, k_y, and k_z coordinates. Figure 4.9 shows the cross section in the $k_y = 0$ plane. Equation (4.61) is an ellipse with semiaxes, $n_e k_0$ and $n_o k_0$ and Eq. (4.62) is a circle with radius $n_o k_0$. With regard to the ellipse, note that the semiaxis $n_e k_0$ lies on the k_x axis, and the semiaxis $n_e k_0$ lies on the k_z axis. Let us now start to dig out as much information as possible from the graph in Fig. 4.9.

Let the wavenormal of the incident light be at an angle θ with respect to the k_z axis, as represented by $\overline{OP'}$ in Fig. 4.9a and $\overline{OP}$ in Fig. 4.9b. The quantities: **D**, **E**, **H**, **s**, **k**, n_{eff}, and the wavenormal or the wavefront are discussed in point form below.

1. There are two **D**'s that are orthogonal to each other, and they are both in the plane perpendicular to **k**. The electric displacement field for the o-wave, labeled as **D'** in Fig. 4.9a, is in the k_y direction. The **D** field associated with the e-wave lies in the k_x–k_z plane in Fig. 4.9b.

2. The direction of **E** for the e-wave is tangent to the ellipse at point P. As this statement may not be immediately evident, a short proof is given.

   Taking the derivative of Eq. (4.61) with respect to k_x gives

$$\frac{2k_z}{n_o^2}\frac{dk_z}{dk_x} + \frac{2k_x}{n_e^2} = 0 \tag{4.70}$$

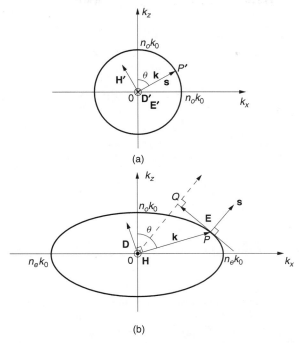

Figure 4.9 Wavevector diagram in a uniaxial crystal with $n_e > n_o$ (positive crystal). (a) o-wave. (b) e-wave.

and hence

$$\frac{dk_z}{dk_x} = -\left(\frac{n_o}{n_e}\right)^2 \frac{k_x}{k_z} \tag{4.71}$$

Using Eq. (4.54) with (4.52) gives

$$\frac{dk_z}{dk_x} = \frac{E_z}{E_x} \tag{4.72}$$

Thus, the tangent of the ellipse is the direction of the **E** field of the e-wave.

3. The direction of **H** is perpendicular to the plane made by **k** and **E** as indicated by Eq. (4.21). The **H** field for the e-wave is in the y direction. The **H**′ field for the o-wave is in the $x-z$ plane and perpendicular to **k**.

4. The Poynting vector **s** is obtained from **E** × **H**. The direction of **s** for the e-wave is normal to **E** and thus normal to the ellipse at P. The Poynting vector **s**′ of the o-wave is in the same direction as **k**.

5. Lastly, it should be mentioned that it is usually the ray direction **s** that is given in the laboratory. In many cases the direction of the wavenormal **k** is not explicitly given. For a given **s**, however, **k** can be found. Let the given direction of **s** be represented by the dotted line $\overline{0Q}$. Find the tangent to the ellipse that is normal to $\overline{0Q}$. The intersection of the tangent line with the ellipse determines the point P. The line $\overline{0P}$ is **k**.

6. $n_{\text{eff}}(\theta)k_0$ is represented by $\overline{0P}$.

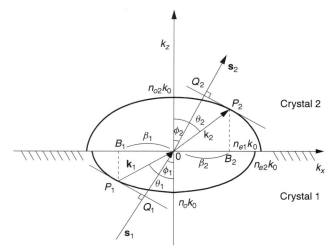

Figure 4.10   Finding the transmitted ray direction s_2 for a given incident ray direction s_1, using the wavevector diagram.

Next, a method of solving the refraction problem using the wavevector diagram will be described by a sequence of steps with brief descriptions for each step.

As indicated in Fig. 4.10, the optic axes of the two crystals are collinear, and the wavevector diagrams of the two crystals share a common origin. Only half of each wavevector diagram is shown in the figure. As described in the previous section, the light ray s_1 is incident from the lower crystal 1 to crystal 2. The problem is to find the direction of the transmitted ray in crystal 2.

Step 1. In order to use Snell's law (phase matching) find the direction θ_1 of the wavenormal $\mathbf{k}_1$ from direction ϕ_1 of s_1, as demonstrated earlier. The tangent line $\overline{P_1 Q_1}$, which is perpendicular to s_1, is drawn. The line drawn from the origin to P_1 is the direction of $\mathbf{k}_1$. The analytic expression relating the two angles is Eq. (4.56).

Step 2. The incident wave normal $\mathbf{k}_1$, is $\overline{OP_1}$. The analytic expression for k_1 is Eq. (4.67).

Step 3. Find the propagation constant β_1 along the interface, which is $\overline{OB_1}$, where B_1 is the normal from P_1 to the boundary surface.

$$\beta_1 = k_1 \sin \theta_1 \tag{4.73}$$

Step 4. In order to satisfy the condition of the phase matching across the boundary, find the point B_2 such that

$$\overline{OB_1} = \overline{OB_2} \tag{4.74}$$

or

$$\beta_1 = \beta_2 \tag{4.75}$$

Step 5. The angle θ_2 of the emergent wavenormal is found. Draw normal $\overline{P_2 B_2}$ from the interface to intersect the wavevector diagram of crystal 2. $\overline{OP_2}$ is the direction of the emergent wavenormal.

The analytical expression for β_2 can be found from Eqs. (4.67) and (4.68):

$$\beta_1 = \beta_2 = \frac{k_0 \sin \theta_2}{\sqrt{\left(\dfrac{\cos \theta_2}{n_{o2}}\right)^2 + \left(\dfrac{\sin \theta_2}{n_{e2}}\right)^2}} \qquad (4.76)$$

which is a quadratic equation in $\sin \theta_2$. The solution of Eq. (4.76) is

$$\sin \theta_2 = \frac{\left(\dfrac{\beta_1}{k_0 n_{o_2}}\right)}{\sqrt{1 + \left(\dfrac{\beta_1}{k_0}\right)^2 \left(\dfrac{1}{n_{o2}^2} - \dfrac{1}{n_{e2}^2}\right)}} \qquad (4.77)$$

Step 6. Finally, $\mathbf{s}_2$, which is perpendicular to the tangent line $\overline{P_2 Q_2}$, is drawn and the emergent ray direction ϕ_2 is found. The alternative is the use of Eq. (4.56).

Next, a description of total internal reflection is added. The condition of total internal reflection is that β_1 starts to exceed $n_{e_2} k_0$;

$$\beta_1 = n_{e_2} k_0 \qquad (4.78)$$

With this condition the critical angle θ_c is obtained as

$$\sin \theta_c = \frac{n_{e2}}{n_{o1}} \frac{1}{\sqrt{1 + n_{e2}^2 \left(\dfrac{1}{n_{o1}^2} - \dfrac{1}{n_{e1}^2}\right)}} \qquad (4.79)$$

4.5.2 Indicatrix Method

Whereas the wavevector method is closely linked to $\mathbf{k}$ space, the indicatrix method [8] utilizes the space formed by the indices of refraction of the crystal. The indicatrix method is an elegant way of handling crystal optics.

Using the vector identity

$$\mathbf{A} \times (\mathbf{B} \times \mathbf{C}) = \mathbf{B}(\mathbf{A} \cdot \mathbf{C}) - \mathbf{C}(\mathbf{A} \cdot \mathbf{B}) \qquad (4.80)$$

Eq. (4.22) is rewritten as

$$k^2 \mathbf{E} - \mathbf{k}(\mathbf{k} \cdot \mathbf{E}) = \omega^2 \mu \mathbf{D} \qquad (4.81)$$

The scalar product of $\mathbf{D}$ with Eq. (4.81) eliminates the second term due to the fact that $\mathbf{D} \cdot \mathbf{k} = 0$, and Eq. (4.81) becomes

$$k^2 \mathbf{E} \cdot \mathbf{D} = \omega^2 \mu \mathbf{D} \cdot \mathbf{D} \qquad (4.82)$$

Equation (4.82) is rewritten as

$$\frac{k^2}{\omega^2 \mu \epsilon_0} \left(\frac{D_x}{\epsilon_{rx}} \hat{\mathbf{i}} + \frac{D_y}{\epsilon_{ry}} \hat{\mathbf{j}} + \frac{D_z}{\epsilon_{rz}} \hat{\mathbf{k}} \right) \cdot \mathbf{D} = D^2 \qquad (4.83)$$

Equation (4.83) can be recast as

$$\left(\frac{x}{n_1}\right)^2 + \left(\frac{y}{n_2}\right)^2 + \left(\frac{z}{n_3}\right)^2 = 1 \tag{4.84}$$

where

$$x = \frac{k}{k_0}\frac{D_x}{D} \qquad n_1^2 = \epsilon_{rx}$$

$$y = \frac{k}{k_0}\frac{D_y}{D} \qquad n_2^2 = \epsilon_{ry} \tag{4.85}$$

$$z = \frac{k}{k_0}\frac{D_z}{D} \qquad n_3^2 = \epsilon_{rz}$$

$$D = |\mathbf{D}|$$

Equation (4.84) is an ellipsoid in the x, y, z coordinates that are the normalized D_x, D_y, D_z coordinates. The principal axes of the ellipsoid, n_1, n_2, and n_3, are the principal refractive indices of the crystal. Such an ellipsoid is called the optical indicatrix, or simply indicatrix, or Fletcher's indicatrix after the scientist who first proposed the method in 1891 [9].

The procedure for obtaining $\mathbf{D}$, $\mathbf{E}$, $\mathbf{k}$, $\mathbf{s}$, and n_{eff} using the indicatrix will be explained in seven steps using the uniaxial case of $n_1 = n_2 = n_o$ and $n_3 = n_e$. Some mathematical verifications will follow.

Step 1. We assume that the direction of the wavenormal is known and is in the x–z plane. Draw line $\overline{OP}$ from the center 0 of the ellipsoid to the direction of the wavenormal, as shown in Fig. 4.11. Figure 4.12 is the cross section of the index ellipsoid in the x–z plane.

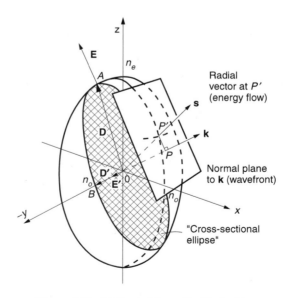

Figure 4.11   Method of the optical indicatrix.

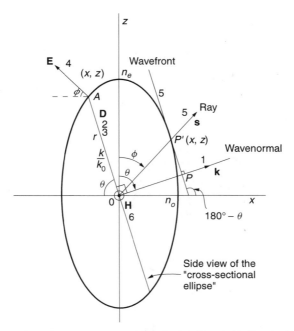

Figure 4.12 The e-wave fields in a uniaxial crystal obtained by the indicatrix. Numbers correspond to the appropriate steps in the text. (The y axis is into the page.)

Step 2. Find the plane that passes through the center of the ellipsoid in the direction perpendicular to $\overline{OP}$. The cross section of the ellipsoid cut by this plane is an ellipse, as indicated by the hatched area in the figure. This "cross-sectional ellipse" is of great importance for investigating propagation in the crystal.

Step 3. The conditions on the polarization direction of **D** in Section 4.4.2, described for the case when **k** is in the $x–z$ plane, are summarized. Only two directions of polarization are allowed for a wave to propagate in an anisotropic medium. One polarization is in the y direction (o-wave) and the other is in the direction defined by the intersection of the plane perpendicular to **k** with the $x–z$ plane (e-wave). With regard to the geometry of the indicatrix, these two allowed directions of polarization are in the directions of the major and minor axes **D** and **D′** of the "cross-sectional ellipse." No other direction of polarization is allowed [10].

Step 4. The normals to the ellipse at A and B are the directions of **E** and **E′**.

Step 5. Find the plane that is tangent to the ellipsoid as well as perpendicular to $\overline{OP}$ or **k**. This tangent plane is the wavefront, and the **D**'s are parallel to this plane. The direction of energy flow is found from the contact point $P′$ of the tangent plane to the ellipsoid. $\overline{OP′}$ is the direction of the ray vector **s**. (The proof is given later in this section.)

Step 6. The directions of **H** and **H′** (not indicated in Fig. 4.11) are the directions perpendicular to the planes made by **k** and **E** and **k** and **E′**.

Step 7. When **k** is in a more general direction, one can always rotate the coordinates so that **k** is in the $x–z$ plane because of the cylindrical symmetry of the indicatrix. However, by expanding the rule for using the indicatrix, there is no need to rotate the

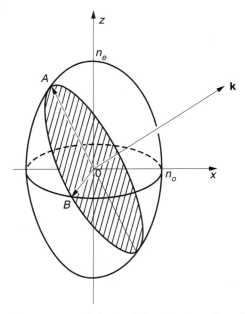

Figure 4.13 The "cross-sectional ellipse" in the indicatrix determines the allowed directions of polarization $\overline{OA}$ (e-wave) and $\overline{OB}$ (o-wave). They are in the directions of the principal axes.

coordinates. The general rule for the use of the indicatrix is as follows. As shown in Fig. 4.13, the intercept of the plane perpendicular to **k** with the ellipsoid generates the "cross-sectional ellipse" with principal semiaxes $\overline{OA}$ and $\overline{OB}$. The only allowed directions of the polarization are $\overline{OA}$ (e-wave) and $\overline{OB}$ (o-wave). The lengths of $\overline{OA}$ and $\overline{OB}$ are the effective refractive indices n_{eff} for the e- and o-waves, respectively.

The proofs of some of the steps will be given.

Let us examine now in more detail the relationship between the quantities shown in Fig. 4.12 and equations derived earlier for the e-wave in a uniaxial crystal. We will now prove that the normal to the ellipse at point A in Fig. 4.12 is the direction of E (as stated in Step 4 of the indicatrix method).

By taking the gradient of the left-hand side of Eq. (4.84) with $y = 0$ and $n_1 = n_o$ and $n_3 = n_e$, the normal vector **N** is obtained. Let (see boxed note)

$$F = \left(\frac{x}{n_o}\right)^2 + \left(\frac{z}{n_e}\right)^2 \tag{4.86}$$

The normal vector **N** is

$$\mathbf{N} = \nabla F \tag{4.87}$$

and

$$\mathbf{N} = \frac{2x}{n_o^2}\hat{\mathbf{i}} + \frac{2z}{n_e^2}\hat{\mathbf{k}} \tag{4.88}$$

Let's examine the meaning of the equation

$$\left(\frac{x}{a}\right)^2 + \left(\frac{y}{b}\right)^2 = F \tag{4.89}$$

If, for example, $F = 1.1^2$, then this equation defines an ellipse that is inflated by 10% compared to the size of the $F = 1$ ellipse, a and b.

$$\frac{x^2}{(1.1a)^2} + \frac{y^2}{(1.1b)^2} = 1$$

The size of the ellipse is determined by the right-hand side F of Eq. (4.89). The component of the gradiant $(\partial F/\partial x)\mathbf{i}$ is the rate of expansion of the ellipse in the x direction.

By replacing x and z at A in Eq. (4.88) by those in Eq. (4.85), the normal vector is expressed as

$$\mathbf{N} = \frac{2k}{k_0 D}\left(\frac{D_x}{n_o^2}\hat{\mathbf{i}} + \frac{D_z}{n_e^2}\hat{\mathbf{k}}\right) \tag{4.90}$$

From Eq. (4.10), $\mathbf{N}$ is rewritten as

$$\mathbf{N} = \frac{2k}{k_0 D}[E_x\hat{\mathbf{i}} + E_z\hat{\mathbf{k}}] \tag{4.91}$$

and

$$\mathbf{N} = \frac{2k}{k_0 D}\mathbf{E} \tag{4.92}$$

Thus, it has been proved that $\mathbf{E}$ is in the direction of the normal.

Referring to Fig. 4.12, we will now prove that $\overline{OP'}$ is perpendicular to $\mathbf{E}$ and hence parallel to $\mathbf{s}$ (as stated in Step 5 of the indicatrix method). P' is the contact point on the ellipse of the tangent line that is perpendicular to $\mathbf{k}$. With the conditions $y = 0$, $n_1 = n_o$, and $n_3 = n_e$, and taking the derivative with respect to x of both sides of Eq. (4.84) at point P', one obtains

$$\frac{dz}{dx} = -\frac{x}{z}\left(\frac{n_e}{n_o}\right)^2 \tag{4.93}$$

The tangent that is perpendicular to $\mathbf{k}$ and hence parallel to $\mathbf{D}$ is

$$\frac{dz}{dx} = \frac{D_z}{D_x} \tag{4.94}$$

Equations (4.11) and (4.94) are put into Eq. (4.93) to obtain

$$\frac{z}{x} = -\frac{E_x}{E_z} \tag{4.95}$$

The line $\overline{OP'}$ connecting the origin and the contact point P' is

$$\overline{OP'} = x\hat{\mathbf{i}} + z\hat{\mathbf{k}} \tag{4.96}$$

From Eqs. (4.95) and (4.96),

$$\overline{OP'} \cdot \mathbf{E} = 0 \tag{4.97}$$

Thus, $\overline{OP'}$ is perpendicular to $\mathbf{E}$ and in the direction of $\mathbf{s}$ because of Eq. (4.47) with $\mathbf{H}$ in the $-y$ direction.

Finally, the value of k is examined. When vector $\mathbf{k}$ is at angle θ with respect to the optic axis, the coordinates (x, z) at point A in Fig. 4.12 are $(-r\cos\theta, r\sin\theta)$. Since this point is on the ellipse expressed by Eq. (4.84), we have

$$r^2 \left[\left(\frac{\cos\theta}{n_o} \right)^2 + \left(\frac{\sin\theta}{n_e} \right)^2 \right] = 1 \tag{4.98}$$

Next, the physical meaning of r is found. The length r in the x, z coordinates can be rewritten using Eq. (4.85) as

$$r = \sqrt{x^2 + z^2} = \frac{k}{k_0} \left(\sqrt{\frac{D_x^2 + D_z^2}{D}} \right) \tag{4.99}$$

Since the value inside the square root is unity, Eq. (4.99) with Eq. (4.98) becomes

$$r = \frac{k}{k_0} = n_{\text{eff}} \tag{4.100}$$

where

$$n_{\text{eff}} = \frac{1}{\sqrt{\left(\dfrac{\cos\theta}{n_o} \right)^2 + \left(\dfrac{\sin\theta}{n_e} \right)^2}} \tag{4.101}$$

which is the same expression as Eq. (4.68). Thus, it has been proved that the optical indicatrix results are identical with the wavevector results.

Both the indicatrix and the wavevector diagram are used for treating refraction at the boundary between anisotropic media. Crystals whose refractive indices change due to electrooptic or acoustooptic effects are characterized by the indicatrix, and the indicatrix method is generally a better choice for this situation. However, the wavevector approach is usually preferred when a knowledge of the phase velocity is required, such as problems involving wave dispersion or the interaction of lightwaves with other types of waves like acoustic waves. Either way, the difference between the two approaches is rather slim. The indicatrix approach to boundary problems will be presented in Section 4.6.

Example 4.3   A light ray is incident from air onto the principal section (Example 4.1) of a calcite crystal whose optic axis is tilted by $48°$ with respect to the front surface, as shown in Fig. 4.14. If the thickness of the crystal is 2 cm, what is the distance between the two spots seen on the emergent surface of the crystal. The indices of refraction are $n_o = 1.658$ and $n_e = 1.486$.

Obtain the qualitative solutions by means of the wavevector method and indicatrix method, and then calculate the distance numerically.

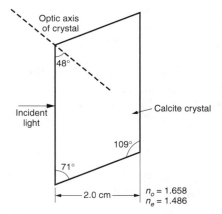

Figure 4.14 A light ray with two orthogonal polarizations is incident onto calcite. The parallelogram represents the principal section shown in Fig. 4.6.

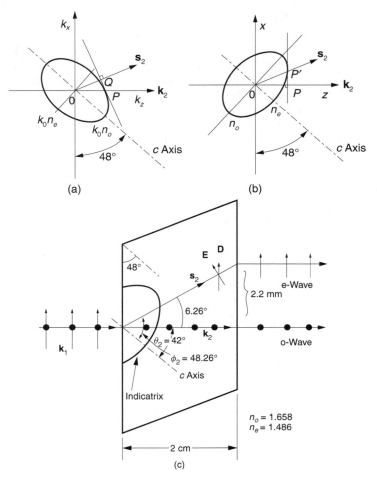

Figure 4.15 Directions of o- and e-waves transversing a calcite crystal. (a) By wavevector diagram. (b) By indicatrix method. (c) By numerical solution.

Solution Graphical solutions are shown in Figs. 4.15a and 4.15b and the analytical solution in Fig. 4.15c. The incident light ray is decomposed into two orthogonal directions of polarization. The o-wave obeys Snell's law, and because it enters the crystal at normal incidence, it propagates straight through the crystal.

The e-wave is slightly different. $\mathbf{k}_1$ and $\mathbf{k}_2$ obey Snell's law, and both are normal to the boundary. The angle of transmittance with respect to the crystal axis is therefore, from Fig. 4.15c, $\theta_2 = 42°$. By Eq. (4.56) the transmitted angle ϕ_2 of the light ray is $\phi_2 = 48.26°$. Thus, the e-wave is refracted at the boundary by $6.26°$ from the normal to the boundary. □

4.6 TREATMENT OF BOUNDARY PROBLEMS BETWEEN ANISOTROPIC MEDIA BY THE INDICATRIX METHOD

When the light encounters a boundary between two media with different optical properties, part of the energy is transmitted into the second medium, and the other part is reflected back into the original medium. In this section, the boundary between uniaxial anisotropic media is treated [5,11,12]. Section 4.6.1 treats the transmitted wave; Section 4.6.2, the reflected wave; and Section 4.6.3, total internal reflection. Before beginning, a few important general remarks made earlier will be repeated here.

On the boundary between the two media with different indices of refraction, the incident light changes its direction of propagation for the sake of phase matching. Recall that it is the direction of the *wavenormal* that obeys Snell's law. It is not the direction of the *energy flow* (ray direction) that obeys Snell's law. This is because Snell's law is the law that synchronizes the phasefronts of the incident, transmitted, and reflected light across the boundary, as explained in Section 2.2.

In dealing with refraction between uniaxial anisotropic media, the incident wave has to be first decomposed into two component waves: one wave is the o-wave and the other is the e-wave. The respective amplitudes of the two waves are found as the projection of the amplitude of the incident wave into these two directions of polarization. The two waves are treated separately and the results are added to reach the final answer.

Since the direction of the wavenormal of the o-wave is identical to the ray direction, the transmitted and reflected waves of the o-wave can be obtained in exactly the same manner as described in Section 2.3 and it will not be repeated here. The explanation in this section is devoted to the e-wave.

4.6.1 Refraction of the e-Wave at the Boundary of Anisotropic Media

For a given incident e-wave ray $\mathbf{s}_1$, the transmitted e-wave ray $\mathbf{s}_2$ will be found [13]. It is the direction of the light ray $\mathbf{s}$ that is usually given. In order to utilize Snell's law, the direction of the light ray $\mathbf{s}$ has to be converted into that of the wavenormal $\mathbf{k}$. After using Snell's law, the direction of the wavenormal is converted back to the ray direction.

It is assumed that the optic axes of the two crystals are collinear, as shown in Fig. 4.16. The indices of refraction of the bottom uniaxial crystal 1 are denoted as

$$\begin{bmatrix} n_{o1}^2 & 0 & 0 \\ 0 & n_{o1}^2 & 0 \\ 0 & 0 & n_{e1}^2 \end{bmatrix}$$

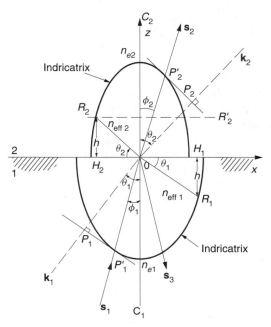

Figure 4.16 Refraction at the boundary of two uniaxial crystals. The optic axes of crystals 1 and 2 are collinear.

and those of the top crystal 2 by

$$
\begin{bmatrix}
n_{o2}^2 & 0 & 0 \\
0 & n_{o2}^2 & 0 \\
0 & 0 & n_{e2}^2
\end{bmatrix}
$$

The indicatrixes of the two crystals share a common origin and only half of each indicatrix is drawn in the figure.

The light ray s_1 is incident from the lower crystal 1 to the upper crystal 2. This ray direction is first converted into the wavenormal k_1. The light ray s_1 intersects the lower ellipse at point P_1'. The tangent to the ellipse at P_1' is drawn. A line is drawn from the origin perpendicular to this tangent line. Let P_1 denote the intersection of the line from the origin with the tangent line. The wavefront of the incident wave is $\overline{P_1 P_1'}$ and the wavenormal k_1 is along $\overline{OP_1}$. The angle k_1 makes with the z axis is θ_1.

Now that the angle θ_1 of the incident wavenormal is found, Snell's law can be applied. The effective index of refraction $n_{\mathrm{eff}\,1}$ of crystal 1 is readily found by drawing line $\overline{OR_1}$ from the origin perpendicular to k_1. Snell's law is

$$
n_{\mathrm{eff}\,1} \sin \theta_1 = n_{\mathrm{eff}\,2} \sin \theta_2 \tag{4.102}
$$

For the special orientation of the optic axis of the crystal perpendicular to the boundary, the angle that $\overline{OR_1}$ makes with the x axis is identical to the angle θ_1 of incidence. Thus, the value of the left-hand side of Eq. (4.102) is graphically represented by the height h of R_1 from the x axis.

Now let us consider the right-hand side of Eq. (4.102). A horizontal line is drawn at a height h above the x axis. The intersection R_2 that the horizontal line $\overline{R_2 R_2'}$ makes with the indicatrix is determined. The length $\overline{OR_2}$ now graphically represents the effective index of refraction $n_{\text{eff}2}$ of the second medium. The direction perpendicular to $\overline{OR_2}$ is the direction of the wavenormal of the transmitted wave. Since the optic axis of the crystal is perpendicular to the boundary, the angle that $\overline{OR_2}$ makes with the x axis is identical to the transmitted angle θ_2 that $\mathbf{k}_2$ makes with the optic axis. Finally, the transmitted ray $\mathbf{s}_2$ is obtained by finding the line $\overline{P_2 P_2'}$, which is tangent to the ellipse as well as perpendicular to $\mathbf{k}_2$. Thus, $\mathbf{s}_2$ was found from $\mathbf{s}_1$.

The above graphical method will be supplemented by a brief description of the analytical method. First, the incident ray direction ϕ_1 is converted into θ_1 of the incident wavenormal using Eq. (4.56).

The value of h of the incident light is

$$h = n_{\text{eff}1} \sin \theta_1 \qquad (4.103)$$

where the effective refractive index $n_{\text{eff}1} = k_1/k_0$ is obtained from Eq. (4.68).

$$h = \frac{\sin \theta_1}{\left[\left(\frac{\cos \theta_1}{n_{o1}} \right)^2 + \left(\frac{\sin \theta_1}{n_{e1}} \right)^2 \right]^{1/2}} \qquad (4.104)$$

The angle θ_2 of the transmitted light is calculated from the fact that h for medium 1 remains the same as that for medium 2. The value of θ_2 has to be calculated for a given h value.

$$h = \frac{\sin \theta_2}{\left[\left(\frac{\cos \theta_2}{n_{o2}} \right)^2 + \left(\frac{\sin \theta_2}{n_{e2}} \right)^2 \right]^{1/2}} \qquad (4.105)$$

This quadratic equation in $\sin \theta_2$ can be solved in the same manner as Eq. (4.77) was obtained from Eq. (4.76). The result is

$$\sin \theta_2 = \frac{\left(\dfrac{h}{n_{o2}} \right)}{\sqrt{1 + h^2 \left(\dfrac{1}{n_{o2}^2} - \dfrac{1}{n_{e2}^2} \right)}}. \qquad (4.106)$$

Finally, the angle θ_2 of the wavenormal of the transmitted light is converted into angle ϕ_2 of the emergent light ray using Eq. (4.56).

Problem 4.3 deals with the case when the two optic axes of the crystal are at an angle.

4.6.2 Reflection of the e-Wave at the Boundary of Anisotropic Media

Regardless of the type of media, the direction of the reflected wave is such that the $\mathbf{k}$ component of the reflected light that is parallel to the boundary is phase matched with that of the incident light. When the optic axis of the crystal is normal to the boundary, as in Fig. 4.16, the angle of the reflected wavenormal is equal to the angle of the incident wavenormal. However, when the optic axes of the crystal are not normal to the boundary, such as shown in Fig. 4.17, the angle of reflection differs from the angle of incidence.

Referring to Fig. 4.17, light is incident from the lower crystal 1. Only the e-wave that is reflected back into crystal 1 is considered here. Let the incident wavenormal be $\mathbf{k}_1$. The angle of the reflected wavenormal has to be such that the wavelength along the boundary matches that of the incident wave. The phase matching conditions (Section 2.2) between the wavefronts of the incident and reflected light are

$$h_1 = n_{\mathrm{eff}\,1} \sin \theta'_1 \tag{4.107}$$

$$h_3 = n_{\mathrm{eff}3} \sin \theta'_3 \tag{4.108}$$

where θ'_1 is the angle between the incident wavenormal to the boundary and the normal to the boundary, and θ'_3 is the angle between the reflected wavenormal and the normal to the boundary. The phase matching condition is satisfied for

$$h_1 = h_3 = h \tag{4.109}$$

The graphical solution for θ'_3 is as follows. Referring to Fig. 4.17, the line $\overline{OR_1}$ perpendicular to $\mathbf{k}_1$, drawn from the origin to the ellipse, is the effective index $n_{\mathrm{eff}\,1}$ of the incident wave. Line $\overline{R_1H_1}$ to the x axis represents h_1. The intersection of the horizontal line $\overline{R_1R_3}$ and the ellipse determines h_3, and $\overline{OR_3}$ represents $n_{\mathrm{eff}\,3}$. The direction normal to $\overline{OR_3}$ gives $\mathbf{k}_3$ of the reflected wave. Thus, the angle of reflection

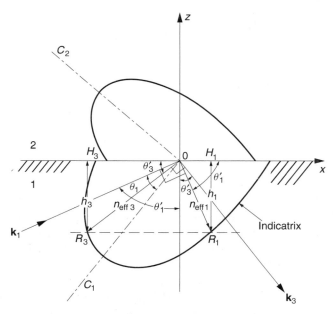

Figure 4.17 Boundary of two crystals where the optic axes are not perpendicular to the boundary. $\overline{C_1 0}$, $\overline{C_2 0}$, and $\mathbf{s}_1$ are all in the $y = 0$ plane.

was obtained from the angle of incidence. It should be noted that $\mathbf{k}_1$ and $\mathbf{k}_3$ are not symmetric with respect to the normal to the boundary when the optic axis of crystal 1 is not normal to the boundary. Transmission across such a boundary is left to Problem 4.3.

Note that the angle of reflection is independent of medium 2. As in the isotropic case, the *amount* of reflection is influenced by medium 2 but not the *angle* of reflection.

4.6.3 Total Internal Reflection of the e-Wave at the Boundary of Anisotropic Media

Total internal reflection takes place when the effective index of refraction of the second medium is not large enough to satisfy Snell's law given by Eq. (4.102). The case shown in Fig. 4.16 is taken as an example. In this case, the indicatrix of the top medium is smaller than that of the bottom medium, and when light is incident from the bottom medium to the top medium, the condition for total internal reflection can exist. The h value of the incident wave is

$$h = n_{\text{eff}\,1} \sin \theta \tag{4.110}$$

The h values of the top and bottom crystals have to be matched but the largest value that h can take in the top crystal 2 is n_{e2}; thus, total internal reflection takes place at an incident angle θ_c such that

$$n_{\text{eff}\,1} \sin \theta_c = n_{e2} \tag{4.111}$$

where θ_c is the critical angle.

Using Eq. (4.68), Eq. (4.111) becomes

$$n_{e2} = \frac{\sin \theta_c}{\left(\dfrac{\cos \theta_c}{n_{o1}} \right)^2 + \left(\dfrac{\sin \theta_c}{n_{e1}} \right)^2}$$

Mere comparison of the form of this equation with that of Eq. (4.105) gives

$$\sin \theta_c = \frac{n_{e2}}{n_{o1}} \cdot \frac{1}{\sqrt{1 + n_{e2}^2 \left(\dfrac{1}{n_{o1}^2} - \dfrac{1}{n_{e1}^2} \right)}} \tag{4.112}$$

where the subscripts 1 and 2 denote crystals 1 and 2. Equation (4.112) is identical to Eq. (4.79), which was obtained by the wavevector method in Section 4.5.1. Equation (4.112) reduces to the familiar expression for the critical angle of the isotropic case when $n_{o1} = n_{e1}$ and $n_{02} = n_{e2}$.

Example 4.4 Lithium niobate (LiNbO$_3$) is deposited over lithium tantalate (LiTaO$_3$) with the optic axes of both crystals normal to the interface. The indicatrixes are shown in Fig. 4.18. A light ray is incident on the boundary from the bottom layer, lithium

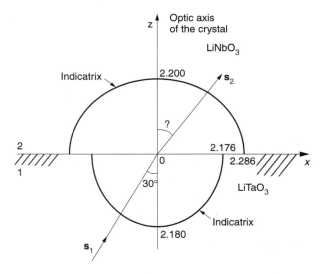

Figure 4.18 Refraction at the boundary between anisotropic media.

tantalate, at an incident angle $\theta_1 = 30°$. Calculate the emergent ray direction of the e-wave.

	Material	n_o	n_e
Medium 2	LiNbO$_3$	2.286	2.200
Medium 1	LiTaO$_3$	2.176	2.180

The wavelength of the incident light is $\lambda = 0.633$ μm.

Solution   The emergent ray direction will be found. First, the ray direction is converted into that of the wavenormal using Eq. (4.56) with $\phi_1 = 30°$:

$$\theta_1 = 30.09°$$

The effective index of refraction k/k_0 at this incident angle is calculated using Eq. (4.68),

$$n_{\text{eff}\,1} = 2.177$$

and

$$h = n_{\text{eff}\,1} \sin \theta_1 \tag{4.113}$$

Equation (4.106) can be used for finding θ_2 by putting $h = 1.092$. The value of θ_2 is

$$\theta_2 = 28.82°$$

Finally, using Eq. (4.56), θ_2 is converted into the emergent ray direction:

$$\phi_2 = 30.72°$$

The results are summarized in Fig. 4.19.   □

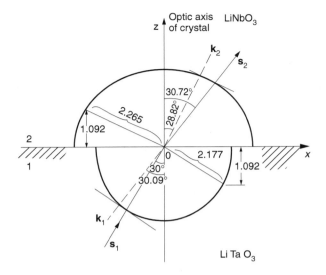

Figure 4.19 Solution to Example 4.4.

PROBLEMS

4.1 What condition enables Eqs. (4.61) and (4.62) to be satisfied simultaneously?

4.2 Obtain the characteristic matrix for the general case including the y components of the wave in an anisotropic medium with

$$\overleftrightarrow{\epsilon}_r = \begin{bmatrix} n_\alpha^2 & 0 & 0 \\ 0 & n_\beta^2 & 0 \\ 0 & 0 & n_\gamma^2 \end{bmatrix}$$

and let

$$\mathbf{k} = k_x \hat{\mathbf{i}} + k_y \hat{\mathbf{j}} + k_z \hat{\mathbf{k}}$$
$$\mathbf{E} = E_x \hat{\mathbf{i}} + E_y \hat{\mathbf{j}} + E_z \hat{\mathbf{k}}$$

4.3 Two crystals have a common boundary in the $z = 0$ plane. The optic axes of the crystals are not normal to the boundary and both are in the $y = 0$ plane with indicatrixes as shown in Fig. P4.3. The light ray $\mathbf{s}_1$ is incident from crystal 1 at the bottom into crystal 2 at the top. The incident light ray is in the $y = 0$ plane. Using the indicatrix, find the following quantities graphically:

(a) The vector $\mathbf{s}_2$ of the transmitted (refracted) ray.
(b) The vector $\mathbf{s}_3$ of the reflected ray when total internal reflection takes place.
(c) The angle of the reflected ray.

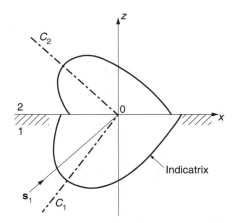

Figure P4.3   Boundary of two crystals.

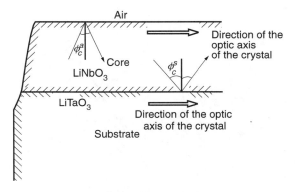

Figure P4.4   Critical angles of an anisotropic optical guide.

4.4   Find the critical angles of the rays of the o- and e-waves inside the core region of an $LiNbO_3$–$LiTaO_3$ optical guide. The optic axes of both the $LiNbO_3$ and $LiTaO_3$ crystals are oriented parallel to the boundaries (Y-cut) as in Fig. P4.4. The indices of refraction of the media are listed in Example 4.4.

4.5 **(a)**   Figure P4.5 shows a contour whose radius r at the ray direction ϕ corresponds to the ray velocity (not phase velocity) in that direction. This ray velocity surface is called Huygens' wavelet of the e-wave ellipsoid. Derive the expression for Huygens' wavelet of the e-wave in terms of ϕ, u, v_e, and v_o, where $v_o = c/n_o$ and $v_e = c/n_e$.

(b)   In the process of second harmonic generation (SHG), the energy of the light at the fundamental frequency is converted into that of the second higher harmonic frequency during the transmission in a nonlinear crystal. It is the ray velocities of the two waves rather than the phase velocities that have to be matched in order to optimize the efficiency of the energy conversion between the two waves. How would you use Huygens' wavelet to find the optimum ray direction in the SHG experiment [7]?

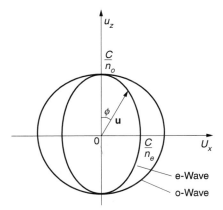

Figure P4.5 Huygens' wavelet. Note the differences of the coordinates among the wavevector diagram, the indicatrix, and Huygens' wavelet.

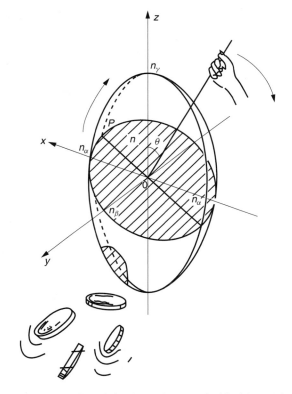

Figure P4.6   Search for the optic axes of a biaxial crystal.

4.6   Find the optic axes of the biaxial crystal shown in Fig. P4.6. The indices of refraction in the x, y, and z directions are n_α, n_β, and n_γ, respectively; assume that $n_\alpha < n_\beta < n_\gamma$. The direction of the optic axis is the direction of propagation such that the elliptic cross-section of the indicatrix made by the intersection with the plane perpendicular to the direction of propagation becomes a circle. In order to find the direction of the optic axes, the direction **k** of propagation is tilted from the z axis in the $x-z$ plane until the elliptic cross section becomes a circle (Fig. P4.6). Prove that the optic axes are at an angle θ from the z axis, which is given by

$$\sin\theta = \pm\frac{n_\gamma}{n_\beta}\sqrt{\frac{n_\beta^2 - n_\alpha^2}{n_\gamma^2 - n_\alpha^2}}$$

REFERENCES

1. C. Kittel, *Introduction to Solid State Physics*, 7th ed., Wiley, New York, 1996.

2. E. M. Purcell, *Electricity and Magnetism*, 2nd ed., McGraw-Hill, New York, 1985.

3. C. R. Wylie and L. C. Barrett, *Advanced Engineering Mathematics*, 6th ed., McGraw-Hill, New York, 1995.

4. A. Sommerfeld, *Optics*, Academic Press, New York, 1967.

5. M. V. Klein and T. E. Furtak, *Optics*, 2nd ed., Wiley, New York, 1986.

6. B. E. A. Saleh and M. C. Teich, *Fundamentals of Photonics*, Wiley, New York, 1991.

7. G. R. Fowles, *Introduction to Modern Optics*, 2nd ed., Dover Publication, New York, 1989.

8. P. Gay, *An introduction to Crystal Optics*, Longman, New York, 1982.

9. L. Fletcher, *The Optical Indicatrix and the Transmission of Light in Crystals*, H. Frowde, London, 1892.

10. K. Iizuka and S. Fujii, "Neural-network laser radar," *Appl. Opt.* **33**(13), 2492–2501 (1994).

11. M. Born and E. Wolf, *Principles of Optics*, 9th ed., Cambridge University Press, London, 1999.

12. P. C. Robinson and S. Bradbury, *Qualitative Polarized-Light Microscopy*, Oxford University Press, Oxford, 1992.

13. M. C. Simon and P. A. Larocca, "Minimum deviation for uniaxial prisms," *Appl. Opt.* **34**(4), 709–715 (1995).

14. A. Yariv, *Optical Electronics*, 3rd ed., Holt, Rinehart and Winston, New York, 1985.

15. M. Mansuripur, "Internal conical refraction," *Opt. and Photonics News*, 43–45 (1997).

5

OPTICAL PROPERTIES OF CRYSTALS UNDER VARIOUS EXTERNAL FIELDS

This chapter describes what happens when an external field is applied to a crystal whose properties are influenced by the external field. The effects of the external field are used to advantage to control integrated optic devices. Applications include spatial light modulators, light displays, switches, light deflectors, isolators, and even optical amplifiers. The fields employed for external control include electric fields, acoustic fields, magnetic fields, or the field of light itself.

An external field in the x direction influences the optical properties not only in the x direction but also in the y and z directions. Figure 5.1 shows an analogy. When a rubber eraser is pressed in the z direction the deformation takes place in all x, y, and z directions. The expression describing the external influence, therefore, becomes a tensor form. Moreover, almost all these materials are anisotropic and the treatment in the previous chapter will be extended to include the influences due to the external field.

5.1 EXPRESSING THE DISTORTION OF THE INDICATRIX

The amount of distortion in the indicatrix due to any external field is rather small, and the distorted indicatrix is still an ellipsoid. The distortions manifest themselves as a change in the lengths of the major and minor axes, as well as a rotation of the axes.

The two-dimensional indicatrix before rotation — that is, before the application of an external field — is given by

$$a_1 x^2 + a_2 y^2 = 1 \tag{5.1}$$

where

$$a_1 = \frac{1}{n_1^2} \quad \text{and} \quad a_2 = \frac{1}{n_2^2}$$

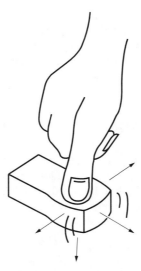

Figure 5.1 Illustration of the tensor property.

The expansion and contraction of the principal axes can be expressed in terms of the changes in a_1 and a_2, but the rotation of the principal axes is possible only after the inclusion of the crossterm $2a_{12}xy$. The expression for the deformed ellipse including the rotation of the principal axes is therefore

$$a_{11}x^2 + a_{22}y^2 + 2a_{12}xy = 1 \qquad (5.2)$$

Generalizing to three dimensions, the indicatrix with no applied field is

$$a_1x^2 + a_2y^2 + a_3z^2 = 1 \qquad (5.3)$$

where

$$a_1 = \frac{1}{n_1^2}, \quad a_2 = \frac{1}{n_2^2}, \quad \text{and} \quad a_3 = \frac{1}{n_3^2}$$

With an applied field, the indicatrix is expressed as

$$a_{11}x^2 + a_{22}y^2 + a_{33}z^2 + 2a_{23}yz + 2a_{31}zx + 2a_{12}xy = 1 \qquad (5.4)$$

Taking the difference between the coefficients in Eqs. (5.4) and (5.3) gives the following vector form Δa_i:

$$\begin{bmatrix} \Delta a_1 \\ \Delta a_2 \\ \Delta a_3 \\ \Delta a_4 \\ \Delta a_5 \\ \Delta a_6 \end{bmatrix} = \begin{bmatrix} a_{11} - 1/n_1^2 \\ a_{22} - 1/n_2^2 \\ a_{33} - 1/n_3^2 \\ a_{23} \\ a_{31} \\ a_{12} \end{bmatrix} \qquad (5.5)$$

Equation (5.5) is used to express the amount of distortion due to various effects [1].

Nonzero Δa_i are generated by various external fields. If the values of Δa_i are proportional to an external electric field, the distortion of the indicatrix is said to be due to the Pockels effect. If the values of Δa_i are proportional to the square of the external electric field, this effect is known as the Kerr effect. If the values of Δa_i are proportional to the strain in the crystal, this effect is called the elastooptic effect.

5.2 ELECTROOPTIC EFFECTS

Recall the simple model used to illustrate the polarization in Figs. 4.1 and 4.2. Strictly speaking, when an E field is applied, the resulting polarization is linearly proportional to E only for low values of E. For high values of E, the relationship is nonlinear. Likewise, for large E, the relationship between E and D is nonlinear. The dynamic dielectric constant

$$\epsilon = \frac{dD}{dE} \tag{5.6}$$

is not constant. It depends on the value of E. The index of refraction n can be changed with the addition of an applied electric field ε. This additional field ε can be a dc or an ac field. The index of refraction may be expressed as

$$n = n_0 + r(E + \varepsilon) + q(E + \varepsilon)^2 + \cdots$$

In this chapter, we are interested in the case where E is much smaller than the external ε, so that n can be approximated as

$$n \doteq n_0 + r\varepsilon + q\varepsilon^2 \tag{5.7}$$

The term $r\varepsilon$ represents the Pockels effect, and the term $q\varepsilon^2$ represents the Kerr effect. The Pockels and Kerr effects are experimentally separable because the Pockels effect depends on the polarity of the applied field while the Kerr effect does not.

5.2.1 Pockels Electrooptic Effect

Pockels coefficients are used to quantitatively describe the Pockels effect. The concept is that the change Δa_i between the coefficients of the undistorted indicatrix and the

---

Notice some coefficients in Table 5.1 are designated by (S) or (T). When the crystal is piezoelectric as well as electrooptic, the applied field creates strain in the crystal, which in turn changes the index of refraction due to the elastooptic effect. This additional effect, however, disappears if the external electric field varies faster than the frequency f_c of the mechanical resonance of the crystal. This condition of no strain is called "clamped," "nondeformed," "strain free," or the "S = 0 condition" and is designated by (S).

If the change of the external field is much slower than f_c, the crystal deforms freely and the elastooptic effect accompanies the electrooptic effect. The crystal is said to be "in an unclamped condition" or "a stress-free T = 0 condition" and is designated by (T). When the external field is a static field, the values marked with (T) are used.

Table 5.1 Electrooptic properties

Name of Substance	Chemical Symbol	Pockels Coefficient (10^{-12} m/V)	Index of Refraction	Wavelength (μm)	Crystal Symmetry	Electrooptic Tensor
Potassium dihydrogen phosphate	KH_2PO_4 or KDP	$r_{41} = 8.6$ $r_{63} = 10.5$ (T)	$n_x = n_y = 1.51$ $n_z = 1.47$	0.63	$\bar{4}2m$	$\begin{bmatrix} 0 & 0 & 0 \\ 0 & 0 & 0 \\ 0 & 0 & 0 \\ r_{41} & 0 & 0 \\ 0 & r_{41} & 0 \\ 0 & 0 & r_{63} \end{bmatrix}$
Ammonium dihydrogen phosphate	$NH_4H_2PO_4$ or ADP	$r_{41} = 23.1$ $r_{63} = 8.5$ (T)	$n_x = n_y = 1.52$ $n_z = 1.48$	0.63	$\bar{4}2m$	
D-KDP	KD_2PO_4	$r_{41} = 8.8$ $r_{63} = 26.4$ (T)	$n_x = n_y = 1.51$ $n_z = 1.47$	0.63	$\bar{4}2m$	
Quartz	SiO_2	$r_{11} = 0.29$ (S) $r_{41} = 0.2$ (T)	$n_x = n_y = 1.546$ $n_z = 1.555$	0.63	32	$\begin{bmatrix} r_{11} & 0 & 0 \\ -r_{11} & 0 & 0 \\ 0 & 0 & 0 \\ r_{41} & 0 & 0 \\ 0 & -r_{41} & 0 \\ 0 & -r_{11} & 0 \end{bmatrix}$
Cinnabar	HgS	$r_{11} = 3.1$ $r_{41} = 1.4$	$n_x = n_y = 2.885$ $n_z = 3.232$	0.63		
Lithium niobate	$LiNbO_3$	$r_{13} = \begin{cases} 10\ (T) \\ 8.6\ (S) \end{cases}$ $r_{33} = \begin{cases} 32.2\ (T) \\ 30.8\ (S) \end{cases}$ $r_{22} = \begin{cases} 6.7\ (T) \\ 3.4\ (S) \end{cases}$ $r_{51} = \begin{cases} 32\ (T) \\ 28\ (S) \end{cases}$	$n_x = n_y = 2.286$ $n_z = 2.200$	0.63	$3m$	$\begin{bmatrix} 0 & -r_{22} & r_{13} \\ 0 & r_{22} & r_{13} \\ 0 & 0 & r_{33} \\ 0 & r_{51} & 0 \\ r_{51} & 0 & 0 \\ -r_{22} & 0 & 0 \end{bmatrix}$
Lithium tantalate	$LiTaO_3$	$r_{13} = 7.0$ (S) $r_{33} = 27$ (S) $r_{22} = 1.0$ (S) $r_{51} = 20$ (S)	$n_x = n_y = 2.176$ $n_z = 2.180$	0.63	$3m$	

Table 5.1 *(Continued)*

Name of Substance	Chemical Symbol	Pockels Coefficient (10^{-12} m/V)	Index of Refraction	Wavelength (μm)	Crystal Symmetry	Electrooptic Tensor
Barium titanate	$BaTiO_3$	$r_{13} = 8$ (S) $r_{33} = 28$ (S) $r_{51} = \begin{cases} 1640 \text{ (T)} \\ 820 \text{ (S)} \end{cases}$	$n_x = n_y = 2.44$ $n_z = 2.37$	0.63	$4mm$	$\begin{bmatrix} 0 & 0 & r_{13} \\ 0 & 0 & r_{13} \\ 0 & 0 & r_{33} \\ 0 & r_{51} & 0 \\ r_{51} & 0 & 0 \\ 0 & 0 & 0 \end{bmatrix}$
Cadmium sulfide	CdS	$r_{13} = 1.1$ (S) $r_{33} = 2.4$ (S) $r_{51} = 3.7$ (T)	$n_x = n_y = 2.46$ $n_z = 2.48$	0.63 0.63 0.59	$6mm$	$\begin{bmatrix} 0 & 0 & r_{13} \\ 0 & 0 & r_{13} \\ 0 & 0 & r_{33} \\ 0 & r_{51} & 0 \\ r_{51} & 0 & 0 \\ 0 & 0 & 0 \end{bmatrix}$
Zinc oxide	ZnO	$r_{13} = 1.4$ (S) $r_{33} = 1.9$ (S) $r_{51} = $ NA	$n_x = n_y = 1.99$ $n_z = 2.015$	0.63		
Lithium iodate	$LiIO_3$	$r_{13} = 4.1$ (S) $r_{33} = 6.4$ (S) $r_{41} = 1.4$ (S) $r_{51} = 3.3$ (S)	$n_x = n_y = 1.88$ $n_z = 1.74$	0.63	6	$\begin{bmatrix} 0 & 0 & r_{13} \\ 0 & 0 & r_{13} \\ 0 & 0 & r_{33} \\ r_{41} & r_{51} & 0 \\ r_{51} & -r_{41} & 0 \\ 0 & 0 & 0 \end{bmatrix}$
Gallium arsenide	GaAs	$r_{41} = 1.2$ (S)	$n_x = n_y = n_z = 3.42$	0.9	$\overline{4}3m$	$\begin{bmatrix} 0 & 0 & 0 \\ 0 & 0 & 0 \\ 0 & 0 & 0 \\ r_{41} & 0 & 0 \\ 0 & r_{41} & 0 \\ 0 & 0 & r_{41} \end{bmatrix}$
Cadmium telluride	CdTe	$r_{41} = 6.8$ (S)	$n_x = n_y = n_z = 2.82$	3.4	$\overline{4}3m$	
Zinc telluride	ZnTe	$r_{41} = 4.2$ (S)	$n_x = n_y = n_z = 3.1$	0.59	$\overline{4}3m$	
Copper chloride	CuCl	$r_{41} = \begin{cases} 6.1 \text{ (T)} \\ 1.6 \text{ (S)} \end{cases}$	$n_x = n_y = n_z = 1.9$	0.54	$\overline{4}3m$	

distorted indicatrix is a linear function of the applied external electric field,

$$\Delta a_i = \sum_{j=1}^{3} r_{ij}\varepsilon_j \tag{5.8}$$

where r_{ij} are the Pockels coefficients and ε_j are the components of the applied electric field. Using Eq. (5.5), this can be expressed in matrix form as

$$\begin{bmatrix} a_{11} - 1/n_1^2 \\ a_{22} - 1/n_2^2 \\ a_{33} - 1/n_3^2 \\ a_{23} \\ a_{31} \\ a_{12} \end{bmatrix} = \begin{bmatrix} r_{11} & r_{12} & r_{13} \\ r_{21} & r_{22} & r_{23} \\ r_{31} & r_{32} & r_{33} \\ r_{41} & r_{42} & r_{43} \\ r_{51} & r_{52} & r_{53} \\ r_{61} & r_{62} & r_{63} \end{bmatrix} \begin{bmatrix} \varepsilon_1 \\ \varepsilon_2 \\ \varepsilon_3 \end{bmatrix} \tag{5.9}$$

where ε_1, ε_2, and ε_3 represent ε_x, ε_y, and ε_z. Table 5.1[3–7] lists Pockels coefficients for a number of electrooptic materials (see boxed note). The pattern of the matrix elements is determined by the symmetries of the crystal structure. Fortunately, many matrix elements are either zero or identical and manipulations are greatly simplified.

The optic axis is often referred to as the Z axis or c axis meaning crystal axis. The "crystal axis" here does not necessarily coincide with "crystal axis" in a crystallographic sense. A biaxial crystal has two optic axes but has only one crystal (crystallographic) axis. The axes perpendicular to the Z axis are referred to as the X and Y axes, and normally capital letters are used.

The manner in which a crystal is sliced is called by its "cut." Z-cut, for instance, means that the crystal is sliced so that it has two parallel flat surfaces, both of which are perpendicular to the Z axis. An illustration of a Y-cut crystal is shown at the top of Fig. 5.2. In diagrams with more than one set of parallel surfaces, the cut is with reference to the most closely spaced set of surfaces.

Example 5.1 Lithium niobate (LiNbO$_3$) is an electrooptic crystal that is widely used for integrated optics devices because of its large-valued Pockels coefficients. Let us investigate the electrooptic effects with various orientations of the external electric field. Find the cross section of the indicatrix cut by a plane perpendicular to the direction of the wavenormal, or in short, the "cross-sectional ellipse" whose major and minor axes are the only allowed directions of polarization (see Section 4.5.2). The incident wavenormal is in the Z direction and a Y-cut LiNbO$_3$ crystal is used. Consider the following external electric fields:

   (a) $\varepsilon = 0$.

   (b) $\varepsilon = \varepsilon_x$.

   (c) $\varepsilon = \varepsilon_y$.

   (d) $\varepsilon = \varepsilon_z$.

Solution Since the direction of the wavenormal is along the z axis, the "cross-sectional ellipse" is in the $z = 0$ plane. The expression for the "cross-sectional ellipse"

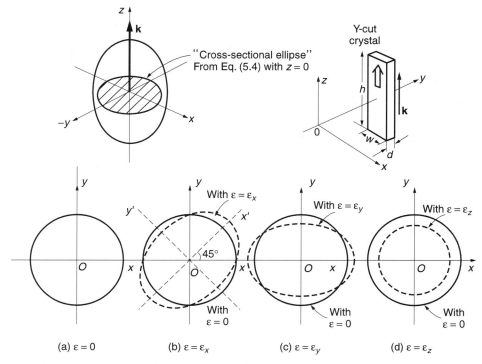

Figure 5.2   The $z = 0$ cross section of the indicatrix of LiNbO$_3$ when an external electric field is applied in various orientations.

is found by setting $z = 0$ in Eq. (5.4). From Table 5.1, the matrix of Pockels coefficients of lithium niobate is

$$\begin{bmatrix} a_{11} - 1/n_0^2 \\ a_{22} - 1/n_0^2 \\ a_{33} - 1/n_e^2 \\ a_{23} \\ a_{31} \\ a_{12} \end{bmatrix} = \begin{bmatrix} 0 & -r_{22} & r_{13} \\ 0 & r_{22} & r_{13} \\ 0 & 0 & r_{33} \\ 0 & r_{51} & 0 \\ r_{51} & 0 & 0 \\ -r_{22} & 0 & 0 \end{bmatrix} \begin{bmatrix} \varepsilon_x \\ \varepsilon_y \\ \varepsilon_z \end{bmatrix} \tag{5.10}$$

(a) $\varepsilon = 0$. When no external field is present, from Eqs. (5.4) and (5.10), we have

$$\frac{x^2}{n_0^2} + \frac{y^2}{n_0^2} = 1 \tag{5.11}$$

and the cross section is a circle.

(b) $\varepsilon = \varepsilon_x$. Only one cross-product term of Eq. (5.4) survives in the $z = 0$ plane, and from Eq. (5.10) we have

$$\frac{x^2}{n_0^2} + \frac{y^2}{n_0^2} - 2r_{22}\varepsilon_x xy = 1 \tag{5.12}$$

The product term creates the rotation of the axes. The product term can be removed from Eq. (5.12) by rotating the coordinates with respect to the z axis. The expression for the rotation of coordinates in two dimensions is

$$\begin{bmatrix} x \\ y \end{bmatrix} = \begin{bmatrix} \cos\theta & -\sin\theta \\ \sin\theta & \cos\theta \end{bmatrix} \begin{bmatrix} x' \\ y' \end{bmatrix} \tag{5.13}$$

where x' and y' are the new coordinates after rotation by θ. Rotation of the coordinates by $45°$ using Eq. (5.13) leads to

$$\left(\frac{1}{n_0^2} - r_{22}\varepsilon_x \right) x'^2 + \left(\frac{1}{n_0^2} + r_{22}\varepsilon_x \right) y'^2 = 1 \tag{5.14}$$

The first term is approximated as follows:

$$\left(\frac{1}{n_0^2} - r_{22}\varepsilon_x \right) x'^2 = \left(\frac{\sqrt{1 - r_{22}n_0^2\varepsilon_x}}{n_0} \right)^2 x'^2$$

$$= \left(\frac{\sqrt{1 - (r_{22}n_0^2\varepsilon_x)^2}}{n_0\sqrt{1 + r_{22}n_0^2\varepsilon_x}} \right)^2 x'^2$$

If $r_{22}n_0^2\varepsilon_x \ll 1$, the first term is approximated as

$$\left(\frac{1}{n_0^2} - r_{22}\varepsilon_x \right) x'^2 \doteq \frac{x'^2}{[n_0(1 + \frac{1}{2}r_{22}n_0^2\varepsilon_x)]^2}$$

$$= \frac{x'^2}{(n_0 + \frac{1}{2}r_{22}n_0^3\varepsilon_x)^2} \tag{5.15}$$

This means the index of refraction in the x' direction is larger than n_0 by $\Delta n = \frac{1}{2}r_{22}n_0^3\varepsilon_x$. A similar approximation is applicable to the second term in Eq. (5.14):

$$\frac{x'^2}{(n_0 + \frac{1}{2}r_{22}n_0^3\varepsilon_x)^2} + \frac{y'^2}{(n_0 - \frac{1}{2}r_{22}n_0^3\varepsilon_x)^2} = 1 \tag{5.16}$$

The major axis is in the x' direction and the minor axis is in the y' direction.

(c) $\varepsilon = \varepsilon_y$. From Eqs. (5.10) and (5.4) with $z = 0$, the ellipse becomes

$$\left(\frac{1}{n_0^2} - r_{22}\varepsilon_y \right) x^2 + \left(\frac{1}{n_0^2} + r_{22}\varepsilon_y \right) y^2 = 1 \tag{5.17}$$

which can be approximated in a manner similar to that used to obtain Eq. (5.15). Equation (5.17) becomes

$$\frac{x^2}{(n_0 + \frac{1}{2}r_{22}n_0^3\varepsilon_y)^2} + \frac{y^2}{(n_0 - \frac{1}{2}r_{22}n_0^3\varepsilon_y)^2} = 1 \tag{5.18}$$

The major and minor axes are in the x and y directions, respectively.

(d) $\varepsilon = \varepsilon_z$. From Eqs. (5.10) and (5.4) with $z = 0$, the ellipse becomes

$$x^2 \left(\frac{1}{n_0^2} + r_{13}\varepsilon_z \right) + y^2 \left(\frac{1}{n_0^2} + r_{13}\varepsilon_z \right) = 1$$

which can be approximated as

$$\frac{x^2}{(n_0 - \frac{1}{2}r_{13}n_0^3\varepsilon_z)^2} + \frac{y^2}{(n_0 - \frac{1}{2}r_{13}n_0^3\varepsilon_z)^2} = 1 \tag{5.19}$$

The cross section is a circle, as in the case of zero field, but the circle has shrunk. All results are summarized in Fig. 5.2. □

Example 5.2   For each of the configurations (b)–(d) in Example 5.1, design a voltage-controlled (I) phase shifter and (II) retarder. With both designs, specify the direction of polarization of the incident wave and find the expressions for the angle of the phase shift of the phase shifter, and the amount of retardation (amount of phase difference between the two orthogonal components of the transmitted light) of the retarder. A Y-cut lithium niobate crystal with dimensions $w \times d \times h$ in the x, y, and z directions is used. The control voltage is V.

Solution   The summary in Fig. 5.3 will be referred to often.

(I) *Phase Shifter.* The directions of polarization of the incident wave have to be in the directions of the major and minor axes of the cross-sectional ellipse.

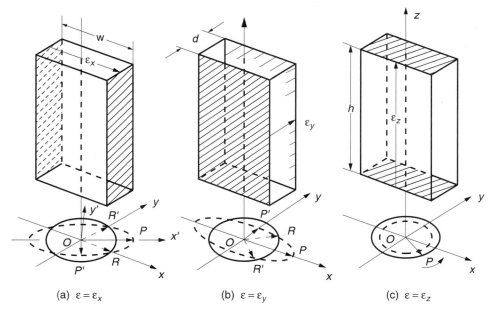

Figure 5.3   The directions of polarization of the incident wave. $\overline{OP}$ and $\overline{OP'}$ correspond to the phase shifter, and $\overline{OR}$ and $\overline{OR'}$ correspond to the retarder.

(b) $\varepsilon = \varepsilon_x$. The axes of the cross-sectional ellipse are rotated by $45°$. The polarization of the incident wave has to be parallel to either the x' or the y' axis.

From Eq. (5.16), the phase shift $\phi_{x',y'}$ is

$$\phi_{x',y'} = \frac{2\pi}{\lambda} \left(n_0 \pm \tfrac{1}{2} r_{22} n_0^3 \frac{V}{w} \right) h \qquad (5.20)$$

The plus sign on the right-hand side of Eq. (5.20) corresponds to polarization in the x' direction, while the minus sign corresponds to the y' direction.

(c) $\varepsilon = \varepsilon_y$. The allowed directions of polarization are in the directions of the x and y axes. The phase shift is, from Eq. (5.18),

$$\phi_{x,y} = \frac{2\pi}{\lambda} \left(n_0 \pm \tfrac{1}{2} r_{22} n_0^3 \frac{V}{d} \right) h \qquad (5.21)$$

(d) $\varepsilon = \varepsilon_z$. The direction of polarization can be any direction and the amount of the phase shift is, from Eq. (5.19),

$$\phi = \frac{2\pi}{\lambda} \left(n_0 - \tfrac{1}{2} r_{13} n_0^3 \frac{V}{h} \right) h \qquad (5.22)$$

It should be noted that the voltage-dependent term in the expression for ϕ is independent of h and does not increase with the length of the crystal as in the previous two cases; it is solely determined by the applied voltage V. Optically transparent electrodes such as tin dioxide (SnO_2) coatings have to be used.

(II) *Retarders.* The retarder is a device that creates a differential phase shift between the two orthogonal components of the transmitted light. It is usually used either to rotate the direction of polarization or to convert linearly polarized light into elliptically polarized light or vice versa (see Chapter 6).

The direction of polarization of the incident wave is most often set at $\pm 45°$ to the allowed directions of polarization so that the amplitude of the incident field is equally decomposed into the two allowed directions of polarization.

(b) $\varepsilon = \varepsilon_x$. In order to equally excite the two allowed polarization directions, the direction of the incident polarization is arranged along either the x or y axis. The amount Δ of retardation of the phase of the E'_y wave with respect to that of E'_x is, from Eq. (5.16),

$$\Delta = -\frac{2\pi}{\lambda} \left(r_{22} n_0^3 \frac{V}{w} \right) h \qquad (5.23)$$

(c) $\varepsilon = \varepsilon_y$. The direction of polarization of the incident wave is at $\pm 45°$ to the x axis. The amount of retardation is, from Eq. (5.18),

$$\Delta = -\frac{2\pi}{\lambda} \left(r_{22} n_0^3 \frac{V}{d} \right) h \qquad (5.24)$$

For both $\varepsilon = \varepsilon_x$ and $\varepsilon = \varepsilon_y$, the amount of retardation is proportional to h as well as V.

(d) $\varepsilon = \varepsilon_z$. Since the elliptic cross section is a circle, the retarder cannot be fabricated with this configuration. $\qquad\square$

An electrooptic amplitude modulator can be fabricated by combining the retarder in (a) with a set of crossed polarizers, as shown in Fig. 5.4. The incident light is polarized in the x direction. The output field E_y is the sum of the y components of E'_x and $E'y$. The variable retarder varies the relative phase between E'_x and E'_y and hence amplitude modulates the output E_y. The quantitative treatment is explained in Example 5.3.

Example 5.3 A Y-cut lithium niobate crystal with an external field in the x direction, such as shown in Fig. 5.3a, is used as an amplitude modulator, as shown in Fig. 5.4. The length of the crystal in the z direction is h. The applied electric field is $\varepsilon_x = \varepsilon_m \cos \omega_m t$, where $\varepsilon_m = V_m/w$. It is assumed that $1/\omega_m$ is much longer than the time that the light propagates through the crystal and the external field does not change appreciably during the transmission. Find the expression for the light intensity at the output of the prism analyzer.

Solution From Eq. (5.23) the retardation angle Δ is

$$\Delta = -\frac{2\pi}{\lambda}(r_{22}n_0^3 \varepsilon_m \cos \omega_m t)h \tag{5.25}$$

The incident lightwave is polarized horizontally and is $45°$ from the x' axis and $E_{x'} = -E_{y'} = (1/\sqrt{2})E_0$, as indicated in Fig. 5.5a. At the output there is a phase retardation of Δ between light amplitudes $E_{x'}$ and $E_{y'}$.

$$E_{x'} = \frac{1}{\sqrt{2}}E_0 \cos(-\omega t) \tag{5.26}$$

$$E_{y'} = -\frac{1}{\sqrt{2}}E_0 \cos(-\omega t + \Delta) \tag{5.27}$$

where ω is the angular frequency of the incident lightwave.

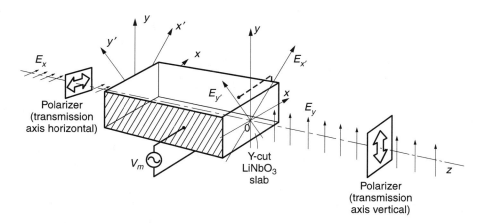

Figure 5.4 Electrooptic amplitude modulator.

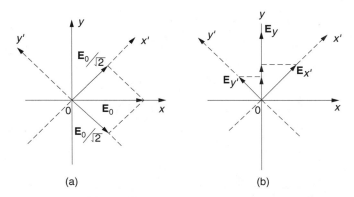

Figure 5.5 Vector diagram of the input and output of an amplitude modulator. (a) Input. (b) Output.

The output from the analyzer is the addition of the **E**-field components parallel to the y direction of the polarizer.

$$E_y = \frac{E_{x'}}{\sqrt{2}} + \frac{E_{y'}}{\sqrt{2}}$$

and

$$E_y = \frac{E_0}{2}[\cos \omega t - \cos(\omega t - \Delta)]$$

$$= E_0 \sin \frac{\Delta}{2} \sin \left(\omega t - \frac{\Delta}{2} \right) \qquad (5.28)$$

Hence, the output amplitude is proportional to $\sin (\Delta/2)$, and the output light intensity I for an input light intensity I_0 is

$$I = I_0 \sin^2 \frac{\Delta}{2} \qquad (5.29)$$

Insertion of Eq. (5.25) into (5.29) gives the final answer as plotted in Fig. 5.6. The output I is highly nonlinear with Δ but this can be minimized by biasing Δ to the point where it is most linear. (See Problem 5.2.) For an in-depth treatment of the retarder see Chapter 6.

It may be added that the curve of Fig. 5.6 lets you conversely find the applied field ε_x from the measured value of I/I_0. Such a LiNbO$_3$ slab can be used as an electrooptic probe tip for measuring the electric field ε_x with minimum disturbance to the original electric field [7]. $\qquad \square$

Example 5.4 The expression for the "cross-sectional ellipse" under an external field often takes the form

$$Ax^2 + Cy^2 + 2Bxy = 1 \qquad (5.30)$$

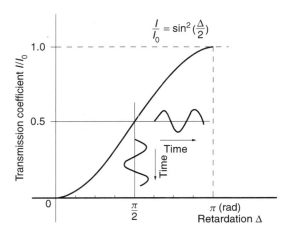

Figure 5.6 Output from the amplitude modulator.

By rotating the coordinates x and y with respect to the z axis, find the directions and amplitudes of the major and minor axes of the ellipse: (a) for the special case of $A = C$, and (b) for the general case.

Solution After rotation of the coordinates, we will find the condition that the $x'y'$ cross-product term vanishes. Insertion of Eq. (5.13) into (5.30) and after some manipulation leads to

$$(A\cos^2\theta + C\sin^2\theta + B\sin 2\theta)x'^2$$
$$+ (A\sin^2\theta + C\cos^2\theta - B\sin 2\theta)y'^2 \qquad (5.31)$$
$$+ [(C - A)\sin 2\theta + 2B\cos 2\theta]x'y' = 1$$

The angle θ of rotation of the coordinates that makes the $x'y'$ term disappear is

$$\tan 2\theta = \frac{2B}{A - C} \qquad (5.32)$$

which determines the orientation of the new coordinates, as shown in Fig. 5.7.

Next, the lengths of the major and minor axes are calculated for the two cases.

(a) For the special case of $A = C$, θ becomes

$$\theta = 45°$$

and Eq. (5.31) becomes

$$(A + B)x'^2 + (A - B)y'^2 = 1 \qquad (5.33)$$

(b) For the general case, by rewriting A and C as

$$A = a + d$$
$$C = a - d \qquad (5.34)$$

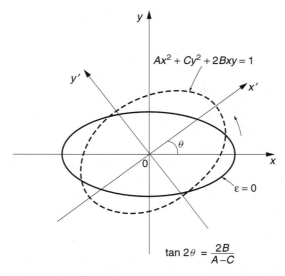

Figure 5.7 The angle of inclination θ of the major (or minor) axis of $Ax^2 + Cy^2 + 2Bxy = 1$ is $\tan 2\theta = 2B/(A - C)$.

or conversely,

$$a = \frac{A + C}{2}$$

$$d = \frac{A - C}{2}$$

both Eqs. (5.31) and (5.32) can be simplified. Equation (5.32) becomes

$$\tan 2\theta = \frac{B}{d} \tag{5.35}$$

Using Eq. (5.34) to express A and C in terms of a and d, and then rewriting $\sin 2\theta$ and $\cos 2\theta$ in terms of B and d by means of Eq. (5.35), a simpler expression for Eq. (5.31) is obtained as

$$(a + \sqrt{B^2 + d^2})x'^2 + (a - \sqrt{B^2 + d^2})y'^2 = 1 \tag{5.36}$$

One can obtain an approximate expression for Eq. (5.36) by putting

$$a = \frac{1}{N^2}$$

and making use of the procedure used to obtain Eq. (5.15):

$$\frac{x'^2}{(N - \frac{1}{2}N^3\sqrt{B^2 + d^2})^2} + \frac{y'^2}{(N + \frac{1}{2}N^3\sqrt{B^2 + d^2})^2} = 1 \tag{5.37}$$

The lengths of the major and minor axes are given by

$$\left(N \pm \frac{1}{2}N^3 \sqrt{B^2 + d^2} \right)$$

The special case $A = C$ corresponds to $d = 0$ in the above expressions.   □

5.2.2 Kerr Electrooptic Effect

The Kerr electrooptic effect is similar to the Pockels effect but the change Δa_i between the coefficients of the undistorted and distorted indicatrix is proportional to the second order of the applied field. There are six different combinations of ε_1, ε_2, and ε_3 that make up the second order field. Equation (5.5) is related to the quadratic electrooptic coefficients or Kerr coefficients as

$$\begin{bmatrix} a_{11} - 1/n_1^2 \\ a_{22} - 1/n_2^2 \\ a_{33} - 1/n_3^2 \\ a_{23} \\ a_{31} \\ a_{12} \end{bmatrix} = \begin{bmatrix} q_{11} & q_{12} & q_{13} & q_{14} & q_{15} & q_{16} \\ q_{21} & q_{22} & q_{23} & q_{24} & q_{25} & q_{26} \\ q_{31} & q_{32} & q_{33} & q_{34} & q_{35} & q_{36} \\ q_{41} & q_{42} & q_{43} & q_{44} & q_{45} & q_{46} \\ q_{51} & q_{52} & q_{53} & q_{54} & q_{55} & q_{56} \\ q_{61} & q_{62} & q_{63} & q_{64} & q_{65} & q_{66} \end{bmatrix} \begin{bmatrix} \varepsilon_1^2 \\ \varepsilon_2^2 \\ \varepsilon_3^2 \\ \varepsilon_2\varepsilon_3 \\ \varepsilon_3\varepsilon_1 \\ \varepsilon_1\varepsilon_2 \end{bmatrix} \qquad (5.38)$$

Examples of materials exhibiting the Kerr effect include such electrooptic crystals as barium titanate ($BaTiO_3$), potassium dihydrogen phosphate (KDP or KH_2PO_4), and ammonium dihydrogen phosphate (ADP or $NH_4H_2PO_4$). Certain liquid materials like benzene (C_6H_6), nitrobenzene ($C_6H_5NO_2$), carbon disulfide (CS_2), and water (H_2O) display the Kerr electrooptic effect. When an electric field is applied, the liquid material behaves like a uniaxial crystal with its optic axis along the applied electric field.

Like Pockels coefficients, many matrix elements vanish and the manipulation is much simpler than it looks. Kerr coefficients of isotropic media such as liquids take on the simple form

$$\begin{bmatrix} q_{11} & q_{12} & q_{12} & 0 & 0 & 0 \\ q_{12} & q_{11} & q_{12} & 0 & 0 & 0 \\ q_{12} & q_{12} & q_{11} & 0 & 0 & 0 \\ 0 & 0 & 0 & \frac{1}{2}(q_{11} - q_{12}) & 0 & 0 \\ 0 & 0 & 0 & 0 & \frac{1}{2}(q_{11} - q_{12}) & 0 \\ 0 & 0 & 0 & 0 & 0 & \frac{1}{2}(q_{11} - q_{12}) \end{bmatrix} \qquad (5.39)$$

and $n_1 = n_2 = n_3 = n$.

Let us derive the expression for the indicatrix when the external field $\varepsilon = \varepsilon_z$ is applied to a Kerr liquid contained in a rectangular cell. Applying Eq. (5.39) to (5.38), the constants a_{ij} for the ellipsoid are found. With $\varepsilon = \varepsilon_z$, we have

$$a_{11} = a_{22} = \frac{1}{n^2} + q_{12}\varepsilon_z^2$$

$$a_{33} = \frac{1}{n^2} + q_{11}\varepsilon_z^2$$

$$a_{23} = a_{31} = a_{12} = 0$$

Table 5.2   Kerr constants of liquids

Substance	Chemical Symbol	Index of Refraction	Kerr Constant K (m/V^2)	Wavelength (μm)
Nitrobenzene	$C_6H_5NO_2$	1.501	2.44×10^{-12}	0.59
Water	H_2O	1.333	5.1×10^{-14}	0.59
Carbon disulfide	CS_2	1.619	3.18×10^{-14}	0.63
Benzene	C_6H_6	1.496	4.14×10^{-15}	0.63
Carbon tetrachloride	CCl_4	1.456	7.4×10^{-16}	0.63

Using the approximation procedure of Eq. (5.15), the expression for the indicatrix becomes

$$\frac{x^2}{(n - \frac{1}{2}q_{12}n^3\varepsilon_z^2)^2} + \frac{y^2}{(n - \frac{1}{2}q_{12}n^3\varepsilon_z^2)^2} + \frac{z^2}{(n - \frac{1}{2}q_{11}n^3\varepsilon_z^2)^2} = 1 \qquad (5.40)$$

Equation (5.40) is the expression for the indicatrix of a uniaxial crystal with the optic axis in the z direction. Thus, the optic axis is in the direction of the external field. The indices of refraction for the ordinary and extraordinary waves are

$$\begin{aligned} n_0 &= n - \tfrac{1}{2}q_{12}n^3\varepsilon_z^2 \\ n_e &= n - \tfrac{1}{2}q_{11}n^3\varepsilon_z^2 \end{aligned} \qquad (5.41)$$

The difference between the indices of refraction for ordinary and extraordinary waves defines the Kerr constant K (not Kerr coefficients) of the liquid. From Eq. (5.41), the difference between refractive indices is

$$n_0 - n_e = K\lambda\varepsilon^2$$

where the Kerr constant is defined as

$$K = \frac{1}{2\lambda}(q_{11} - q_{12})n^3 \qquad (5.42)$$

and where λ is the wavelength of the light in vacuum. Table 5.2 shows the Kerr constants of various liquids. For example, when the external field $\varepsilon_z = 3 \times 10^6$ V/m is applied to a nitrobenzene Kerr cell of thickness 1 cm, the retardation is

$$\frac{2\pi}{\lambda}l(n_0 - n_e) = 2\pi l K\varepsilon^2$$

$$= 2\pi(0.01)(2.44 \times 10^{-12})(3 \times 10^6)^2$$

$$= 1.38 \text{ rad}$$

5.3   ELASTOOPTIC EFFECT

A change in the index of refraction takes place when a crystal is physically deformed. Such an effect is called the elastooptic effect. When the strain is created by an acoustic wave, it is sometimes called the acoustooptic effect.

So that the effect can be quantitatively expressed, the strains must clearly be defined. There are two kinds of strain — principal and shearing strains. As shown in Fig. 5.8a, when an elemental volume $dx\, dy\, dz$ is deformed by dU in the positive x direction by an external force, the strain in the x direction is defined as

$$S_{xx} = \frac{\partial U}{\partial x} \tag{5.43}$$

and for deformation dV in the positive y direction, the strain is

$$S_{yy} = \frac{\partial V}{\partial y} \tag{5.44}$$

Similarly, for deformation dW in the positive z direction, the strain is

$$S_{zz} = \frac{\partial W}{\partial z} \tag{5.45}$$

These three strains are called the principal strains.

Shearing strain causes a change in the angle $x0y$ of an edge with the application of a force. Referring to Fig. 5.8b, the changes in angles θ_1 and θ_2 are

$$\tan \theta_1 = \frac{\partial V}{\partial x}$$
$$\tan \theta_2 = \frac{\partial U}{\partial y} \tag{5.46}$$

The shearing strain S_{xy} is defined as

$$S_{xy} = \tfrac{1}{2}(\theta_1 + \theta_2)$$

and is approximated as

$$S_{xy} = \frac{1}{2}\left(\frac{\partial V}{\partial x} + \frac{\partial U}{\partial y}\right) \tag{5.47}$$

It is important to remember that dU, dV, and dW are positive when they are in the positive x, y, and z directions.

For instance, when the square is simply rotated as shown in Fig. 5.8c, the shearing strain is absent. In this case, note that $\partial U/\partial y$ is a negative quantity and cancels $\partial V/\partial x$ and $S_{xy} = 0$.

Similarly, the two other shearing strains are defined as

$$S_{yz} = \frac{1}{2}\left(\frac{\partial V}{\partial z} + \frac{\partial W}{\partial y}\right) \tag{5.48}$$

$$S_{xz} = \frac{1}{2}\left(\frac{\partial U}{\partial z} + \frac{\partial W}{\partial x}\right) \tag{5.49}$$

S_{xx}, S_{yy}, S_{zz}, S_{yz}, S_{xz}, and S_{xy} are represented by subscripts 1 to 6 as S_1, S_2, S_3, S_4, S_5, and S_6, respectively.

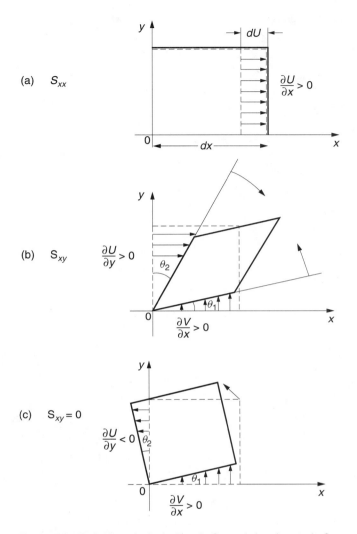

Figure 5.8   Definition of principal strain S_{xx} and shearing strain S_{xy}.

The change in the indicatrix, Eq. (5.5), is now represented using elastooptic constants as

$$
\begin{bmatrix}
a_{11} - 1/n_1^2 \\
a_{22} - 1/n_2^2 \\
a_{23} - 1/n_3^2 \\
a_{23} \\
a_{31} \\
a_{12}
\end{bmatrix}
=
\begin{bmatrix}
p_{11} & p_{12} & p_{13} & p_{14} & p_{15} & p_{16} \\
p_{21} & p_{22} & p_{23} & p_{24} & p_{25} & p_{26} \\
p_{31} & p_{32} & p_{33} & p_{34} & p_{35} & p_{36} \\
p_{41} & p_{42} & p_{43} & p_{44} & p_{45} & p_{46} \\
p_{51} & p_{52} & p_{53} & p_{54} & p_{55} & p_{56} \\
p_{61} & p_{62} & p_{63} & p_{64} & p_{65} & p_{66}
\end{bmatrix}
\begin{bmatrix}
S_1 \\
S_2 \\
S_3 \\
S_4 \\
S_5 \\
S_6
\end{bmatrix}
\tag{5.50}
$$

Matrices of the elastooptic constants for commonly used materials are tabulated in Table 5.3.

Table 5.3 Elastooptic properties

Name of Substance	Chemical Symbol	Photoelastic Constant	Index of Refraction	Wavelength (μm)	Crystal Symmetry	Elastooptic Tensor
Fused silica	SiO_2	$p_{11} = 0.121$ $p_{12} = 0.270$ $p_{44} = p_{55} = p_{66}$ $= \frac{1}{2}(p_{11} - p_{12})$	$n = 1.457$	0.63	Isotropic	$\begin{bmatrix} p_{11} & p_{12} & p_{12} & 0 & 0 & 0 \\ p_{12} & p_{11} & p_{12} & 0 & 0 & 0 \\ p_{12} & p_{12} & p_{11} & 0 & 0 & 0 \\ 0 & 0 & 0 & p_{44} & 0 & 0 \\ 0 & 0 & 0 & 0 & p_{55} & 0 \\ 0 & 0 & 0 & 0 & 0 & p_{66} \end{bmatrix}$
Water	H_2O	$p_{11} = 0.31$ $p_{12} = 0.31$ $p_{44} = p_{55} = p_{66}$ $= \frac{1}{2}(p_{11} - p_{12})$	$n = 1.33$	0.63	Isotropic	
Gallium arsenide	GaAs	$p_{11} = -0.165$ $p_{12} = -0.140$ $p_{44} = -0.061$	$n_x = n_y = n_z = 3.42$	1.15	$\bar{4}3m$	$\begin{bmatrix} p_{11} & p_{12} & p_{12} & 0 & 0 & 0 \\ p_{12} & p_{11} & p_{12} & 0 & 0 & 0 \\ p_{12} & p_{12} & p_{11} & 0 & 0 & 0 \\ 0 & 0 & 0 & p_{44} & 0 & 0 \\ 0 & 0 & 0 & 0 & p_{44} & 0 \\ 0 & 0 & 0 & 0 & 0 & p_{44} \end{bmatrix}$
Zinc sulfide	β-ZnS	$p_{11} = 0.091$ $p_{12} = -0.01$ $p_{44} = 0.075$	$n_x = n_y = n_z = 2.352$	0.63		
Lithium niobate	$LiNbO_3$	$p_{11} = -0.02$ $p_{12} = 0.08$ $p_{13} = 0.13$ $p_{14} = -0.08$ $p_{31} = 0.17$ $p_{33} = 0.07$ $p_{41} = -0.15$ $p_{44} = 0.12$ $p_{66} = \frac{1}{2}(p_{11} - p_{12})$	$n_x = n_y = 2.286$ $n_z = 2.20$	0.63	$3m$	$\begin{bmatrix} p_{11} & p_{12} & p_{13} & p_{14} & 0 & 0 \\ p_{12} & p_{11} & p_{13} & -p_{14} & 0 & 0 \\ p_{31} & p_{31} & p_{33} & 0 & 0 & 0 \\ p_{41} & -p_{41} & 0 & p_{44} & 0 & 0 \\ 0 & 0 & 0 & 0 & p_{44} & p_{41} \\ 0 & 0 & 0 & 0 & p_{14} & p_{66} \end{bmatrix}$

$$\begin{bmatrix} p_{11} & p_{12} & p_{13} & 0 & 0 & 0 \\ p_{12} & p_{11} & p_{13} & 0 & 0 & 0 \\ p_{31} & p_{31} & p_{33} & 0 & 0 & 0 \\ 0 & 0 & 0 & p_{44} & 0 & 0 \\ 0 & 0 & 0 & 0 & p_{44} & 0 \\ 0 & 0 & 0 & 0 & 0 & p_{66} \end{bmatrix}$$

Material	Coefficients	Refractive index		Symmetry
Lithium tantalate (LiTaO₃)	$p_{11} = -0.08$ $p_{12} = -0.08$ $p_{13} = 0.09$ $p_{14} = -0.03$ $p_{31} = 0.09$ $p_{33} = -0.044$ $p_{41} = -0.085$ $p_{44} = 0.02$ $p_{66} = \frac{1}{2}(p_{11} - p_{12})$	$n_x = n_y = 2.176$ $n_z = 2.180$	0.63	$3m$
Rutile (TiO₂)	$p_{11} = -0.011$ $p_{12} = 0.172$ $p_{13} = -0.168$ $p_{31} = -0.096$ $p_{33} = -0.058$ $p_{44} = 0.0095$ $p_{66} = \pm0.072$	$n_x = n_y = 2.585$ $n_z = 2.875$	0.63 0.51 0.63	$\bar{4}2m$
Potassium dihydrogen phosphate (KDP) (KH₂PO₄)	$p_{11} = 0.251$ $p_{12} = 0.249$ $p_{13} = 0.246$ $p_{31} = 0.225$ $p_{33} = 0.221$ $p_{44} = -0.019$ $p_{66} = -0.058$	$n_x = n_y = 1.51$ $n_z = 1.47$	0.63 0.59 0.63	$\bar{4}2m$

Table 5.3 *(Continued)*

Name of Substance	Chemical Symbol	Photoelastic Constant	Index of Refraction	Wavelength (μm)	Crystal Symmetry	Elastooptic Tensor
Ammonium dihydrogen phosphate (ADP)	$NH_4H_2PO_4$ or ADP	$p_{11} = 0.302$ $p_{12} = 0.246$ $p_{13} = 0.236$ $p_{31} = 0.195$ $p_{33} = 0.263$ $p_{44} = -0.058$ $p_{66} = -0.075$	$n_x = n_y = 1.52$ $n_z = 1.48$	0.63 0.59 0.59	$\bar{4}2m$	$\begin{bmatrix} p_{11} & p_{12} & p_{13} & 0 & 0 & 0 \\ p_{12} & p_{11} & p_{13} & 0 & 0 & 0 \\ p_{31} & p_{31} & p_{33} & 0 & 0 & 0 \\ 0 & 0 & 0 & p_{44} & 0 & 0 \\ 0 & 0 & 0 & 0 & p_{44} & 0 \\ 0 & 0 & 0 & 0 & 0 & p_{66} \end{bmatrix}$
Tellurium dioxide	TeO_2	$p_{11} = 0.0074$ $p_{12} = 0.187$ $p_{13} = 0.340$ $p_{31} = 0.090$ $p_{33} = 0.240$ $p_{44} = -0.17$ $p_{66} = -0.046$	$n_x = n_y = n_z = 2.35$	0.63	$\bar{4}2m$	

Example 5.5   A transverse acoustic wave is launched in a crystal in the x, y, and z directions as shown in Fig. 5.9.

(a) Identify the predominant shearing strains in each of the three cases, and designate this strain by one of $S_1, S_2, \ldots, S_6$.

(b) Are there any other directions of propagation of the transverse acoustic wave that create the same shearing strains as those in Fig. 5.9?

Solution

(a) A useful mnemonic for identifying shearing strain is pinching a rubber eraser with a cross mark on its side, as shown in Fig. 5.10. The deviation of θ from 90° identifies the presence of a shearing strain. The shearing strain is in the plane of the cross of the eraser. Now imagine the pinched eraser to coincide with the crest of an acoustic wave in Fig. 5.9. The shearing strain in Fig. 5.9a takes place in the y–z plane and it is S_4. The strain in Fig. 5.9b is S_5 and that in Fig. 5.9c is S_6.

(b) The propagation of the transverse acoustic wave in the z direction, which is obtained by rotating the wave in Fig. 5.9a by 90° around the x axis, also creates S_4. The propagation in the x direction, which is obtained by rotating the wave

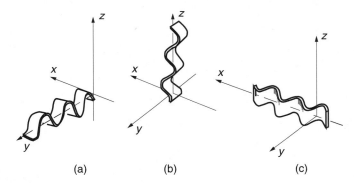

| (a) | (b) | (c) |

Figure 5.9   Identification of shearing strains due to transverse acoustic waves in a crystal.

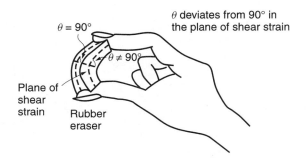

Figure 5.10   Identification of shearing strain at your fingertips.

in Fig. 5.9b by 90° around the y axis, also creates S_5. The propagation in the y direction, which is obtained by rotating the wave in Fig. 5.9c by 90° around the z axis, also creates S_6.   □

Example 5.6   When an acoustic wave is launched into a crystal that displays the elastooptic effect, the index of refraction of the crystal is spatially modulated by the periodic strains caused by the acoustic wave. The incident light into the crystal is diffracted by the modulation of the index of refraction. The direction of diffraction is changed by changing the period of modulation. Such a light deflector is called an acoustooptic light deflector. If the acoustic wave is launched on the surface of the crystal, the device is called a surface acoustic wave (SAW) light deflector. The surface acoustic wave is generated by interdigital electrodes, such as shown in Fig. 5.11. Strain is generated by the electric field set up by a pair of electrodes because of the piezo-electric effect. The generated longitudinal surface acoustic wave propagates along the crystal.

If a Y-cut tellurium dioxide (TeO$_2$) crystal is used for the SAW light deflector, what are the directions of polarization and propagation of the incident light that require minimum power to the interdigital transducer? For the ideal situation, assume that the lightwave and the surface acoustic wave are launched in the same plane.

Solution   Referring to Table 5.3, the largest elastooptic coefficient of TeO$_2$ is $p_{13} = 0.34$. From the matrix, Eq. (5.50), S_3 has to be nonzero to make use of p_{13}. Thus, a longitudinal acoustic wave has to be launched in the z direction of the crystal.

The change in the indicatrix due to S_3 is calculated using Table 5.3 and Eq. (5.50) with $S = (0, 0, S_3, 0, 0, 0)$. With approximations as were used in Eq. (5.15), the

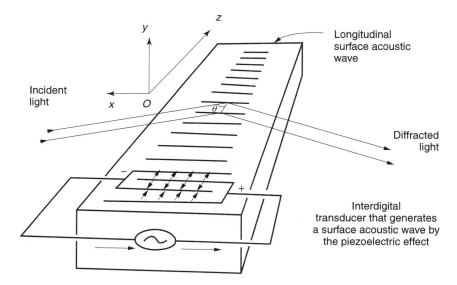

Figure 5.11   SAW light deflector.

indicatrix becomes

$$\frac{x^2}{(n_0 - \frac{1}{2}n_0^3 p_{13}S_3)^2} + \frac{y^2}{(n_0 - \frac{1}{2}n_0^3 p_{13}S_3)^2} + \frac{z^2}{(n_e - \frac{1}{2}n_e^3 p_{33}S_3)^2} = 1 \qquad (5.51)$$

All three major axes of the indicatrix change with S_3, but the x and y axis change the most. Referring to the indicatrix shown in Fig. 5.12, various polarizations and propagation directions are considered for a Y-cut crystal.

When the light is polarized in the x direction, the direction of the light propagation has to be in the y–z plane. This means the light has to propagate along the z direction, otherwise the light would pass through the crystal. For z propagation, θ in Fig. 5.11 must be 0°.

When the light is polarized in the y direction, the direction of the propagation of the light is in the x–z plane. In this case the deflected light stays in the x–z plane. This polarization gives more flexibility in reflection angle in that θ in Fig. 5.11 need not be 0°.

It is also possible to use the e-wave, which is polarized in the x–z plane, as shown in Fig. 5.12, but this is less desirable because the effective index of refraction is dependent on the angle of deflection.

The acoustic power needed to produce S is approximately

$$P \propto |S|^2 \qquad (5.52)$$

(stress is proportional to S and power is the product of the force and displacement). Thus, if one were to use the next biggest coefficient $p_{33} = 0.24$, the driving acoustic power has to be $(0.34/0.24)^2 = 2$ times larger. □

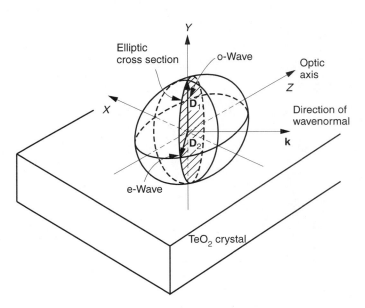

Figure 5.12   Indicatrix of a TeO₂ SAW light deflector.

5.4 MAGNETOOPTIC EFFECT

Magnetooptic effects cause the optical properties of a medium to change when the medium is subjected to a magnetic field. The Faraday effect is the most widely used magnetooptic effect. Another magnetooptic effect is the Cotton–Mouton effect. In order to gain some insight into how magnetic fields can alter the optical properties of a material, the Faraday effect will be explained using a classical model.

5.4.1 Faraday Effect

Using the classical model of precession of electron spin due to an external magnetic field, the Faraday effect will be phenomenologically explained [8]. As an electron orbits the nucleus, it spins on its own axis. Since the electron is charged, the spinning electron creates a small current loop and possesses a magnetic moment **m**. If the magnetic moment is under the influence of an external dc magnetic field, the magnetic moment tends to line up with the external magnetic field and a force is applied to the magnetic moment as indicated by **F** in Fig. 5.13. Due to the applied force **F**, the spinning electron with an angular momentum starts to precess around the external magnetic field $\mathbf{H}_{dc}$. It is analogous to a precessing top. When the axis of a top is tilted and the gravitational force attempts to change the direction of the angular momentum of the top, the top starts precessing. The precession of the magnetic moment of the spinning electron is the source of the Faraday effect.

When a circularly polarized wave is incident into a medium, the magnetic permeability that the wave sees depends on the sense of the circular polarization. The propagation constant k for a circularly polarized wave that propagates along $\mathbf{H}_{dc}$ is [8]

$$k^2 = \omega^2 \epsilon (\mu \pm K) \tag{5.53}$$

where ϵ is the dielectric constant, μ is the magnetic permeability in the absence of the precession or in the absence of $\mathbf{H}_{dc}$, and K is the difference in magnetic permeability

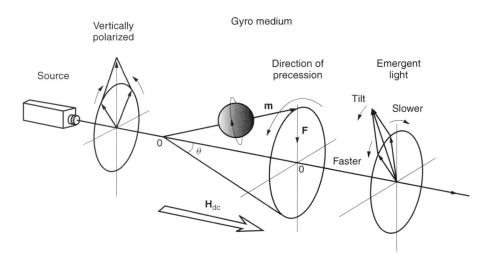

Figure 5.13   Gyro effect of the electron spin due to an external field $\mathbf{H}_{dc}$.

caused by the precession. K in general is a complex number. The positive sign corresponds to circular polarization whose sense is in the same direction as that of the precession, and the negative sign means that the sense of the circular polarization is opposite to that of the precession.

When the incident light field is linearly polarized, it can be decomposed into two circularly polarized waves of equal magnitude but having the opposite sense of circular polarization. Any effect causing the propagation constants for these two circularly polarized component waves to differ results in a rotation of the direction of the emergent linearly polarized wave, as illustrated in Fig. 5.13. This rotation of the emergent linearly polarized wave is called the *Faraday effect*. An interesting and useful property of the Faraday effect is that *the direction of the rotation of the linearly polarized light depends only on the direction of precession*, which is determined by the direction of the applied magnetic field; it does not depend on whether the light is traveling from left to right, or right to left. The rotation of the linearly polarized light is always in the same direction as the precession. To be more specific, in the example shown in Fig. 5.14, the direction of polarization rotates from vertically upward to the horizontal direction after passing through the Faraday medium of an appropriate length from the left to the right in Fig. 5.14a. If, however, the direction of propagation is reversed and the light passes from the right to the left, as in Fig. 5.14b, the direction of polarization points downward rather than upward. This is called the nonreciprocity of the Faraday effect. This property is used for fabricating optical isolators that will be explained further in the next section.

The quantitative expression for the rotation angle Φ of the polarization is

$$\Phi = V\mathbf{H}_{dc} \cdot \mathbf{l} \tag{5.54}$$

where V is the Verdet constant, usually expressed in units of degrees/(oersted-cm), $\mathbf{H}_{dc}$ is the external dc magnetic field, and $\mathbf{l}$ is the path of light inside the medium. Both $\mathbf{H}_{dc}$ and $\mathbf{l}$ are vector quantities.

5.4.2 Cotton-Mouton Effect

As indicated by Eq. (5.54), the Faraday effect disappears when the direction of the dc magnetic field is perpendicular to that of the light propagation, but birefringence can be observed under these conditions with some substances like nitrobenzene. This birefringence observed in a liquid like nitrobenzene is attributed to the Cotton–Mouton effect and is proportional to the square of the applied dc magnetic field. The substance behaves like a uniaxial crystal with its optic axis along the external magnetic field. When the light is polarized in the direction of H_{dc} it behaves like an e-wave, and when it is polarized perpendicular to the plane set by the direction of the light propagation and the external magnetic field, the wave behaves like an o-wave. This resembles the Kerr effect with a liquid substance.

5.5 OPTICAL ISOLATOR

An optical isolator is a device that allows transmission of light in one direction while it suppresses light transmission in the opposite direction. Optical isolators are used to

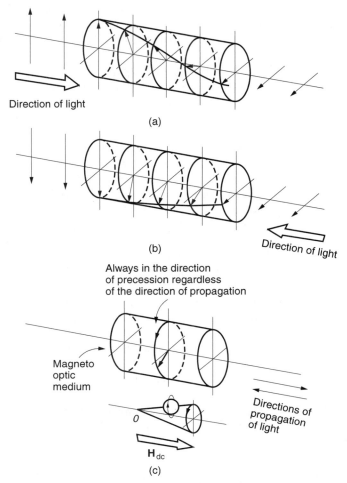

Figure 5.14 Rotation of the direction of polarization due to the Faraday effect. (a) Light from the left to the right. (b) Light from the right to the left. (c) Direction of the rotation.

suppress the effects of reflected light. For example, the insertion of an isolator between the laser diode and the rest of the system cuts down on the light reflected back into the laser, and not only stabilizes the laser operation but also substantially reduces the laser noise.

There are two types of optical isolators: one whose performance is influenced by the direction of polarization of the incident light, and one whose performance is independent of the polarization of the incident light.

5.5.1 Polarization-Dependent Optical Isolator

Figure 5.15a shows the layout of the polarization-dependent optical isolator. The magnetooptic material is sandwiched between the polarizer and analyzer prisms. The direction of the polarization of the analyzer is tilted by 45° from that of the polarizer. Referring to Fig. 5.15b, the vertically polarized incident wave can pass through the

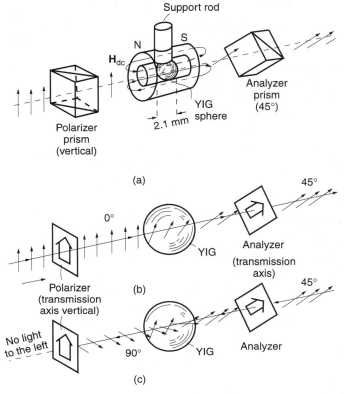

Figure 5.15 Optical isolator. (a) Construction of an optical isolator. (b) Light in the forward direction. (c) Light in the backward direction. The arrows in the polarizer and analyzer indicate the transmission axis.

polarizer with minimum attenuation and enters the magnetooptic material. The Faraday effect causes the direction of the polarization to rotate by 45°. Since the analyzer is also set at 45°, the emergent light from the magnetooptic material can proceed through the analyzer with minimum attenuation into the optical system. Referring to Fig. 5.15c, the reflected light from the optical system passes through the analyzer and enters the magnetooptic material, where the Faraday effect again causes the direction of polarization to rotate by 45°. As was discussed in Section 5.4.1, the sense of rotation of the polarization is independent of the direction of light propagation so that light passing through twice (once in the forward direction and once in the reverse direction) experiences a total rotation of 90°.

On reaching the polarizer, the reflected light has its polarization in the direction perpendicular to the transmission axis of the polarizer and cannot go through the polarizer toward the source. Thus, the light source is isolated from the reflected light from the outside system.

Ferromagnetic substances like yttrium iron garnet (YIG), $Y_3Fe_5O_{12}$, display a strong Faraday effect because, as a characteristic of ferromagnetism, the electron spins are all pointed in the same direction, and the effect of the precession is significantly accentuated. The Verdet constant is much higher than in other magnetooptic substances like arsenic trisulfide (AsS_3), soda lime silicate, or terbium (Tb)-doped paramagnetic glass. Another feature of YIG is that the Faraday effect is saturated at 1780 gauss* and there is no need to accurately control the magnetic field as long as the field is stronger than this value. The Verdet constant with a saturated magnetic field is expressed in units of degrees/cm and the value for YIG is 220 degrees/cm. A YIG sphere with a diameter as small as 2.1 mm is enough to give a 45° rotation at a wavelength of $\lambda = 1.3$ μm [9]. It should, however, be noted that the Verdet constant is a strong function of wavelength and decreases as the square of the wavelength.

The optical isolator shown in Fig. 5.15 uses a YIG sphere. The YIG sphere is housed in a doughnut-shaped permanent magnet that can provide a magnetic field much stronger than the saturation magnetization. A polymer-bonded rare earth magnet is conveniently used because it is easily machined.

5.5.2 Polarization-Independent Optical Isolator

During light transmission in an ordinary optical fiber, the direction of polarization fluctuates. When an optical isolator [10,11] is to be used in such a system, either the isolator has to be the polarization-independent type, or the polarization has to be stabilized by means of polarization-maintaining fiber or polarization controllers. In this section, the polarization-independent optical isolator is explained.

The polarization-independent isolator combines the birefringence of prisms and the nonreciprocity of the Faraday rotator. Figure 5.16a shows the layout of the components. The Selfoc lens (graded index lens, or GRIN lens) converts the output of the fiber into a plane wave. The input as well as output prisms P_1 and P_2 are made out of birefringent crystal. The prisms have the same shape, but the directions of their optic axes are tilted from each other by 45°. The Faraday rotator in the center rotates the polarization directions of both forward and backward waves by 45°. The direction of rotation is the same for forward and backward waves.

Let us first follow the path of the forward wave propagating from the left to the right in Fig. 5.16b. The plane wave from the Selfoc lens is arbitrarily polarized. Let it be decomposed into the e-wave and o-wave components of crystal prism P_1. The prism being birefringent, the e-wave sees the refractive index n_e and the o-wave sees n_0, and each is refracted into different directions by prism P_1. The Faraday rotator rotates the polarization directions of both the o-wave and the e-wave by 45°, and the direction of rotation is the same for both waves. Since the optic axis of P_2 is also rotated by 45°, the o-wave still sees refractive index n_0 in P_2, and likewise the e-wave sees refractive index n_e in P_2. Both waves emerge from P_2 parallel to the axis of the Selfoc lens and are focused properly into the center of the core of the output fiber F_2 for transmission.

The blockage of the backward wave, as illustrated in Fig. 5.16c, will now be explained. The backward wave is also decomposed into components parallel and perpendicular to the tilted optic axis of prism P_2, which are the e-wave and o-wave, respectively. The Faraday rotator rotates the polarization directions of both waves further by 45°. For the forward wave, the Faraday rotation compensates for the tilt of the optic axis of prism P_2; but for the backward wave, the tilt and the Faraday rotation do not compensate each other. The backward o-wave from P_2 is rotated 90°

---

* The magnetic field of a small magnet used for posting memo papers on a refrigerator door is about 1000 gauss.

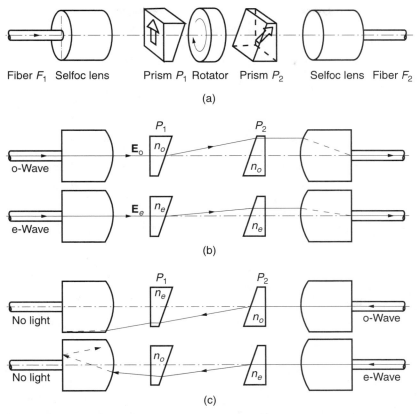

Figure 5.16 Principle of operation of a polarization-independent isolator. (a) General layout. (b) Passage of forward light. (c) Blockage of backward light.

with respect to the optical axis of P_1 and sees the refractive index n_e in P_1. Similarly, the backward e-wave from P_2 sees the refractive index n_0 in P_1. Upon emerging from P_1, the backward waves are not parallel to the axis of the Selfoc lens and cannot be focused properly onto the core of fiber F_1. The decrease in coupling into fiber F_1 for slanted rays is significant. A $1°$ slant angle from the optical axis causes a 45-dB decrease in the coupling power. A ratio of forward to backward waves of over 35 dB is obtainable by this type of isolator.

5.6 PHOTOREFRACTIVE EFFECT

When a fringe pattern of sufficient intensity impinges on a photorefractive medium, a phase grating is formed in the medium with an index of refraction profile that follows the intensity distribution of the fringe pattern. This grating persists even after the incident fringe pattern has been removed, as long as the crystal is kept in low light levels. The phase grating is erasable with a floodlight illuminating all directions. After it is erased, the same procedure can be repeated with minimum fatigue. Such an effect is called the photorefractive effect and is used as a means of recording real-time holograms, for producing phase conjugate mirrors [12,13] (see Chapter 8), and for optical communication applications such as an energy converter of an optical amplifier. Photorefractive crystals include lithium niobate (LiNbO$_3$), barium titanate (BaTiO$_3$), gallium arsenide (GaAs), bismuth silicon oxide (BSO, Bi$_{12}$SiO$_{20}$), bismuth germanium oxide (BGO, Bi$_{12}$GeO$_{20}$), and some kinds of liquid crystals.

A heuristic explanation of the photorefractive effect is as follows. The photorefractive effect is due to the local electrooptic effect. The illuminating light frees electrons from the orbits of the atoms in the crystal, and the freed electrons disturb the electrical neutrality and locally establish an electric field. The electric field, in turn, causes changes in the index of refraction due to the electrooptic effect.

A slightly more detailed explanation will be attempted using Figs. 5.17 and 5.18. When two coherent light beams R and S interfere inside a photorefractive crystal, the two beams generate a fringe pattern as shown in Fig. 5.17 that is responsible for writing in the phase grating. Figure 5.18 illustrates the sequence of events in the crystal after illumination by the fringe pattern. Unlike electrons that were freed out of the orbits of the atoms, the atoms (ions) are locked in the crystal and are immobile. Thus, atoms whose electrons have been lost becomes centers of $\oplus$ charges. The fact that these charges are immobile plays an important role. Even though the freed electrons move through the crystal (under a diffusion force), their movement is not completely free because of the electric field established by the immobile $\oplus$ charges. The immobile $\oplus$ charges attract the electrons and try to pull the electrons back. Such a force of attraction is called a *drift force. The drift force and the diffusion force are opposed.* The free electrons redistribute themselves to an equilibrium where these two opposing forces balance. The equilibrium distribution of the mobile negative charges and the immobile $\oplus$ charges is something like that shown in Fig. 5.18a. (Immobile charges are circled).

Below the picture of Fig. 5.18a, the corresponding spatial distributions of various quantities are presented. The curve in Fig. 5.18b shows the intensity distribution of the incident light fringe pattern. The curve in Fig. 5.18c represents the spatial distribution of the charge in the crystal. Electrons are piled up in the dark regions. This is as if the electrons are swept away into the dark regions by a fly swatter, just as cockroaches hide themselves in the shade. The curve in Fig. 5.18d represents the electric field

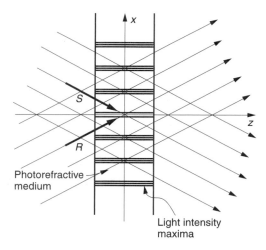

Figure 5.17   Fringe pattern projected into a photorefractive substance by R and S beams of light.

Cockroach theory of photorefractivity. This theory was conceived by a scientist who watched how cockroaches hide themselves in the shade. The cockroaches represent the electrons in a photorefractive crystal.

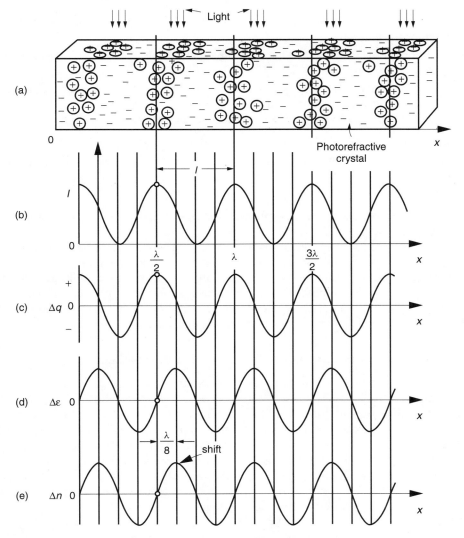

Figure 5.18 Illustration of the photorefractive effect.

distribution $\Delta\varepsilon$, which is caused by the spatial modulation of the charge. Finally, the curve in Fig. 5.18e represents the change in the index of refraction Δn, resulting from the modulated $\Delta\varepsilon$ field in Fig. 5.18d due to the electrooptic effect. This Δn persists as a phase grating. Comparing graphs of Figs. 5.18b and 5.18e, one sees that the spatial modulation Δn of the index of refraction is shifted from that of the light intensity pattern by one-quarter period of the light intensity pattern or by $\lambda/8$.

The following is a summary of the observed properties of the photorefractive effect.

1. The photorefractive effect increases with an increase in the spatial frequency of the light fringe pattern.
2. The photorefractive effect is at its maximum when the crystal axis is normal to the planes of constant index of refraction; namely, the crystal axis is along the x direction in Fig. 5.18a.
3. The longer the wavelength of the floodlight is, the longer time it takes to erase the phase grating.
4. A dc external electric field in the direction of the crystal axis enhances the photorefractive effect.
5. Even when the intensities of the R and S beams are made equal at the input to the crystal, in general, the intensities are not equal at the output of the crystal. This is the result of a transfer of energy between the two beams inside the crystal.

Observed Properties 4 and 5 are the principle of operation of an optical amplifier based on the photorefractive effect and will be explained in more detail in the next section.

5.7 OPTICAL AMPLIFIER BASED ON THE PHOTOREFRACTIVE EFFECT

There are two major types of optical amplifiers: one is based on the transfer of light energy in a photorefractive medium, while the other is an erbium-doped fiber amplifier or a laser diode at the verge of lasing. The former type will be explained here while the latter in Chapter 14. The optical amplifier described here uses the enhanced photorefractive effect produced by an external field combined with the transfer of light energy in the crystal. These properties will be explained separately and then combined afterward.

5.7.1 Enhanced Photorefractive Effect by an External Electric Field

An external electric field is applied to the photorefractive medium shown in Fig. 5.19a. The diffusion force is not influenced by the external field but the drift force is. As a result, the equilibrium distribution shown in Fig. 5.18a is upset. Since the $\oplus$ charges are immobile, the electrons are pushed by the external field, and the distribution of the charges will become somewhat like that shown in Fig. 5.19a. Note the spacings between $\oplus$ and $-$ in Fig. 5.19a. This results in an increase in the local electric field $\Delta\varepsilon$ and enhances the amount of Δn. The drift of free electrons due to the external field is primarily responsible for the formation of the phase grating. When an external field aids the drift force as in Fig. 5.19a, the positions of the maxima of Δn coincide very nearly with those of the light fringe pattern, as indicated in Figs. 5.19b and 5.19c. The quarter-period shift that was seen in Fig. 5.18, however, starts to diminish and results in a change in the efficiency of the energy transfer.

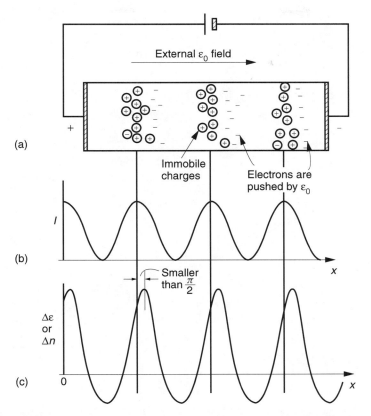

Figure 5.19 Photorefractive effect with an external dc electric field.

5.7.2 Energy Transfer in the Crystal

The amount of energy transfer between the R and S beams within the crystal critically depends on the relative position between the fringe pattern of the incident beams and the photorefractive phase grating in the crystal. In order to study the effect of the relative position, let us first assume that the maxima of the fringe pattern coincide with the established Δn grating as shown in Fig. 5.20a. The origin of the coordinates is taken at the fringe maximum where the two waves R_0 and S_0 are in phase. A portion of the incident beam is reflected by the fringe and the other portion is transmitted through the fringe. The reflected portion of R_0 merges into $S(x)$, and similarly the reflected portion of S_0 merges into $R(x)$. The waves reflected by the fringe undergo a $(\pi/2)$-radian phase shift [14–16], while the waves transmitted through the fringe undergo no phase shift. Referring to Fig. 5.20a, the $S(x)$ and $R(x)$ beams are

$$R(x) = tR_0 + jrS_0 \tag{5.55}$$

$$S(x) = jrR_0 + tS_0 \tag{5.56}$$

where r and t are reflection and transmission coefficients and are assumed to be real numbers.

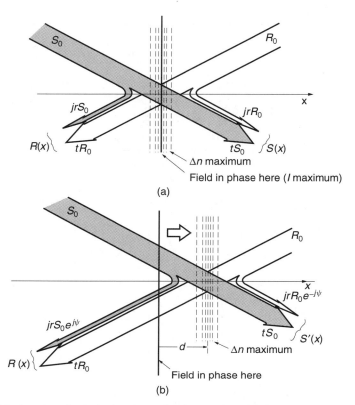

Figure 5.20 Explanation of gain in the photorefractive medium. (a) The Δn maximum matches that of the light fringe. (b) The Δn maximum is shifted from that of the light fringe.

When $R_0 = S_0$ and the beams are like polarized, the location of the fringe maximum lines up with the Δn grating maximum, and from Eqs. (5.55) and (5.56), $R(x) = S(x)$. There is no net energy transfer between the R and S beams.

Next, let us shift the location of the maximum of Δn to the right by d, as shown in Fig. 5.20b, in order to analyze the case of the shifted Δn in Fig. 5.18e or Fig. 5.19c. This shift creates a phase shift of the reflected wave but not of the transmitted wave. Equations (5.55) and (5.56) become

$$R(x) = tR_0 + jrS_0e^{j\psi} \tag{5.57}$$

$$S(x) = jrR_0e^{-j\psi} + tS_0 \tag{5.58}$$

where $\psi = 2d(2\pi/\lambda)$. When $d = \lambda/8$ and $\psi = \pi/2$ radians, then

$$R(x) = tR_0 - rS_0 \tag{5.59}$$

$$S(x) = rR_0 + tS_0 \tag{5.60}$$

and the magnitude of $S(x)$ is larger than $R(x)$. This is interpreted as a net transfer of energy from the R beam into the S beam. On the other hand, when $\psi = +\pi/2$ radians, the magnitude of $R(x)$ becomes larger than $S(x)$ and the direction of the transfer of energy is reversed.

Figure 5.21 illustrates the above description by a phasor diagram. Outputs $R(x)$ and $S(x)$ from the crystal are expressed in terms of the inputs R_0, S_0, $S_0 e^{j\psi}$, and $R_0 e^{-j\psi}$ so as to demonstrate the influence of ψ on the transfer of energy between $R(x)$ and $S(x)$. The following three cases are considered:

Case (a) is for $\psi = 0$.
Case (b) is for $0 < \psi < \pi/2$.
Case (c) is for $\psi = \pi/2$.

For all cases, the amplitudes R_0 and S_0 are assumed identical. From the figure, the contribution to the energy transfer from $R_0 e^{-j\psi}$ into $S_0 e^{j\psi}$ increases as ψ approaches $\psi = \pi/2$ radians. This phase shift is essential for the energy transfer. The phasor representation of Fig. 5.21 is easily extended to the case where R_0 and S_0 are real but unequal. For $S_0 \ll R_0$, the energy transfer from the R beam into the S beam can result in $S(x)$ being much larger than its initial value S_0.

So far, the crystal was assumed to be isotropic and the value of ψ was assumed not to depend on the choice of the direction of x in Fig. 5.18. In reality, the value of ψ, which determines which way the energy flows, critically depends on the orientation of the crystal axis.

The direction of the energy flow in the case of barium titanate (BaTiO$_3$) is shown in Fig. 5.22. When the incident light R_0 and S_0 of equal amplitude pass through the crystal, the angle ψ is established in such a way that the energy of R_0 flows into S_0

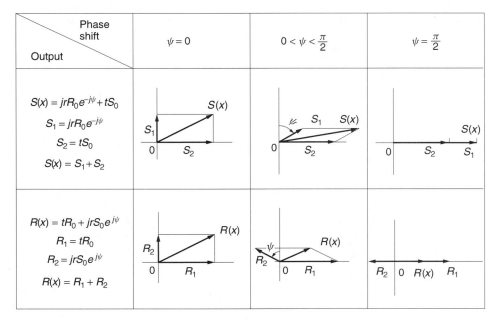

Figure 5.21 Phasor diagram of the energy transfer for various values of ψ.

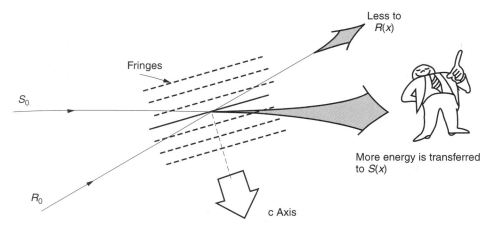

Figure 5.22 Direction of the energy flow. In a BaTiO$_3$ crystal, the energy flows as if the light is bent toward the c axis.

and $S(x) > R(x)$. It behaves as if the light energy is bent toward the direction of the crystal axis.

5.7.3 Optical Amplifier Structure

Now the results of the previous two sections are combined to explain the principle of operation of an optical amplifier. Most of the discussion up until now dealt with the case where $R_0 = S_0$. However, the more common situation in the use of this optical amplifier is to start with one of the beams being much smaller in intensity than the other, namely, $S_0 \ll R_0$. In this case, as the intensity S_0 increases, the intensity fringe contrast increases, which increases the magnitude Δn of the phase modulation, which causes the reflectivity of the phase grating fringe to increase, which causes more energy transfer from the R beam into the $S(x)$ beam. Thus, the output $S(x)$ increases with an increase in input S_0, and the device functions as an optical amplifier. The phase grating plays the role of a gate to control the energy flow from R_0 into $S(x)$.

Figure 5.23 shows the structure of an optical amplifier using a BaTiO$_3$ optical guide whose crystal axis is periodically reversed [17]. The periodic reversal of the crystal

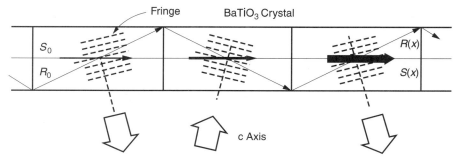

Figure 5.23 Periodically arranged BaTiO$_3$ photorefractive amplifier. (After F. Ito, K. Kitayama, and O. Nakao [17].)

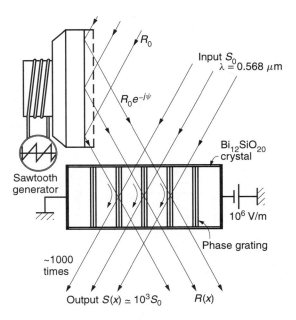

Figure 5.24 Optical amplifier based on the photorefractive effect.

axis makes it possible to fabricate the optical amplifier in a thin layer. The R wave (pump wave) changes its direction of propagation after each reflection from the wall, but the direction of the crystal axis is also reversed so that the light energy always flows from the pump wave into the signal wave following the rule explained in Fig. 5.22.

Figure 5.24 shows a model that uses an external electric field. An external electric field is applied because, as discussed in Section 5.5.1, it significantly increases the magnitude Δn of the phase grating modulation. Unfortunately, however, the phase grating formed under an external electric field loses the $\pi/2$-radian shift that is essential for the transfer of the light energy from R_0 into $S(x)$. A moving mirror mounted on a piezoelectric crystal changes the phase of the R beam and artificially provides the necessary $\pi/2$-radian phase shift. The speed of the movement of the mirror critically depends on the speed of formation of the phase grating by $R_0 e^{-j\psi}$ and $S_0 e^{j\psi}$ [14].

5.8 PHOTOREFRACTIVE BEAM COMBINER FOR COHERENT HOMODYNE DETECTION

A homodyne detection scheme has significantly higher sensitivity than that of a direct detection scheme (see Chapter 12). With the homodyne detection scheme, a local oscillator light is added collinearly with the signal light by means of a beam coupler and fed into the mixer diode as shown in Fig. 12.9. The phase as well as the frequency of the two lightwaves have to be kept identical at all times. If there exists an α-radian phase difference between these two lightwaves, the output signal current from the mixer reduces by $\cos\alpha$. One way of keeping α zero is the elaborate Costas loop mentioned in Chapter 12. Another way of alleviating the stringent requirement on α is by using a photorefractive crystal as a combiner for these two lightwaves.

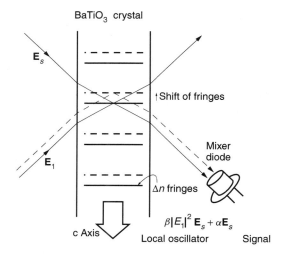

Figure 5.25 Beam combiner made out of BaTiO$_3$ for homodyne detection.

Figure 5.25 explains the operating principle of the photorefractive crystal combiner [18–20]. As soon as $\mathbf{E}_1$, which eventually becomes the local oscillator light, and the signal light $\mathbf{E}_s$ enter into the photorefractive crystal, a refractive index fringe pattern starts to establish itself in the crystal as shown by the solid lines.

For the sake of simplicity, the phase shift ψ is assumed zero and the photorefractive crystal will be treated as an ordinary holographic plate. The refractive index fringe is then expressed as

$$\Delta n = \beta[(E_s + E_1)(E_s^* + E_1^*)$$
$$= \beta[|E_s|^2 + |E_1|^2 + E_sE_1^* + E_s^*E_1]$$
(5.61)

The third term in Eq. (5.61) is responsible for reconstructing $\mathbf{E}_s$ when the reconstruction beam $\mathbf{E}_1$ illuminates the fringe.

$$\mathbf{E}_r = \beta\mathbf{E}_s|E_1|^2$$
(5.62)

Another field reaching the detector is due to the direct transmission with some attenuation:

$$\mathbf{E}_d = \alpha\mathbf{E}_s$$
(5.63)

The output $\mathbf{E}_t$ from the lower right of the crystal is the sum of Eqs. (5.62) and (5.63)

$$\mathbf{E}_t = \beta|E_1|^2\mathbf{E}_s + \alpha\mathbf{E}_s$$
(5.64)

$\mathbf{E}_t$ is detected by the mixer photodiode. The frequency and phase of the first term are identical with those of $\mathbf{E}_s$, constituting an ideal local oscillator light for homodyne detection.

If the crystal is a photorefractive material with $\psi \neq 0$, and if the direction of the crystal axis c is downward, $\mathbf{E}_1$ is even more efficiently refracted toward the signal $\mathbf{E}_s$ because of the transfer of light energy in the direction of the crystal axis.

Let us consider the stability of the system in the event of a phase fluctuation. Let's say the phase fluctuation of $\mathbf{E}_s$ is a delay. The delay in the phase of the signal $\mathbf{E}_s$ causes a shift of fringe pattern. The newly established fringe pattern moves upward, as drawn by the dashed lines in Fig. 5.25. Now, the light $\mathbf{E}_1$ has to travel a longer distance to reach the mixer diode because of the upward shift of the fringe pattern and $\mathbf{E}_1$ is also delayed. Thus, the fringe pattern reestablishes itself such that the fluctuation of the signal light is compensated.

In the field of satellite-to-satellite free-space optical communication, such a combiner automatically compensates for phase fluctuations. A common occurrence in these optical communication systems is that the direction of the signal light fluctuates. If a beamsplitter such as the one shown in Fig. 12.13 is used as the combiner, then the physical orientation of the beamsplitter has to follow the direction of fluctuation so as to align the signal and local oscillator light to be collinear at all times. The fringe pattern in the photorefractive crystal, however, can reestablish itself to compensate for the phase fluctuation originating from the directional fluctuation.

5.9 OPTICALLY TUNABLE OPTICAL FILTER

Figure 5.26a shows an example where the refractive index pattern in a photorefractive crystal itself is used as a tunable filter. The index of refraction grating is generated by illuminating a $BaTiO_3$ crystal with the optical fringe pattern generated by the interference of two coherent writing beams. The period of the refractive index grating can be changed by changing the angle 2θ between the two writing beams. The results in Fig. 5.26 [21] was obtained by scanning θ and thus demonstrate the capability of resolving the combined outputs from two laser diodes whose wavelengths are closely spaced.

5.10 LIQUID CRYSTALS

A striking difference between a liquid crystal and a solid crystal is that the molecular orientation of a liquid crystal [22–25] is easily altered by low-amplitude electric or magnetic fields and by small changes in temperature or mechanical pressures. This high degree of susceptibility to external influences comes from the fact that the molecular orientation of the crystal is governed by van der Waals forces. Van der Waals forces arise from dipole–dipole interactions and are proportional to $1/r^6$, where r is the distance between the molecules. The interatomic force of a simple solid ionic crystal like NaCl is a coulombic force and is proportional to $1/r^2$.

The necessary control voltages of liquid crystal devices are significantly lower than similar devices based on the Pockels electrooptic effect. An electrically controlled birefringence cell, such as shown in Fig. 5.3, normally requires a voltage close to an arcing potential, while liquid crystal devices of the same function typically need just a few volts.

5.10.1 Types of Liquid Crystals

Liquid crystals are classified into the following four types: (1) cholesteric, (2) smectic, (3) nematic, and (4) discotic. Each of these types is discussed individually in the sections to follow.

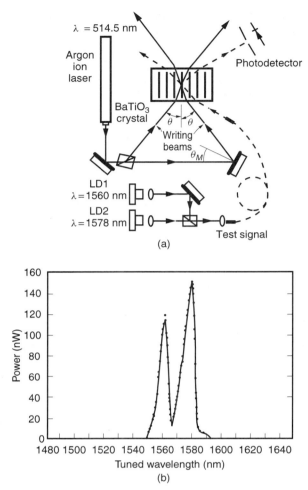

Figure 5.26 Optically tunable optical filter. (a) Experimental setup for wavelength division multiplexing using an optically tunable filter. (b) Tunable photorefractive filter output: resolving simultaneous signals from laser diodes LD 1 and LD 2. (After R. James et al. [21].)

5.10.1.1 Cholesteric

The molecular orientation of the cholesteric-type liquid crystal is shown in Fig. 5.27a. The molecules are arranged in parallel planes that are perpendicular to the optic axis. The planes rotate helically along the optical axis. To light whose wavelength is much shorter than the pitch of the helix, the cholesteric liquid crystal film behaves like a rotator described in Section 6.7. Linearly polarized incident light rotates its direction of polarization as it propagates through the film. The light that is transmitted through the film remains linearly polarized, but the direction of linear polarization is rotated.

When the light wavelength is comparable in magnitude to the pitch of the helix, there will be a reflected wave as well as a transmitted wave under the right conditions. Light whose wavelength matches with the pitch of the helix multiplied by the index of refraction is reflected from the internal layers and comes back to the entry surface as

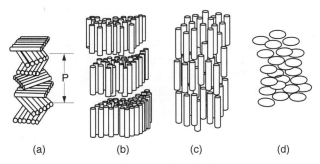

Figure 5.27   Types of liquid crystals. (a) Cholesteric crystal. (b) Smectic liquid crystal (*A* type). (c) Nematic liquid crystal. (d) Discotic liquid crystal in nematic-like orientation.

the reflected wave. As a result, the cholesteric liquid displays a vivid color associated with the pitch of the helix.

It should be added that a linearly polarized wave can be decomposed into two oppositely rotating circularly polarized waves. Only the circularly polarized wave whose sense of rotation matches that of the helix participates in creating the color. The pitch of the cholesteric crystal changes with temperature, and hence the color of the reflected light also changes with temperature. This fact makes the crystal film very useful for mapping the spatial distribution of temperature. Combined with a resistive film, the cholesteric liquid crystal film is also used for spatial mapping of microwave radiation [26].

Since the pitch of the cholesteric crystal is pressure sensitive, the change in color of the cholesteric liquid crystal can also be used for spatial pressure mapping.

5.10.1.2 Smectic

As shown in Fig. 5.27b, molecules of the smectic liquid crystal are structured in layers. In each layer, the molecules are stacked with their axes parallel to each other. The stacks are aligned, nearly perpendicular (smectic *A* type) to the layers or slightly tilted (smectic *C* type) to the layers. The spacing between the layers is more or less one molecular length long.

This type of liquid crystal is turbid white in color and slimy in texture. The name smectic is derived from the Greek *smektis*, meaning soap.

Some smectic liquid crystals are ferroelectric and are used as fast response time display devices.

5.10.1.3 Nematic

Figure 5.27c shows the orientation of the molecules of the nematic liquid crystal. The molecules are aligned in parallel lines, but not in layers. Vector **n**, which represents the statistically preferred direction, is called the director. The orientation of the director in Fig. 5.27c is vertical. Under the microscope, the nematic liquid crystal looks like combed threads. The name nematic is derived from the Greek *nemat*, meaning thread.

The optical properties of well-aligned nematic liquid crystals are similar to those of uniaxial crystals.

5.10.1.4 Discotic

The discotic liquid crystal is the latest addition to the family of liquid crystals. For the liquid crystals shown in Figs. 5.27a–5.27c, the molecules are all cigar–shaped. In the discotic liquid crystal, the molecules are shaped like disks. Figure 5.27d shows a discotic liquid crystal in a nematic-like orientation. Besides this orientation, there are discotic liquid crystals in a smectic-like orientation and in a cholesteric-like orientation.

5.10.2   Molecular Orientations of the Nematic Liquid Crystal Without an External Field

The molecular orientation of the nematic liquid crystal is influenced by its container or, more specifically, by the surface of the inner walls of the cell used to contain the liquid crystal. When the molecules are oriented parallel to the liquid crystal cell as shown in Fig. 5.28a, this orientation is called the homogeneous orientation. Such cell walls are coated with a thin film of polymers such as polyvinyl alcohol (PVA) or polyimide (PI). On the polymer surface, fine streak marks are made by rubbing the surface with a nylon cloth. The fine streaks act as anchors to the molecules in the direction of the rubbing.

When the molecules are oriented perpendicular to the cell surface as shown in Fig. 5.28b, this orientation is called the homeotropic orientation. Such an orientation is possible by chemically treating the surface with a coupling agent [24,27,28]. The treated surface anchors one end of the molecules, and a perpendicular orientation of the molecules results.

The third orientation is the twisted nematic orientation. With the homogeneous orientation, when the rubbing directions of the facing cell walls are not parallel, the direction of the molecules is twisted between the two walls. If the twist is gradual and the thickness d between the walls is thick enough to satisfy

$$4d > \lambda(n_e - n_0)$$

then the direction of polarization of the light follows the twisted molecular axis. The cell displays the property of a rotator. The direction of polarization is rotated by the same angle as the twist of the rubbing directions.

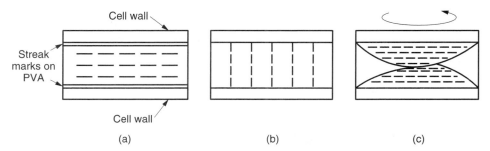

Cell wall

Streak marks on PVA

Cell wall

(a)　　　　　(b)　　　　　(c)

Figure 5.28   Molecular orientations of nematic liquid crystal. (a) Homogeneous. (b) Homeotropic. (c) Twisted (180°).

5.10.3 Molecular Reorientation of the Nematic Liquid Crystal with an External Electric Field

The usefulness of the nematic liquid crystal resides in the controllability of the molecular orientation by an external electric field. The change in molecular orientation produced by an external electric field depends on the polarity of the birefringence of the particular nematic crystal used.

The molecular energy of the nematic liquid crystal is made up of polarization and elastic energies. The static energy of the induced polarization is given by [29]

$$U = -\int \mathbf{P} \cdot \mathbf{E}/dv \qquad (5.65)$$

and the larger the induced polarization $\mathbf{P}$ is, the lower the polarization energy U is. Let us consider a nematic liquid crystal with a positive birefringence (see Section 4.2)

$$n_e > n_0 \qquad (5.66)$$

where n_e is the refractive index in the direction of the director (in the molecular axis) and n_0 is that in the direction perpendicular to the director. From Eqs. (4.3) and (4.5), with a positive birefringence, the induced polarization $\mathbf{P}$ is larger; hence, the polarization energy U is lower when the director is parallel to the external field. Thus, the director of a positive birefringent nematic crystal is reoriented to the same direction as the external electric field. On the other hand, the molecular axis of a negative birefringent liquid crystal is reoriented to the direction perpendicular to the external electric field.

The transition between homogeneous and homeotropic orientation takes place only when the external electric field exceeds a critical value. In other words, there is a threshold value for the external field. The threshold value is reached when the energy due to the induced polarization exceeds the elastic energy of the molecules. This is called the Freédericksz effect.

Example 5.7 An electrically controlled birefringence (ECB) cell is to be constructed using a nematic liquid crystal. Proper combinations of the liquid crystal type and the zero-field molecular orientation are important. Figure 5.29 summarizes possible combinations of the polarity of the birefringence and zero-field molecular orientation. Select the workable combinations.

Solution The molecules have to be reoriented each time the on and off states are switched.

With the homogeneous zero-field orientation, a positive birefringent nematic liquid crystal has to be used because the positive birefringent liquid crystal reorients to the homeotropic configuration when the external field is applied. The negative birefringent liquid crystal tends even more to the homogeneous orientation at the on state.

For the same reason, with the homeotropic zero-field orientation, a negative birefringent nematic liquid crystal has to be used so that the molecules are reoriented to the homogeneous configuration when the external field is applied. The proper selections are Cases 1 and 4 in Fig. 5.29. □

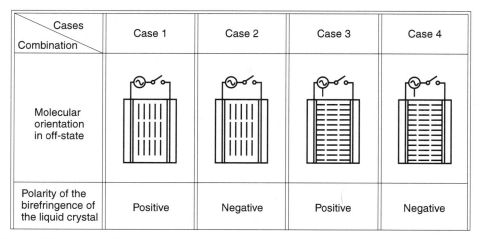

Cases Combination	Case 1	Case 2	Case 3	Case 4
Molecular orientation in off-state				
Polarity of the birefringence of the liquid crystal	Positive	Negative	Positive	Negative

Figure 5.29 Combinations of the molecular orientation and polarity of the birefringence of a liquid crystal for constructing an ECB.

5.10.4 Liquid Crystal Devices

Devices utilizing liquid crystals will be described in this section. These devices include phase shifters, variable focus lenses, rotators, waveplates, spatial light modulators, general displays, and liquid crystal television. The attractive features of liquid crystal devices are their low control voltage, low power consumption, light weight, small size, and long life.

5.10.4.1 Liquid Crystal Fabry–Pérot Resonator
The structure of the liquid crystal Fabry–Pérot resonator has already been shown in Fig. 3.12.

The maximum change in the refractive index is equal to the birefringence of the nematic liquid crystal. Typical values are $n_0 = 1.5$ and $n_e = 1.7$, which correspond to a change of about 0.2.

With a 10-μm long cavity, the obtainable tuning range is approximately 200 nm. The switching speed from one wavelength to another is of the order of tens of milliseconds.

5.10.4.2 Liquid Crystal Rotatable Waveplate
A nematic liquid crystal displays a birefringence like a uniaxial solid crystal and can be used as a waveplate [30]. By applying a rotating external electric field, the director (molecular axis) of the nematic liquid crystal can be rotated and an electrically rotatable waveplate can be realized.

Figure 5.30 shows the electrode geometry for a rotatable waveplate. By applying a sinusoidal electric field with consecutive phase delays of 45°, a rotating electric field is established in the center.

Do not confuse this device with a rotator. A rotator simply rotates the direction of polarization of the passing light. On the other hand, in a waveplate, also called a retarder or retardation plate, the incident polarized light is resolved into the two allowed directions of polarization inside the crystal. Because of the different refractive indices, one component lags the other. The phase difference between the two components

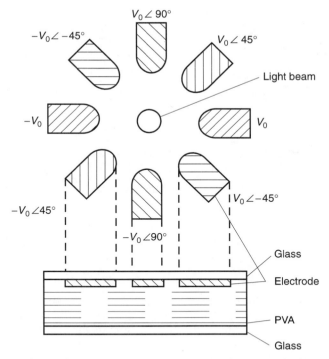

Figure 5.30 Top and side views of a rotatable waveplate. (After T. Chiba, Y. Ohtera, and S. Kawakami [30].)

is the retardation. When the two components recombine in the emergent light, the polarization of the emergent light is changed depending on the amount of retardation. More quantitative discussions on rotators and waveplates can be found in Chapter 6. Figure 6.19 shows conventional waveplates being used to convert the polarization of the incident light. In this figure, human fingers rotate the direction of the fast axis of the waveplates. In the liquid crystal rotatable waveplate, the human fingers are replaced by electronic control.

The speed of rotation is limited by the response of the liquid crystal and is slower than about 100 revolutions per second. The values of the retardation are determined by the thickness of the liquid crystal cell, and any desired value can be designed.

Such a device is useful for the automatic compensation for the fluctuations in the state of polarization of the received signal for coherent detection in fiber-optic communication systems (see Chapter 12).

5.10.4.3 Liquid Crystal Microlens

A nonuniform electric field is established in a nematic liquid crystal (NLC) cell when one of the cell's electrodes has a circular hole in its center. This nonuniform electric field produces a semiquadratic spatial distribution of the refractive index in the nematic liquid crystal. This refractive index distribution acts as a microlens [31].

Figure 5.31a shows the geometry of the liquid crystal lens using a positive birefringent nematic crystal. Without the external electric field, the molecular axes are in the homogeneous orientation. With the electric field on, the strength of the electric

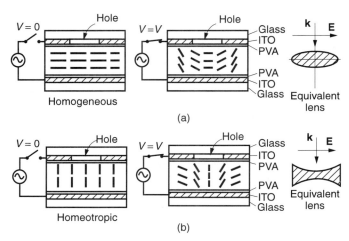

Figure 5.31 Liquid crystal microlenses. (a) With a positive birefringent nematic liquid crystal. (b) With a negative birefringent nematic liquid crystal.

field is increased as the rim of the circular hole is approached. Because the nematic liquid crystal is the positive birefringent type, the orientation of the molecular axes becomes homeotropic as the rim of the circular hole of the electrode is approached.

Figure 5.32b shows the equiphase lines of light emergent from the microlens whose geometry is shown in Fig. 5.32a. Typical parameters for a liquid crystal microlens are given in Table 5.4.

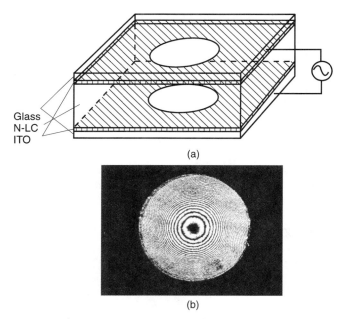

Figure 5.32 Geometry and interference fringes of a liquid crystal microlens. (a) Geometry of two-hole type. (b) Interference fringe. (After M. Homma, T. Nose, and S. Sato [31].)

Table 5.4 Typical parameters for a liquid crystal microlens

Parameter	Range
Cell thickness	50–200 μm
Hole diameter	20–80 μm
Control voltage	1–3 V
Focal length	0.2–0.5 mm

Arrays of liquid crystal microlenses are used in optical displays, optical interconnects, and optical signal processing.

5.10.4.4 Twisted Nematic (TN) Liquid Crystal Spatial Light Modulator (TNSLM)

The twisted nematic (TN) liquid crystal cell is constructed by inserting a positive birefringent liquid crystal between polyimide film sheets rubbed in two different directions, which determine the amount of the desired twist.

Figure 5.33 illustrates the principle of a TN liquid crystal rotator functioning as a light on–off panel. In Fig. 5.33a, the TN liquid crystal cell is designed to give a 90° twist. The incident light from the left is first filtered to be horizontally polarized light

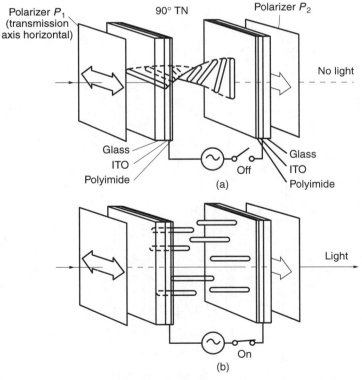

Figure 5.33 Principle of filtering light by twisted nematic liquid crystal. (a) No voltage applied. (b) Voltage applied.

by polarizer P_1. The long liquid crystal molecules are twisted and the direction of polarization of the incident light is twisted as it propagates, as if trying to follow the direction of the higher refractive index, which is along the axis of the molecule. The direction of polarization is rotated from horizontal to vertical. The light is blocked by the output polarizer P_2 and there is no emergent light. This corresponds to the "off" state of the pannel.

When an external electric field (either ac or dc) is applied to the electrodes, the twisted positive birefringent liquid crystal molecules are straightened in the direction of the applied electric field and the rotation of the polarization ceases, as shown in Fig. 5.33b. The transmitted light is thus horizontally polarized and emerges from polarizer P_2, thereby creating the "on"state of the pannel. Such a panel is widely used in display devices.

5.10.4.5 *Electrically Addressed Spatial Light Modulator (EASLM)*

A typical example of the electrically addressed spatial light modulator (EASLM) is the liquid crystal television (TV). The liquid crystal TV is made by arranging microsize ($\sim 50 \times 50 \ \mu m^2$) TN liquid crystal cells in a matrix form, such as shown in Fig. 5.34.

With color TV, microcolor filters of red, green, and blue (R, G, B) are placed so that the white light from the back panel light source is converted into color.

Each pixel is equipped with a thin film transistor (TFT), which can be addressed by way of the data and scanning electrode arrays. The light transmission is varied by varying the control voltage to the TFT. The capacitance C_p of each pixel holds

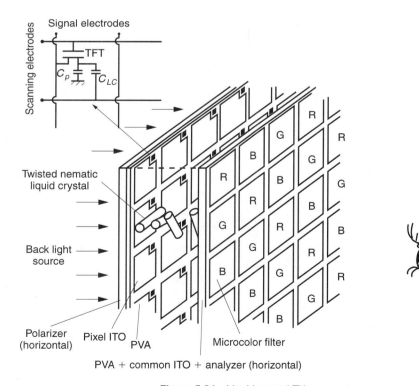

Figure 5.34 Liquid crystal TV.

the charge and maintains the voltage until the subsequent signal is addressed. This holding mechanism significantly enhances the brightness of the pixel. The additional capacitance C_{LC} is for further enhancement in the brightness.

5.10.4.6 *Optically Addressed Spatial Light Modulator (OASLM)*

The construction of the optically addressed spatial light modulator (OASLM) [32] is shown in Fig. 5.35. Unlike the EASLM, the OASLM directly uses the light image as an input signal. It is not divided into pixels.

A pair of ITO electrodes provide an external electric field to the two major components: the photoconductor layer and the TN liquid crystal cell. These two major components are optically isolated by a mirror layer in the midle.

At the locations where the light intensity from the cathode ray tube is low, the impedance of the photoconductor is high, and no external electric field is applied to the TN liquid crystal. However, at the locations of high-intensity light, the impedance of the photoconductor is low and an external electric field that exceeds the Freédericksz threshold is applied to the liquid crystal layer. The twisted orientation of the nematic liquid crystal converts into the homeotropic orientation.

Next, referring to Fig. 5.35, we will explain how the external image is extracted. The extraction of the external image makes use of a polarization beamsplitter (PBS), such as described in Section 6.5.2. The PBS reflects the *s* wave (vertically polarized wave) and transmits the *p* wave (horizontally polarized wave).

Light *ab* with arbitrary polarization is used as the readout light beam of the system. Light *bd* emergent from the PBS toward the TN liquid crystal cell is vertically polarized and is used as the input light to the OASLM. At location *e* where the light intensity from the cathode tube is low, the molecular orientation is twisted. As the incident light *bd* penetrates into the TN liquid crystal cell, the direction of polarization rotates with the twist of the liquid crystal molecule and reaches the mirror. Then, the direction of polarization of the light reflected from the mirror rerotates back the same amount. The emergent light *ef* from the TN liquid crystal cell recovers the original vertical

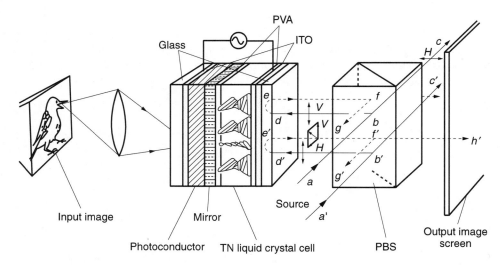

Figure 5.35 Optically addressed spatial light modulator.

polarization and is reflected by the PBS toward the original source as fg and does not reach the output image plane.

On the other hand, for light $a'b'$, which is incident at location e' where the intensity of the image light from the cathode ray tube is high, the molecular orientation tends to be homeotropic and the molecular axis is no longer parallel to the direction of polarization of the incident light. The incident light starts to have components both parallel and perpendicular to the molecular axis. Light $e'f'$ emergent from the SLM is no longer linearly polarized. Even though the vertically polarized component is reflected to the source as $f'g'$ by the PBS, the horizontally polarized component passes through the PBS and is projected onto the output screen as $f'h'$ to form the output image.

The applications of the OASLM are multifold:

1. *Image Intensifer.* The intensity of the output image can be increased to any desired level by increasing the power of the read light source. Large-scale image projectors are made.

2. *Incoherent–Coherent Converter.* The input image can be either coherent or incoherent light. Many types of optical signal processors need coherent light images. The output from the OASLM is a coherent optical image if coherent light is used for readout.

3. *Wavelength Converter.* A far-infrared image can be converted into a visible image because the nematic liquid crystal works beyond the infrared light region.

5.10.4.7 Polymer-Dispersed Liquid Crystal (PDLC)-Type Spatial Light Modulator (SLM)

A polymer-dispersed liquid crystal (PDLC) is a homogeneous mixture of polymer and liquid crystal microdroplets [33,34]. Figure 5.36 shows examples of spatial light modulators using the PDLC. The microdroplets are positive birefringent nematic liquid crystals with diameters less than a micron. The refractive index n_p of the polymer is set close to that of n_0 of the liquid crystal and

$$n_0 = n_p$$

$$n_e > n_p$$

When the external electric field is absent, the orientation of the molecular axes of the nematic liquid crystal in the microdroplet is arbitrary and the microdroplets become scattering centers, as shown in Fig. 5.36a. The transmission of light is low.

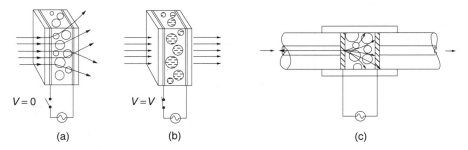

Figure 5.36 Polymer-dispersed liquid crystal (PDLC) light modulator. (a) Off state. (b) On state. (c) Light modulator of the fiber optic type. (After K. Takizawa et al. [33].)

On the other hand, when the external field is applied, the crystal axis is aligned because of the positive birefringence. Incident light with a normal incident angle but with an arbitrary state of polarization sees the refractive index n_0. The light transmits through the modulator with minimum attenuation, as shown in Fig. 5.36b.

The advantages of the PDLC-type SLM compared to the TN-type SLM is that neither polarizer nor analyzer is needed, the efficiency is higher, and the operation is polarization independent.

In Fig. 5.36c, the PDLC is used as a fiber-optic light modulator. PDLC film can be also used as an erasable volume hologram [35]. The reconstructed image from such a hologram is shown in Fig. 5.37.

5.10.5 Guest–Host Liquid Crystal Cell

A dichroic dye is a dye whose wavelength absorption characteristics depend on the direction of the molecular axis with respect to the incident light polarization. When a small amount of elongated dichroic dye is mixed as a guest molecule into a host of nematic or cholesteric liquid crystal, the dye molecules are aligned in the liquid crystal matrix. The dye molecules follow the reorientation of the liquid crystal when an external field is applied. Such an effect is called the *guest–host effect* and can be used to make a color display cell. Figure 5.38 shows a double guest–host (DGH) cell, which concatenates two GH cells. The DGH cell does away with the polarizer sheet, resulting in a higher brightness. The DGH cell uses a doped negative birefringent nematic liquid crystal oriented homeotropically in the off state, but with a slight tilt angle from the normal, as shown in Fig. 5.38a. The tilted homeotropic orientation is achieved by an oblique angle evaporation of SiO followed by treatment with a homeotropic surface coupling agent, N, N-dimethyl-N-octadecyl-3-aminopropyltrimethoxysilyl chloride (DMOAP) [27].

The guest dye is a positive-type dye, which means the dye absorbs light when the direction of polarization of the incident light is parallel to the axis of the dye molecule.

In the off state, as shown in Fig. 5.38a, both horizontal and vertical components of the incident light are polarized perpendicular to the dye molecule, and only minimum interaction takes place. The emergent light is either untinted or slightly tinted in color.

Figure 5.37 Reconstructed image of a hologram recorded in a PDLC film. (Courtesy of V. P. Tondiglia et al. [35].)

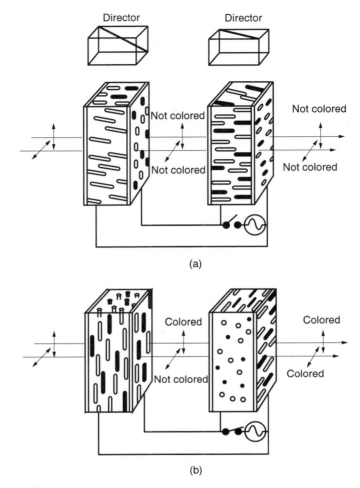

Figure 5.38 Double guest host cell. (a) Off state. (b) On state. (After T. Uchida, and M. Wada [27].)

In the on state, as shown in Fig. 5.38b, the dye molecules are reoriented to be homogeneous together with the liquid crystal molecules. The direction of the dye molecules is parallel to the vertical polarization in the first cell, and parallel to the horizontal polarization in the second cell. Significant interaction takes place for both polarizations and the emergent light is colored.

The GH cell is superior to cells with microcolor filters in the vividness and uniformity of the color as well as a wider viewing angle.

5.10.6 Ferroelectric Liquid Crystal

While the switching speed of a nematic crystal display is of the order of $1-10$ milliseconds, that of a ferroelectric crystal is much faster and is of the order of $1-10$ microseconds. A ferroelectric liquid crystal [23,37,38] display uses a chiral smectic C phase liquid crystal. The term "smectic C" means a smectic crystal with the molecular axis tilted from the molecular layers. *Chiral* means handed in Greek. A chiral molecule is a molecule that is not superimposable on its mirror image [22], for example, a helix whose mirror image can never be superimposed on its original. Here the molecular structure is spiral.

Figure 5.39a shows the off-state structure of a chiral smectic C liquid crystal in a display cell. While the molecules in the same layer are pointing in the same direction,

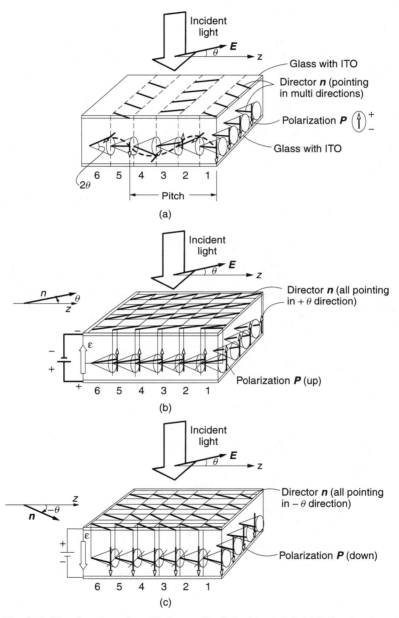

Figure 5.39 Switching function of a chiral smectic C liquid crystal. (a) Without external field **E**. (b) With upward external field **E**. (c) With downward external field **E**.

molecules between layers 1,2,3,..., are consecutively rotating like a cholesteric crystal with the helical axis z perpendicular to the boundary planes of the layers (shown in fine vertical dotted lines) and with the apex angle θ. The refractive index of the crystal in the θ direction is the same as that in the $-\theta$ direction.

Each molecule has a spontaneous dipole moment perpendicular to the molecular axis, as indicated by the white head arrow. Do not confuse the *spontaneous* dipole moment with the dipole moment induced by an external field. The spontaneous polarization is present regardless of the external field ε.

This is the very source of the difference between the properties of the ferroelectric liquid crystal and the properties of the nonferroelectric liquid crystal. Unlike the induced polarization, the direction of the spontaneous polarization is unchanged by the external field and it is fixed perpendicular to the molecular axis or the director. (Visualize a person with open arms. The body represents the director, the arms represent the dipole moment.)

The effect of the torque of the dipole moment on the molecular orientation is illustrated in Fig. 5.40 and explained in the boxed note.

As soon as the external electric field is applied, the torque generated by the spontaneous dipole moment flips both the director and the dipole moment so that the dipole moment lines up with the external electric lines of force, as illustrated in the boxed note. Figure 5.39b shows the case when the direction of the external electric lines of force is upward and all molecules (directors) are lined up in the direction tilted θ degrees from the helix axis z.

The switching mechanism of the chiral smectic C liquid crystal will be illustrated using an analogy. One of the ends of a banana is pierced by a metal fork (Fig. 5.40). The fork is assumed magnetized. The banana represents the elongated molecule or the director, and the fork, the spontaneous polarization. By loosely holding the other end of the banana, if the banana is placed under a strong external magnetic field with its N pole at the bottom, the banana will be flipped such that the magnetic moment of the fork lines up with the external magnetic field. As a consequence, the banana will be oriented in a plane perpendicular to the external magnetic field and to the left-hand side of the figure.

When, however, the polarity of the external magnetic field is reversed, the banana will flip to the right-hand side of the figure.

Thus, by switching the polarity of the external magnetic field, the orientation of the banana (director) can be flipped between the two binary horizontal orientations.

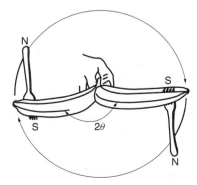

Figure 5.40 Explanation of the switching action of a ferroelectric liquid crystal.

On the other hand, when the polarity of the external electric field is switched to downward, all molecules (directors) are switched to the direction tilted $-\theta$ degrees from the helix axis z as shown in Fig. 5.39c.

Thus, an electrically controlled birefringent (ECB) cell can be made out of the chiral smectic C cell. The change in the direction of the director is $2\theta = 20°-30°$. This is large enough to fabricate a light modulator in collaboration with a polarizer and an analyzer.

In order to enhance the performance of the cell, it is important to keep the cell thickness shorter than the pitch of the helix, which is $2-3$ μm, so that the surface force may assist in aligning the molecules parallel to the surface. This cell has the merits of memory and a fast switching speed. The change between two discrete states is binary.

5.11   DYE-DOPED LIQUID CRYSTAL

The dye-doped nematic liquid crystal [39] displays a "photorefractive-like" nonlinearity in the sense that its refractive index changes when exposed to light.

When a nematic liquid crystal is exposed to the fringe pattern created by laser light, a spatial variation of the charges is established in small amounts just like the case of the photorefractive crystal shown in Fig. 5.18.

The amount of charge is significantly increased when the nematic liquid crystal is doped with dyes like rhodamine-6G, fullerene C_{60}, and especially methyl-red. The electric field established by this spatial charge distribution contributes to the reorientation of the directors of the nematic liquid crystal. The reorientation of the directors results in a change in the index of refraction. With regard to the mechanisms responsible for changing the refractive index, the difference between the photorefractive crystal and the dye-doped liquid crystal is that, with the former, the Pockels effect

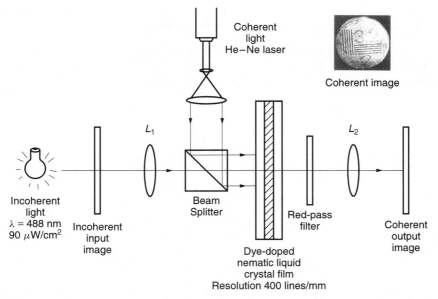

Figure 5.41   An incoherent to coherent image converter using a dye-doped nematic liquid crystal. (After I. C. Khoo et al. [39].)

changes the refractive index, but with the latter, the reorientation of the directors changes the refractive index.

Another difference to be noted is that the sensitivity of the photorefractive crystal improves when a dc E field is applied because of the shift in the charge pattern, as mentioned Section 5.7.1. The sensitivity of the doped nematic liquid crystal is also increased when a dc E field is applied, but the mechanism of the enhancement of the sensitivity is slightly different. The spatial variation of the orientation of the directors that has already been established creates a spatial variation of the conductivity. When a dc E field is applied, a spatially varying electric field is established. This varying field further assists the spatial orientation of directors of the nematic liquid crystal.

Figure 5.41 shows an incoherent to coherent image converter using the doped nematic liquid crystal. Compared to the optically addressed spatial light modulator mentioned in Section 5.10.4.6, the fabrication is simpler because the deposition of the photoconductive film has become unnecessary.

PROBLEMS

5.1 In Example 5.2, the direction of propagation was taken as the z direction. For this direction of propagation, the retardation was calculated for the three cases $\varepsilon = \varepsilon_x$, $\varepsilon = \varepsilon_y$, and $\varepsilon = \varepsilon_z$. In this problem, we remove the restriction that light must propagate in the z direction. Of the following three possibilities for the applied field direction — $\varepsilon = \varepsilon_x$, $\varepsilon = \varepsilon_y$, or $\varepsilon = \varepsilon_z$ — determine which applied field direction best takes advantage of the largest Pockels coefficient r_{33} of lithium niobate. Find the expression for the indicatrix. Determine the propagation direction that maximizes the retardation Δ, and find an expression for Δ.

5.2 Referring to Fig. 5.6, find the bias Δ_b that best makes use of the linear portion of the I versus Δ curve. Find an expression for the modulated light intensity, assuming that the phase changes produced by the modulation are much less than Δ_b.

5.3 Sinusoidal external electric fields are applied to a lithium niobate crystal in both the x and y directions, and the direction of the incident wave is along the z axis as shown in Fig. P5.3. Prove that the cross-sectional ellipse is rotated if the external

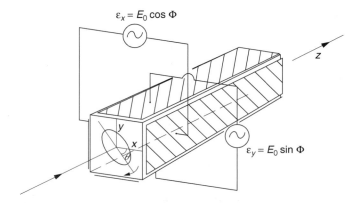

Figure P5.3 Application of a rotating external electric field to LiNbO$_3$.

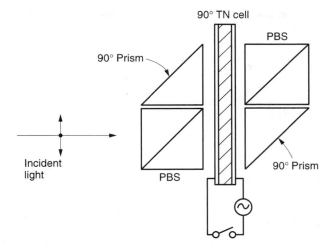

Figure P5.6 Ninety-degree TN cell combined with polarizing beamsplitters and 90° prisms.

fields are

$$\varepsilon_x = E_0 \cos \Phi$$

$$\varepsilon_y = E_0 \sin \Phi$$

5.4 In the discussion of the photorefractive effect, an intensity fringe pattern was used to create a phase grating. Draw the intensity fringe pattern generated by the two intersecting beams

$$R = R_0 e^{j(k_x x + k_z z + \psi)}$$

$$S = S_0 e^{j(-k_x x + k_z z)}$$

To simplify the problem, let $R_0 = S_0 = 1$, and consider only the $z = 0$ plane. Draw the intensity pattern as a function of x for the following three ψ values: $\psi = 0$, $\psi = \pi/2$, $\psi = -\pi/2$.

5.5 Referring to Fig. 5.21, find the value of ψ that makes the maximum transfer from $S(x)$ to $R(x)$.

5.6 Ninety-degree TN cells are combined with polarizing prisms and 90° prisms as shown in Fig. P5.6. Draw the path of light when the electric circuit switch is on and when it is off [40].

REFERENCES

1. J. F. Nye, *Physical Properties of Crystals*, Clarendon Press Oxford University Press, New York, 1985.
2. A. Yariv and P. Yeh, *Optical Waves in Crystals*, Wiley, New York, 1984.

3. M. Bass, E. W. Van Stryland, D. R. Williams, and W. L. Wolfe, *Handbook of Optics*, Vol. 2, McGraw-Hill, New York, 1995.

4. D. A. Pinnow, "Elastooptical materials," in *Handbook of Lasers with Selected Data on Optical Technology*, R. J. Pressley (Ed.), Chemical Rubber Co., Cleveland, OH, 1971, pp. 478–488.

5. I. P. Kaminow and E. H. Turner, "Linear electrooptical material" in *Handbook of Lasers with Selected Data on Optical Technology*, R. J. Pressley (Ed.), Chemical Rubber Co., Cleveland, OH, 1971, pp. 447–459.

6. M. J. Weber, *Handbook of Laser Science and Technology*, Chemical Rubber Co., Cleveland, Ohio, 1995.

7. W.-K. Kuo, W.-H. Chen, Y.-T. Huang, and S.-L. Huang, "Two-dimensional electric-field vector measurement by a $LiTaO_3$ electro-optic probe tip," *Appl. Opt.* **39**(27), 4985–4993 (2000).

8. S. Ramo, J. R. Whinnery, and T. Van Duzer, *Fields and Waves in Communication Electronics*, 3rd ed., Wiley, New York, 1994.

9. T. Sugie and M. Saruwatari, "An effective non-reciprocal circuit for semiconductor laser to optical fiber coupling using a YIG sphere," *J. Lighwave Technol.* **LT-1**(1), 121–130 (1983).

10. M. Shirasaki and K. Asama, "Compact optical isolator fibers using birefringent wedges," *Appl. Opt.* **21**(23), 4296–4299 (1982).

11. T. Tamaki, "Optical isolators," Essentials in the design and applications of optical integrated circuits," G. Nishikawa (Ed.), Nippon Kogyo Gizitsu Center, Tokyo, 1985, Chap. 8, pp. 203–234.

12. R. Yeh, "Photorefractive phase conjugators," *Proc. IEEE* **80**, 436–450 (1992).

13. J. Feinberg, "Photorefractive nonlinear optics," *Phys. Today*, 46–52 (Oct. 1988).

14. Ph. Refregier, L. Solymar, H. Rajbenbach, and J. P. Huignard, "Two-beam coupling in photorefractive $Bi_{12}SiO_{20}$ crystals with moving grating — theory and experiments," *J. Appl. Phys.* **58**(1), 45–57 (1985).

15. V. Kondilenko, V. Markov, S. Odulov, and M. Soskin, "Diffraction of coupled waves and determination of phase mismatch between holographic grating and fringe pattern," *Opt. Acta* **26**(2), 239–251 (1979).

16. H. Kogelnik, "Coupled wave theory for thick hologram gratings," *Bell Syst. Tech. J.* **48**(9), 2909–2947 (1969).

17. F. Ito, K. Kitayama, and O. Nakao, "Enhanced two-wave mixing in a photorefractive waveguide having a periodically reversed c-axis by electrical poling technique," *Appl. Phys. Lett.* **60**(7), 793–795 (Feb. 1992).

18. R. T. B. James, *Polarization Independent Coherent Optical Communication Systems*, Ph.D. thesis, University of Toronto, Toronto, 1993.

19. R. T. B. James and K. Iizuka, "Polarization insensitive homodyne detection with all optical processing based on the photorefractive effect," *J. Lightwave Technol.* **11**(4), 633–638 (1993).

20. F. M. Davidson and C. T. Field, "Coherent homodyne optical communication receivers with photorefractive optical beam combiners," *J. Lightwave Technol.* **12**(7), 1207–1223 (1994).

21. R. T. B. James, C. Wah, K. Iizuka, and H. Shimotahira, "Optically tunable optical filter," *Appl. Opt.* **34**(35), 8230–8235 (1995).

22. P. J. Collings and M. Hird, *Introduction to Liquid Crystals Chemistry and Physics*, Taylor and Francis, London, 1997.

23. S. Chandrasekhar, *Liquid Crystals*, 2nd ed., Cambridge University Press, Cambridge, 1992.

24. S. Sato, *A World of Liquid Crystals*, Sangyotosho Publishing Company, Tokyo, 1994.

25. A. Sasaki, *Liquid Crystal Electronics*, Fundamentals and Applications, Ohm-sha, Tokyo, 1979.

26. K. Iizuka, "Subtractive microwave holography and its application to plasma studies," *Appl. Phys. Lett.* **20**(1), 27–29 (1972).

27. T. Uchida and M. Wada, "Guest–host type liquid crystal displays," *Mol. Cryst. Liq. Cryst.* **63**(1), 19–44 (1981).

28. C. J. Newsome, M. O'Neill, R. J. Farley, and G. P. Bryan-Brown, "Azimuthal anchoring energy of liquid crystals aligned using laser ablated grating on polyimide," Conference on Liquid Crystal Materials, Devices and Applications VI, San Jose, California, January 1998, *Proc. SPIE* **3297**, 19–25 (1998).

29. J. A. Stratton, *Electromagnetic Theory*, McGraw-Hill, New York, 1941.

30. T. Chiba, Y. Ohtera, and S. Kawakami, "Polarization stabilizer using liquid crystal rotatable waveplates," *J. Lightwave Technol.* **17**(5), 885–890 (1999).

31. M. Honma, T. Nose, and S. Sato, "Optimization of device parameters for minimizing spherical aberration and astigmatism in liquid crystal microlenses," *Opt. Rev.* **6**(2), 139–143 (1999).

32. A. Tanone and S. Jutamulia, "Liquid-crystal spatial light modulators," *Optical Pattern Recognition*, F. T. S. Yu and S. Jutamulia (Eds.), Cambridge University Press, New York, 1998, Chap. 14.

33. K. Takizawa, H. Kikuchi, H. Fujikake, Y. Namikawa, and K. Tada, "Polymer-dispersed liquid crystal light valves for projection display application," *SPIE* **1815**, 223–232 (1992).

34. R. L. Sutherland, V. P. Tondiglia, L. V. Natarajan, T. J. Bunning, and W. W. Adams, "Electrically switchable volume grating in polymer-dispersed liquid crystals," *Appl. Phys. Lett.* **64**(9), 1074–1076 (1994).

35. V. P. Tondiglia, L. V. Natarajan, R. L. Sutherland, T. J. Bunning, and W. W. Adams, "Volume holographic image storage and electro-optical readout in a polymer-dispersed liquid-crystal film," *Opt. Lett.* **20**(11), 1325–1327 (1995).

36. B. Bahadur, "Current status of dichroic liquid crystal displays," *Mol. Cryst. Liq. Cryst.*, **209**, 39–61, 1991.

37. S. T. Lagerwall, M. Matuszczyk, and T. Matuszczyk, "Old and new ideas in ferroelectric liquid crystal technology. Liquid crystals: physics, technology and applications," Zakopane, Poland, 3–8 March, 1997, *Proc. SPIE*, **3318**, 2–38 (1997).

38. N. Collings, W. A. Crossland, P. J. Ayliffe, D. G. Vass, and I. Underwood, "Evolutionary development of advanced liquid crystal spatial light modulators," *Appl. Opt.* **28**(22), 4740–4747 (1989).

39. I. C. Khoo, M. -Y. Shih, M. V. Wood, B. D. Guenther, P. H. Chen, F. Simoni, S. S. Slussarenko, O. Francescangeli, and L. Lucchetti, "Dye-doped photorefractive liquid crystals for dynamic and storage holographic grating formation and spatial light modulation," *Proc. IEEE* **87**(11), 1897–1911 (1999).

40. R. E. Wagner and J. Cheng, "Electrically controlled optical switch for multimode fiber applications," *Appl. Opt.* **19**(17), 2921–2925 (1980).

6

POLARIZATION OF LIGHT

Clever uses for polarized light are not restricted to just the field of photonics. Devices for manipulating polarized light can be found in a wide range of settings, from advanced research laboratories to the common household [1,2]. Perhaps one of the most familiar household polarizers is a pair of sunglasses, a necessity for many car drivers on a sunny day. Sunglasses filter the light specularly reflected from a flat paved surface. The reflected light from the flat pavement is predominantly horizontally polarized, so that a polarizer with a vertical transmission axis rejects the specular reflection.

Another example of polarizing glasses are those worn for viewing a stereoscopic motion picture. In this case, the transmission axis of the polarizer covering the right eye is orthogonal to that of the polarizer for the left eye. Likewise, the motion picture scenes for the right and left eyes are projected using orthogonally polarized light. The right eye polarizer passes the light for the right eye scene and rejects the left eye scene. Similarly, the left eye polarizer passes the left eye scene and rejects the right eye scene. The viewer enjoys a stereoscopic picture.

While the polarizing glasses of the previous examples are normally constructed with linearly polarizing material, antiglare screens frequently employ circularly polarizing sheets. Displays, such as radar screens, use these circularly polarizing sheets to suppress glare. Light that enters the circular polarizer, and subsequently undergoes reflection at some other surface, is blocked from reemerging from the circular polarizer because of the reversal of handedness, while the light generated by the screen passes through.

Scientists in many disciplines use polarized light as a tool for their investigations. Physicists are still trying to unfold the mysteries of the invariance of the state of polarization of a photon before and after collisions with high-speed particles. Polarization also presents puzzles such as: "What happens when only one photon polarized at 45° with respect to the birefringent crystal axis enters the crystal?" Is the photon, which is considered the smallest unit of light, further split into horizontally and vertically polarized half-photons?

While puzzles such as these make polarization itself an interesting study, applications of polarization devices and phenomena to other disciplines are extensive. Spectroscopists

High definition 3D television.

use the Lyot–Ohman filter [1] made out of a combination of polarizers and retarders for their work. With this filter, resolving powers as high as 0.01 nm can be achieved.

Astrophysicists study the pattern of magnetic fields in nebulae by mapping the pattern of perturbation of the state of polarization of light from a nebula. These perturbations result from the Faraday rotation caused by the magnetic field of the nebula.

Many organic materials rotate the direction of light polarization as light passes through them. Chemists use this fact for analyzing the structure of new organic molecules. One of the most familiar examples is the determination of the sugar content of a sugar solution by measuring the rotation of the polarization.

Mechanical engineers use the strain birefringence pattern of a plastic model as an aid to strain analysis. Colorful strain patterns in the plastic model can be viewed under a polariscope.

Biologists are certainly beneficiaries of polarization microscopes [3] that enable them to observe microbes that are transparent and invisible under normal light. The polarization microscope sees the pattern of the retardance that the microbes create. Biologists also know that the direction of polarization of the illuminating light controls the direction of growth of some fungi is used for navigation by certain animals such as bees and horseshoe crabs.

Principles of operation of many liquid crystal displays are based on the manipulation of the polarized light as detailed in Chapter 5.

In the field of fiber-optic communication, many electrooptic devices are polarization dependent. Coherent optical communication systems detect the received light by mixing it with local oscillator light. Fluctuations in the state of polarization of the received light or the local oscillator light will cause the output power of the intermediate frequency IF signal to fluctuate. Countermeasures have to be exploited. The concepts in this chapter establish the foundation for understanding polarization. In Chapter 12, we will deal with issues such as countermeasures for polarization jitter in coherent communication systems.

6.1 INTRODUCTION

The types of waves that have so far appeared in this book have been linearly polarized waves. The **E** field component did not change direction as the wave propagated. As shown in Fig. 6.1a, this type of wave is called a linearly polarized wave, and the direction of the **E** field is called the direction of polarization.

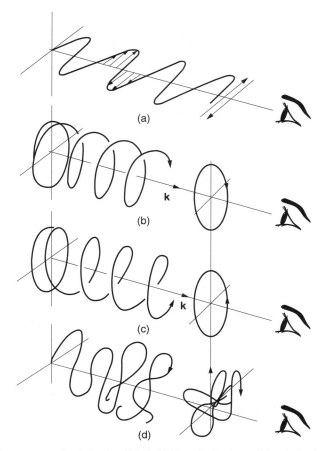

Figure 6.1 Various states of polarization (SOP). (a) Linearly (horizontally) polarized. (b) Right-handed circularly polarized. (c) Left-handed circularly polarized. (d) Depolarized.

In this chapter, waves whose directions of polarization rotate as the waves propagate will be described. The **E** vector rotates around the propagation direction **k**, as the wave propagates, as shown in Figs. 6.1b and 6.1c. When the cross section of the helix is an ellipse, the wave is said to be elliptically polarized. When the cross section is circular as in Fig. 6.1b, it is naturally called a circularly polarized wave. If the **E** vector rotates in a clockwise sense when observed at a distant location in the propagation path while looking toward the light source, as in Fig. 6.1b, the handedness of the polarization is right-handed rotation. Similarly, if the **E** vector rotates in a counterclockwise direction, as in Fig. 6.1c, the rotation is left-handed. If there is no repetition in the pattern of the **E** field as the wave propagates, as shown in Fig. 6.1d, the wave is said to be unpolarized or depolarized.

For handedness to be meaningful, both the direction of observation and the direction of propagation have to be specified. By convention, the handedness is specified by looking into the source of light.

Information describing the pattern and orientation of the polarized light is called the state of polarization. Any given state of polarization can be decomposed into two linearly polarized component waves in perpendicular directions. The state of polarization is determined by the relative amplitude and difference in phase between the two component waves. This relative phase difference is termed retardance.

The three most basic optical components that are used for manipulating or measuring the state of polarization are the (1) retarder, (2) linear polarizer, and (3) rotator.

In this chapter, the prime emphasis is placed on how to use these optical components. The circle diagrams are predominantly used for explaining the operation. In the next chapter, however, the Poincaré sphere will be used for explaining the operation.

6.2 CIRCLE DIAGRAMS FOR GRAPHICAL SOLUTIONS

Graphical and analytical methods for finding the state of polarization complement each other. The graphical method is fail-safe and is often used to confirm the results obtained by analytical methods. The graphical method helps visualize the state of polarization for a given set of parameters and also makes it easier to visualize intermediate stages. On the other hand, analytical methods provide higher accuracy and are easier to generalize. This chapter begins with a look at graphical solutions to common polarization problems.

6.2.1 Linearly Polarized Light Through a Retarder

A retarder can be made from any birefringent material, that is, any material whose refractive index depends on direction. As an example, let us take the uniaxial crystal characterized by refractive indices n_e and n_o as described in Chapter 4. The orthogonal linearly polarized component waves are the e-wave and the o-wave. It is further assumed that the front and back surfaces of the retarder are parallel to the optic axis of the crystal, and the propagation direction of the incident light is normal to the front surface of the retarder. In this situation, the directions of the component e-wave and o-wave do not separate as they propagate through the retarder; rather, they emerge

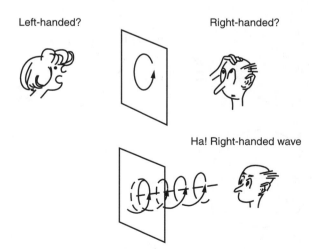

Right- or left-handedness can only be determined after both the direction of propagation and the direction of observation have been specified.

together. Depending on which is smaller, n_e or n_o, one of the component waves moves through the retarder faster than the other. The relative phase difference is the retardance Δ. The polarization direction of the faster component wave is called the fast axis of the retarder, and the polarization direction of the slower component wave is called the slow axis. The emergent state of polarization is the superposition of the two component waves and will depend on the relative amplitudes of the two component waves, as well as the retardance.

A circle diagram will be used to find the state of polarization as the incident linearly polarized light transmits through the retarder. Figure 6.2a shows the configuration. A linearly polarized wave at azimuth $\theta = 55°$ is incident onto a retarder with retardance

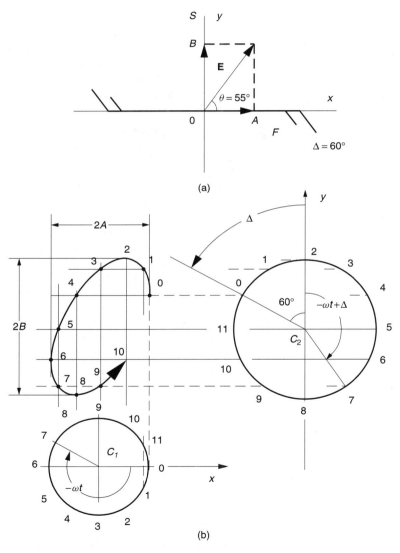

Figure 6.2 Graphical solution. (a) Geometry. (b) Circle diagram.

> The difference in the usage of the words retardance and retardation is analogous to that between transmittance and transmission. Retardance is a measurable quantity representing the difference in phase angles.

$\Delta = 60°$. The direction of the fast axis of the retarder is designated by an elongated F and in this case is oriented in the x direction. The direction of the slow axis is perpendicular to that of the fast axis and is taken as the y direction. The z direction is the direction of propagation.

The incident light $\mathbf{E}$ is decomposed into the directions of the fast and slow axes, that is, in the x and y directions. In complex notation, the component waves are

$$E_x = A e^{j(-\omega t + \beta z)} \tag{6.1}$$

$$E_y = B e^{j(-\omega t + \beta z + \Delta)} \tag{6.2}$$

with

$$\mathbf{E} = E_x \hat{\mathbf{i}} + E_y \hat{\mathbf{j}}$$

$$A = |\mathbf{E}| \cos 55°$$

$$B = |\mathbf{E}| \sin 55°$$

and the corresponding real expressions are

$$E_x = A \cos(-\omega t + \beta z) \tag{6.3}$$

$$E_y = B \cos(-\omega t + \beta z + \Delta) \tag{6.4}$$

The phasor circle C_1 in Fig. 6.2 represents Eq. (6.1) and C_2 represents Eq. (6.2). As time progresses, both phasors rotate at the same angular velocity as $e^{-j\omega t}$ (for now a fixed z), or clockwise as indicated by 0, 1, 2, 3, . . . , 11. The phase of E_y, however, lags by $\Delta = 60°$ because of the retarder. The projection from the circumference of circle C_1 onto the x axis represents E_x, and that from the C_2 circle onto the y axis represents E_y. It should be noted that the phase angle $-\omega t$ in C_1 is with respect to the horizontal axis and $-\omega t + \Delta$ in C_2 is with respect to the vertical axis.

By connecting the cross points of the projections from 0, 1, 2, 3, . . . , 11 on each phasor circle, the desired vectorial sum of E_x and E_y is obtained. The emergent light is elliptically polarized with left-handed or counterclockwise rotation.

Next, the case when the fast axis is not necessarily along the x axis will be treated. For this example, a retarder with $\Delta = 90°$ will be used. Referring to the geometry in Fig. 6.3a, the fast axis F is at azimuth Θ with respect to the x axis, and linearly polarized light with field $\mathbf{E}$ is incident at azimuth θ. Aside from the new value of Δ and the azimuth angles, the conditions are the same as the previous example.

The only difference in the procedure from that in the previous case is that the fast axis is no longer in the x direction and the incident field has to be decomposed into components parallel to the fast and slow axes, rather than into x and y components. Figure 6.3b shows the circle diagram for this case.

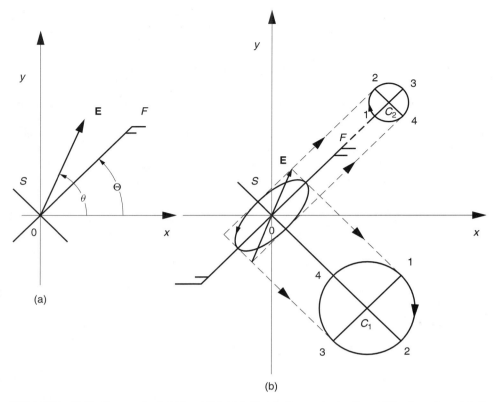

Figure 6.3 Circle diagram for linearly polarized light entering a retarder ($\Delta = 90°$), where the azimuth angle of the retarder is given by Θ. (a) Geometry. (b) Circle diagram.

It is important to recall that the only allowed directions of polarization inside the crystal are along the fast and slow axes; no other directions in between the two axes are allowed. This is the reason why the incident field is decomposed into components along the fast and slow axes.

6.2.2 Sign Conventions

As mentioned in Chapter 1, this book has employed the convention of

$$e^{-j\omega t} \tag{6.5}$$

rather than

$$e^{j\omega t} \tag{6.6}$$

which appears in some textbooks. This section attempts to clarify some of the confusion surrounding signs and the choice of Eq. (6.5) or (6.6). Let us take the example of the retarder in Fig. 6.2 as the basis for discussion. The expression for the state of polarization depends critically on the difference between ϕ_x and ϕ_y of the E_x and E_y component waves. With the convention of Eq. (6.5), the phases of E_x and E_y for the positive z direction of propagation are

$$\phi_x = \beta z - \omega t \tag{6.7}$$

$$\phi_y = \beta z - \omega t + \Delta \tag{6.8}$$

where $\Delta = 60°$. We will now explain why ϕ_y was expressed as $\beta z - \omega t + \Delta$ rather than $\beta z - \omega t - \Delta$. The example was defined such that the x direction is the direction of the fast axis, which means E_x advances faster than E_y, and hence E_x leads E_y. Let us examine Eqs. (6.7) and (6.8) more closely to see if it is indeed the case that E_x leads E_y. For simplicity, the observation is made on the $z = 0$ plane. Both ϕ_x and ϕ_y are becoming large negative values as time elapses. At the time when $\phi_x = 0$, ϕ_y is still a positive number, namely, $\phi_y = 60°$ and ϕ_y lags ϕ_x by $60°/\omega$ seconds in the movement toward large negative values. Hence, one can say that E_y is lagging E_x by $60°$ or E_x is leading E_y by $60°$. This confirms that $\phi_y = \beta z - \omega t + \Delta$ was the correct choice to represent E_x leading E_y, for the convention of Eq. (6.5). When E_x leads E_y, a left-handed polarization results for $\Delta = 60°$, as shown in Fig. 6.2.

Now, let us look at the other convention of using $e^{j\omega t}$ instead of $e^{-j\omega t}$. The same example of the retarder in Fig. 6.2 will be used. When E_x and E_y are propagating in the positive z direction, the signs of the βz and ωt terms are opposite (Chapter 1). Let ϕ'_x and ϕ'_y denote the phases of the x and y components using the $e^{j\omega t}$ convention.

$$\phi'_x = -\beta z + \omega t \tag{6.9}$$

$$\phi'_y = -\beta z + \omega t - \Delta \tag{6.10}$$

where $\Delta = 60°$. As the x direction was specified as the direction of the fast axis, Eqs. (6.9) and (6.10) have to represent the case where E_x leads E_y by $60°$. Let us verify that this is true. Taking $z = 0$ as the plane of observation, both ϕ'_x and ϕ'_y become large positive numbers as time elapses. At the time $\phi'_x = 0$, the phase of ϕ'_y is a negative number and is behind ϕ'_x by $60°/\omega$ seconds in the movement toward large positive numbers. This is consistent with E_x leading E_y by $60°$.

To conclude this example, if E_x leads E_y by $60°$, the resulting polarization is left-handed, regardless of the choice of convention of Eq. (6.5) or (6.6). However, for this previous statement to be true, the *sign of Δ does depend on the choice of convention*.

In Problem 6.1, the same reasoning is applied to the geometry of Fig. 6.3.

Next, the state of polarization of the emergent wave will be investigated as a function of retardance. For simplicity, the amplitudes of E_x and E_y are kept the same, that is, $B/A = 1$, and the fast axis is kept along the x axis. The case of $B/A \neq 1$ is left for Problem 6.2. A series of circle diagrams were drawn to obtain the states of polarization with Δ as a parameter. The results are summarized in Fig. 6.4.

With $\Delta = 0°$ ($360°$), the state of polarization is linear with azimuth $\theta = 45°$. As Δ is increased from $0°$ to $90°$, the shape becomes elliptical, growing fatter and fatter while keeping the major axis always at $45°$, and the rotation of polarization always left-handed. When Δ reaches $90°$, the wave becomes a circularly polarized wave, still with left-handed rotation. As soon as Δ passes $90°$, the radius in the $45°$ direction starts to shrink while that in the $135°$ direction expands, and the state of polarization becomes elliptically polarized with its major axis at $135°$, but still with left-handed rotation. This trend continues until $\Delta = 180°$.

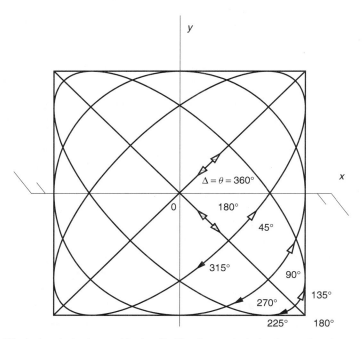

Figure 6.4 Elliptical polarizations with $A = B$. The fast axis is in the x direction, and Δ is the parameter. ⊸▷ represents left-handedness, ⟶ represents right-handedness, and ⊸▷ represents linear polarization.

Table 6.1 Summary of states of polarization with fixed $B/A = 1$ for various retardance values

Retardance Δ	0°		90°		180°		270°		360°
Shape									
Inclination θ		45°		135°		135°		45°	
Sign of sin Δ			+				−		
Handedness			Left				Right		

With $\Delta = 180°$, the wave becomes linearly polarized again, but this time the direction of polarization is at 135°. As soon as Δ exceeds 180°, the wave starts to become elliptically polarized but with right-handed rotation. As Δ increases between 180° and 360°, the state of polarization changes from linear (135°) to right-handed elliptical (major axis at 135°) to right-handed circular to right-handed elliptical (major axis at 45°) to linear (45°). In this region of Δ, the handedness is always right-handed. The results are summarized in Table 6.1.

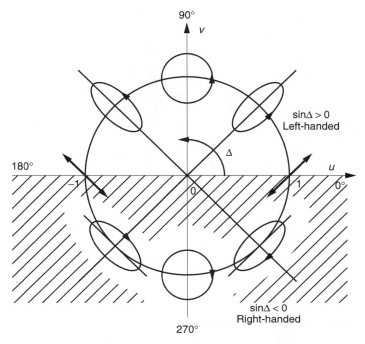

Figure 6.5 Summary of elliptical polarization with $A = B$ and Δ as a parameter on the E_y/E_x complex plane.

Next, a method of classification other than Table 6.1 will be considered. As one may have already realized, it is the combination of two numbers — B/A and Δ — that determines the state of polarization of the emergent light from the retarder. These two numbers, however, are obtainable from the quotient of Eqs. (6.1) and (6.2), namely,

$$\frac{E_y}{E_x} = \left(\frac{B}{A}\right) e^{j\Delta} \tag{6.11}$$

Each point on the complex number plane of E_y/E_x corresponds to a state of polarization. As a matter of fact, this representation will be extensively used in the next chapter. Figure 6.5 shows such a complex plane. For this example, $B/A = 1$ and the states of polarization are drawn on a unit circle for various values of Δ. Figure 6.5 summarizes the results in Fig. 6.4.

Example 6.1 A quarter-waveplate, commonly written as a $\lambda/4$ plate, is a retarder with $\Delta = 90°$. As shown in Fig. 6.6, horizontally linearly polarized light is intercepted by a $\lambda/4$ plate whose orientation Θ is rotated. Draw the sequence of elliptically polarized waves of the emergent light as the fast axis of the $\lambda/4$ plate is rotated at $\Theta = 0°$, 22.5°, 45°, 67.5°, 90°, 112.5°, 135°, 157.5°, 180°, and 202.5°.

Solution The series of circles is drawn in Fig. 6.7. As the correct numbering of the circles C_1 and C_2 is crucial to the final result, a few tips are given here on how to set up the numbering. The convention of $e^{-j\omega t}$ is being used, so that numbering of both circles C_1 and C_2 is in a clockwise sense. Refer to the drawings with $\Theta = 22.5°$

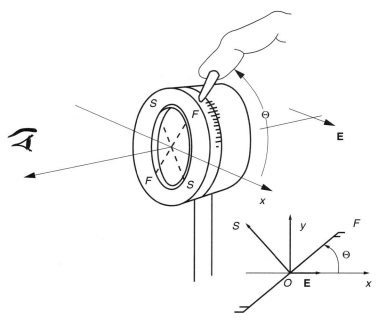

Figure 6.6 The state of polarization of the emergent light is observed as the quarter-waveplate is rotated.

and $\Theta = 157.5°$ as examples. Point P on the horizontal axis represents the tip of the incident light polarization at $t = 0$. P is decomposed into points P_1 and P_2 on circles C_1 and C_2, respectively, as shown in Fig. 6.7. If the retardance had been zero ($\Delta = 0°$), then P_1 and P_2 would correspond to point 1 for each of the circles C_1 and C_2. Because of the retardance of the $\lambda/4$ plate, C_2 is delayed 90° with respect to C_1, which corresponds to a rotation of C_2 by 90° in the counterclockwise direction. P_2 now lines up with point 2 of C_2. Observe in the case of the $\lambda/4$ plate, for all diagrams in Fig. 6.7, the line drawn from point 1 of C_1 and the line drawn from point 2 of C_2 intersect along the horizontal axis at P. This is a good method for obtaining the correct numbering.

With $\Theta = 0°$, the radius of C_2 becomes zero, and with $\Theta = 90°$, that of C_1 becomes zero. The emergent light is identical to the incident light for these cases.

The results are summarized in Fig. 6.8. The major or minor axis is always along the direction of the fast axis. This is a characteristic of a quarter-waveplate when the incident light is linearly polarized. First, the major axis follows the fast axis, and then the minor axis, and then the major axis. They alternate at every 45°. In the region $0 < \Theta < \pi/2$, the emergent light is right-handed, while in the region $\pi/2 < \Theta < \pi$ the emergent light is left-handed. It is worthwhile remembering that the handedness of the emergent circularly polarized wave alternates every 90° of rotation of the retarder. At

---

With the case of $\theta = 157.5°$, two circles C_2 are drawn, one on each side. Either circle C_2 can be used, as long as one makes sure that point 1 on circle C_1 as well as on circle C_2 correspond to point P if the retardance is momentarily reduced to zero.

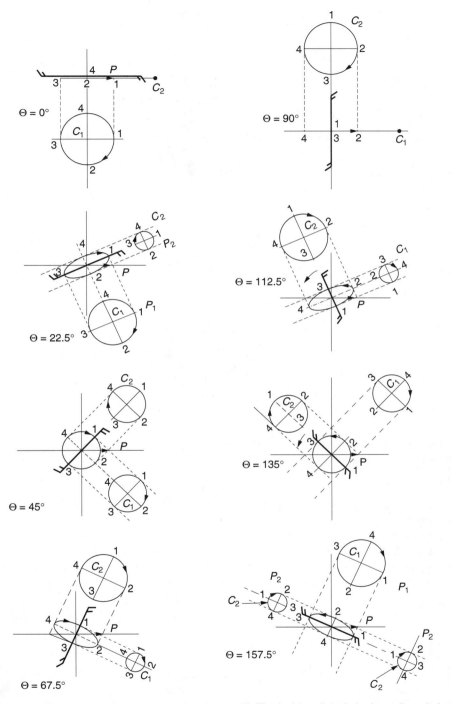

Figure 6.7 Circle diagrams as the $\lambda/4$ plate is rotated. The incident light is horizontally polarized. *Note*: Handedness changes at the azimuth of the retarder, $\Theta = 90°$ and $\Theta = 180°$.

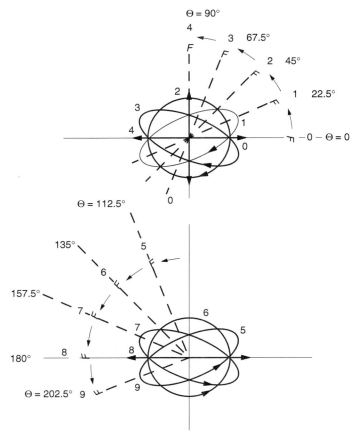

Figure 6.8 Transitions of the state of polarization as the $\lambda/4$ plate is rotated. The incident light is horizontally linearly polarized.

$\Theta = 45°$, right-handed circular polarization is obtained, and at $\Theta = 135°$ left-handed circular polarization is obtained. This is a quick convenient way of obtaining a right circularly or left circularly polarized wave.

At every 90°, the emergent light becomes identical with the incident light and is horizontally linearly polarized. Note that the orientation with $\Theta = 180°$ is identical to that of $\Theta = 0°$, and the orientation with $\Theta = 202.5°$ is identical to that of $\Theta = 22.5°$.   □

6.2.3 Handedness

The question of how the direction of the handedness is determined will be resolved. The instantaneous value $\theta(t)$ of the direction of polarization with respect to the x axis is

$$\theta(t, z) = \tan^{-1}\left(\frac{E_y(t)}{E_x(t)}\right) \tag{6.12}$$

From Eqs. (6.3) and (6.4), $\theta(t, z)$ is expressed as

$$\theta(t, z) = \tan^{-1}\left(\frac{B}{A}(\cos \Delta + \tan(\omega t - \beta z)\sin \Delta)\right) \tag{6.13}$$

For $\sin \Delta = 0$, the azimuth $\theta(t, z)$ becomes independent of time and location, and a linearly polarized wave results. With other values of $\sin \Delta$, $\theta(t, z)$ depends on time and location.

The direction of the movement of $\theta(t, z_0)$ at a fixed point $z = z_0$ is found from the derivative of Eq. (6.13):

$$\frac{d\theta(t, z_0)}{dt} = \left[\frac{\omega \sec^2(\omega t - \beta z)}{1 + \left(\frac{B}{A}[\cos \Delta + \tan(\omega t - \beta z_0)\sin \Delta]\right)^2}\right] \sin \Delta \tag{6.14}$$

The factor in the large parentheses of Eq. (6.14) is always positive, and $d\theta(t, z_0)/dt$ is the same sign as $\sin \Delta$. When $\sin \Delta$ is positive, the azimuth $\theta(t, z_0)$ increases with time and if negative, $\theta(t, z_0)$ decreases. These results are summarized in Eq. (6.15):

$$\begin{array}{lll}
\text{Counter clockwise (left-handed)} & \text{when} & \sin \Delta > 0 \\
\text{Linearly polarized} & \text{when} & \sin \Delta = 0 \\
\text{Clockwise (right-handed)} & \text{when} & \sin \Delta < 0
\end{array} \tag{6.15}$$

For the case of circularly polarized emergent light, that is,

$$B/A = 1 \quad \text{and} \quad \sin \Delta = \pm 1 \tag{6.16}$$

the derivative simplifies to

$$\frac{d\theta(t, z_0)}{dt} = \pm\omega \tag{6.17}$$

The angular velocity of the rotation of the polarization is the same as that of the component wave. The sense of rotation of $d\theta(t, z_0)/dt$ also matches the sign of $\sin \Delta$.

It should be noted that Eq. (6.15) is true only when the incident wave is linearly polarized.

6.2.4 Decomposition of Elliptically Polarized Light

The graphical method for constructing an elliptical polarization has been described. Up to this point, the incident light has been decomposed into components along the fast and slow axes of the retarder. In this section, the graphical method will be generalized to allow decomposition of a given elliptical polarization into an arbitrary set of mutually perpendicular component waves.

The values B'/A' and Δ' of the newly decomposed waves, however, vary according to the desired orientation of the decomposed component waves. An example will be used for explaining the decomposition.

Example 6.2 Graphically decompose the elliptically polarized wave with $B/A = 1$ and $\Delta = 45°$ shown in Fig. 6.4 into E'_x polarized at 22.5° and E'_y polarized at 112.5° and then determine the values of B'/A' and Δ' of the newly decomposed component waves.

Solution Referring to Fig. 6.9, the decomposition is performed as follows:

1. Draw coordinates $x' - y'$ in the desired directions.
2. Determine the radius of circles C_1 and C_2 from the points of the extrema on the ellipse in the x' and y' directions.
3. Extend a line downward parallel to the y' axis from point 1 on the ellipse to point 1 on circle C_1. Similarly, extend a line to the right parallel to the x' axis from the same point on the ellipse to intersect point 1 and point 1' on the circle C_2 (set the point 1' aside for now).

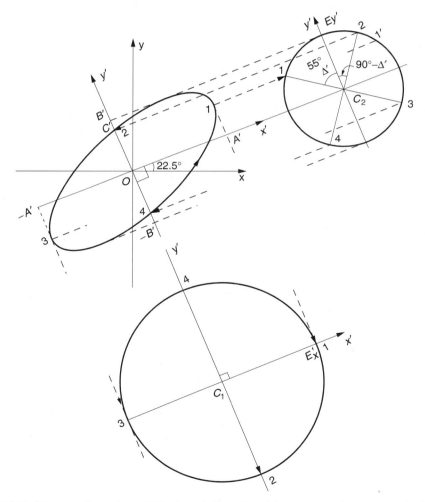

Figure 6.9 Decomposition of an elliptically polarized wave into component waves along arbitrary orthogonal coordinates.

4. From point 1 on the ellipse O, follow the ellipse in the direction of rotation of polarization to point 2. Point 2 is the intersection of the ellipse O with the y' axis. Draw the extension line from point 2 on the ellipse parallel to the x' axis to make point 2 on circle C_2.

5. Find point 3, which is the tangent to the ellipse parallel to the y' axis. Draw an extension line parallel to the x' axis from point 3 on the ellipse O to point 3 on circle C_2.

6. Find point 4, which is the intersection of the ellipse O with the y' axes, and draw an extension parallel to the x' axis to point 4 on circle C_2.

7. Now, B'/A' and Δ' can be obtained from this graph. The ratio of the radii of C_1 and C_2 gives

$$B'/A' = 0.59$$

and the angle Δ' on circle C_2 gives

$$\Delta' = 55°$$

The E_y phasor rotates clockwise from the y' axis, so that E'_y is lagging by $\Delta' = 55°$ from that of E'_x, consistent with the left-handed rotation of the ellipse given by Eq. (6.15). As a matter of fact, examination of Fig. 6.9 shows that Δ' can be calculated directly from the intersections of B' and C' of the ellipse on the y' axis:

$$\cos(90° - \Delta') = \frac{OC'}{OB'}$$

and

$$\sin \Delta' = \frac{OC'}{OB'}$$

8. Note that if the phasor on E'_y starts from point $1'$ and both E'_x and E'_y rotate in the clockwise sense $(-\omega t)$, then the intersections will not form the original ellipse, and point $1'$ has to be discarded. □

6.2.5 Transmission of an Elliptically Polarized Wave Through a λ/4 Plate

Previous sections dealt with the transmission of linearly polarized light through a retarder. This section treats the more general case of transmission of an elliptically polarized incident wave through a retarder.

As shown in Fig. 6.10, let the azimuth and ellipticity of the incident wave be θ and ϵ, respectively. The retarder is again a λ/4 plate. The circle diagram method starts with the decomposition of the incident elliptic field into the field parallel to the fast axis of the λ/4 plate and that parallel to the slow axis. The former component field is represented by phasor circle C_1 and the latter by phasor circle C_2.

Next, the retardance is considered. The endpoint P of the phasor vector of the incident wave is projected onto point 1 of circle C_1 and projected onto point 0 of circle C_2. Point 0 on circle C_2 is delayed by 90° with respect to point 1 in order to account for transmission through the λ/4 plate. The circumference of the phasor circles

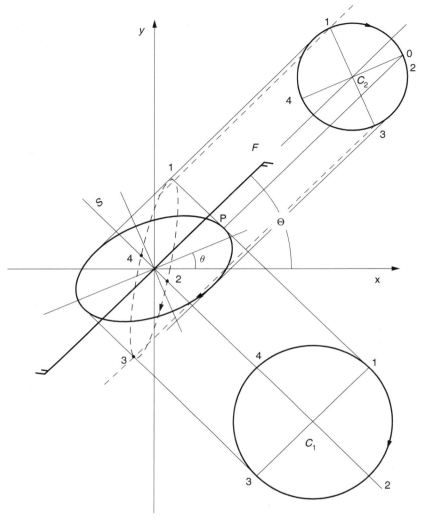

Figure 6.10 An elliptically polarized wave (solid line) is incident onto a $\lambda/4$ plate at $\Theta = 45°$. The emergent ellipse (dotted line) is obtained by circle diagrams.

is divided into four and numbered 1, 2, 3, and 4 sequentially. The intercepts of the projections from each of the circles parallel to the fast and slow axes are numbered 1, 2, 3, and 4. The emergent ellipse, shown as a dotted line in Fig. 6.10, is completed by connecting 1, 2, 3, 4 and the handedness is in the direction of 1, 2, 3, and 4.

6.3 VARIOUS TYPES OF RETARDERS

A waveplate is a retarder with a fixed retardance. Waveplates providing retardances of 360°, 180°, and 90° are called full-waveplates, half-waveplates, and quarter-waveplates, respectively. They are also written simply as λ, $\lambda/2$, and $\lambda/4$ plates. Retarders with adjustable retardance are called compensators.

6.3.1 Waveplates

Waveplates can be fabricated either from a single piece of birefringent crystal or from a combination of two pieces of crystal. The difficulty with fabricating a single crystal waveplate is that the plate has to be made extremely thin. The thickness d for a $\lambda/4$ plate is calculated as

$$\frac{2\pi}{\lambda} d|n_e - n_o| = \frac{\pi}{2} \tag{6.18}$$

Taking $\lambda = 0.63$ μm, the values of d for typical birefringent crystals are:

> For calcite, $d = 0.92$ μm ($n_e = 1.4864$, $n_o = 1.6584$).
> For quartz, $d = 17.3$ μm ($n_e = 1.5443$, $n_o = 1.5534$).
> For mica, $d = 31.5$ μm ($n_e = 1.594$, $n_o = 1.599$).

The thickness is in the range of tens of microns.

Even though calcite has a cleavage plane and need not be polished, its brittleness makes it hard to handle thin pieces. Quartz is not as brittle but requires polishing because it does not have a cleavage plane. Mica has more favorable properties. It is not only flexible, but also possesses cleavage planes; however, there is some difficulty in cleaving at exactly the right thickness. The plate with the desired thickness is selected among many cleaved pieces.

The stringent requirement of excessively small thicknesses can be avoided by taking advantage of the rollover of the retardance at every 2π radians. The retardance of a $\lambda/4$ plate, for instance, is designed as

$$\frac{2\pi}{\lambda}|n_e - n_o| d = \left(2\pi N + \frac{\pi}{2}\right) \tag{6.19}$$

where the value of N is normally a few hundred.

With $N = 100$, the thickness of a quartz $\lambda/4$ plate is 7 mm. The drawbacks of a retarder with a large N are a higher sensitivity to temperature and to the angle of incidence. The increase in the path of a ray with incident angle θ compared to the ray normal to the plate of thickness d is

$$\Delta d \doteq \left(\frac{1}{\cos\theta} - 1\right) \doteq d\frac{\theta^2}{2} \tag{6.20}$$

The value of θ that creates a retardance error of $\pi/2$ radians is

$$\frac{2\pi}{\lambda}|n_e - n_o| d\frac{\theta^2}{2} = \frac{\pi}{2} \tag{6.21}$$

Inserting Eq. (6.19) into (6.21) gives

$$\theta \doteq \frac{1}{\sqrt{2N}} \text{ rad} \tag{6.22}$$

With $N = 100$, the cone of the allowed angle of incidence is narrower than $4°$.

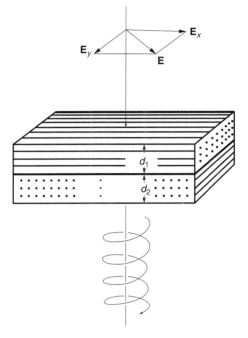

Figure 6.11 Structure of a quarter-waveplate.

The second approach to alleviating the thinness requirement is the combination of two plates with their optical axes perpendicular to each other. Figure 6.11 shows the construction of a waveplate of this kind. The optical paths ϕ_x and ϕ_y for E_x and E_y are

$$\phi_x = \frac{2\pi}{\lambda}(d_1 n_e + d_2 n_o) \tag{6.23}$$

$$\phi_y = \frac{2\pi}{\lambda}(d_1 n_o + d_2 n_e) \tag{6.24}$$

and the retardance $\Delta = \phi_y - \phi_x$ is

$$\Delta = \frac{2\pi}{\lambda}(d_1 - d_2)(n_o - n_e) \text{ rad} \tag{6.25}$$

What matters in Eq. (6.25) is the difference in thickness, rather than the total thickness so that the thickness of each plate can be comfortably large to facilitate polishing.

6.3.2 Compensator

Figure 6.12a shows the structure of a Babinet compensator. Two wedge-shaped birefringent crystals are stacked such that their optical axes are perpendicular to each other. The apex angle, however, is made small so that the separation of the o- and e-waves is negligibly small. They can be slid against one another so that the difference in thicknesses $d_1 - d_2$ is adjustable.

Depending on the location of the incident ray, the retardance is varied. At the location where $d_1 = d_2$, regardless of the values of n_e and n_o, both waves E_x and E_y

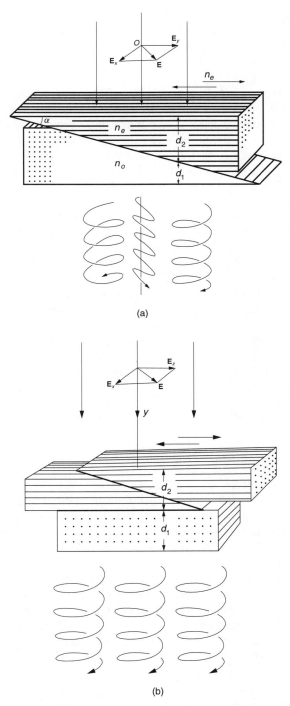

(a)

(b)

Figure 6.12 Comparison between Babinet-type and Soleil-type compensators. (a) Babinet compensator. Depending on the horizontal locations of the incident light, the emergent state of polarization varies. (b) Soleil compensator. For a given d_1 and d_2, the emergent state of polarization does not change as the incident light position is varied.

go through the same amount of phase shift and the retardance is zero. However, at the locations where $d_1 \neq d_2$, the two waves do not go through the same phase shift. If the crystal is a positive uniaxial crystal and $n_e > n_o$ like quartz and if $d_1 < d_2$, then the phase of E_y lags behind that of the E_x wave. The amount of phase lag, or retardation, can continuously be adjusted by either shifting the location of the incident light while keeping the relative positions of the crystal stack fixed, or sliding one crystal against the other by means of a micrometer while keeping the location of the incident light fixed.

The compensator can be used at any wavelength provided that the light at that wavelength is not significantly absorbed by the crystal. The compensator also provides a full range of retardation so that it can be used as a zero-wave, quarter-wave, half-wave, or full-wave retarder. The sense of circular polarization can also be changed by changing plus $\lambda/4$ to minus $\lambda/4$ of retardation. In some applications, the compensator is used to measure the retardation of a sample material. Light of a known state of polarization enters the sample, which causes a change in the polarization of the emergent light. The emergent light is passed through the compensator and the thickness d_2 is adjusted to regain the initial state of polarization. From this adjustment of d_2, which is precalibrated in terms of retardance, the retardation of the sample can be determined.

Figure 6.12b shows the structure of a Soleil compensator. The difference between the Babinet and Soleil compensators is that the Soleil compensator has another block of crystal so that the thickness d_2 is independent of the location of the incident wave. The top two blocks consist of two wedge-shaped crystals. They can be slid by a micrometer against one another so that the thickness d_2 is adjustable. The thickness d_1 of the bottom block is fixed. With the Soleil compensator, the emergent state of polarization is independent of the location of incidence.

As mentioned earlier, when N is large, there are stringent requirements on the collimation of the light entering the retarder, as well as a larger temperature dependence. However, for the Soleil or Babinet compensator, it is possible to reduce N to zero.

6.3.3   Fiber-Loop Retarder

Bending an optical fiber creates birefringence within the fiber. This is the basis of the fiber-loop retarder whereby a controlled amount of bend-induced birefringence is used to change the retardance. An advantage of the fiber-loop retarder controller over conventional optical devices, such as quarter-waveplates and half-waveplates, is that the polarization control is achieved without interrupting transmission of light in the fiber.

When an optical fiber is bent, the fiber is compressed in the radial direction of the bend and is expanded in the direction perpendicular to it, as shown in Fig. 6.13. The refractive index of glass is lowered where it is compressed and raised where it is expanded. In the coordinates shown in Fig. 6.13, the difference between the change Δn_x in the x direction before and after the bending, and the change Δn_y in the y direction before and after the bending is calculated as [4,5]

$$\Delta n_y - \Delta n_x = -0.0439 n^3 \left(\frac{r}{R}\right)^2 \tag{6.26}$$

where n is the index of refraction of the core, $2r$ is the outer diameter of the fiber (diameter of the cladding), and R is the radius of curvature of the loop as indicated in

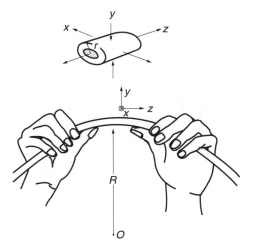

Figure 6.13 Bending of a fiber to create birefringence.

Fig. 6.14. The coefficient 0.0439 was calculated from Poisson's ratio and the strain-optical coefficients for a silica glass fiber.

Figure 6.15 shows the structure of a fiber-loop polarization controller based on the bend-induced birefringence in the fiber [6]. It combines both a $\lambda/4$ and $\lambda/2$ fiber-loop retarder and a polarizer loop. The retarders are made of ordinary single-mode fibers. The radius of the left-hand side retarder spool is designed such that the phase of the y-polarized wave is $\pi/2$ radians behind that of the x-polarized wave. This spool is the fiber equivalent of a "quarter-waveplate" and it will be referred to as the $\lambda/4$ loop. The right-hand side retarder spool is designed to create a π-radian phase shift, analogous to a "half-waveplate", and will be referred to as the $\lambda/2$ loop. Usually, the two spools have the same radius, and the $\lambda/2$ spool has twice the number of fiber turns as the $\lambda/4$ spool.

The orientation of the fiber loops can be changed as indicated by the arrows in Fig. 6.15. When conversion of elliptic to linear polarization is desired, the $\lambda/4$ loop is oriented so that the elliptically polarized wave going into the loop is converted into a linearly polarized wave upon exiting the loop (see Section 6.4.3.2). The $\lambda/2$ loop is then oriented to rotate the direction of the linear polarization to the desired direction.

Example 6.3 Find the loop radius of a fiber-loop polarization controller for a $\lambda = 1.30$ μm wavelength system. The number of turns is $N = 4$ for the $\lambda/4$ loop and $N = 8$ for the $\lambda/2$ loop. The diameter of the fiber is 125 μm, and the index of refraction of the core is $n = 1.55$.

Solution The radius R of the $\lambda/4$ loop is first calculated. The difference $\Delta\beta$ in propagation constants in the x and y directions is

$$\Delta\beta = k_0(n + \Delta n_x) - k_0(n + \Delta n_y) \tag{6.27}$$

where k_0 is the free-space propagation constant. Combining Eq. (6.26) with (6.27),

$$\Delta\beta = 0.0439 \, k_0 n^3 \left(\frac{r}{R}\right)^2 \text{ rad/m} \tag{6.28}$$

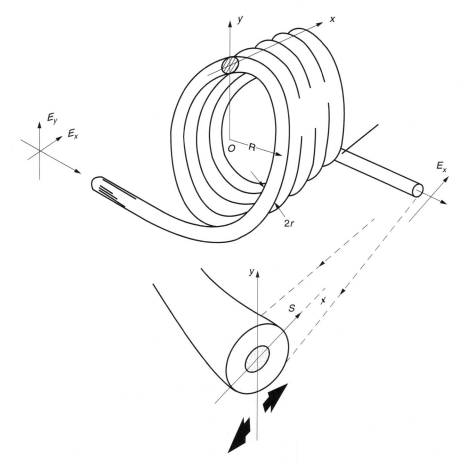

Figure 6.14   Fiber-loop-type retarder using a single-mode fiber.

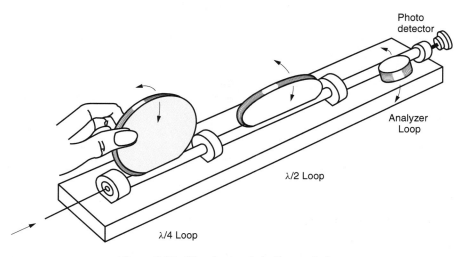

Figure 6.15   Fiber-loop polarization controller.

The $\lambda/4$ condition requires

$$2\pi RN \cdot \Delta\beta = \frac{\pi}{2} \qquad (6.29)$$

Equations (6.28) and (6.29) give

$$R = 0.176\, k_0 n^3 r^2 N \qquad (6.30)$$

With the parameters provided, the loop radius is

$$R = 4.95 \;\; \text{cm}$$

A choice of radius smaller than $R = 1.5$ cm is not recommended, because the transmission loss (as will be mentioned in Section 6.5.3) becomes increasingly significant as R is decreased. $\qquad \square$

6.4 HOW TO USE WAVEPLATES

The waveplate is one of the most versatile optical components for manipulating the state of polarization. Various applications of the waveplate will be summarized from the viewpoint of laboratory users.

6.4.1 How to Use a Full-Waveplate

A full-wave plate (λ plate) combined with an analyzer (polarizer) functions like a wavelength filter. The retardance of any thick waveplate critically depends on the wavelength of light. This can be seen from Eq. (6.19), which gives the expression for the retardance of a thick quarter-waveplate. The expression for the retardance of the thick full-waveplate can be obtained by substituting 2π for $\pi/2$ on the right-hand side of Eq. (6.19).

The operation of the wavelength filter is described as follows. Multiwavelength linearly polarized light is incident onto the full-waveplate. The azimuth of the incident light is chosen so that both fast and slow component waves propagate through the full-waveplate. Only those wavelengths that satisfy the full-wave retardance condition will emerge as linearly polarized waves. The transmission axis of the analyzer is oriented in the same direction as the azimuth of the incident light, so that the emergent linearly polarized waves experience the least amount of attenuation on passing through the analyzer. For all other wavelengths, the state of polarization is changed by going through the full-waveplates, and these will be attenuated on passing through the analyzer.

By adding additional full-waveplate and analyzer pairs, a narrower linewidth wavelength filter is obtainable. Such filters are the Lyot–Ohman and Šolic filters [1,13].

6.4.2 How to Use a Half-Waveplate

The half-waveplate can be used to change the orientation and/or the handedness of a polarized wave. The case of a linearly polarized incident wave is first considered. Let the vector $\overline{OP}$ of the incident E wave be in the vertical direction, and let the direction of propagation be into the page. The plane of the half-waveplate is in the plane of the page. Let the fast axis of the half-waveplate ($\lambda/2$ plate) be oriented at an angle ϕ from the direction of polarization, as shown in Fig. 6.16a. The incident vector $\overline{OP}$ is decomposed into $\overline{OQ}$ and $\overline{QP}$ in the directions of the fast and slow axes, respectively. After transmission through the $\lambda/2$ plate, the direction of the vector $\overline{QP}$ is reversed to $\overline{QP}'$, while the vector $\overline{OQ}$ remains unchanged. Alternatively, $\overline{OQ}$ is reversed and $\overline{PQ}$ is unchanged. The resultant vector for the transmitted wave becomes the vector $\overline{OP}'$. The emergent vector $\overline{OP}'$ is a mirror image of the incident vector $\overline{OP}$ with respect to the fast axis. Another way of looking at the emergent wave is to say that the incident vector has been rotated toward the fast axis by 2ϕ. For instance, a vertically polarized incident wave can be converted into a horizontally polarized wave by inserting the $\lambda/2$ plate at an azimuth angle of 45°.

Next, the case of elliptically polarized incident light is considered. Let us say that the direction of the major axis is vertical and the handedness is right. The direction of the fast axis is ϕ degrees from the major axis of the ellipse. As illustrated in Fig. 6.16b, the incident elliptically polarized wave is decomposed into components parallel to the fast and slow axes, represented by circles C_1 and C_2, respectively. The points 1,2,3,4 on the circle C_1 correspond to points 1,2,3,4 on the circle C_2.

After transmission through the $\lambda/2$ plate, the points on circle C_2 (slow axis) lag the points on circle C_1 (fast axis) by 180°. This is shown in Fig. 6.16b by moving the points 1,2,3,4 on C_2 diametrically opposite, as indicated by $1'$, $2'$, $3'$, and $4'$, while the points 1,2,3, and 4 on C_1 remain unchanged. The emergent wave is drawn with the new combination. The emergent wave is elliptically polarized with the same shape but rotated from that of the incident wave. The emergent wave looks like a mirror image with respect to the fast axis of the $\lambda/2$ plate. The handedness is reversed. Alternatively, the emergent ellipse also looks as if it were made by rotating the incident ellipse toward the fast axis by 2ϕ, then reversing the handedness.

The $\lambda/2$ plate does not change the shape of the ellipse, only the orientation and handedness. When just a simple change of handedness is desired, one of the axes of the ellipse is made to coincide with the fast axis of the $\lambda/2$ plate.

6.4.3 How to Use a Quarter-Waveplate

The quarter-waveplate is probably the most popular of the waveplates. It has a variety of uses including polarization converter, handedness interrogator, and retardance measuring tool.

6.4.3.1 Conversion from Linear to Circular Polarization by Means of a λ/4 Plate

Figure 6.17 shows three configurations for converting a linearly polarized wave into either a circularly or elliptically polarized wave by means of a quarter-waveplate. In the figure, the incident light is vertically polarized and propagating from left to right. In Fig. 6.17a, looking into the source of light, the direction of the polarization of the incident light is 45° to the left of the fast axis of the quarter-waveplate. With this configuration, the emergent light is a left-handed circularly polarized wave (Problem 6.3).

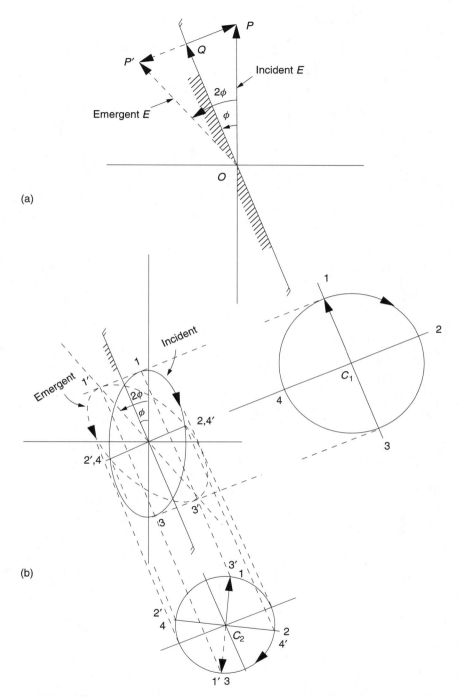

Figure 6.16 Usage of a $\lambda/2$ plate. (a) Incident wave is linearly polarized. (b) Incident wave is elliptically polarized.

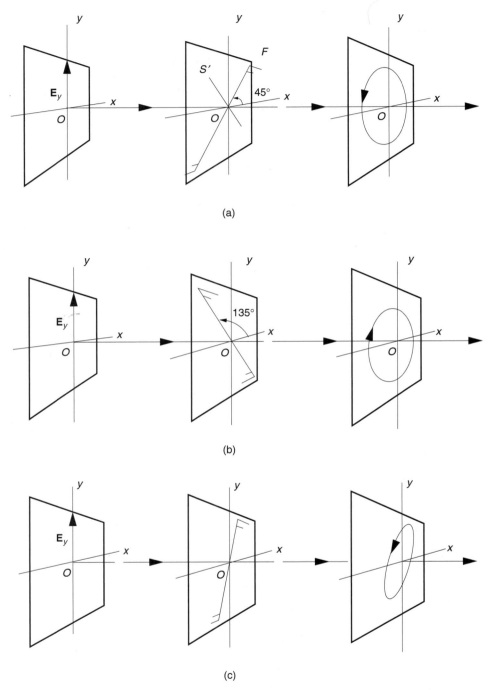

Figure 6.17 Converting a linearly polarized wave by means of a $\lambda/4$ plate. (a) Linear to left-handed circular polarization. (b) Linear to right-handed circular polarization. (c) Linear to elliptical polarization.

Figure 6.17b is similar to Fig. 6.17a except that the direction of polarization of the incident wave is 45° to the right of the fast axis of the quarter-waveplate. The emerging light is a right-handed circularly polarized wave.

In summary, looking toward the direction of the source, if the incident light polarization is oriented 45° to the left of the fast axis, the handedness is also left-handed. On the other hand, if the incident light polarization is oriented 45° to the right of the fast axis, the handedness becomes also right-handed. Thus, for the same linearly polarized incident wave, a change in the handedness of the emergent light is achieved by just rotating the quarter-waveplate by 90° in either direction.

An elliptically polarized wave is generated by orienting the fast axis at an azimuth angle other than 45° with respect to the direction of the incident light polarization, as shown in Fig. 6.17c.

Inspection of the results shown in Fig. 6.17 and the answers to Problem 6.3 reveals a shortcut method to drawing the emergent wave from a $\lambda/4$ plate when the incident light is linearly polarized. Figure. 6.18 illustrates this shortcut method, and the steps are explained below.

Step 1. Draw in the azimuth θ of the input light and Θ of the fast axis of the $\lambda/4$ plate.

Step 2. Draw the line $\overline{ef}$ perpendicular to the fast axis of the $\lambda/4$ plate.

Step 3. Complete the rectangle $efgh$. The center of the rectangle coincides with the origin.

Step 4. The ellipse that is tangent to this rectangle represents the polarization of the emergent light. If **E** is to the left of the fast axis, the emergent wave has left-handed elliptical polarization.

6.4.3.2 Converting Light with an Unknown State of Polarization into Linearly Polarized Light by Means of a $\lambda/4$ Plate

Figure 6.19 shows an arrangement for converting an elliptically polarized wave into a linearly polarized wave. The incident light beam goes through a $\lambda/4$ plate, a $\lambda/2$ plate, and an analyzer and finally reaches the photodetector. If the fast axis of the quarter-waveplate is aligned to the major (or minor) axis of the elliptically polarized incident light, the output from the quarter-waveplate will be linearly polarized. This fact is detailed in Fig. 6.20, where elliptically polarized light is decomposed into two component fields perpendicular and parallel to the major axis. The phase of the perpendicular component is delayed from that of the parallel component by 90° (see also Problem 6.4). If the fast axis of the quarter-waveplate is aligned with the delayed perpendicular component, the two component waves become in phase and the emergent wave is a linearly polarized wave as indicated by the vector $\overline{OP}$.

If the fast axis of the $\lambda/4$ plate is further rotated by 90° in either direction, again the emergent wave is a linearly polarized wave as indicated by vector $\overline{OP'}$.

A method for aligning the $\lambda/4$ plate in the desired location will now be explained. Besides the $\lambda/4$ plate, a $\lambda/2$ plate and an analyzer are added as shown in Fig. 6.19. The transmission axis (major principal axis) of the analyzer is for now set in the vertical direction. The output of the analyzer is monitored with a photodetector. The function of the $\lambda/2$ plate is to rotate the direction of the light polarization emerging from the $\lambda/4$ plate.

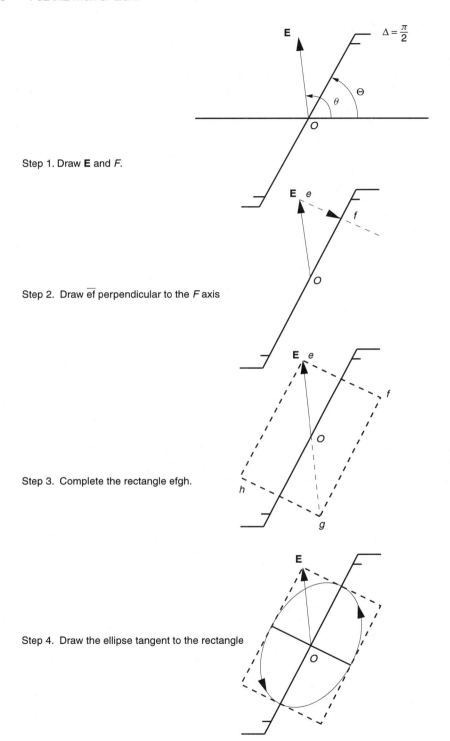

Step 1. Draw **E** and _F_.

Step 2. Draw $\overline{ef}$ perpendicular to the _F_ axis

Step 3. Complete the rectangle efgh.

Step 4. Draw the ellipse tangent to the rectangle

Figure 6.18   A shortcut method of finding the state of polarization when linearly polarized light is incident onto a $\lambda/4$ plate.

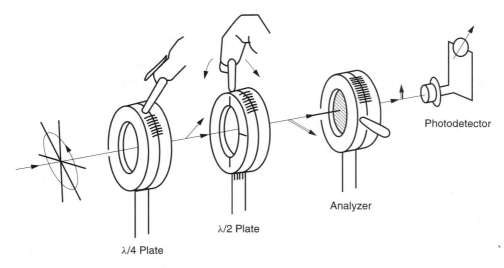

Figure 6.19 Converting from elliptical to linear polarization.

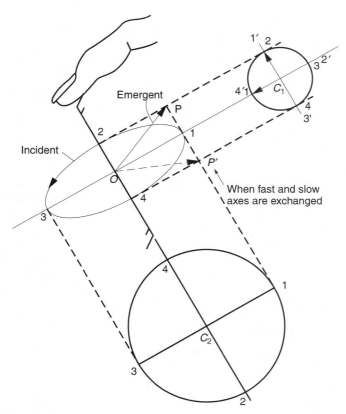

Figure 6.20 Converting an elliptically polarized wave into a linearly polarized wave by means of a λ/4 plate.

First, with an arbitrary orientation of the $\lambda/4$ plate, the $\lambda/2$ plate is rotated. As the $\lambda/2$ plate is rotated, the value of the minimum output from the photodetector is noted. This is the beginning of an iterative procedure aimed at producing a sharp null in the photodetector output. As the next step, the $\lambda/4$ plate is rotated by a small amount in one direction. The $\lambda/2$ plate is again rotated. The new minimum output from the photodetector is compared to the previous minimum. If the new minimum is smaller, the $\lambda/4$ plate was turned in the correct direction. The procedure of rotating the $\lambda/4$ plate by a small amount followed by a rotation of the $\lambda/2$ plate is repeated until the absolute minimum has been found. Only when the input to the $\lambda/2$ plate is linearly polarized is its output linearly polarized, and the photodetector output shows sharp nulls where the linearly polarized output from the $\lambda/2$ plate is perpendicularly polarized to the transmission axis of the analyzer.

Once a linearly polarized wave and a sharp null are obtained, the direction of polarization can be changed to the desired direction by rotating the $\lambda/2$ plate.

The fiber equivalent of Fig. 6.19 is shown in Fig. 6.15. The fiber-loop $\lambda/4$ plate can be treated as a conventional $\lambda/4$ plate whose surface is perpendicular to the plane of the fiber loop. The alignment procedure for the conventional waveplates in Fig. 6.19 applies equally to the fiber-loop waveplates in Fig. 6.15. The direction of the fast axis of the $\lambda/2$ and $\lambda/4$ loops lies in the plane of the fiber loop as explained in Section 6.3.3.

6.4.3.3 Measuring the Retardance of a Sample

Figure 6.4 summarizes the sequential change in the state of polarization as the retardance Δ is increased from $0°$ to $360°$ for a linearly polarized initial state with $B/A = 1$. By the same token, if linearly polarized light with $B/A = 1$ is incident on a birefringent sample of known orientation but unknown retardance, the retardance can be determined from the state of polarization of the emergent light.

Figure 6.21 shows an arrangement of Senarmont's method for measuring the retardance of a sample. The incident light is linearly polarized at $\theta = 45°$ with respect to the $x-y$ axes and the amplitudes of the x and y components are the same, namely, $B/A = 1$. Either the fast or slow axis of the crystal is aligned to the x axis (a method for locating the fast or slow axis can be found in Problem 6.10). The emergent light is an elliptically polarized wave with a $45°$ (or $135°$) azimuth angle. The ellipticity $\tan \beta$, which is the ratio of the major axis to the minor axis, depends on the value of the

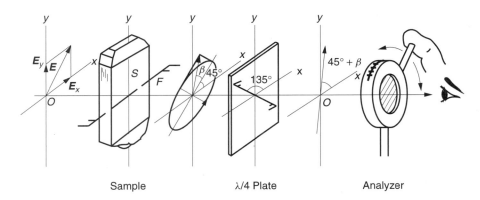

Figure 6.21   Measurement of the retardance Δ of a crystal by Senarmont's method.

retardance and, as obtained in Problem 6.11,

$$\Delta = \tfrac{1}{2}\beta \tag{6.31}$$

The elliptically polarized wave further enters a $\lambda/4$ plate whose fast axis orientation is set at 45° or 135°. The emergent light from the $\lambda/4$ plate is linearly polarized because the fast axis is aligned to the major or minor axis of the ellipse as mentioned in Section 6.4.3.2. The azimuth angle of the light emergent from the $\lambda/4$ plate is $\theta = 45° + \beta$. The value of β can be found from the direction of the sharp null when rotating the analyzer. Finally, Δ is found from β by Eq. (6.31).

It is important to realize that the direction of the major or minor axis of the elliptically polarized light incident to the $\lambda/4$ plate is always at $\theta = 45°$ or 135° if $B/A = 1$, and the fast axis of the sample is along the x axis, regardless of Δ. Once the fast axis of the $\lambda/4$ plate is set to $\theta = 45°$ or 135°, the $\lambda/4$ plate need not be adjusted during the measurement. The only adjustment needed is the direction of the analyzer. This is a noble feature of Senarmont's method.

6.4.3.4 *Measurement of Retardance of an Incident Field*

The previously mentioned Senarmont's method used a priori knowledge of the azimuth angle of 45° of the emergent light from the sample, but in this case, the retardance between the x and y directions of an incident wave with an arbitrary state of polarization will be measured.

The measurement consists of three steps using a polarizer, a $\lambda/4$ plate, and a photodetector. Let an arbitrary incident wave be represented by Eqs. (6.1) and (6.2). The arrangement is similar to the one shown in Fig. 6.21, but the sample is removed.

Step 1. First, only the polarizer, which is used as an analyzer, and the photodetector are installed. With the transmission axis of the analyzer along the x axis, the transmitted power is measured. The transmitted power I_x is expressed as

$$I_x = \tfrac{1}{2}|E_x|^2 = \tfrac{1}{2}A^2$$

where an ideal analyzer is assumed (i.e., lossless transmission of the through polarization and complete rejection of the cross polarization). Next, the transmission axis of the analyzer is rotated along the y axis, and the transmitted power is measured. The transmitted power I_y is expressed as

$$I_y = \tfrac{1}{2}|E_y|^2 = \tfrac{1}{2}B^2$$

The total transmitted power I_0 is

$$I_0 = I_x + I_y$$

Next, the transmission axis of the analyzer is rotated at a 45° azimuth angle to the x axis and the transmitted light power is measured. Both E_x and E_y contribute to the component along the analyzer transmission axis:

$$I_1 = \frac{1}{2}\left|\frac{1}{\sqrt{2}}E_x + \frac{1}{\sqrt{2}}E_y\right|^2$$

$$= \tfrac{1}{4}|A + Be^{j\Delta}|^2$$

$$= \tfrac{1}{4}(A + Be^{j\Delta})(A + Be^{-j\Delta})$$

$$= \tfrac{1}{4}(A^2 + B^2 + 2AB\cos\Delta)$$

Step 2. The $\lambda/4$ plate is inserted in front of the analyzer. The fast axis of the $\lambda/4$ plate is parallel to the y axis, and the analyzer is kept with its transmission axis at $45°$ to the x axis. The transmitted power is

$$I_2 = \frac{1}{2}\left|\frac{1}{\sqrt{2}}E_x + \frac{1}{\sqrt{2}}E_y e^{-j90°}\right|^2$$

$$= \tfrac{1}{4}(A^2 + B^2 + 2AB\sin\Delta)$$

From these measured values of I_0, I_1, and I_2, the retardance is

$$\Delta = \tan^{-1}\left(\frac{2I_2 - I_0}{2I_1 - I_0}\right)$$

6.5   LINEAR POLARIZERS

A linear polarizer favorably passes the component of light polarized parallel to the transmission axis and suppresses the component polarized parallel to the extinction axis. The extinction axis is perpendicular to the transmission axis.

The three major types of linear polarizers are the (1) dichroic polarizer, (2) birefringence polarizer, and (3) polarizer based on Brewster's angle and scattering. Each type has merits and demerits and a choice has to be made considering such parameters as transmission loss, power of extinction, wavelength bandwidth, bulkiness, weight, durability, and cost.

6.5.1   Dichroic Polarizer

Figure 6.22 shows an oversimplified view of the molecular structure of a dichroic sheet polarizer. It is analogous to a lacy curtain suspending an array of long slender conducting molecules.

The dichroic sheet is quite thin and is normally laminated on a transparent substrate for strength. Transmission through the dichroic sheet depends on the direction of polarization of the incident wave [7,8].

When the axis of a conducting molecule is parallel to the **E** field, the situation is similar to a linear dipole antenna receiving a radio signal. A current is induced in the axial direction and can flow freely along the molecule except at both ends. At the ends, the axial current has to be zero and the direction of the current has to be

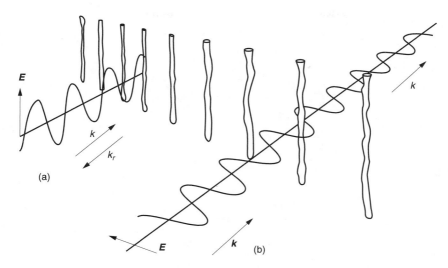

Figure 6.22 Dichroic linear polarizer sheet. (a) Extinction. (b) Transmission.

The "dichroic" polarizer has to do with "two colors." Historically [1], certain crystals displaying polarizing properties were observed to change color when they were held up to sunlight and were viewed with a polarizer. The color changes occur when the polarizer is rotated.

Although somewhat of a misnomer, the term dichroic has persisted. At present, any sheet whose absorption depends on the direction of polarization of the incident light is called a dichroic sheet. Another example is the dichroic filter. It reflects at a specified wavelength and transmits at another specified wavelength, while maintaining a nearly zero coefficient of absorption for all wavelengths of interest.

reversed, resulting in a current standing wave I_z with a sinusoidal distribution along the molecular axis as shown in Fig. 6.23a.

When, however, the direction of the **E** field is perpendicular to the axis of the molecule, the current is induced in a diametrical direction. The current has to be zero at the left and right edges of the molecule. The magnitude of the excited current cannot be large because the zero current boundary conditions are located so close to each other. The distribution of the current I_x has a quasitriangular shape with a short height. The magnitude of the induced current for a perpendicular orientation of the **E** field is small compared to that for a parallel orientation of the **E** field, as shown in Fig. 6.23b.

Regardless of the conductivity of the molecule, the transmitted light is attenuated as long as the direction of the **E** field is parallel to the molecular axis. For resistive-type slender molecules, the induced current is converted into heat and there is no reflected wave. For slender molecules that are conductors, the induced current sets up a secondary cylindrical wave whose amplitude is identical to that of the incident wave but whose phase is shifted by 180° in order that the resultant field on the surface of the molecule vanishes, thereby satisfying the boundary condition of a perfect conductor. In the region beyond the molecule, both the transmitted and the 180° out-of-phase secondary wave propagate in the same direction but in opposite phase, and they cancel each other. When the incident wave is from the left to the right, there is no emergent wave in the region to the right of the molecule, as illustrated in Fig. 6.24. In the region

Fleming's shaking rope.

Sir John A. Fleming (1849–1946) used to explain electromagnetic wave phenomena by making analogies with waggling a rope, and it is tempting to apply a rope analogy to the case of the slender molecule polarizer. Imagine that a rope is stretched horizontally through a set of vertical bars. If on one side of the bars, the rope is shaken up and down to produce a wave propagating with its crests in the vertical direction, the wave will pass through the bars unhindered. On the other hand, if the rope is shaken left and right so that its crests are in the horizontal direction, the propagating wave is blocked by the bars. This analogy is opposite to reality in the case of light transmission through the slender molecule polarizer. The transmission axis is perpendicular to the bars (axis of slender molecules), and the extinction axis is parallel to the bars. The shaking rope analogy is shaky in this case.

in front of the molecule, these two waves are propagating in opposite directions, and there exists a standing wave in the region to the left of the molecule in Fig. 6.24.

As mentioned earlier, when the **E** field is perpendicular to the molecule axis, the degree of excitation of the induced current I_x on the molecule is small and the wave can propagate through the molecule curtain with minimum attenuation.

The quality of a polarizer is characterized by two parameters: the major principal transmittance k_1 and the minor principal transmittance k_2. k_1 is the ratio of the intensity of the transmitted light to that of the incident light when the polarizer is oriented to maximize the transmission of linearly polarized incident light. k_2 is the same ratio but when the polarizer is oriented to minimize transmission. The values of k_1 and k_2 are defined when the incident light direction is perpendicular to the surface of the polarizer. The performance of the polarizer is optimum at this angle of incidence.

The value of k_2 can be reduced by increasing the density of the slender molecules, but always with a sacrifice of a reduction in k_1. Figure 6.25 shows the characteristic curves of k_1 and k_2 for a typical dichroic sheet polarizer. Even though *the transmission ratio* defined as $R_t = k_1/k_2$ can be as large as 10^5, it is hard to obtain the ideal value of $k_1 = 1$ with a dichroic polarizer sheet. On the other hand, the birefringent-type polarizer can provide both a large transmission ratio and a value of k_1 very close to unity.

The advantages of the dichroic sheet are that it is thin, lightweight, and low-cost, but the disadvantages are low k_1 values (70% is common) and relatively low

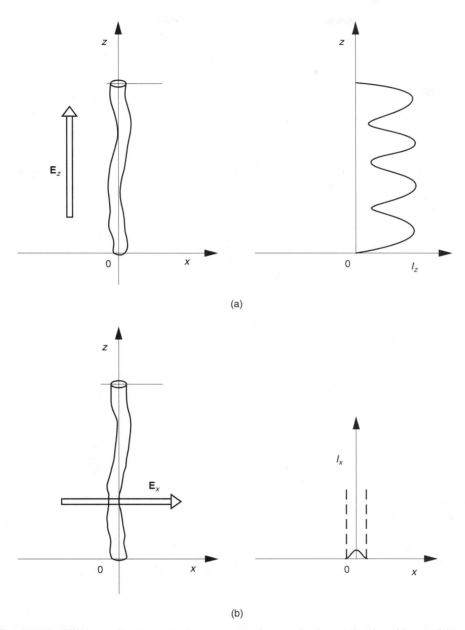

Figure 6.23 Difference in the excited current on the conducting molecule with parallel and perpendicular **E**. (a) **E** is parallel to the axis. (b) **E** is perpendicular to the axis.

power handling capability due to absorption. The transmission ratio deteriorates in the ultraviolet region, $\lambda < 300$ nm.

Example 6.4   Find the state of polarization of the emergent wave E_i for the following combinations of incident field E_0 and polarizer.

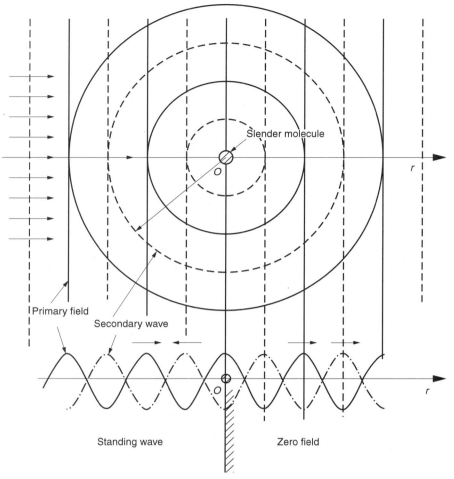

Slender molecule

Primary field

Secondary wave

Standing wave

Zero field

Figure 6.24 Top view of the fields incident onto and scattered from a slender perfectly conducting molecule.

(a) The incident light is linearly polarized, and a poor-quality polarizer is used with major and minor transmittances $\sqrt{k_1} = 1$ and $\sqrt{k_2} = 0.5$, respectively.

(b) This situation is the same as case (a) but with an elliptically polarized incident wave.

(c) The incident wave is elliptically polarized in the same way as case (b) but the polarizer has ideal characteristics, namely, $\sqrt{k_1} = 1$ and $\sqrt{k_2} = 0$. Draw the locus of the major axis of the emergent light as the polarizer is rotated in its plane.

Solution

The solutions are shown in Fig. 6.27.

(a) Figure 6.27a shows the configuration. The transmission axis of the polarizer is shown by an extended T. The incident field E_0 is decomposed into components E_{01}

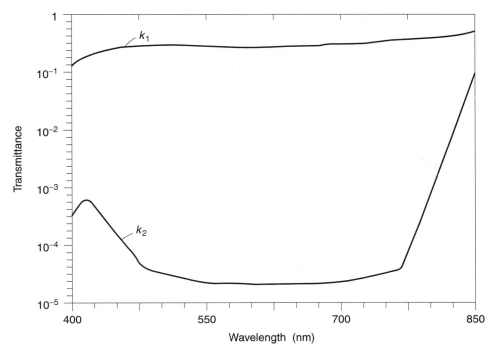

Figure 6.25 Characteristics of type HN38 Polaroid polarizer. (From Polaroid Corporation Catalog).

The concept of an absorption indicatrix is shown in Fig. 6.26. It is used to analyze the transmission of a polarized wave through a bulk medium that possesses an absorboanisotropy like a dichroic crystal. The method for using this indicatrix is similar to that used for the refraction indicatrix presented in Section 4.5.2. Referring to Fig. 6.26, consider the case when light propagating along the direction *ON* is incident onto an absorboanisotropic crystal. The intercept of the plane containing the origin and perpendicular to *ON* with the ellipsoid generates the "cross-sectional ellipse." The lengths of the vectors $\mathbf{a}_1$ and $\mathbf{a}_2$ of the major and minor axes represent absorbancies in these two directions of polarization.

If the direction of polarization of the incident light is arbitrary, the $\mathbf{E}$ field of the incident light is decomposed into components parallel to $\mathbf{a}_1$ and $\mathbf{a}_2$, which suffer absorbancies a_1 and a_2, where a_1 and a_2 are the major and minor axes of the ellipse. The amplitude of the emergent light is the vectorial sum of these two components [9].

The shape of the ellipsoid of the absorption indicatrix is significantly more slender than that of the refraction indicatrix in Section 4.5.2.

and E_{02}, which are parallel and perpendicular to the transmission axis of the polarizer. Their phasor circles are C_1 and C_2. The incident field being linear, the phasors are in phase and points 1, 2, 3, and 4 are numbered accordingly. The E_{02} component suffers an attenuation of $\sqrt{k_2} = 0.5$, which is represented in Fig. 6.27a by shrinking circle C_2. On C_2, the points 1, 2, 3, and 4 shrink to $1'$, $2'$, $3'$, and $4'$. Successive intersections of

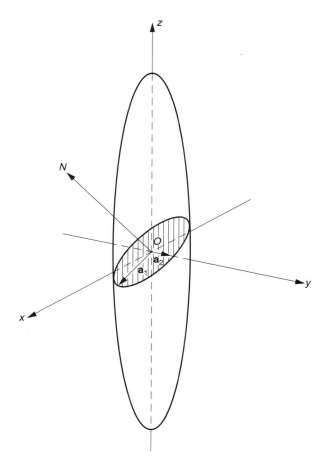

Figure 6.26 Absorption indicatrix used in finding the transmission through an absorboanisotropic crystal.

points 1, 2, 3, and 4 of circle C_1 and points $1'$, $2'$, $3'$, and $4'$ of the shrunken circle C_2 produce the state of polarization of the emergent wave.

The emergent wave is linearly polarized but the azimuth angle is not the same as that of the incident wave.

(b) Circles C_1 and C_2 are set up in a similar fashion to that of part (a), the only difference being that points 1, 2, 3, and 4 of the linear incident light are replaced by points 1, 2, 3, and 4 of the incident ellipse, as shown in Fig. 6.27b. The emergent light polarization is formed from points 1, 2, 3, and 4 of C_1 and points $1'$, $2'$, $3'$, and $4'$ of the shrunken C_2.

The azimuth angle of the emergent wave is closer to the azimuth of the polarizer than the incident wave.

(c) Figure 6.27c explains the case with an ideal polarizer. Since $k_2 = 0$, the radius of the circle C_2 shrinks to zero, and the amplitude of the emergent wave is determined solely by the radius of circle C_1. The emergent light is linearly polarized and the direction of the emergent E field is always along the direction of the transmission axis. Referring to Fig. 6.27c, the solid line ellipse represents the incident ellipse. When the

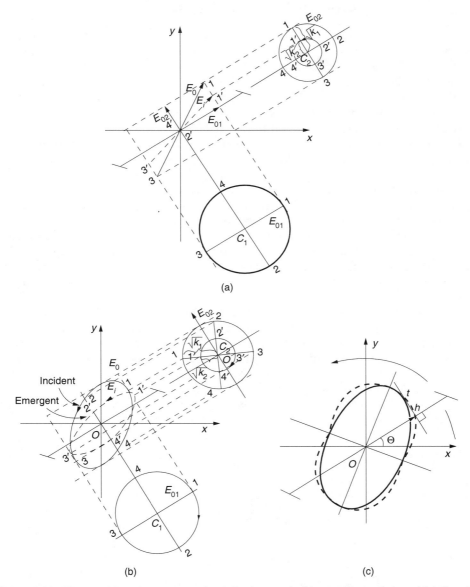

Figure 6.27 Waves with various states of polarization are incident onto a polarizer. (a) A linearly polarized wave is incident onto a polarizer with $k_1 = 1.0$ and $k_2 = 0.5$. (b) An elliptically polarized wave is incident onto a polarizer with $k_1 = 1.0$ and $k_2 = 0.5$. (c) Locus of the amplitude of the emergent wave when an ideal polarizer is rotated.

azimuth angle of the polarizer is at Θ, the direction of the emergent linear polarization is also at $\theta = \Theta$, and the amplitude is represented by $\overline{Oh}$, which is perpendicular to the tangent to the incident ellipse. The dashed line in the figure shows the locus of the field vector as the polarizer is rotated. The position of h in Fig. 6.27c corresponds to $1'$ when $k_2 = 0$ in Fig. 6.27b. It should be noted that the answer is *not* an ellipse.

□

6.5.2 Birefringence Polarizer or Polarizing Prism

As mentioned in Chapter 4, there are only two possible directions of polarization (e- and o-waves) inside a uniaxial crystal. No other directions of polarization are allowed. A birefringence polarizer, which is sometimes called a prism polarizer, creates a linearly polarized wave by eliminating one of the two waves. One type makes use of the difference in the critical angles of total internal reflection for e- and o-waves. Another type makes use of the difference in the angle of refraction for the two waves. A significant advantage of the birefringence polarizer over a simple dichroic polymer polarizing sheet is its high transmission coefficient of 90% to 95% or better compared to 70% for the polarizing sheet. Moreover, the polarizing beamsplitter gives access to beams of each polarization.

Figure 6.28 shows a cut-away view of the Nicol prism. A calcite crystal is sliced diagonally and is cemented back together with Canada balsam cement whose index of refraction is $n = 1.55$. Since the indices of refraction of calcite are $n_e = 1.486$ and $n_0 = 1.658$ at $\lambda = 0.58$ μm, the e-wave does not encounter total internal reflection and exits through the crystal to the right. The crystal is sliced at such an angle that total internal reflection of the o-wave takes place at the interface between the calcite crystal and the Canada balsam cement. The reflected o-wave is absorbed by the surrounding dark coating.

Even though the Nicol prism is one of the best-known polarizers because of its long history, the Nicol prism has the following disadvantages: the Canada balsam cement absorbs in the ultraviolet region of the spectrum, the power handling capability is limited by the deterioration of the cement, and the emergent beam is laterally displaced from the position of the incident beam. A favorable characteristic of the Nicol prism is its reasonable field of view of 28°.

The Glan–Foucault or Glan–Air polarizing prism is shown in Fig. 6.29a. This type eliminates the use of Canada balsam cement so as to avoid absorption in the

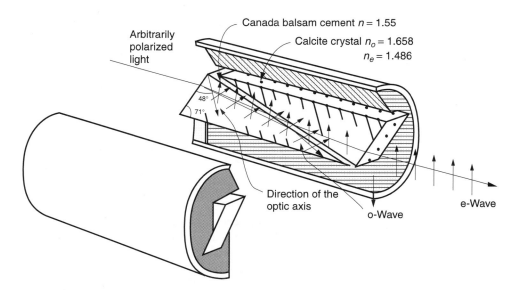

Figure 6.28 Cutaway view of the Nicol prism.

ultraviolet region and the limitation on the power handling capability of the prism. As shown in Fig. 6.29a, the front surface of the prism is cut parallel to the optic axis and perpendicular to the incident beam. The angle of the slanted airgap is chosen such that total internal reflection takes place for the o-wave. A polarizing prism normally discards the o-wave by the use of an absorptive coating. However, if desired, the side surface of the polarizer can be polished to allow the o-wave to exit. The o-wave from this polarizing beamsplitter can be used for monitoring purposes or for providing an additional source with an orthogonal direction of polarization to

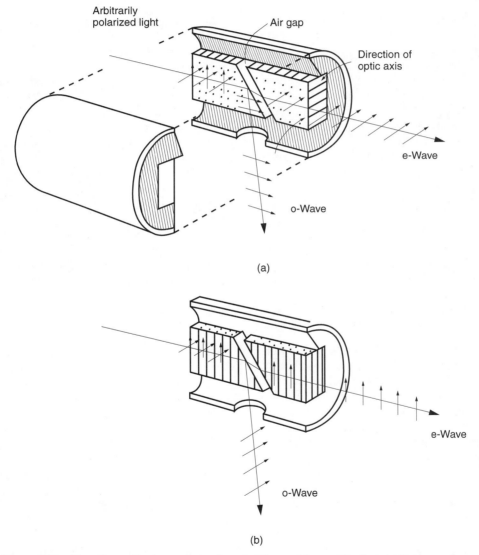

(a)

(b)

Figure 6.29 Glan–Foucault prism polarizer/beamsplitter and its modification. (a) Cutaway view of the Glan–Foucault prism polarizer/beamsplitter. (b) Same as (a) but modified by Taylor for better transmission.

the e-wave. An additional merit of this prism is its short longitudinal dimension. The Glan–Air prism, however, suffers from the demerits of a narrow acceptance angle of 15° to 17° and multiple images caused by multiple reflections in the airgap.

The Glan–Faucault prism was modified by Taylor who rotated the crystal axis by 90°, as illustrated in Fig. 6.29b. With this orientation Brewster's angle can be used to minimize the reflection. The value of k_1 was significantly increased.

The Glan–Thomson polarizing prism has the same geometry as the Glan–Foucault prism but uses Canada balsam cement in place of the airgap in order to increase the viewing angle to 25° to 28° at the cost of the aforementioned drawbacks of Canada balsam cement.

Several other types of polarizing prisms are similar and all are shown for comparison in Fig. 6.30. The direction of refraction is determined by assuming that a negative birefringent crystal ($n_e < n_o$) like calcite is used. The angular separation of the o- and e-waves is made by different arrangements of the optic axes of two pieces of the same crystal material. The Rochon, Senarmont, and Ahrens polarizing prisms do not deviate the direction of one of the transmitted lightwaves from the direction of the incident light. With reference to Fig. 6.30, the deviated transmitted light from the Rochon prism is vertically polarized while that of the Senarmont prism is horizontally polarized. The Wollaston polarizing prism maximizes the angular separation between the two beams because what is labeled the output o-wave is, in fact, the e-wave in the first prism and both waves are refracted at the interface. The geometry of the Cotton-type prism is almost the same as the bottom piece of the Ahrens type, except for the larger apex angle of the prism for optimization of operation.

Prisms based on refraction create aberrations when they are introduced in a convergent beam. This is because the vertical geometry is not the same as the horizontal geometry, and the angle of refraction from the boundaries in the vertical direction is different from that of the horizontal direction just like a cylindrical lens.

6.5.3   Birefringence Fiber Polarizer

Next, the fiber-loop birefringence polarizer will be explained. When an optical fiber is bent too tightly, the light in the core starts to leak out. The amount of leakage, however, depends on the direction of polarization of the light because the change in the refractive index caused by the bending is not isotropic. The fiber-loop-type polarizer makes use of this property. Needless to say, the fiber-loop polarizer is especially advantageous for use in fiber-optic communications because polarization control is achieved without having to exit the fiber and transmission of light in the fiber is uninterrupted.

When an optical fiber is bent, the fiber is compressed in the radial direction of the bend and is expanded in the direction perpendicular to it, as shown at the top of Fig. 6.13. The refractive index of glass is lowered where it is compressed and raised where it is expanded. The differential stress creates anisotropy in the refractive indices in the two aforementioned directions in the fiber.

Both single-mode and polarization-preserving fibers can be used for fabricating a polarizer, but better results are obtained with polarization-preserving fibers which already have birefringence even before bending the fibers. With ordinary single-mode

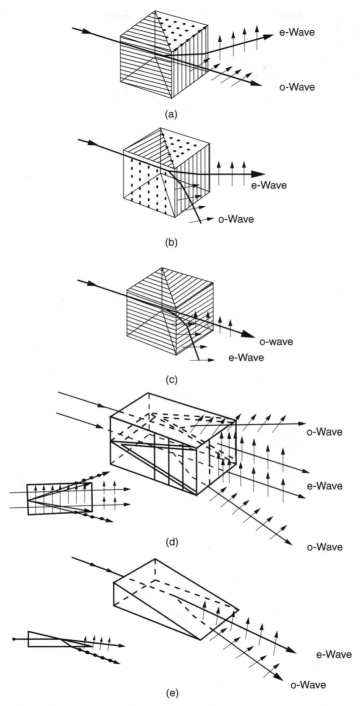

e-Wave

o-Wave

(a)

e-Wave

o-Wave

(b)

o-wave

e-Wave

(c)

o-Wave

e-Wave

o-Wave

(d)

e-Wave

o-Wave

(e)

Figure 6.30 Various types of birefringence polarizers (polarizing prisms) using calcite. (a) Rochon. (b) Wollaston. (c) Senarmont. (d) Ahrens. (e) Cotton.

fibers, the fiber has to be bent over a much tighter radius to achieve the desired effect and, consequently, is prone to breakage. Birefringence in the Panda-type polarization-preserving fiber, shown in the inset in Fig. 6.31, is produced by contraction of the glass with a higher thermal expansion coefficient in the "eyes" region when the drawn fiber solidifies. The index of refraction n_x seen by the wave polarized in the direction of the "eyes" is raised due to the expansion of the core and that of n_y seen by the wave polarized in the direction of the "nose" is lowered due to the contraction of the core. The slow axis is in the direction of the "eyes" and the fast axis is in the direction of the "nose."

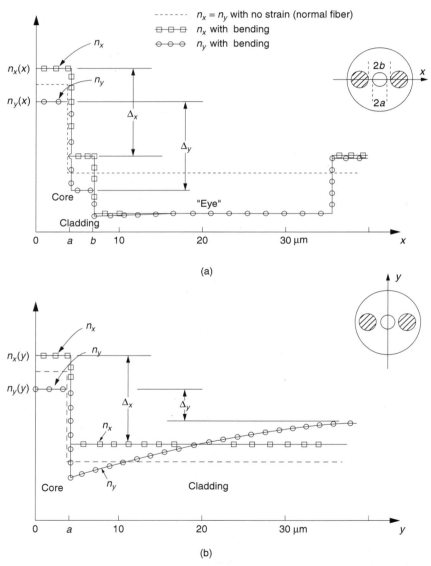

(a)

(b)

Figure 6.31   Refractive index profiles of a Panda fiber. (a) Profile along the x axis. (b) Profile along the y axis. (After K. Okamoto, T. Hosaka, and J. Noda [10].)

By comparing the inset in Fig. 6.31 with Fig. 6.13, one soon notices that the same effect caused by the contraction of the "nose" can be generated by just bending the fiber in the y direction. As a result, the birefringence of the Panda fiber is even more enhanced when the Panda fiber is bent in the y direction.

Figure 6.31 shows the calculated profile of the indices of refraction when the Panda fiber is bent [10]. The distribution along the "eyes" direction of the Panda fiber is shown in Fig. 6.31a while that along the "nose" direction is shown in Fig. 6.31b.

As long as one stays on the line connecting the centers of the "eyes" (x axis) the strains inside the core and the cladding are identical and the difference between n_x and n_y, which is directly related to the strain, is the same in the core and cladding, as shown in Fig. 6.31a.

Along the "nose" direction (y axis), however, the amount of strain varies significantly with the distance from the center of the fiber, and the difference between n_x and n_y also varies with the radius in the y direction. As shown in Fig. 6.31b, even though n_y is smaller than n_x inside and on the periphery of the core, n_y grows bigger than n_x in the region beyond 20 μm. The difference Δ_y between n_y in the core and n_y in the cladding also decreases with y, whereas the difference Δ_x between n_x in the core and n_x in the cladding stays the same with y. The evanescent wave exists in these regions and a slight decrease in Δ_y significantly increases the bending loss of the E_y component [10]. Thus, the emergent light is predominantly E_x polarized in the direction of the "eye." It is this anisotropic strain distribution that makes the fiber polarizer work.

The tensile stress due to the Panda "eyes" can be enhanced further by increasing the bending of the fiber in the y direction. Excessive bending, however, starts incurring the transmission loss of the x-polarized wave. The amount of bending has to be determined from a compromise between the transmittance k_1 and the *extinction ratio* (defined as the inverse of the transmission ratio) R. A transmission loss of 0.5 dB with an extinction ratio $R = -30$ dB is obtainable for a wide wavelength range by 10 turns of a 3-cm-diameter Panda fiber loop [11].

It is difficult to know the orientation of the Panda eyes unless its cross section is examined under a microscope. The use of a special Panda fiber whose cross section is oval shaped to indicate the orientation of the Panda "eyes" makes it easier to bend the fiber into a coil while maintaining the right bending orientation [10].

6.5.4 Polarizers Based on Brewster's Angle and Scattering

Brewster's angle is another phenomenon that depends on the direction of polarization of light and can be utilized to design a polarizer. Brewster's angle of total transmission exists only for a lightwave whose direction of polarization is in the plane of incidence.

Figure 6.32 shows a pile-of-plates polarizer that is based on Brewster's angle. Brewster's condition is

$$\tan \theta_B = \frac{n_1}{n_0}$$

The wave polarized in the plane of incidence transmits through totally without reflection. The wave polarized perpendicular to the plane of incidence also transmits through, with some loss due to reflection. To be effective as a polarizer, several plates are necessary in order to increase the loss due to reflection of the wave polarized perpendicular to the plane of incidence.

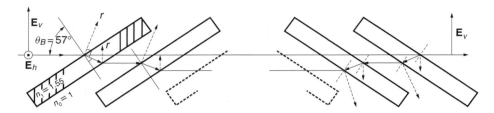

Figure 6.32 Pile-of-plates polarizer.

Light with high purity of linear polarization is obtainable from an external cavity-type gas laser such as shown in Fig. 14.1. This type of laser uses a Brewster window, and in the case of the He−Ne laser, the light goes back and forth more than 2000 times before exiting the cavity. This is equivalent to a pile of 2000 plate polarizers, and light with very pure linear polarization is obtained.

6.5.5 Polarization Based on Scattering

A rather unconventional polarizer makes use of the nature of Rayleigh scattering. Scattering from a particle smaller than the wavelength of light creates polarized light. Referring to Fig. 6.33, the light scattered in the direction normal to the incident ray is linearly polarized. The vertically polarized component of the incident light cannot

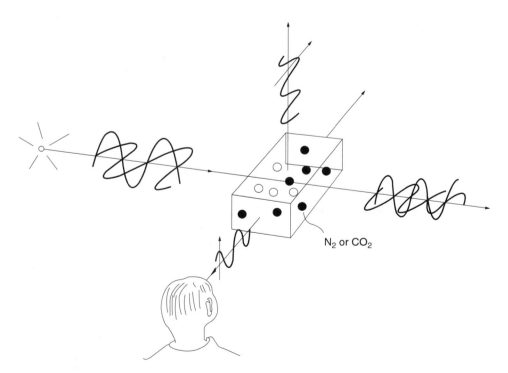

Figure 6.33 Polarizer based on Rayleigh scattering.

be scattered in the vertical direction because the **E** field would become parallel to the direction of propagation. The light scattered toward the vertical direction is highly horizontally polarized light, and the light scattered toward the horizontal direction is highly vertically polarized light.

A chamber filled with either N_2 or CO_2 molecules makes a polarizer. Even though the amount of the scattered light is small, the purity of the polarization is good. The direction of polarization is perpendicular to the plane containing the path of the light from the source to the observer by way of the scatterer, as indicated in Fig. 6.33.

6.6 CIRCULARLY POLARIZING SHEETS

A polarizer sheet laminated with a $\lambda/4$ plate sheet is sometimes marketed as a circularly polarizing sheet. Its usages are presented here.

6.6.1 Antiglare Sheet

In this section, a method of preventing glare using a circularly polarizing sheet will be described. Figure 6.34 shows a circularly polarizing sheet being used as an antiglare cover for a radar screen. Figure 6.35 explains the function of the sheet, and for purposes of the explanation, the polarizer and the $\lambda/4$ plate sheet are separated. Figure 6.35a shows the state of polarization of the light incident on to the radar surface, and Fig. 6.35b shows the state of polarization of the reflected wave from the radar screen. In Fig. 6.35a, the direction of the polarization is 45° to the left of the fast axis, and left-handed circularly polarized light is incident onto the radar screen.

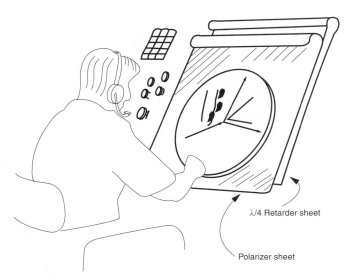

λ/4 Retarder sheet

Polarizer sheet

Figure 6.34 A circularly polarizing sheet, which is a lamination of polarizer and $\lambda/4$ plate sheets, is used for prevention of glare on a radar screen.

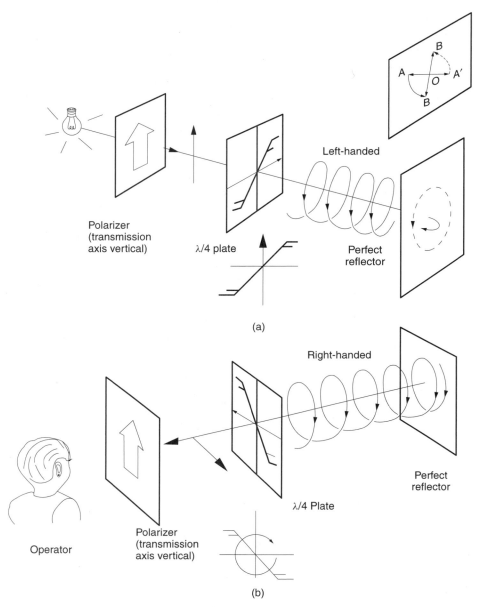

Figure 6.35 Step-by-step explanation of the antiglare circularly polarizing sheet. (a) Incident wave. (b) Reflected wave.

The very right top inset in the Figure shows what happens on reflection. If the surface is assumed to be a perfect reflector, at the moment when the field vector of the incident light points in the direction *OA*, the field vector of the induced field should point in the opposite direction *OA'* to satisfy the boundary condition that the resultant tangential *E* field is zero on the surface of a perfect conductor. At the next moment, when the incident vector moves to *OB*, the induced vector moves to *OB'*. Although the incident vector and the induced vector always point in opposite directions, they always rotate in the same direction.

The reflected wave is the expansion of the induced wave. Figure 6.35b shows how the reflected wave propagates toward the operator. Recall that the observer looks toward the source of light, and the reflected wave is right-handed circularly polarized. Likewise, the azimuth angle of the fast axis of the $\lambda/4$ plate now looks to the observer like $\Theta = 135°$.

The light transmitted through the $\lambda/4$ plate is found by the circle diagram to be horizontally polarized. The light cannot go through the polarizer, and the light reflected from the radar surface does not reach the radar operator. The blips originating from the radar screen, which are randomly polarized, reach the operator's eye with some attenuation.

6.6.2 Monitoring the Reflected Light with Minimum Loss

A reflectometer gathers information from reflected light. One of the simplest ways to sample the reflected light is to use a nonpolarizing beamsplitter (NPBS) in the manner shown in Fig. 6.36a. With this configuration, however, the reflected as well as the

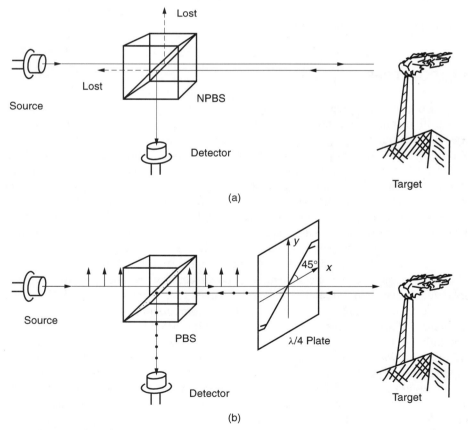

Figure 6.36 Comparison between two types of reflectometers. (a) Using a nonpolarizing beamsplitter. (b) Using a polarizing beamsplitter and a $\lambda/4$ plate.

incident beam will be split by the splitter, resulting in light being lost to the system. If a beamsplitter with reflectance R is used, the intensity of the light collected by the system is $I_{in}TR\sigma$, where I_{in}, T, and σ are incident light intensity, the transmittance of the beamsplitter, and the reflectivity of the target, respectively. Because of the constraint $T + R = 1$, the optimum intensity of the collected light occurs when $R = 0.5$, and the collected light intensity is at best $0.25\, I_{in}\sigma$. Only one-quarter of the incident light intensity is useful.

The reflected light is often weak, as, for instance, in a system for remotely analyzing the gas contents from a smokestack. The system shown in Fig. 6.36b can be used to maximize the sensitivity. This reflectometer uses the combination of a polarizing beamsplitter (PBS) and a quarter-waveplate.

The arrangement is quite similar to that for preventing the glare explained in Fig. 6.35, where the vertically polarized incident light is converted into a horizontally polarized reflected light after passing through the $\lambda/4$ plate twice. In the reflectometer of Fig. 6.36b, the vertically polarized light transmits through the PBS and is converted into a left-handed circularly polarized wave by the $\lambda/4$ plate whose azimuth is $45°$. The light reflected from the target is a right-handed circularly polarized wave, which in turn is converted into a horizontally polarized light by the same $\lambda/4$ plate. The horizontally polarized wave is reflected by the PBS to the detector.

The power loss due to the transmission loss of the optical components is $10^{-3}-10^{-5}$, depending on the quality of the components.

6.7 ROTATORS

When a linearly polarized light propagates in quartz along its optic axis, the direction of polarization rotates as it propagates. Similar phenomena can be observed inside other crystals like cinnabar (HgS) and sodium chlorate ($NaClO_3$), as well as solutions like sucrose ($C_{12}H_{22}O_{11}$), turpentine ($C_{10}H_{16}$), and cholesteric liquid crystals. Even some biological substances like amino acids display this effect. This phenomenon of rotation of the direction of polarization is called *optical activity*. A substance that displays optical activity is called an *optically active substance*. Each optically active substance has a particular sense of rotation. Media in which the rotation of polarization is right-handed looking toward the source are called dextrorotary (*dextro* in Latin means right). Media in which the rotation of polarization is left-handed are called levorotary (*levo* in Latin means left). There are both *d*- and *l*-rotary varieties of quartz.

Fresnel explained the mechanism of optical activity by decomposing a linearly polarized wave into circularly polarized waves. As shown in Fig. 6.37, linearly polarized incident light can be considered as a combination of right- and left-handed circularly polarized waves with equal amplitudes. If these two oppositely rotating circularly polarized waves rotate at the same speed, the direction of the polarization of the resultant wave remains unchanged. However, if the rotation speeds are different, the direction of polarization of the resultant wave rotates as the two waves propagate.

In explaining the difference in rotation speeds of the left and right circular component waves, Fresnel attributed this to the rotational asymmetry of the molecular structure of the optically active medium.

A birefringent material is a material that is characterized by two indices of refraction. If, for example, the refractive indices are n_x and n_y, corresponding to x and y linearly

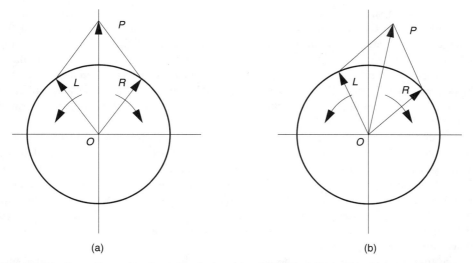

(a)   (b)

Figure 6.37 Fresnel's explanation of optical activity. (a) Incident light. (b) Inside an optically active substance.

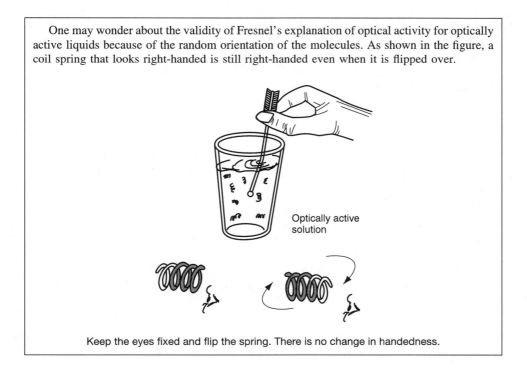

One may wonder about the validity of Fresnel's explanation of optical activity for optically active liquids because of the random orientation of the molecules. As shown in the figure, a coil spring that looks right-handed is still right-handed even when it is flipped over.

Optically active solution

Keep the eyes fixed and flip the spring. There is no change in handedness.

polarized component waves, the material is said to be linearly birefringent. Retarders are examples of linearly birefringent devices. If the refractive indices are n_l and n_r, corresponding to left and right circularly polarized component waves, the material is said to be circularly birefringent. Optically active substances are examples of circular birefringence.

Let us compare the emergent polarization for these two different types of birefringence. In the case of linear birefringence, the shape and/or orientation of the emergent polarization may differ from that of the incident light, as illustrated in the retarder examples shown in Figs. 6.7 and 6.8. On the other hand, for a circularly birefringent medium, the orientation of the emergent polarization changes, but the shape remains the same. For example, linearly polarized light incident on an optically active medium will remain linearly polarized, but the direction will rotate, as illustrated in Fig. 6.37.

The angle of rotation in an optically active medium is proportional to the distance of propagation in the medium. The angle of rotation per unit distance is called the rotary power. The rotary power of quartz, for instance, is $27.71°/mm$ at the D line of the sodium spectrum ($\lambda = 0.5893$ μm) and at $20°C$. In the case of a liquid substance like natural sugar dissolved in water, the angle of rotation is proportional to both the length of transmission and the concentration of the solute. The saccharimeter detailed in Section 6.7.1 determines the concentration of an optically active sugar solution by measuring the angle of rotation.

The rotary power depends on the wavelength of the light as well as the temperature of the substance. If an optically active medium is placed between the orthogonally oriented polarizer and analyzer shown in Fig. 6.38, and white light is used as the incident light, then the optical spectrum is attenuated for wavelengths whose angles of rotation are an integral multiple of π radians and their complimentary colors appear. A beautiful color pattern is observed. This phenomenon is called rotary dispersion.

The Faraday effect causes a substance to behave like an optically active medium when an external magnetic field is applied. This induced optical activity exists only when an external magnetic field is applied. The sense of rotation is solely determined

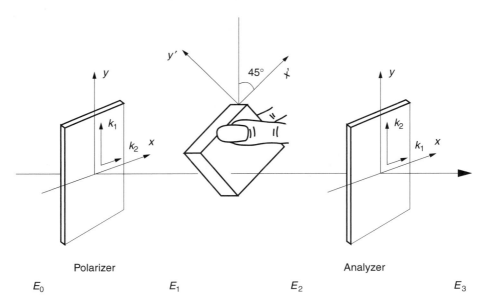

Figure 6.38   In the polariscope, an optical component is inserted between two crossed polarizers for inspection. The transmission axis for the polarizer on the left is in the y direction, and the transmission axis for the polarizer on the right is in the x direction.

by the direction of the magnetic field and does not depend on the direction of light propagation. This is an important distinction between ordinary optical activity (a reciprocal phenomenon) and the Faraday effect (a nonreciprocal phenomenon). In the case of natural optical activity, if the direction of light propagation is reversed in an l-rotary material, the rotation is still l-rotary. In the case of the Faraday effect, if a material is l-rotary when the light propagates in the direction parallel to the magnetic field, the material is d-rotary when the light propagates antiparallel to the magnetic field.

Example 6.5   As shown in Fig. 6.38, the polariscope consists of two polarizer sheets arranged with their transmission axes perpendicular to each other. Find the amplitude E_3 of the emergent light when the following components are inserted between the polarizers. Assume $k_1 = 1$ and $k_2 = 0$.

(a) No component.

(b) A polarizer sheet with transmission axis along x' and an azimuth angle of $45°$.

(c) A $\lambda/4$ plate with an azimuth angle of $45°$ (fast axis along x' axis in Fig. 6.38).

(d) A $\lambda/2$ plate with azimuth angle of $45°$.

(e) A full-waveplate with azimuth angle of $45°$.

(f) A $90°$ rotator.

Solution

This time, the solutions are found without resorting to circle diagrams.

(a) Nothing is inserted. $E_3 = 0$.

(b) A polarizer is inserted at $45°$. As shown in Fig. 6.39a, $\mathbf{E}_1$ is decomposed into $\mathbf{E}_{x'}$ and $\mathbf{E}_{y'}$. $\mathbf{E}_{y'}$ is extinguished. Only the horizontal component of $\mathbf{E}_{x'}$ is transmitted through the analyzer.

$$E_3 = E_1 \times \frac{1}{\sqrt{2}} \times \frac{1}{\sqrt{2}} = \frac{E_1}{2}$$

(c) A $\lambda/4$ plate is inserted at $45°$. There are two ways to solve this problem.

(1) Decompose $\mathbf{E}_1$ into components $\mathbf{E}_{x'}$ parallel to the x' axis and $\mathbf{E}_{y'}$ parallel to the y' axis. The components are

$$E_{x'} = \frac{E_1}{\sqrt{2}} \quad \text{and} \quad E_{y'} = \frac{E_1}{\sqrt{2}} e^{j90°}$$

These two waves are further decomposed into both horizontal and vertical (x and y) components, but one needs to be concerned only with the x component because only the horizontal component passes through the analyzer. The horizontal component of the incident wave to the analyzer is

$$E_x = \frac{1}{\sqrt{2}} \frac{E_1}{\sqrt{2}} - \frac{1}{\sqrt{2}} \frac{E_1}{\sqrt{2}} e^{j90°}$$

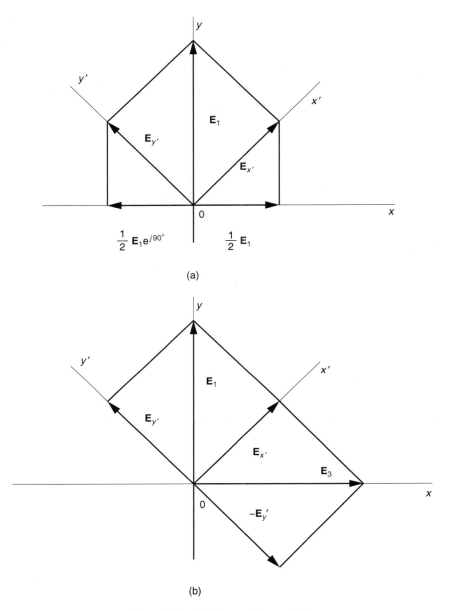

(a)

(b)

Figure 6.39 Solutions of Example 6.5.

where the second term is from $\mathbf{E}_{y'}$ as shown in Fig. 6.39a and the first term is from $\mathbf{E}_{x'}$, and

$$E_3 = \frac{E_1}{2}(1 - e^{j90°}) = \frac{E_1}{\sqrt{2}}e^{-j45°}$$

(2) Note that emergent wave $\mathbf{E}_2$ is a circularly polarized wave with radius $E_1/\sqrt{2}$. The magnitude of the horizontally polarized wave $\mathbf{E}_3$ is $E_1/\sqrt{2}$.

(d) E_1 is decomposed into $E_{x'}$ and $E_{y'}$. The vector $E_{y'}$ is reversed in direction because of the $\lambda/2$ retarder, as shown in Fig. 6.39b. The resultant of $E_{x'}$ and $-E_{y'}$ becomes E_3.

$$E_3 = 2\frac{E_1}{\sqrt{2}}\cos 45°$$
$$= E_1$$

Another way of obtaining the same result is to make use of the fact that a $\lambda/2$ plate rotates the polarization by 2θ. The vertical polarization becomes the horizontal polarization.

(e) The full-waveplate does not disturb the state of polarization and the answer is the same as (a):

$$E_3 = 0$$

(f) If a vertical vector pointing toward the $+y$ direction is rotated by $90°$, the result is a horizontal vector pointing toward the $-x$ direction.

$$E_3 = -E_1 \qquad \qquad \square$$

6.7.1 Saccharimeter

As a sugar solution is an optically active substance, the concentration of sugar can be determined by measuring the angle of rotation of the transmitted light polarization. The Lausent-type saccharimeter such as shown in Fig. 6.40 is widely used to monitor the sugar concentration of grapes in a vineyard. This is a pocketable outdoor type and uses white light. The combination of a wavelength filter F and a polarizing beamsplitter (PBS) converts the incident white light into quasimonochromatic linearly polarized light. In the left half of the field, the light passes through a thin quartz rotator R, while in the right half of the field, the light misses the rotator. Thus, a slight difference in the direction of polarization is created between the light E_L passing through in the left field and E_R in the right field. This slight difference in the direction of polarization is for the purpose of increasing the accuracy of reading the azimuth of the analyzer A through which the incident light is viewed.

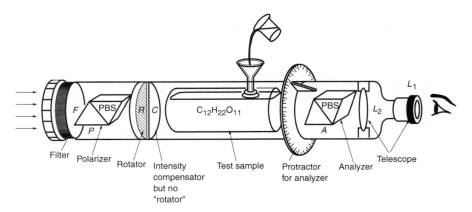

Figure 6.40 Lausent-type saccharimeter.

The first step is the calibration without solution. When the direction k_2 of the extinction axis is adjusted at exactly the midpoint of the angle between the two directions of polarization, the contrast in the intensities between the left and right fields diminishes. The azimuth angle θ_1 of the analyzer of diminishing contrast is noted.

Next, the solution under test is poured into the chamber. The directions of polarization in both left and right fields will rotate by an amount that is proportional to the concentration of the sugar.

The analyzer is again rotated so that the direction k_2 of the extinction axis lies at the midpoint of the rotated directions of polarizations, and the contrast between the left and right field diminishes. This new azimuth angle θ_2 of the analyzer is noted.

The difference $\theta = \theta_2 - \theta_1$ of the azimuth angles of the analyzer is the angle of rotation caused by the optical activity of the sugar solution.

The explanation of the operation of the saccharimeter will be repeated referring to Fig. 6.41. As shown in Fig. 6.41a, when the analyzer is not exactly adjusted such that k_2 is at the midpoint of $\mathbf{E}_L$ and $\mathbf{E}_R$, a contrast between the left and right field intensities can be seen (the right side is darker). As soon as k_2 of the analyzer is adjusted to the midpoint, as shown in Fig. 6.41b, the contrast disappears. The azimuth angle θ_1 is noted. In the field, this calibration is performed prior to introducing the sample, as the power of rotation of the quartz rotator R is temperature dependent.

As the second step, the test sample is introduced into the chamber. Both $\mathbf{E}_L$ and $\mathbf{E}_R$ rotate by the same amount due to the optical activity of the sample, as shown in Fig. 6.41c, and a contrast in field intensities appears again (the left side is darker). The analyzer is rotated to find the azimuth θ_2 that diminishes the contrast, as shown in Fig. 6.41d. The rotation is computed as $\theta = \theta_2 - \theta_1$.

The concentration P of sugar in grams per 100 cc of solution is given by the formula

$$\theta = [\theta]_\lambda^t \, l \, \frac{P}{100}$$

where $[\theta]_\lambda^t$ is the specific rotary power of the substance at a temperature $t°C$ and a light wavelength λ μm. For sugar, $[\theta]_{0.5893 \ \mu m}^{20} = 66.5°$ (per length in decimeter × concentration in grams per 100 cc). The quantity l is the length of the chamber in units of 10 cm.

The high-accuracy performance of this type of saccharimeter is attributed to the following:

1. The contrast of two adjacent fields rather than the absolute value of the transmitted light through the analyzer was used. The eyes are quite sensitive to detecting differences in intensities between adjacent fields.

2. The region of minimum rather than maximum light transmission through the analyzer was used. The sensitivity of the eyes to detecting a change in the transmitted light is greater near the minimum of transmission.

Another approach is to eliminate both the quartz rotator and the intensity compensator. The incident light to the sample is not divided. The direction of the polarization of the emergent light is directly measured by a split-field polarizer. The split-field polarizer is made up of two analyzers side by side with a 5° to 10° angle between the extinction axes, as shown in Fig. 6.41e. When the split-field analyzer is rotated so that the $d'-d'$ axis aligns with the direction of polarization of the light, the contrast between the left and right sections disappears. The split-field polarizer again

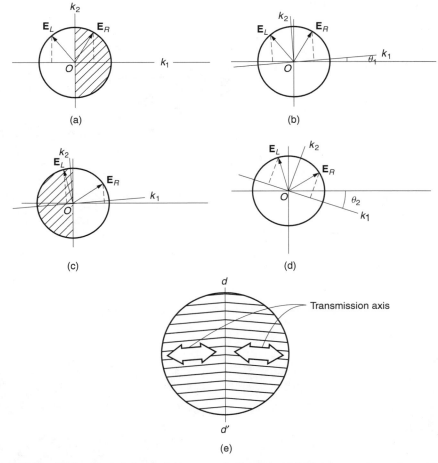

Figure 6.41 Field view of Lausent-type saccharimeter. (a) Without sample and unadjusted. (b) Without sample and adjusted to eliminate contrast. (c) With sample and unadjusted. (d) With sample and adjusted to eliminate contrast. (e) Split-field polarizer.

uses the contrast of the fields near the minimum of transmission and the precision reaches $0.001°$.

6.7.2 Antiglare TV Camera

It is often difficult for a TV reporter to videotape a passenger inside a car due to the light reflected from the surface of the car window.

The geometry of an antiglare camera [12] is shown in Fig. 6.42. When the incident angle to the car window is in the vicinity of Brewster's angle ($56°$ for glass), reflection of the p-polarized light is suppressed, but the s-polarized light is not, and thus the light reflected from the car window is strongly linearly polarized. Removal of this particular component of the light minimizes glare to the TV camera. One way of accomplishing this is by means of a liquid crystal rotator such as the one shown in the display pannel in Fig. 5.33, but without the input polarizer P_1.

Figure 6.42 Operation of an antiglare TV camera.

(a) off (b) on

(c) off (d) on

Figure 6.43 Demonstration of antiglare TV camera. (Courtesy of H. Fujikake et al. [12].)

The total amount of light into the camera is monitored and used as an electrical servosignal. The amount of light for the same scene should be at a minimum when the glare light has successfully been removed. The electrical servosignal rotates the polarization direction of the incident glare light until it becomes blocked by polarizer sheet P_2. The servosignal is minimized when the required amount of polarization rotation is achieved. In order to construct a variable rotator, TN liquid crystal rotators of 45° and 90° are combined. By selecting the appropriate combination of the applied electric field to the two TN liquid crystal rotators, the amount of rotation can discretely be varied from 45° to 135° at intervals of 45°. The photographs in Fig. 6.43 demonstrate the effectiveness of the antiglare TV camera. The photographs on the left were taken with an ordinary camera, while those on the right are the same scenes taken with the antiglare camera. With the antiglare camera, the passengers in the car, and the fish in the pond, are clearly visible.

6.8   THE JONES VECTOR AND THE JONES MATRIX

A method of analysis based on 2×2 matrices was introduced by R. Clark Jones [13,14] of Polaroid Corporation to describe the operation of optical systems. Each component of the system has an associated Jones matrix, and the analysis of the system as a whole is performed by multiplication of the 2×2 component matrices. Moreover, the state of polarization at each stage of the multiplication is easily known.

The state of polarization is described by the Jones vector whose vector components are E_x and E_y. From Eqs. (6.1) and (6.2), the Jones vector is

$$\begin{bmatrix} E_x \\ E_y \end{bmatrix} = e^{j(\beta z - \omega t)} \begin{bmatrix} A \\ Be^{j\Delta} \end{bmatrix} \tag{6.32}$$

The common factor is generally of no importance and is omitted. Eliminating the common factor $e^{j(\beta z - \omega t)}$, Eq. (6.32) is written as

$$\mathbf{E} = \begin{bmatrix} A \\ Be^{j\Delta} \end{bmatrix} \tag{6.33}$$

Representative states of polarizations expressed by the Jones vector are shown in Fig. 6.44.

If only the relative phase between E_x and E_y is important, the common factor $e^{j\Delta/2}$ can be removed, and Eq. (6.33) becomes

$$\mathbf{E} = \begin{bmatrix} Ae^{-j\Delta/2} \\ Be^{j\Delta/2} \end{bmatrix} \tag{6.34}$$

If one interprets an optical component as a converter of the state of polarization from $[E_x \quad E_y]$ into $[E'_x \quad E'_y]$, then the function of the optical component is represented by the 2×2 matrix that transforms $[E_x \quad E_y]$ into $[E'_x \quad E'_y]$.

$$\begin{bmatrix} E'_x \\ E'_y \end{bmatrix} = [2 \times 2] \begin{bmatrix} E_x \\ E_y \end{bmatrix} \tag{6.35}$$

Such a 2×2 matrix is called the Jones matrix.

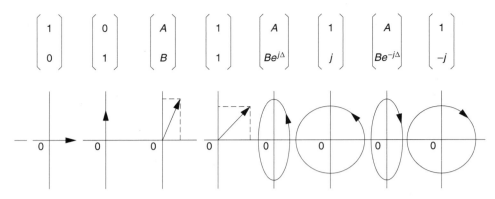

Figure 6.44 Jones vector and the state of polarization.

6.8.1 The Jones Matrix of a Polarizer

The Jones matrix of a polarizer whose major principal transmission axis is along the x axis is

$$P = \begin{bmatrix} k_1 & 0 \\ 0 & k_2 \end{bmatrix} \qquad (6.36)$$

For an ideal polarizer, $k_1 = 1$ and $k_2 = 0$.

Next, the case when the polarizer is rotated in its plane will be considered. Let the direction of the transmission axis be rotated by Θ from the x axis. The incident field **E** is expressed in $x-y$ coordinates.

In this case, the incident field **E** has to be decomposed into E_{x1}, which is along the major principal axis of the polarizer, and E_{y1}, which is along the minor principal axis. Referring Fig. 6.45,

$$E_{x1} = E \cos(\theta - \Theta)$$
$$= E \cos\theta \cos\Theta - E \sin\theta \sin\Theta$$
$$= E_x \cos\Theta - E_y \sin\Theta$$

Similarly,

$$E_{y1} = -E_x \sin\Theta + E_y \cos\Theta$$

E_{x_1} and E_{y_1} can be rewritten in a matrix form as

$$\begin{bmatrix} E_{x1} \\ E_{y1} \end{bmatrix} = \begin{bmatrix} \cos\Theta & \sin\Theta \\ -\sin\Theta & \cos\Theta \end{bmatrix} \begin{bmatrix} E_x \\ E_y \end{bmatrix} \qquad (6.37)$$

The matrix in Eq. (6.37) is a rotation by Θ degrees from the original coordinates; the incident field **E** is expressed in coordinates x_1 and y_1 that match the directions of the principal axes of the polarizer.

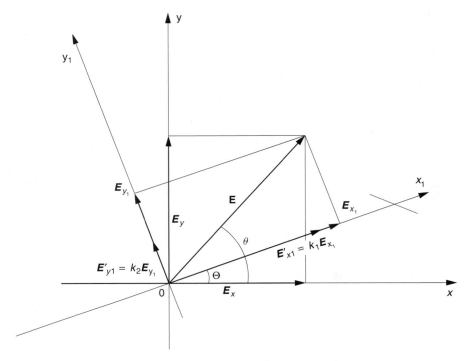

Figure 6.45 Finding the Jones matrix for a polarizer rotated by Θ.

The light emergent from the polarizer is now

$$
\begin{bmatrix} E'_{x1} \\ E'_{y1} \end{bmatrix} = \begin{bmatrix} k_1 & 0 \\ 0 & k_2 \end{bmatrix} \begin{bmatrix} E_{x1} \\ E_{y1} \end{bmatrix}
\tag{6.38}
$$

The emergent wave, however, is in x_1 and y_1 coordinates, and needs to be expressed in the original x and y coordinates.

Referring Fig. 6.45, the sum of the projections of E'_{x1} and E'_{y1} to the x axis provides E'_x. A similar projection to the y axis provides E'_y.

$$
E'_x = E'_{x1} \cos \Theta - E'_{y1} \sin \Theta
$$
$$
E'_y = E'_{x1} \sin \Theta + E'_{y1} \cos \Theta
$$

which again can be rewritten in a matrix form as

$$
\begin{bmatrix} E'_x \\ E'_y \end{bmatrix} = \begin{bmatrix} \cos \Theta & -\sin \Theta \\ \sin \Theta & \cos \Theta \end{bmatrix} \begin{bmatrix} E'_{x1} \\ E'_{y1} \end{bmatrix}
\tag{6.39}
$$

This is a rotation by $-\Theta$ degrees from the x_1 and y_1 coordinates. The emergent wave is expressed in the original x and y coordinates.

Combining Eqs. (6.37) to (6.39), the Jones matrix P_Θ for a polarizer rotated by Θ is given by

$$\begin{bmatrix} E'_x \\ E'_y \end{bmatrix} = P_\Theta \begin{bmatrix} E_x \\ E_y \end{bmatrix} \tag{6.40}$$

$$P_\Theta = \begin{bmatrix} \cos\Theta & -\sin\Theta \\ \sin\Theta & \cos\Theta \end{bmatrix} \begin{bmatrix} k_1 & 0 \\ 0 & k_2 \end{bmatrix} \begin{bmatrix} \cos\Theta & \sin\Theta \\ -\sin\Theta & \cos\Theta \end{bmatrix} \tag{6.41}$$

$$P_\Theta = \begin{bmatrix} k_1 \cos^2\Theta + k_2 \sin^2\Theta & (k_1 - k_2)\sin\Theta\cos\Theta \\ (k_1 - k_2)\sin\Theta\cos\Theta & k_1 \sin^2\Theta + k_2 \cos^2\Theta \end{bmatrix}$$

If the polarizer is ideal and $k_1 = 1$ and $k_2 = 0$, Eq. (6.41) becomes

$$P_\Theta = \begin{bmatrix} \cos^2\Theta & \sin\Theta\cos\Theta \\ \sin\Theta\cos\Theta & \sin^2\Theta \end{bmatrix} \tag{6.42}$$

6.8.2 The Jones Matrix of a Retarder

The Jones matrix of a retarder whose fast axis is oriented along the x axis is

$$R = \begin{bmatrix} 1 & 0 \\ 0 & e^{j\Delta} \end{bmatrix} \tag{6.43}$$

$$R = \begin{bmatrix} e^{-j\Delta/2} & 0 \\ 0 & e^{j\Delta/2} \end{bmatrix} \tag{6.44}$$

The Jones matrix of the half-waveplate is

$$H = \begin{bmatrix} -j & 0 \\ 0 & j \end{bmatrix} \tag{6.45}$$

and that of the quarter-waveplate is

$$Q = \begin{bmatrix} e^{-j\pi/4} & 0 \\ 0 & e^{j\pi/4} \end{bmatrix} \tag{6.46}$$

When a retarder is rotated, the treatment is similar to that of the rotated polarizer. The Jones matrix whose fast axis is rotated by Θ from the x axis is

$$R_\Theta = \begin{bmatrix} \cos\Theta & -\sin\Theta \\ \sin\Theta & \cos\Theta \end{bmatrix} \begin{bmatrix} e^{-j\Delta/2} & 0 \\ 0 & e^{j\Delta/2} \end{bmatrix} \begin{bmatrix} \cos\Theta & \sin\Theta \\ -\sin\Theta & \cos\Theta \end{bmatrix} \tag{6.47}$$

Noting that Eq. (6.47) becomes the same as Eq. (6.40) if k_1 and k_2 are replaced by $e^{-j\Delta/2}$ and $e^{j\Delta/2}$, respectively, the product of the matrix Eq. (6.47) is obtained as

$$R_\Theta = \begin{bmatrix} e^{-j\Delta/2}\cos^2\Theta + e^{j\Delta/2}\sin^2\Theta & -j2\sin\dfrac{\Delta}{2}\sin\Theta\cos\Theta \\ -j2\sin\dfrac{\Delta}{2}\sin\Theta\cos\Theta & e^{-j\Delta/2}\sin^2\Theta + e^{j\Delta/2}\cos^2\Theta \end{bmatrix} \tag{6.48}$$

The Jones matrix of a retarder with retardance Δ rotated by $\pm 45°$ is, from Eq. (6.48),

$$R_{\pm 45°} = \begin{bmatrix} \cos \dfrac{\Delta}{2} & \mp j \sin \dfrac{\Delta}{2} \\ \mp j \sin \dfrac{\Delta}{2} & \cos \dfrac{\Delta}{2} \end{bmatrix} \tag{6.49}$$

When a half-waveplate is rotated by $\Theta = \pm 45°$, the Jones matrix is

$$H_{\pm 45°} = \begin{bmatrix} 0 & 1 \\ 1 & 0 \end{bmatrix} \tag{6.50}$$

where a common factor of $e^{\mp j\pi/2}$ which appears after inserting $\Delta/2 = \pi/2$ and $\Theta = \pm 45°$ into Eq. (6.48), is suppressed.

When a quarter-waveplate is rotated by $\Theta = \pm 45°$, the Jones matrix is

$$Q_{\pm 45°} = \frac{1}{\sqrt{2}} \begin{bmatrix} 1 & \mp j \\ \mp j & 1 \end{bmatrix} \tag{6.51}$$

6.8.3 The Jones Matrix of a Rotator

A rotator changes the azimuth angle without disturbing all other parameters of the state of polarization.

Let the incident linearly polarized field **E** be converted into **E**′ by rotation as shown in Fig. 6.46. Noting that $|\mathbf{E}'| = |\mathbf{E}|$,

$$\begin{aligned} E'_x &= E \cos(\theta_0 + \theta) = E \cos \theta_0 \cos \theta - E \sin \theta_0 \sin \theta \\ E'_y &= E \sin(\theta_0 + \theta) = E \cos \theta_0 \sin \theta + E \sin \theta_0 \cos \theta \end{aligned} \tag{6.52}$$

Since

$$\begin{aligned} E_x &= E \cos \theta_0 \\ E_y &= E \sin \theta_0 \end{aligned}$$

Eq. (6.52) is equivalent to

$$\begin{bmatrix} E'_x \\ E'_y \end{bmatrix} = \begin{bmatrix} \cos \theta & -\sin \theta \\ \sin \theta & \cos \theta \end{bmatrix} \begin{bmatrix} E_x \\ E_y \end{bmatrix} \tag{6.53}$$

which is the same expression as that for rotating the coordinates by $-\theta$.

While the above explanation dealt with the rotation of a linearly polarized incident light, the same holds true for elliptically polarized incident light. For elliptical polarization, each decomposed wave rotates by the same amount and the elliptical shape does not change, but the azimuth of the axes rotates by θ.

Regardless of the orientation of a light wave incident onto a rotator, the amount of rotation is the same.

Example 6.6 Find the answers to Example 6.5 using the Jones matrix.

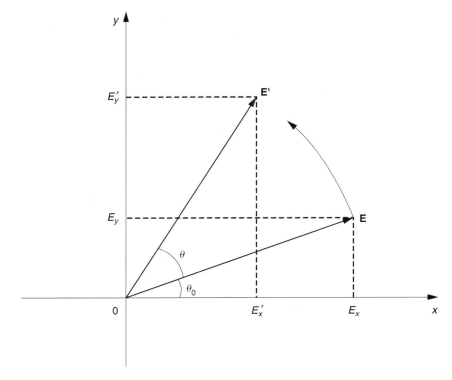

Figure 6.46 Rotation of a field vector by a rotator.

Solution

(a) No plate is inserted. From Eq. (6.42) with $\Theta = 90°$ and then $0°$, the Jones matrix expression is

$$\begin{bmatrix} E'_x \\ E'_y \end{bmatrix} = \begin{bmatrix} 1 & 0 \\ 0 & 0 \end{bmatrix} \begin{bmatrix} 0 & 0 \\ 0 & 1 \end{bmatrix} \begin{bmatrix} E_x \\ E_y \end{bmatrix}$$

$$= \begin{bmatrix} 0 & 0 \\ 0 & 0 \end{bmatrix} \begin{bmatrix} E_x \\ E_y \end{bmatrix}$$

There is no output.

(b) A polarizer is inserted at $\Theta = 45°$. The Jones matrix expression from Eq. (6.42) is

$$\begin{bmatrix} E'_x \\ E'_y \end{bmatrix} = \frac{1}{2} \begin{bmatrix} 1 & 0 \\ 0 & 0 \end{bmatrix} \begin{bmatrix} 1 & 1 \\ 1 & 1 \end{bmatrix} \begin{bmatrix} 0 & 0 \\ 0 & 1 \end{bmatrix} \begin{bmatrix} E_x \\ E_y \end{bmatrix}$$

$$= \frac{1}{2} \begin{bmatrix} 1 & 0 \\ 0 & 0 \end{bmatrix} \begin{bmatrix} E_y \\ E_y \end{bmatrix}$$

In the above expression, the vector just after the inserted polarizer is linearly polarized at 45°. The advantage of the Jones matrix is that the state of polarization can be known at each stage of manipulation. Performing the final matrix multiplication gives

$$\begin{bmatrix} E'_x \\ E'_y \end{bmatrix} = \frac{1}{2} \begin{bmatrix} E_y \\ 0 \end{bmatrix}$$

which is a linearly polarized wave along the x direction.

(c) A quarter-waveplate is inserted at $\Theta = 45°$. From Eqs. (6.42) and (6.51), the Jones matrix expression is

$$\begin{bmatrix} E'_x \\ E'_y \end{bmatrix} = \frac{1}{\sqrt{2}} \begin{bmatrix} 1 & 0 \\ 0 & 0 \end{bmatrix} \begin{bmatrix} 1 & -j \\ -j & 1 \end{bmatrix} \begin{bmatrix} 0 & 0 \\ 0 & 1 \end{bmatrix} \begin{bmatrix} E_x \\ E_y \end{bmatrix}$$

$$= \frac{e^{-j\pi/2}}{\sqrt{2}} \begin{bmatrix} 1 & 0 \\ 0 & 0 \end{bmatrix} \begin{bmatrix} E_y \\ E_y e^{j\pi/2} \end{bmatrix}$$

The intermediate state of polarization after passing through the polarizer and quarter-waveplate is left-handed circular polarization from Fig. 6.44. The emergent wave is

$$\begin{bmatrix} E'_x \\ E'_y \end{bmatrix} = \frac{e^{-j\pi/2}}{\sqrt{2}} \begin{bmatrix} E_y \\ 0 \end{bmatrix}$$

(d) A half-waveplate is inserted at $\Theta = 45°$. From Eqs. (6.42) and (6.50), the Jones matrix expression is

$$\begin{bmatrix} E'_x \\ E'_y \end{bmatrix} = \begin{bmatrix} 1 & 0 \\ 0 & 0 \end{bmatrix} \begin{bmatrix} 0 & 1 \\ 1 & 0 \end{bmatrix} \begin{bmatrix} 0 & 0 \\ 0 & 1 \end{bmatrix} \begin{bmatrix} E_x \\ E_y \end{bmatrix}$$

$$= \begin{bmatrix} 1 & 0 \\ 0 & 0 \end{bmatrix} \begin{bmatrix} E_y \\ 0 \end{bmatrix}$$

The light leaving the half-wave plate is linearly polarized along the x direction. The emergent wave is

$$\begin{bmatrix} E'_x \\ E'_y \end{bmatrix} = \begin{bmatrix} E_y \\ 0 \end{bmatrix}$$

(e) A full-waveplate is inserted with azimuth 45°. From Eqs. (6.42) and (6.49) with $\Delta = 2\pi$, the Jones matrix expression is

$$\begin{bmatrix} E'_x \\ E'_y \end{bmatrix} = - \begin{bmatrix} 1 & 0 \\ 0 & 0 \end{bmatrix} \begin{bmatrix} 1 & 0 \\ 0 & 1 \end{bmatrix} \begin{bmatrix} 0 & 0 \\ 0 & 1 \end{bmatrix} \begin{bmatrix} E_x \\ E_y \end{bmatrix}$$

$$= - \begin{bmatrix} 1 & 0 \\ 0 & 0 \end{bmatrix} \begin{bmatrix} 0 \\ E_y \end{bmatrix}$$

$$= \begin{bmatrix} 0 \\ 0 \end{bmatrix}$$

(f) A 90° rotator is inserted. From Eqs. (6.42) and (6.53), the Jones matrix expression is

$$\begin{bmatrix} E'_x \\ E'_y \end{bmatrix} = \begin{bmatrix} 1 & 0 \\ 0 & 0 \end{bmatrix} \begin{bmatrix} 0 & -1 \\ 1 & 0 \end{bmatrix} \begin{bmatrix} 0 & 0 \\ 0 & 1 \end{bmatrix} \begin{bmatrix} E_x \\ E_y \end{bmatrix}$$

$$= \begin{bmatrix} 1 & 0 \\ 0 & 0 \end{bmatrix} \begin{bmatrix} -E_y \\ 0 \end{bmatrix}$$

$$= \begin{bmatrix} -E_y \\ 0 \end{bmatrix} \qquad\qquad \square$$

Example 6.7 Apply Jones matrices to Senarmont's method for measuring the retardance Δ of a crystal plate.

Solution   As shown in Fig. 6.21, a linearly polarized wave inclined at 45° is incident onto the crystal under test. The light emergent from the crystal further goes through a quarter-waveplate at −45°, where the wave is converted into a linearly polarized wave whose azimuth angle determines the retardance of the sample under test.

The output field **E** is from Eqs. (6.44) and (6.51)

$$\begin{bmatrix} E_x \\ E_y \end{bmatrix} = \frac{1}{\sqrt{2}} \begin{bmatrix} 1 & j \\ j & 1 \end{bmatrix} \begin{bmatrix} e^{-j\Delta/2} & 0 \\ 0 & e^{j\Delta/2} \end{bmatrix} \begin{bmatrix} 1 \\ 1 \end{bmatrix}$$

$$= \sqrt{2} e^{j45°} \begin{bmatrix} \cos\left(\dfrac{\Delta}{2} + 45°\right) \\[2mm] \sin\left(\dfrac{\Delta}{2} + 45°\right) \end{bmatrix}$$

The emergent wave from the quarter-waveplate is linearly polarized with azimuth angle $\Delta/2 + 45°$. $\square$

6.8.4   Eigenvectors of an Optical System

With most optical systems, if the state of polarization of the incident wave is varied, the state of polarization of the emergent wave also varies. However, one may find a particular state of polarization that does not differ between the incident and emergent waves, except for a proportionality constant. The field vector that represents such an incident wave is called an eigenvector and the value of the proportionality constant is called an eigenvalue of the given optical system. For instance, a lasing light beam (see Section 14.2.3) bouncing back and forth inside the laser cavity has to be in the same state of polarization after each trip, over and above the matching of the phase, so that the field is built up as the beam goes back and forth. When the laser system is expressed in terms of the Jones matrix, the eigenvector of such a matrix provides the lasing condition and the eigenvalue, the gain or loss of the system. [15]

Let $\begin{bmatrix} E_x \\ E_y \end{bmatrix}$ be an eigenvector of the optical system, and let λ be its eigenvalue. The relationship between incident and emergent waves in Jones matrix representation is

$$\lambda \begin{bmatrix} E_x \\ E_y \end{bmatrix} = \begin{bmatrix} a_{11} & a_{12} \\ a_{21} & a_{22} \end{bmatrix} \begin{bmatrix} E_x \\ E_y \end{bmatrix} \qquad (6.54)$$

Equation (6.54) is rewritten as

$$\begin{bmatrix} a_{11} - \lambda & a_{12} \\ a_{21} & a_{22} - \lambda \end{bmatrix} \begin{bmatrix} E_x \\ E_y \end{bmatrix} = 0 \tag{6.55}$$

The eigenvalues and corresponding eigenvectors of the system will be found by solving Eq. (6.55).

For nontrivial solutions for E_x and E_y to exist, the determinant of Eq. (6.55) has to vanish:

$$(a_{11} - \lambda)(a_{22} - \lambda) - a_{12}a_{21} = 0 \tag{6.56}$$

Equation (6.56) is a quadratic equation in eigenvalue λ and the solution is

$$\begin{aligned} \lambda_1 &= \tfrac{1}{2}\left[a_{22} + a_{11} - \sqrt{(a_{22} - a_{11})^2 + 4a_{12}^2} \right] \\ \lambda_2 &= \tfrac{1}{2}\left[a_{22} + a_{11} + \sqrt{(a_{22} - a_{11})^2 + 4a_{12}^2} \right] \end{aligned} \tag{6.57}$$

The convention of choosing $\lambda_1 < \lambda_2$ will become clear as the analysis progresses (see the discussion surrounding Eq. (6.86)).

Next, the eigenvectors will be found. Inserting λ_1 into either the top or bottom row of Eq. (6.55) gives

$$E_y = \frac{a_{11} - \lambda_1}{-a_{12}} E_x \tag{6.58}$$

or

$$E_y = \frac{-a_{21}}{a_{22} - \lambda_1} E_x \tag{6.59}$$

The equality of Eqs. (6.58) and (6.59) is verified from Eq. (6.56). One has to be careful whenever $a_{12} = 0$ or $a_{22} - \lambda_1 = 0$, as explained in Example 6.6.

Eigenvector $\mathbf{v}_1$, whose components E_x and E_y are related by either Eq. (6.58) or (6.59), is rewritten as

$$\mathbf{v}_1 = \begin{bmatrix} E_{x1} \\ E_{y1} \end{bmatrix} = \begin{bmatrix} -a_{12} \\ a_{11} - \lambda_1 \end{bmatrix} \tag{6.60}$$

and similarly for λ_2

$$\mathbf{v}_2 = \begin{bmatrix} E_{x2} \\ E_{y2} \end{bmatrix} = \begin{bmatrix} -a_{12} \\ a_{11} - \lambda_2 \end{bmatrix} \tag{6.61}$$

By taking the inner product of the eigenvectors given by Eqs. (6.60) and (6.61), we will find the condition that makes the eigenvectors orthogonal. Simplification of the product using Eq. (6.57) leads to

$$\begin{bmatrix} E_{x1} & E_{y1} \end{bmatrix} \begin{bmatrix} E_{x2} \\ E_{y2} \end{bmatrix} = a_{12}(a_{12} - a_{21}) \tag{6.62}$$

Thus, these vectors are orthogonal if

$$a_{12} = a_{21} \tag{6.63}$$

meaning Eq. (6.54) is a symmetric matrix.

Example 6.8   Find the eigenvalues and eigenvectors of a quarter-waveplate with its fast axis along the x axis.

Solution   The Jones matrix expression of a quarter-waveplate is, from Eq. (6.46),

$$\lambda \begin{bmatrix} E_x \\ E_y \end{bmatrix} = \begin{bmatrix} e^{-j\pi/4} & 0 \\ 0 & e^{j\pi/4} \end{bmatrix} \begin{bmatrix} E_x \\ E_y \end{bmatrix} \tag{6.64}$$

Comparing Eq. (6.64) with (6.54) gives

$$a_{11} = e^{-j\pi/4}$$
$$a_{22} = e^{j\pi/4} \tag{6.65}$$
$$a_{12} = a_{21} = 0$$

Inserting Eq. (6.65) into (6.57) gives

$$\lambda_{1,2} = \frac{1}{\sqrt{2}}(1 \mp j) = e^{\mp j\pi/4} \tag{6.66}$$

As mentioned earlier, if a_{12} or $a_{22} - \lambda_1$ is zero, one has to be careful. Here, the original equation Eq. (6.55) is used,

$$(a_{11} - \lambda)E_x + a_{12}E_y = 0$$
$$a_{21}E_x + (a_{22} - \lambda)E_y = 0 \tag{6.67}$$

and is combined with Eqs. (6.65) and (6.66) with $\lambda = \lambda_1$ to give

$$0E_x + 0E_y = 0$$
$$0E_x + \left(2j \sin\frac{\pi}{4}\right) E_y = 0$$

The above two equations are simultaneously satisfied if $E_y = 0$ and E_x is an arbitrary number, meaning a horizontally polarized wave. The output is $\lambda_1 E_x$.

Similarly, with $\lambda = \lambda_2$, Eq. (6.55) becomes

$$\left(-2j \sin\frac{\pi}{4}\right) E_x + 0E_y = 0$$
$$E_x + 0E_y = 0$$

which leads to $E_x = 0$ and E_y can be any number. The eigenvector is a vertically polarized wave. The output is $\lambda_2 E_y$.

The magnitude of the transmitted light is $|\lambda_{1,2}| = 1$. If the phase of the output is important, the phase factor $(e^{j\Delta/2})$ that was discarded from Eq. (6.34) should be retained in Eq. (6.64).   □

6.9   STATES OF POLARIZATION AND THEIR COMPONENT WAVES

Relationships existing among ellipticity, azimuth of the major axes of the ellipse, E_x and E_y component waves, and retardance will be derived. Such relationships will help to convert the expression for an elliptically polarized wave into that of E_x and E_y component waves.

6.9.1   Major and Minor Axes of an Elliptically Polarized Wave

The lengths of the major and minor axes will be found from the expressions for the E_x and E_y component waves.

Letting $\phi_0 = -\omega t + \beta z$, Eqs. (6.3) and (6.4) can be rewritten for convenience as

$$\frac{E_x}{A} = \cos \phi_0 \tag{6.68}$$

$$\frac{E_y}{B} = \cos \phi_0 \cos \Delta - \sin \phi_0 \sin \Delta \tag{6.69}$$

In order to find an expression that is invariant of time and location, ϕ_0 is eliminated by putting Eq. (6.68) into (6.69):

$$\frac{E_y}{B} = \left(\frac{E_x}{A}\right) \cos \Delta - \sqrt{1 - \left(\frac{E_x}{A}\right)^2} \sin \Delta \tag{6.70}$$

Rearranged, Eq. (6.70) is

$$\left(\frac{E_x}{A}\right)^2 + \left(\frac{E_y}{B}\right)^2 - 2\frac{E_x}{A}\frac{E_y}{B} \cos \Delta = \sin^2 \Delta \tag{6.71}$$

In order to facilitate the manipulation, let's rewrite Eq. (6.71) as

$$g(X, Y) = \sin^2 \Delta \tag{6.72}$$

$$g(X, Y) = a_{11}X^2 + a_{22}Y^2 + 2a_{12}XY \tag{6.73}$$

where

$$X = E_x, \qquad Y = E_y$$

$$a_{11} = \frac{1}{A^2}, \qquad a_{22} = \frac{1}{B^2}, \qquad a_{12} = -\frac{\cos \Delta}{AB} \tag{6.74}$$

Let us express Eq. (6.73) in matrix form so that various rules [16] associated with the matrix operation can be utilized. As shown in Fig. 6.47, let the position vector $\mathbf{v}$ of a point (X, Y) on the ellipse be represented by

$$\mathbf{v} = \begin{bmatrix} X \\ Y \end{bmatrix} \qquad \mathbf{v}^t = [X \quad Y] \tag{6.75}$$

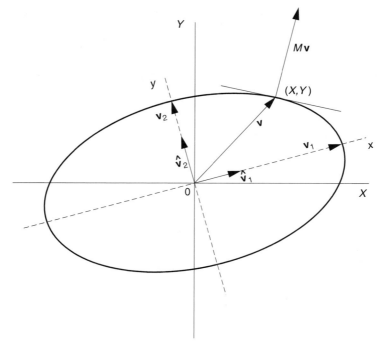

Figure 6.47 The normal to the ellipse is $M\mathbf{v}$, and the directions of the major and minor axes are those $\mathbf{v}$ that satisfy $M\mathbf{v} = \lambda\mathbf{v}$. $X = E_x$ and $Y = E_y$.

and define a symmetric matrix M as

$$M = \begin{bmatrix} a_{11} & a_{12} \\ a_{12} & a_{22} \end{bmatrix} \tag{6.76}$$

Realize that an equality exists as

$$\mathbf{v}^t M \mathbf{v} = a_{11}X^2 + a_{22}Y^2 + 2a_{12}XY \tag{6.77}$$

The normal $\mathbf{v}_N$ from the circumference of the ellipse is obtained by taking the gradient of Eq. (6.73) (see Eq. (4.89)) and is expressed in vector form as

$$\mathbf{v}_N = 2\begin{bmatrix} a_{11}X + a_{12}Y \\ a_{12}X + a_{22}Y \end{bmatrix} \tag{6.78}$$

Hence,

$$\mathbf{v}_N = 2M\mathbf{v} \tag{6.79}$$

As shown in Fig. 6.47, if the vector $\mathbf{v}$ were to represent the direction of the major or minor axis of the ellipse, $\mathbf{v}_N$ should be parallel to $\mathbf{v}$ or $\mathbf{v}_N = \lambda\mathbf{v}$, and hence, the condition for $\mathbf{v}$ to be along the major or minor axis is, from Eq. (6.79),

$$M\mathbf{v} = \lambda\mathbf{v} \tag{6.80}$$

where the factor 2 was absorbed in λ. Thus, the eigenvectors of matrix M provides the directions of the major and minor axes, and the eigenvectors are given by Eqs. (6.60) and (6.61).

Next, the actual lengths a and b of the major and minor axes of the ellipse will be found. The position vector $\mathbf{v}$ in Fig. 6.47 of a point (X,Y) on the ellipse in the $X-Y$ coordinates is expressed in the new $x-y$ coordinates taken along the major and minor axes as

$$\mathbf{v} = x\hat{\mathbf{v}}_1 + y\hat{\mathbf{v}}_2 \tag{6.81}$$

where $\hat{\mathbf{v}}_1$ and $\hat{\mathbf{v}}_2$ are the unit vectors of $\mathbf{v}_1$ and $\mathbf{v}_2$.

Inserting Eq. (6.82) into (6.77) gives

$$\mathbf{v}^t M\mathbf{v} = (x\hat{\mathbf{v}}_1^t + y\hat{\mathbf{v}}_2^t)M(x\hat{\mathbf{v}}_1 + y\hat{\mathbf{v}}_2)$$
$$= \lambda_1 x^2 + \lambda_2 y^2 \tag{6.82}$$

where use was made of

$$M\hat{\mathbf{v}}_1 = \lambda_1\hat{\mathbf{v}}_1 \tag{6.83}$$

$$M\hat{\mathbf{v}}_2 = \lambda_1\hat{\mathbf{v}}_2 \tag{6.84}$$

$$\hat{\mathbf{v}}_1^t \cdot \hat{\mathbf{v}}_2 = 0$$

Combining Eqs. (6.72), (6.73), (6.77) and (6.83) gives

$$\frac{x^2}{(\sin\Delta)^2/\lambda_1} + \frac{y^2}{(\sin\Delta)^2/\lambda_2} = 1 \tag{6.85}$$

Thus, in the new $x-y$ coordinates along $\mathbf{v}_1$ and $\mathbf{v}_2$, the major and minor axes of the ellipse a and b are

$$a = \frac{|\sin\Delta|}{\sqrt{\lambda_1}} \quad \text{and} \quad b = \frac{|\sin\Delta|}{\sqrt{\lambda_2}} \tag{6.86}$$

Since a is conventionally taken as the length of the major axis, the smaller eigenvalue is taken for λ_1. That is, the negative sign of Eq. (6.57) will be taken for λ_1, and the positive sign for λ_2.

In summary, the eigenvalues and eigenvectors of M have given the lengths as well as the directions of the major and minor axes.

Before going any further, we will verify that Eq. (6.71) is indeed the expression of an ellipse, and not that of a hyperbola, as both formulas are quite alike. Note that if $\lambda_1\lambda_2 > 0$, then Eq. (6.85) is an ellipse, but if $\lambda_1\lambda_2 < 0$, it is hyperbola. From Eq. (6.57), the product $\lambda_1\lambda_2$ is

$$\lambda_1\lambda_2 = a_{11}a_{22} - a_{12}^2 \tag{6.87}$$

Note that Eq. (6.87) is exactly the determinant of M. Thus, the conclusions are

$$\det M > 0, \quad \text{ellipse}$$
$$\det M < 0, \quad \text{hyperbola} \tag{6.88}$$

The value of the determinant is, from Eqs. (6.74) and (6.76),

$$\lambda_1 \lambda_2 = \det \begin{bmatrix} \dfrac{1}{A^2} & -\dfrac{\cos \Delta}{AB} \\[2ex] -\dfrac{\cos \Delta}{AB} & \dfrac{1}{B^2} \end{bmatrix} = \frac{\sin^2 \Delta}{A^2 B^2} \tag{6.89}$$

Thus,

$$\lambda_1 \lambda_2 > 0 \tag{6.90}$$

and Eq. (6.71) is indeed the expression of an ellipse.

6.9.2   Azimuth of the Principal Axes of an Elliptically Polarized Wave

Figure 6.48 shows the general geometry of an ellipse. Capital letters will be used for the quantities associated with the X and Y components of the $\mathbf{E}$ field, and lowercase letters

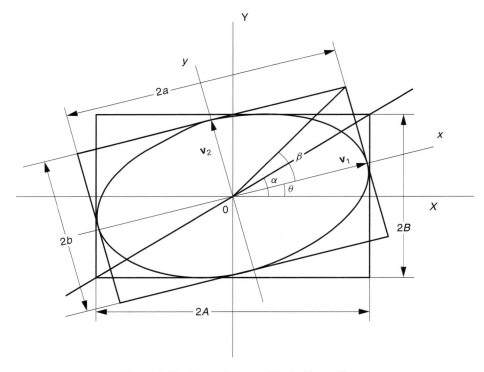

Figure 6.48   Parameter associated with an ellipse.

for those quantities expressed in $x-y$ coordinates. The $x-y$ coordinates correspond to the directions of the major and minor axes of the ellipse. E_X does not exceed A and E_Y does not exceed B, and the ellipse is always bordered by a rectangle $2A \times 2B$. The ratio B/A is often expressed in terms of the angle α as

$$\tan \alpha = \frac{B}{A} \tag{6.91}$$

Since the right-hand side of Eq. (6.91) is a positive quantity, α must lie in the range

$$0 \leq \alpha \leq \pi/2 \tag{6.92}$$

The vector $\mathbf{v}_1$ points in the direction of the x axis. The azimuth angle θ of the major axis with respect to the X axis will be found.

From Eq. (6.60), $\tan \theta$ is expressed as

$$\tan \theta = -\frac{(a_{11} - \lambda_1)}{a_{12}} \tag{6.93}$$

Inserting Eqs. (6.57) and (6.74) into Eq. (6.93) gives

$$\tan \theta = (-t + \sqrt{t^2 + \cos^2 \Delta}) \frac{1}{\cos \Delta} \tag{6.94}$$

where

$$t = \frac{A^2 - B^2}{2AB} \tag{6.95}$$

Equations (6.94) and (6.95) will be simplified further. Using the double-angle relationship of the tangent function given by

$$\tan 2\theta = \frac{2 \tan \theta}{1 - \tan^2 \theta} \tag{6.96}$$

Eq. (6.94) is greatly simplified as

$$\tan 2\theta = \frac{\cos \Delta}{t} \tag{6.97}$$

Applying the double-angle relationship of Eq. (6.96) to the angle α, and making use of Eq. (6.91), Eq. (6.95) becomes

$$t = \frac{1}{\tan 2\alpha} \tag{6.98}$$

The final result is obtained from Eqs. (6.97) and (6.98):

$$\tan 2\theta = \tan 2\alpha \cos \Delta \tag{6.99}$$

As seen from Fig. 6.48, all configurations of the principal axes can be expressed by $0 \le \theta \le 180°$.

The following conclusions are immediately drawn from Eq. (6.99):

1. If the amplitudes A and B are identical and $\alpha = 45°$, then the azimuth θ can only be $45°$ or $135°$, regardless of the value of Δ:

$$\theta = 45° \quad \text{for} \ \cos \Delta > 0$$

$$\theta = 135° \quad \text{for} \ \cos \Delta < 0$$

This agrees with previous discussions involving Fig. 6.4.

2. For any value of A and B, if $\Delta = 90°$, the azimuth θ is either $0°$ or $90°$.

6.9.3 Ellipticity of an Elliptically Polarized Wave

Ellipticity is another quantity that describes the shape of an ellipse. The ellipticity ϵ is defined as

$$\epsilon = \frac{b}{a} \tag{6.100}$$

where a is the length of the major axis, and b is the length of the minor axis of the ellipse.

From Eqs. (6.57), (6.86), and (6.100) the ellipticity is

$$\epsilon = \sqrt{\frac{1 - Y}{1 + Y}} \tag{6.101}$$

where

$$Y = \sqrt{\left(\frac{a_{11} - a_{22}}{a_{11} + a_{22}}\right)^2 + \left(\frac{2a_{12}}{a_{11} + a_{22}}\right)^2} \tag{6.102}$$

From Eq. (6.74), the quantities under the square root of Eq. (6.102) are simplified as

$$\frac{a_{11} - a_{22}}{a_{11} + a_{22}} = -\cos 2\alpha$$

$$\frac{2a_{12}}{a_{11} + a_{22}} = -\sin 2\alpha \cos \Delta \tag{6.103}$$

where the trigonometric relationships

$$\cos 2\alpha = \frac{1 - \tan^2 \alpha}{1 + \tan^2 \alpha} \tag{6.104}$$

$$\sin 2\alpha = \frac{2 \tan \alpha}{1 + \tan^2 \alpha} \tag{6.105}$$

were used.

Insertion of Eq. (6.103) into (6.102) gives

$$Y = \sqrt{1 - \sin^2 2\alpha \sin^2 \Delta} \tag{6.106}$$

Thus, insertion of Eq. (6.106) into (6.101) gives the ellipticity. A few more manipulations will be made on the expression for ϵ, but first, observe the following behavior of ϵ for given values of B/A and Δ:

1. With zero retardance Δ, the value ϵ is always zero and the wave is linearly polarized.
2. Only when $B/A = 1$ and $\Delta = 90°$, can the wave be circularly polarized.

Returning to the manipulations on the ellipticity expression, ϵ will be rewritten further in terms of trigonometric functions. Referring to Fig. 6.48, ϵ can be represented by the angle β:

$$\tan \beta = \epsilon \tag{6.107}$$

Since ϵ is a quantity between 0 and 1

$$0 \le \beta \le \pi/4 \tag{6.108}$$

The trigonometric relationship

$$\sin 2\beta = \frac{2 \tan \beta}{1 + \tan^2 \beta} \tag{6.109}$$

is applied to Eqs. (6.101) and (6.107) to obtain

$$\sin 2\beta = \sqrt{1 - Y^2} \tag{6.110}$$

Insertion of Eq. (6.106) into (6.110) gives

$$\sin 2\beta = \sqrt{\sin^2 2\alpha \sin^2 \Delta} \tag{6.111}$$

$$\sin 2\beta = \sin 2\alpha |\sin \Delta| \tag{6.112}$$

Because of the restrictions imposed on α and β in Eqs. (6.92) and (6.108), both $\sin 2\beta$ and $\sin 2\alpha$ are positive and the absolute value of $\sin \Delta$ has to be taken.

6.9.4   Conservation of Energy

When the state of polarization is converted, the light power neither increases nor decreases, aside from the loss due to nonideal optical components. Conservation of energy dictates that

$$a^2 + b^2 = A^2 + B^2 \tag{6.113}$$

Equation (6.113) will now be verified. From Eq. (6.86), $a^2 + b^2$ is expressed in terms of the eigenvalues λ_1 and λ_2 and Δ as

$$a^2 + b^2 = \left(\frac{1}{\lambda_1} + \frac{1}{\lambda_2} \right) \sin^2 \Delta \tag{6.114}$$

From Eqs. (6.57) and (6.74), the sum of the eigenvalues is

$$\lambda_1 + \lambda_2 = \frac{1}{A^2} + \frac{1}{B^2} \tag{6.115}$$

Insertion of Eqs. (6.89) and (6.115) into Eq. (6.114) finally proves the equality of Eq. (6.113).

Next, area relationships will be derived from Eq. (6.86). The product ab is

$$ab = \frac{\sin^2 \Delta}{\sqrt{\lambda_1 \lambda_2}} \tag{6.116}$$

With Eq. (6.89), a substitution for $\sqrt{\lambda_1 \lambda_2}$ is found and

$$ab = AB|\sin \Delta| \tag{6.117}$$

where the absolute value sign was used because all other quantities are positive. Note area πab of the ellipse becomes zero when $\Delta = 0$, and a maximum when $\Delta = \pm\pi/2$, for given values of A and B.

Furthermore, the difference $a^2 - b^2$ will be calculated. From Eq. (6.86), $a^2 - b^2$ is

$$a^2 - b^2 = \left(\frac{1}{\lambda_1} - \frac{1}{\lambda_2} \right) \sin^2 \Delta \tag{6.118}$$

With Eq. (6.89), Eq. (6.118) becomes

$$a^2 - b^2 = (\lambda_2 - \lambda_1)A^2B^2 \tag{6.119}$$

In the following, $\lambda_2 - \lambda_1$ will be calculated. From Eq. (6.57), the difference $\lambda_2 - \lambda_1$ is

$$\lambda_2 - \lambda_1 = (a_{22} - a_{11})\sqrt{1 + \left(\frac{2a_{12}}{a_{22} - a_{11}} \right)^2} \tag{6.120}$$

Manipulation of Eqs. (6.74), (6.95), (6.98), and (6.99) gives

$$\frac{2a_{12}}{a_{22} - a_{11}} = \tan 2\theta \tag{6.121}$$

From Eq. (6.120) and (6.121), the difference becomes

$$\lambda_2 - \lambda_1 = \frac{A^2 - B^2}{A^2B^2} \frac{1}{\cos 2\theta} \tag{6.122}$$

Inserting Eq. (6.122) back into (6.119) gives the final result of

$$(a^2 - b^2)\cos 2\theta = A^2 - B^2 \tag{6.123}$$

This relationship is used later on in converting between X–Y and x–y components.

6.9.5 Relating the Parameters of an Elliptically Polarized Wave to Those of Component Waves

So far, parameters such as $\epsilon(=\tan\beta)$ and θ have been derived from $B/A(=\tan\alpha)$ and Δ. In this section, α and Δ will conversely be obtained from β and θ.

Using the trigonometric identity of Eq. (6.104) for β instead of α and using Eqs. (6.107), (6.113), and (6.123), $\cos 2\beta$ is expressed as

$$\cos 2\beta = \frac{A^2 - B^2}{A^2 + B^2} \cdot \frac{1}{\cos 2\theta} \tag{6.124}$$

Dividing both numerator and denominator by A^2, and using the trigonometric relationship Eq. (6.104) for $\tan\alpha$, Eq. (6.124) becomes

$$\cos 2\alpha = \cos 2\theta \cos 2\beta \tag{6.125}$$

Next, the expression for the retardance Δ will be derived. The derivation makes use of Eq. (6.112), which contains $|\sin\Delta|$, and the absolute value cannot be ignored.

$$|\sin\Delta| = \sin\Delta \qquad \text{for } \sin\Delta > 0; \text{ left-handed}$$
$$|\sin\Delta| = -\sin\Delta \qquad \text{for } \sin\Delta < 0; \text{ right-handed} \tag{6.126}$$

where the handedness information is given by Eq. (6.15). The ratio between Eqs. (6.99) and (6.112), and the use of Eqs. (6.125) and (6.126) leads to

$$\tan\Delta = \pm\operatorname{cosec} 2\theta \tan 2\beta \tag{6.127}$$

The plus and minus signs are for left-handed and right-handed elliptical polarization, respectively.

6.9.6 Summary of Essential Formulas

The formulas derived in the last few sections are often used for calculating the state of polarization and will be summarized here.

$$\Delta = \phi_y - \phi_x$$
$$\Delta > 0, \quad y \text{ component is lagging} \tag{6.7} \text{ and } (6.8)$$
$$\Delta < 0, \quad y \text{ component is leading}$$
$$\tan\alpha = \frac{B}{A} \quad \left(0 \le \alpha \le \frac{\pi}{2}\right) \tag{6.91}$$

$$\tan \beta = \frac{b}{a} = \epsilon \quad \left(0 \le \beta \le \frac{\pi}{4}\right) \qquad \text{(6.100) and (6.107)}$$

$$\tan \theta = \left(-t + \sqrt{t^2 + \cos^2 \Delta}\right) / \cos \Delta \quad (0 \le \theta < \pi) \qquad \text{(6.94)}$$

$$t = \frac{A^2 - B^2}{2AB} \qquad \text{(6.95)}$$

$$t = 1/\tan 2\alpha \qquad \text{(6.98)}$$

$$\tan 2\theta = \tan 2\alpha \cos \Delta \quad (0 \le 2\theta < 2\pi) \qquad \text{(6.99)}$$

$$\sin 2\beta = \sin 2\alpha |\sin \Delta| \qquad \text{(6.112)}$$

$$a^2 + b^2 = A^2 + B^2 \qquad \text{(6.113)}$$

$$ab = AB |\sin \Delta| \qquad \text{(6.117)}$$

$$(a^2 - b^2) \cos 2\theta = A^2 - B^2 \qquad \text{(6.123)}$$

$$(a^2 - b^2) \sin 2\theta = 2AB \cos \Delta \qquad \text{(Prob. 6.12a)}$$

$$(a^2 - b^2) \cos 2\theta = (A^2 + B^2) \cos 2\alpha \qquad \text{(Prob. 6.12b)}$$

$$\cos 2\alpha = \cos 2\theta \cos 2\beta \qquad \text{(6.125)}$$

$$\sin \Delta > 0 \qquad \text{and } |\sin \Delta| = \sin \Delta; \text{ left-handed}$$

$$\sin \Delta < 0 \qquad \text{and } |\sin \Delta| = -\sin \Delta; \text{ right-handed} \qquad \text{(6.126)}$$

$$\tan \Delta = \pm \text{cosec } 2\theta \tan 2\beta \quad (+ \text{ is for left-handed and } - \text{ for right-handed)} \qquad \text{(6.127)}$$

Example 6.9   A linearly polarized wave is incident onto a retarder whose fast axis is along the x axis. The retardance Δ is 38° and the amplitudes of the E_x and E_y components are 2.0 V/m and 3.1 V/m, respectively. Calculate the azimuth θ and the ellipticity ϵ of the emergent elliptically polarized wave. Also, determine the lengths a and b of the major and minor axes. Find the solution graphically as well as analytically.

Solution

For the given parameters,

$$A = 2.0 \text{ V/m}$$

$$B = 3.1 \text{ V/m}$$

$$\Delta = 38°$$

θ and ϵ will be found.

$$\tan \alpha = \frac{B}{A} = 1.55$$

$$\alpha = 57.2°$$

From Eq. (6.99), the angle θ is obtained:

$$\tan 2\theta = \tan 2\alpha \cos \Delta$$

$$= (-2.2)(0.788)$$

$$= -1.733$$

$$\theta = -30° \text{ or } 60°$$

Since $0 \le \theta \le \pi$, $\theta = 60°$ is the answer.
From Eq. (6.112)

$$\sin 2\beta = \sin 2\alpha |\sin \Delta|$$

$$= (0.910)(0.616)$$

$$= 0.560$$

$$\beta = 17.0°$$

$$\epsilon = \tan \beta = 0.31$$

Next, a and b are calculated from Eqs. (6.113) and (6.123):

$$a^2 + b^2 = A^2 + B^2 = 13.61$$

$$a^2 - b^2 = \frac{1}{\cos 2\theta}(A^2 - B^2) = \frac{5.61}{0.5}$$

$$= 11.22$$

$$a = 3.52$$

$$b = 1.09$$

The circle diagram is shown in Fig. 6.49a and the calculated results are summarized in Fig. 6.49b.   □

Example 6.10

The parameters of an elliptically polarized wave are $A = 10$ V/m, $B = 8$ V/m, $a = 12.40$ V/m, and $b = 3.22$ V/m.

(a) Find the azimuth θ and the retardance Δ.
(b) For the given values of A and B, what is the maximum ellipticity ϵ that can be obtained by manipulating the retardance?
(c) For the given values of A and B, is it possible to obtain an ellipse with $\epsilon = 0.26$ and azimuth $\theta = 50°$ by manipulating the retardance Δ?

Solution   From A and B, $\tan \alpha$ is obtained:

$$A = 10 \text{ V/m}$$

$$B = 8 \text{ V/m}$$

$$\tan \alpha = 0.8$$

From a and b, $\tan \beta$ is obtained:

$$a = 12.4 \text{ V/m}$$

$$b = 3.20 \text{ V/m}$$

$$\tan \beta = 0.26$$

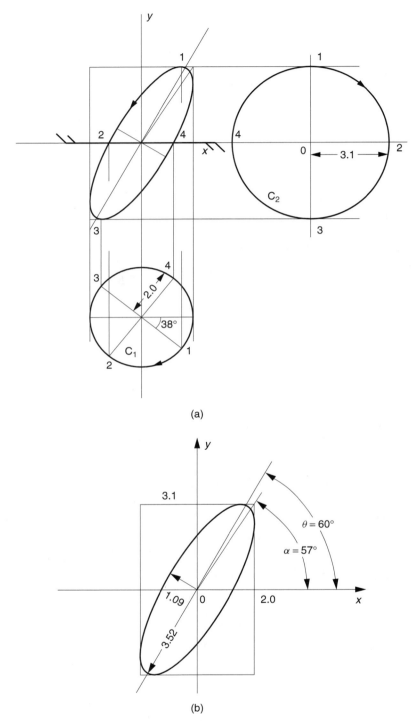

(a)

(b)

Figure 6.49 Solutions of Example 6.9 obtained graphically as well as analytically. (a) Graphical solution. (b) Summary of calculated results.

(a) θ and Δ are computed as follows. From Eq. (6.123), θ is calculated:

$$(a^2 - b^2)\cos 2\theta = A^2 - B^2$$

$$\cos 2\theta = \frac{36}{143.52} = 0.25$$

$$\theta = 37.8°$$

From Eq. (6.117), Δ can be found:

$$ab = AB|\sin \Delta|$$

$$|\sin \Delta| = \frac{ab}{AB} = 0.496$$

$$\Delta = \pm 29.7°$$

(b) From Eq. (6.112), the value of Δ that maximizes 2β for a given value of α is $\Delta = 90°$. The maximum value of ϵ is

$$\epsilon = \tan \beta = \tan \alpha = 0.8$$

(c) Let us see if Eq. (6.125) is satisfied:

$$\cos 2\alpha = \cos 2\theta \cos 2\beta$$

with $\tan \alpha = 0.8$,

$$\alpha = 38.7°$$

and with $\tan \beta = 0.26$

$$\beta = 14.6°$$

$$\theta = 50°$$

$$0.218 = -1.74 \times 0.873$$

$$0.218 \neq -1.52$$

For given A and B, the value of ϵ and θ are mutually related, and one cannot arbitrarily pick the two values.   □

Example 6.11   A right-handed elliptically polarized wave with $a = \sqrt{3}$ V/m and $b = 1$ V/m and $\theta = 22.5°$ is incident onto a $\lambda/4$ plate with its fast axis oriented at $\Theta = 45°$ with respect to the X axis. Find the state of polarization of the wave emergent from the $\lambda/4$ plate.

The circle diagram was used to solve the same question in Section 6.2.5, Fig. 6.10.

Solution
The following steps will be taken:

1. Calculate A, B, and Δ of the incident wave.

2. Find B'/A', and Δ' of the emergent wave from the $\lambda/4$ plate by means of the Jones matrix.
3. Convert B'/A', and Δ' into a', b', and θ'.

Step 1. The given parameters are

$$a = \sqrt{3} \text{ V/m}$$

$$b = 1 \text{ V/m}$$

$$\theta = 22.5°$$

$$\tan \beta = b/a = \frac{1}{\sqrt{3}}$$

$$\beta = 30°$$

From Eq. (6.125), α is calculated:

$$\cos 2\alpha = \cos 2\theta \cos 2\beta$$

$$= 0.354$$

$$\alpha = 34.6°$$

Next, the retardance Δ will be found using Eq. (6.127):

$$\tan \Delta = \pm \, \text{cosec} 2\theta \tan 2\beta$$

$$= \pm 2.45$$

where the $+$ sign is for left-handed and the $-$sign is for right-handed. Since the problem specifies right-handed, $\tan \Delta = +2.45$ is eliminated.

The two possibilities for Δ are

$$\Delta = -67.8° \text{ or } + 112.2°$$

From Eq. (6.126), the correct choice of Δ is

$$\Delta = -67.8°$$

From Eq. (6.113), A and B are found:

$$a^2 + b^2 = A^2 + B^2$$

$$= A^2(1 + \tan^2 \alpha)$$

$$A^2 = \frac{1 + 3}{1 + \tan^2 34.6°}$$

$$A = 1.65$$

$$B = 1.14$$

Step 2. The Jones matrix of the $\lambda/4$ plate whose fast axis azimuth angle Θ is $45°$ is used to calculate emergent wave.

$$\begin{bmatrix} E_x \\ E_y \end{bmatrix} = \frac{1}{\sqrt{2}} \begin{bmatrix} 1 & -j \\ -j & 1 \end{bmatrix} \begin{bmatrix} 1.65 \\ 1.14e^{-j67.8°} \end{bmatrix}$$

$$= \frac{1}{\sqrt{2}} \begin{bmatrix} 1 & e^{-j90°} \\ e^{-j90°} & 1 \end{bmatrix} \begin{bmatrix} 1.65 \\ 1.14e^{-j67.8°} \end{bmatrix}$$

$$= \frac{1}{\sqrt{2}} \begin{bmatrix} 1.65 + 1.14e^{-j157.8°} \\ 1.65e^{-j90°} + 1.14e^{-j67.8°} \end{bmatrix}$$

$$= \frac{1}{\sqrt{2}} \begin{bmatrix} \sqrt{[1.65 + 1.14\cos(-157.8°)]^2 + [1.14\sin(-157.8°)]^2}\,e^{j\phi_x} \\ \sqrt{[1.14\cos(-67.8°)]^2 + [-1.65 + 1.14\sin(-67.8°)]^2}\,e^{j\phi_y} \end{bmatrix}$$

$$= \frac{1}{\sqrt{2}} \begin{bmatrix} \sqrt{(0.595)^2 + (-0.43)^2}\,e^{j\phi_x} \\ \sqrt{(0.43)^2 + (-2.71)^2}\,e^{j\phi_y} \end{bmatrix}$$

$$= \frac{1}{\sqrt{2}} \begin{bmatrix} 0.73\ e^{-j35.5°} \\ 2.74\ e^{-j81.0} \end{bmatrix} = \frac{e^{-j35.5°}}{\sqrt{2}} \begin{bmatrix} 0.733 \\ 2.74e^{-45.5°} \end{bmatrix}$$

$$\Delta' = -45.5°$$

$$\tan\alpha = \frac{2.74}{0.733} = 3.74$$

$$\alpha' = 75.0°$$

Step 3. From Eq. (6.99), θ is found:

$$\tan 2\theta' = \tan 2\alpha' \cos\Delta'$$

$$= (-0.566)(0.72) = -0.407$$

$$2\theta' = -22.2°\ \text{or} 157.8°$$

Since $0 \le \theta' \le 180°$, and $0 \le 2\theta' \le 360°$, the negative value is rejected and $\theta' = 79°$. From Eq. (6.112), β' is calculated:

$$\sin 2\beta' = \sin 2\alpha' |\sin\Delta'|$$

$$= 0.5 \times |-0.707|$$

$$= 0.354$$

$$\beta' = 10.36°$$

$$\epsilon = \tan\beta' = 0.18$$

$$\sin\Delta' = -0.707 < 0$$

The emergent wave is right-handed.

$$a'^2 + b'^2 = a'^2(1 + \epsilon^2)$$

From Eq. (6.113), the above equation is rewritten

$$A^2 + B^2 = a'^2(1 + \epsilon^2)$$

and conservation of energy gives

$$A^2 + B^2 = a^2 + b^2 = a'^2 + b'^2$$

$$a'^2 = \frac{\sqrt{3}^2 + 1}{1 + 0.18^2} = 3.87$$

$$a' = 1.96$$

$$b' = 0.35$$

Compare with the answer shown in Fig. 6.10 using a circle diagram.

PROBLEMS

6.1 A linearly polarized wave with $A = \sqrt{3}$ V/m, and $B = 1$ V/m is incident onto a $\lambda/4$ plate with its fast axis along the x axis. A is the amplitude of the E_x component wave, and B is the amplitude of the E_y component wave. Obtain the emergent elliptically polarized wave by using the two different conventions of $E_- = e^{j\omega t - j\beta z}$ and $E_+ = e^{-j\omega t + j\beta z}$ representing a forward wave, and demonstrate that both results are the same.

6.2 A linearly polarized wave with azimuth $\theta_1 = 63.4°$ is incident onto a retarder with $\Delta = 315°$ whose fast axis is oriented along the x axis. Graphically determine the azimuth angle θ_2, which is the angle between the major axis of the emergent ellipse and the x axis, and the ellipticity ϵ of the ellipse.

6.3 In Example 6.1, the direction of polarization of the incident light was fixed and the fast axis of the $\lambda/4$ plate (retarder with $\Delta = 90°$) was rotated. The results were drawn in Fig. 6.8. Draw the results (analogous to Fig. 6.8) for

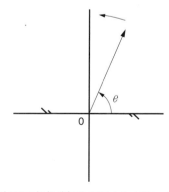

Figure P6.3   A linearly polarized wave is incident onto a quarter-waveplate with its fast axis oriented horizontally.

the case when the direction of the fast axis is fixed horizontally, and the direction of the incident linear polarization is rotated as shown in Fig. P6.3 at $\theta = 0°, 22.5°, 45°, 67.5°, 90°, 112.5°, 135°, 157.5°,$ and $180°$:

6.4 Decompose graphically an elliptically polarized wave into component waves that are parallel and perpendicular to the major or minor axis, and verify that the phase difference between the component waves is $90°$.

6.5 Obtain graphically the state of polarization with the same configuration as that shown in Fig. 6.10, but with the opposite handedness of rotation of the incident wave, that is, left-handed.

6.6 A linearly polarized light wave is incident normal to a pair of polarizers P_1 and P_2 whose transmission axes are oriented at θ_1 and θ_2 (Fig. P6.6). Assume $k_1 = 1$ and $k_2 = 0$ for both polarizers.

    (a) The light is incident from P_1 to P_2. What is the orientation θ of the linearly polarized eigenvector?

    (b) What is the orientation θ of the linearly polarized eigenvector when the light is incident from P_2 to P_1?

6.7 The horseshoe crab's eyes are known to be polarization sensitive. It is believed that this sensitivity is used as a means of navigating in sunlight, and the principle involved is that of the polarization of sunlight by scattering from particles in the water. Referring to the configuration in Fig. P6.7, how can the horseshoe crab orient itself along a north–south line in the early morning? What is the direction of polarization that the horseshoe crab sees when facing south in the early morning?

6.8 Linearly polarized laser light ($\lambda = 0.63$ µm) is transmitted through a quartz crystal along its optical axis. Due to Rayleigh scattering from minute irregularities in the crystal, one can observe a trace of the laser beam from the side of the crystal. One may even notice a spatial modulation of the intensity along the

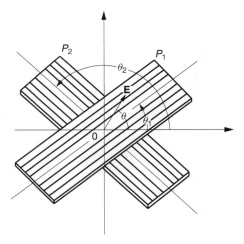

Figure P6.6 Direction of the eigenvector (looking from the source).

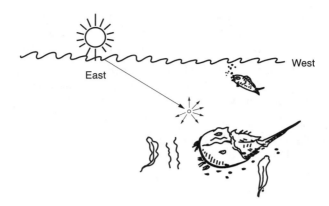

Figure P6.7   The horseshoe crab is known to be sensitive to polarized light.

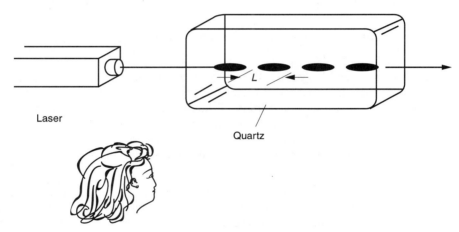

Figure P6.8   Modulation of the intensity of Rayleigh scattering inside a quartz crystal due to the optical activity of the crystal.

trace, as illustrated in Fig. P6.8. If one assumes that the spatial modulation is due to the rotary power, what is the period of the modulation? The rotary power of quartz is 19.5 deg/mm at $\lambda = 0.63$ μm and at a temperature of 20°C.

6.9   Figure 6.11 shows a diagram of a $\lambda/4$ plate. If one assumes $d_1 > d_2$, is the birefringence of the crystal in the figure positive or negative?

6.10   Devise a scheme to determine the directions of the fast and slow axes of a retarder.

6.11   The ellipse shown in Fig. P6.11 was made with $B/A = 1$ and $\Theta = 0$, just like the ellipses shown in Fig. 6.4. Prove that the retardance Δ is identical to the angle $\angle ABC = 2\beta$ on the ellipse.

6.12   Prove the following equalities:

(a) $(a^2 - b^2)\sin 2\theta = 2AB\cos\Delta$.

(b) $(a^2 - b^2)\cos 2\theta = (A^2 + B^2)\cos 2\alpha$.

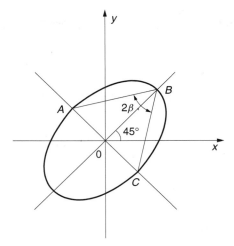

Figure P6.11 Prove that 2β is identical to the retardance Δ when $B/A = 1$ and $\Theta = 0$.

REFERENCES

1. W. A. Shurcliff, *Polarized Light*, Harvard University Press, Cambridge, MA, 1962. A condensed version was published with the same title, with S. S. Ballard as coauthor, Book #7, Van Nostrand Momentum, Princeton, 1964.

2. H. G. Gerrard, "Transmission of light through birefringent and optically active media; the Poincaré sphere," *J. Opt. Soc. Am.* **44**(8), 634–640 (1954).

3. P. C. Robinson and S. Bradbury, *Quantitative Polarized-Light Microscopy*, Oxford University Press, Oxford, 1992.

4. R. Ulrich, S. C. Rashleigh, and W. Eickhoff, "Bending-induced birefringence in single-mode fibers," *Opt. Let.* **5**(6), 273–275 (1980).

5. Y. Imai, M. A. Rodorigues, and K. Iizuka, "Temperature insensitive fiber coil sensor for altimeters," *Appl. Opt.* **29**(7), 975–978 (1990).

6. H. C. Lefevre, "Single mode fiber fractional wave devices and polarization controllers," *Electron. Let.* **16**(20), 778–780 (1980).

7. K. Shiraishi, H. Hatakeyama, H. Matsumoto, and K. Matsumura, "Laminated polarizers exhibiting high performance over a wide range of wavelength," *J. Lightwave Technol.* **15**(6), 1042–1050 (1997).

8. K. Baba, Y. Sato, T. Yoshitake and M. Miyagi, "Fabrication and patterning of submicrometer-thick optical polarizing films for 800 nm using streched silver island multilayers," *Opt. Rev.*, **7**(3), 209–215 (2000).

9. C. Viney, *Transmitted Polarized Light Microscopy*, McCrone Research Institute, Chicago, 1990.

10. K. Okamoto, T. Hosaka, and J. Noda, "High birefringence fiber with flat cladding," *J. Lightwave Technol.* **LT-3**(4), 758–762 (1985).

11. M. Takagi, Y. Kubo, E. Sasaoka, and H. Suganuma, "Ultra-wide bandwidth fiber polarizer," *Symposium Digest of the 8th Fiber Sensor Conference*, Monterey, CA, pp. 284–287, 1992.

12. H. Fujikake, K. Takizawa, T. Aida, H. Kikuchi, T. Fujii, and M. Kawakita, "Electrically-controllable liquid crystal polarizing filter for eliminating reflected light," *Opt. Rev.* **5**(2), 93–98 (1998).

13. A. Yariv and P. Yeh, *Optical Waves in Crystals*, Wiley, New York, 1984.

14. D. S. Kliger, J. W. Lewis, and C. E. Randall, *Polarized Light in Optics and Spectroscopy*, Academic Press, Boston, 1990.

15. T. J. Kane and R. L. Byer, "Monolithic unidirectional single-mode Nd:YAG ring laser," *Opt. Let.* **10**(2), 65–67 (1985).

16. G. James, *Advanced Modern Engineering Mathematics*, 2nd ed. Addison-Wesley, Reading, MA, 1999.

HOW TO CONSTRUCT AND USE THE POINCARÉ SPHERE

In this chapter, the Argand diagram and the Poincaré sphere will be introduced as additional graphical methods for dealing with polarization.

The Poincaré sphere was proposed as early as 1892 by the French scientist Henrie Poincaré [1–6], but only in recent years has the Poincaré sphere been given the attention it really deserves [6]. The Poincaré sphere is a projection of the Argand diagram onto a spherical surface to make the diagram spherically symmetric [3]. Spherical symmetry eliminates the step of rotating and rerotating the coordinates, which is necessary when using the Argand diagram.

The Poincaré sphere can be used in:

1. Problems associated with any retardance or orientation of the fast axis.
2. Problems associated with a polarizer with any orientation.
3. Determining the Stokes parameters, which are the projections of a point on the sphere to the equatorial plane.

Although primarily used for polarized waves, the Poincaré sphere can be extended to partially polarized waves.

A special feature of the Poincaré sphere is the simplicity of manipulation. Regardless of whether the incident wave is linearly polarized or elliptically polarized, whether the desired quantities are the optical parameters of a lumped element or a distributed element like the twist rate of an optical fiber [7], or whether the axis of the optical element is horizontally oriented or tilted at an arbitrary angle, the procedure remains the same. Multiple retarders undergo the same kind of manipulation. The answer after each stage is provided as the manipulation continues.

Moreover, when parameters of optical components have to be selected to achieve a particular state of polarization, the Poincaré sphere becomes even more valuable.

This chapter uses many results of Chapter 6 and should be considered as an extension of Chapter 6.

7.1 COMPONENT FIELD RATIO IN THE COMPLEX PLANE

Let the complex number representation of the x and y component fields be

$$E_x = Ae^{j(-\omega t + \beta z + \phi_x)} \tag{7.1}$$

$$E_y = Be^{j(-\omega t + \beta z + \phi_y)} \tag{7.2}$$

A new quantity, which is the ratio of these complex fields, is defined in polar form as

$$\frac{E_y}{E_x} = \left(\frac{B}{A}\right) e^{j\Delta} \tag{7.3}$$

where Δ is the retardance. The retardance Δ is the phase of the y component with respect to that of the x component and is expressed as

$$\Delta = \phi_y - \phi_x \tag{7.4}$$

As mentioned in Chapter 6, Eq. (7.3) is called the component field ratio. The component field ratio is represented as a point P on the u–v complex plane, as shown in Fig. 7.1. The magnitude $0P$ represents the amplitude ratio $B/A (= \tan \alpha)$ and the phase angle represents the retardance Δ.

The real and imaginary parts of the component field ratio are

$$\left(\frac{B}{A}\right) \cos \Delta + j \left(\frac{B}{A}\right) \sin \Delta = u + jv \tag{7.5}$$

$$u = \tan \alpha \cos \Delta$$

$$v = \tan \alpha \sin \Delta \tag{7.6}$$

with

$$B/A = \tan \alpha \tag{7.7}$$

Hence, the ratio v/u is simply

$$\frac{v}{u} = \tan \Delta \tag{7.8}$$

Each point on the u–v complex plane of the component field ratio corresponds to a state of polarization because Eqs. (6.99) and (6.112) give the values of θ and β from given α and Δ values. This correspondence between states of polarization and points in the u–v complex plane is illustrated in Fig. 7.2. Some observations will be made concerning Figs. 7.1 and 7.2.

1. In the upper half-plane, $0 < \Delta < \pi$ and $\sin \Delta > 0$. From Eq. (6.126), all states in this region are left-handed. In the lower half-plane, $\pi < \Delta < 2\pi$ and $\sin \Delta < 0$, and here the states are all right-handed.

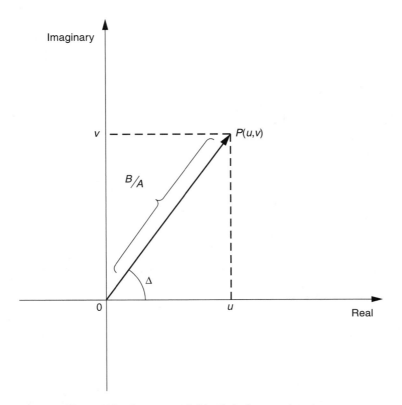

Figure 7.1 Component field ratio in the complex plane.

2. Along the positive u axis, $\Delta = 0$ and there is no phase difference between E_x and E_y. The states are all linearly polarized, as seen from Eq. (6.112). According to Eq. (6.99), the azimuth θ of the linear polarization is equal to α and increases with u from horizontally polarized at $u = 0$ to vertically polarized at $u = \infty$.

3. Since $B/A > 0$, the sign of u and v are determined solely by the value of the retardance Δ. Along the negative u axis, $\Delta = 180°$. The retardance $\Delta = 180°$ means that the direction of E_y is reversed from that of E_y in the corresponding positive u direction as shown in Fig. 7.3. The azimuth θ along the negative u axis is the mirror image of that along the positive u axis.

4. Along the positive v axis, $\Delta = 90°$ and with an increase in v, $\epsilon (= \tan \beta)$ increases within the range given by Eq. (6.108). According to Eq. (6.112), $\beta = \alpha$. In the region above the u axis, the states are all left-handed elliptically polarized waves from Eq. (6.126). Their major axis is horizontal in the region $v < 1$ and their minor axis is horizontal in the region $v > 1$.

5. Along the negative v axis, $\Delta = -90°$. With an increase in $|v|$, ϵ increases within the range given by Eq. (6.108). As with positive v, $\beta = \alpha$. Below the u axis all states of polarization are right-handed elliptically polarized waves with their major axis horizontal when $|v| < 1$ and with their minor axis horizontal when $|v| > 1$.

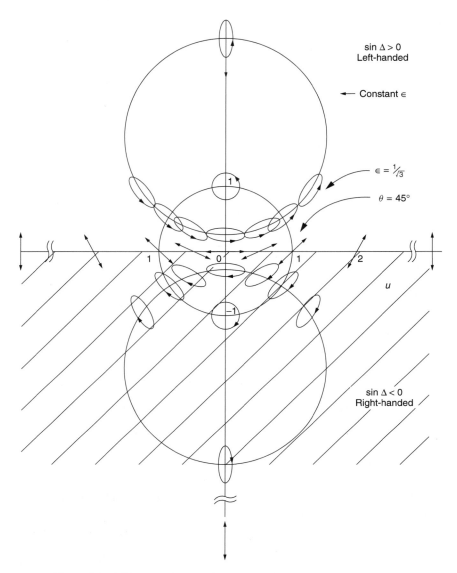

Figure 7.2   Ellipses on the complex plane of the complex amplitude ratio.

6. On the unit circle, $B/A = 1$ and α is either $45°$ or $135°$. The results shown in Fig. 6.4 are arranged along the unit circle in Figs. 6.5 and 7.2. From Eq. (6.112), the relationship between β and Δ is

$$\beta = \tfrac{1}{2}|\Delta| \tag{7.9}$$

The intercepts of the unit circle with the positive and negative v axes are left-handed and right-handed circularly polarized waves, respectively.

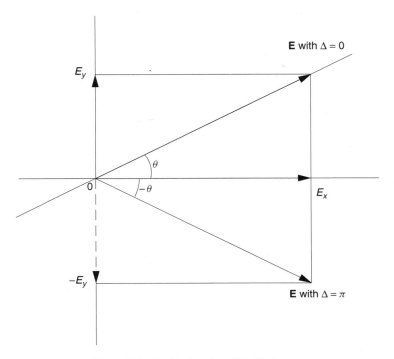

Figure 7.3   **E** with $\Delta = 0$ and **E** with $\Delta = \pi$.

7.2   CONSTANT AZIMUTH θ AND ELLIPTICITY ϵ LINES IN THE COMPONENT FIELD RATIO COMPLEX PLANE

In Fig. 7.2, the states of polarization were drawn in the component field ratio plane. The values of θ and ϵ were superposed on the values of α and Δ. In this section the points having the same θ values will be connected together, as well as the points having the same ϵ values. The two sets of θ and ϵ loci will be put together in one component field ratio complex plane. This plane is referred to as the Argand diagram.

7.2.1   Lines of Constant Azimuth θ

In this section, a set of curves of constant θ will be generated. The constant θ line can be generated if θ is expressed in terms of u and v. Equations (6.99) and (7.6) are used for this purpose.

$$\tan 2\theta = \frac{\left| 2\dfrac{u}{\cos \Delta} \right|}{1 - \left(\dfrac{u}{\cos \Delta} \right)^2} \cos \Delta \qquad (7.10)$$

where the double-angle relationship of Eq. (6.96) was applied to the angle α.

Inverting both sides of Eq. (7.10) gives

$$2u \cot 2\theta = 1 - \left(\frac{u}{\cos \Delta}\right)^2 \tag{7.11}$$

Now, an alternate way of expressing Eq. (7.8) is

$$\frac{1}{\cos^2 \Delta} = 1 + \left(\frac{v}{u}\right)^2 \tag{7.12}$$

Inserting Eq. (7.12) into (7.11) gives

$$u^2 + v^2 + 2u \cot 2\theta - 1 = 0 \tag{7.13}$$

which can be further rewritten as

$$(u + \cot 2\theta)^2 + v^2 = \text{cosec}^2 2\theta \tag{7.14}$$

Equation (7.14) is the expression of a circle with radius $\text{cosec}\, 2\theta$ centered at $(-\cot 2\theta, 0)$. This is the contour of constant azimuth θ. To be more exact, it is the contour of constant $\tan 2\theta$.

The intersections of the circle with the u and v axes are investigated. From Eq. (7.13) with $u = 0$, the intersects P_1 and P_2 with the v axis are found:

$$v = \pm 1 \tag{7.15}$$

As a matter of fact, since the θ-dependent third term on the left-hand side of Eq. (7.13) vanishes with $u = 0$, the curves for all values of θ pass through the points P_1 and P_2.

From Eq. (7.14) with $v = 0$, the intersects with the u axis are found:

$$\begin{aligned} u &= -\cot 2\theta \pm \frac{1}{\sin 2\theta} \\ &= \frac{-\cos^2 \theta + \sin^2 \theta \pm 1}{2 \cos \theta \sin \theta} \end{aligned} \tag{7.16}$$

The intersections are

$$u_1 = \tan \theta \tag{7.17}$$

$$u_2 = -\cot \theta = \tan(\theta + 90°) \tag{7.18}$$

A series of circles for different values of θ are drawn in Fig. 7.4.

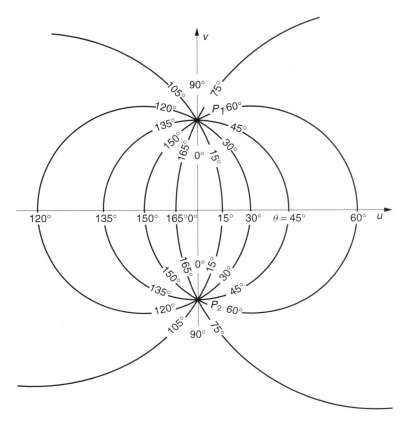

Figure 7.4 Constant θ curves.

A word of caution should be added. As mentioned at the beginning of this section, the above curves are not curves of constant θ but curves of constant $\tan 2\theta$. The multivalue problem of $\tan 2\theta$ has to be resolved. There is more than one value of θ that gives the same value of $\tan 2\theta$. For example, both $\theta = 30°$ and $\theta = 120°$ give the same value of $\tan 2\theta$. The value of $\tan 2\theta$ is the same for θ and $\theta \pm 90°n$, where n is an integer and the correct value of θ has to be selected. Equation (6.123) is useful for making this selection:

$$(a^2 - b^2)\cos 2\theta = A^2 - B^2 \qquad (6.123)$$

If a is chosen as the length of the major axis, then $a^2 - b^2$ is always a positive number. Inside the unit circle in the u–v plane, $B/A < 1$ and $A^2 - B^2$ is also positive. Only the θ that satisfies $\cos 2\theta > 0$ is permitted inside the unit circle.

On the other hand, outside the unit circle, $A^2 - B^2 < 0$ and, hence, only the θ that satisfies $\cos 2\theta < 0$ is permitted outside the unit circle. The values of θ indicated in Fig. 7.4 are based on this selection.

7.2.2 Lines of Constant Ellipticity ϵ

Similar to the lines of constant azimuth θ, the lines of constant ellipticity ϵ can be obtained by expressing $\epsilon(=\tan\beta)$ in terms of u and v. First, the case of the left-handed rotation, or $|\sin\Delta| = \sin\Delta$, is treated. Equation (6.112) will be used to find the constant ϵ line. Using the trigonometric relationship of Eq. (6.105), and then expressing $\tan\alpha$ in terms of $v/(\sin\Delta)$ by means of Eq. (7.6), Eq. (6.112) is rewritten as

$$\sin 2\beta = \frac{2v}{1 + \left(\dfrac{v}{\sin\Delta}\right)^2} \tag{7.19}$$

Now, $\sin^2\Delta$ is converted into $\tan^2\Delta$ so that Eq. (7.8) can be used. Equation (7.19) becomes

$$u^2 + v^2 - 2v\operatorname{cosec} 2\beta + 1 = 0 \tag{7.20}$$

which can be rewritten further as

$$u^2 + (v - \operatorname{cosec} 2\beta)^2 = \cot^2 2\beta \tag{7.21}$$

Equation (7.21) is the equation of a circle with radius $\cot 2\beta$ centered at $(0, \operatorname{cosec} 2\beta)$. The constant ϵ lines are plotted in Fig. 7.5.

Specific points on the circle will be investigated. The intersection with the u axis is examined first. Setting $v = 0$ in Eq. (7.20) gives $u^2 + 1 = 0$, and there is no intersection with the u axis. The intersections with the v axis are found from Eq. (7.20) with $u = 0$ as

$$v^2 - 2v\operatorname{cosec} 2\beta + 1 = 0 \tag{7.22}$$

The two solutions of Eq. (7.22) are

$$v = \operatorname{cosec} 2\beta \pm \sqrt{\operatorname{cosec}^2 2\beta - 1}$$
$$= \frac{1 \pm \cos 2\beta}{\sin 2\beta} \tag{7.23}$$

The solutions corresponding to the positive and negative signs in Eq. (7.23) are, respectively,

$$v_1 = \cot\beta = \frac{1}{\epsilon} \tag{7.24}$$

$$v_2 = \tan\beta = \epsilon \tag{7.25}$$

Since $\epsilon < 1$, and hence $v_1 > v_2$, the lower intersection explicitly represents ϵ. From this information, the constant ϵ lines for the left-handed rotation are drawn. These circles all stay in the upper half of the u–v plane.

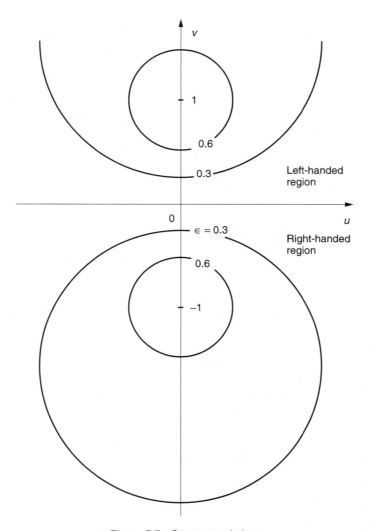

Figure 7.5 Constant ϵ circles.

Exactly the same procedure is repeated for the right-handed rotation except that $|\sin \Delta| = -\sin \Delta$ is used in Eq. (6.112). The result is

$$u^2 + (v + \operatorname{cosec} 2\beta)^2 = \cot^2 2\beta \qquad (7.26)$$

The circles represented by Eq. (7.26) fill the lower half-plane.

7.3 ARGAND DIAGRAM

The loci of the constant θ and those of the constant ϵ are combined together on the same plane in Fig. 7.6. This diagram is called an Argand diagram [3,6] because its shape resembles the spherical Argand gas lamps commonly used for street lighting in

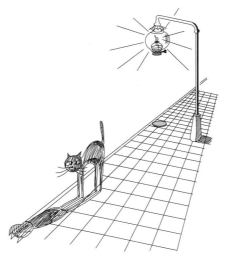

Argand lamp.

days of old [1]. The Argand diagram graphically represents ϵ and θ on the α and Δ plane. It permits us to find any two parameters in terms of the other two.

Just like the circle diagram and the Jones matrix described in Chapter 6, the Argand diagram not only simplifies the procedure of the calculation but also provides the states of polarization after each stage of the optical system.

The Poincaré sphere is nothing but a projection of the Argand diagram onto a unit sphere. Knowledge about the Argand diagram is essential to a clear understanding of the operation of the Poincaré sphere.

7.3.1   Solution Using a Ready-Made Argand Diagram

There are two ways of using the Argand diagram. One way is to draw in the lines on the fully completed Argand diagram. The other way is to construct the constant θ and ϵ curves only as needed. The advantage of the former way is simplicity, and that of the latter is accuracy. Both ways will be explained using examples. For the first two examples, the incident light is linearly polarized. Following these, two examples are given where the incident light is elliptically polarized.

Example 7.1   Using the Argand diagram, answer the questions in Example 6.9, which concerned the emergent state of polarization from a retarder with $\Delta = 38°$ and its fast axis in the x direction. The parameters of the incident linear polarization were $E_x = 2.0$ V/m and $E_y = 3.1$ V/m.

Solution   With the given values of $B/A = 1.55$ and $\Delta = 38°$, point P is picked as shown in Fig. 7.6. The answer is read from the diagram as $\theta = 60°$ and $\epsilon = 0.3$, which matches the results that were given in Example 6.9.   □

Example 7.2   Linearly polarized light with azimuth $\theta = 165°$ is incident on a $\lambda/4$ plate. Find the state of polarization of the emergent wave when the fast axis of the $\lambda/4$ plate is (a) along the x axis and (b) along the y axis.

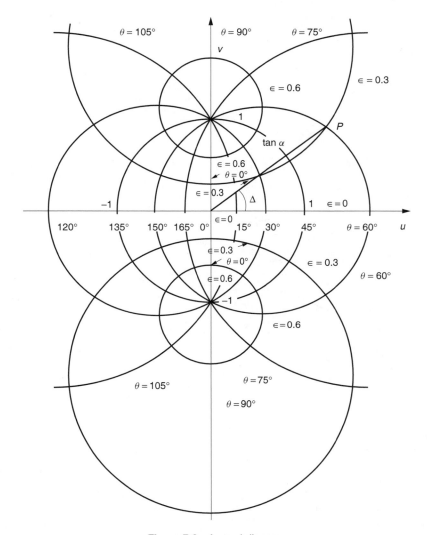

Figure 7.6 Argand diagram.

Solution

(a) The Argand diagram facilitates the calculation. Figure 7.7 shows the answer. Since the incident light is linearly polarized, $\epsilon = 0$. The point P_1, which represents the incident light, is $\theta = 165°$, $\epsilon = 0$; or $\theta = 165°$ on the u axis. From the measurement of the length $0P_1 = 0.27$, the incident wave is

$$\left(\frac{B}{A}\right) e^{j\Delta} = 0.27 \ e^{j180°}$$

The introduction of the $\lambda/4$ plate delays the phase of the y component by $\Delta = 90°$ more than that of the x component and the component field ratio becomes $0.27 \ e^{j270°}$. Point P_1 moves to point P_2. The answer is right-handed elliptical polarization with $\theta = 0°$ and $\epsilon = 0.27$.

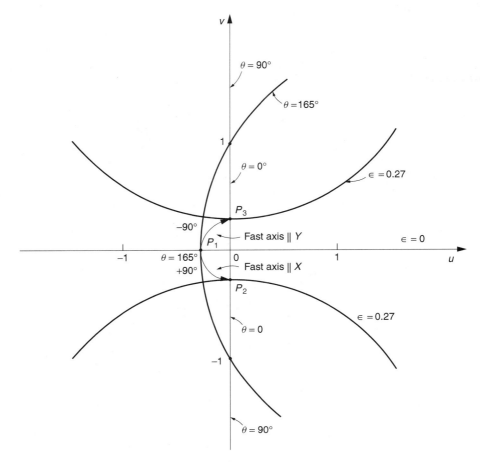

Figure 7.7 Solution of Example 7.2.

(b) This situation is similar to (a) but with $\Delta = -90°$. The point P_1 moves to point P_3 and the answer is the left-handed elliptical polarization with $\theta = 0°$ and $\epsilon = 0.27$. ☐

Example 7.3 Example 6.1 looked at how the state of polarization of a horizontally linearly polarized wave changes as the azimuth of the fast axis of the $\lambda/4$ plate is rotated. Repeat this exercise using the Argand diagram. Solve only for $\Theta = 22.5°$, for $\Theta = 67.5°$, and for $\Theta = 112.5°$.

Solution It should be remembered that Eqs. (6.1) and (6.2) are based on the condition that the x axis is parallel to the fast axis of the retarder. Just as was done analytically in Section 6.8.1, the incident field **E** has to be represented in the coordinates rotated by Θ so that the new x' axis aligns with the fast axis of the retarder. After this has been done, the retardation is accounted for and, finally, the coordinates are rotated back by $-\Theta$ to express the result in the original coordinates.

Let us start with the case $\Theta = 22.5°$. The incident light is $(\theta, \epsilon) = (0, 0)$ and is represented by P_0 in Fig. 7.8. The coordinates are rotated by $+22.5°$ so that the new x'

axis aligns with the fast axis. The azimuth of the incident light in the new coordinates decreases by the same amount and in the rotated coordinates, $\theta = -22.5°$, which is equivalent to $\theta = 157.5°$. Recall that $\theta = 157.4°$ rather than $\theta = -22.5°$ has to be used because of the restriction of $0 \leq \theta < 180°$. This point is represented by P_1^1 at $(\theta, \epsilon) = (157.5°, 0)$. The $\lambda/4$ plate rotates P_1^1 by $90°$ as indicated by the dotted line to P_2^1 at $(\theta, \epsilon) = (0, 0.42)$. The coordinates are then rotated back by $-22.5°$ in order to express the emergent light in the original coordinates. The azimuth of the emergent light in the rotated back coordinates increases the same amount. P_2^1 moves along the $\epsilon = 0.42$ line to point P_3^1 at $(\theta, \epsilon) = (22.5°, 0.42)$. The handedness is right-handed.

Next, the case with $\Theta = 67.5°$ will be solved. The incident light is again at P_0. The coordinates are rotated by $+67.5°$ to match the new x' axis to the fast axis. The azimuth of the incident light in the new coordinates decreases by the same amount and becomes $\theta = -67.5°$, which is equivalent to $\theta = 112.5$, which meets the restriction $0 \leq \theta < 180°$. The incident light is represented by P_1^3 at $(\theta, \epsilon) = (112.5°, 0)$. The retardance of $90°$ brings the point P_1^3 as indicated by the dotted line to P_2^3 at $(\theta, \epsilon) = (90°, 0.42)$.

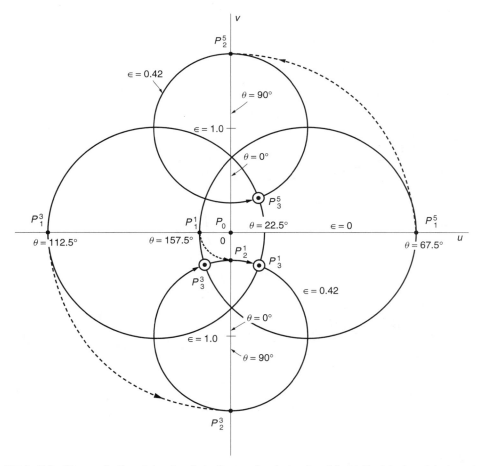

Figure 7.8 Change in the state of polarization as the fast axis of the $\lambda/4$ plate is rotated, as in Example 6.1.

Back-rotation of the coordinates moves P_2^3 along the constant ϵ line to P_3^3 at $(\theta, \epsilon) = (157.5°, 0.42)$. The handedness is also right-handed.

Finally, the case the $\Theta = 112.5°$ will be solved. The coordinates are rotated by $+112.5°$ and as a result the azimuth of the incident light in the new coordinates is $\theta = -112.5°$, which is equivalent to $\theta = 67.5°$ to meet the restriction of $0 \leq \theta < 180°$. The incident light is indicated by P_1^5 at $(\theta, \epsilon) = (67.5°, 0)$. The retardance rotates P_1^5 as indicated by the dotted line to P_2^5, where $(\theta, \epsilon) = (90°, 0.42)$. Back-rotation of the coordinates by $-112.5°$ along $\epsilon = 0.42$ brings P_2^5 to P_3^5 at $\theta = 202.5°$, which is equivalent to $\theta = 22.5°$, which meets the restriction $0 \leq \theta < 180°$. The final result is at P_3^5 with $(\theta, \epsilon) = (22.5, 0.42)$. The handedness is left-handed. $\square$

The incident waves of the next two examples are elliptically polarized.

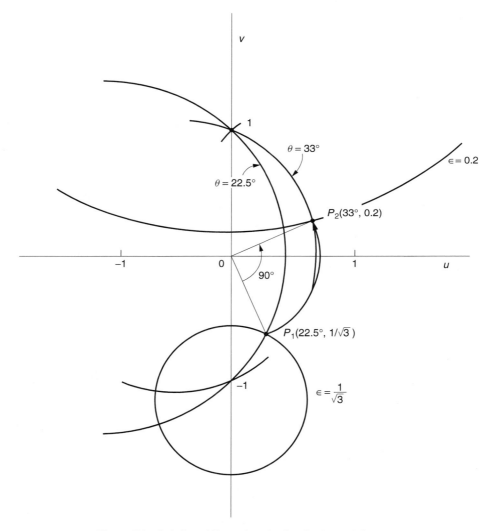

Figure 7.9   Solution of Example 7.4 using the Argand diagram.

Example 7.4   A right-handed elliptically polarized wave with $\theta = 22.5°$ and $\epsilon = 1/\sqrt{3}$ is incident on a $\lambda/4$ plate whose fast axis is along the x axis. Find the state of polarization of the emergent light using the Argand diagram.

Solution   Point P_1, which represents the state of polarization of the incident wave, is plotted at $(\theta, \epsilon) = (22.5°, 1/\sqrt{3})$ in the lower half-plane as in Fig. 7.9. The $\lambda/4$ plate with its fast axis along the x axis moves P_1 by $90°$ counterclockwise along the circumference with radius $0P_1$ to point P_2. The coordinates of P_2 represent the state of polarization of the emergent wave. Point P_2 is read from the Argand diagram at approximately $(\theta, \epsilon) = (33°, 0.2)$ in the upper half-plane. The emergent wave therefore has left-handed elliptical polarization with $\theta = 33°$ and $\epsilon = 0.2$.   □

Example 7.5   The same elliptically polarized wave is incident onto a $\lambda/4$ plate but this time its fast axis is not along the x axis but is tilted from the x axis by $\Theta = 45°$. Find the state of polarization of the emergent wave.

Solution   Now that the fast axis is not along the x axis, the coordinates have to be rotated by Θ; then the retardance is accounted for, followed by a rotation back by $-\Theta$. The state of polarization of the incident wave is represented by P_1 at $(\theta, \epsilon) = (22.5°, 1/\sqrt{3})$ in the lower half-plane in Fig. 7.10. Due to the rotation of the coordinates, point P_1 moves along the $\epsilon = 1/\sqrt{3}$ line to point P_2 at $(\theta, \epsilon) = (-22.5°, 1/\sqrt{3}) = (157.5°, 1/\sqrt{3})$. The $\lambda/4$ plate rotates the point by $+90°$ to point P_3 at $(\theta, \epsilon) = (34.2°, 0.2)$. Finally, the coordinates are rotated back by $-\Theta = 45°$. The azimuth of the emergent light in the rotated back coordinates increases the same amount. P_3 is brought to the final point P_4 at $(\theta, \epsilon) = (79.2°, 0.2)$ in the lower half-plane. The emergent wave has right-handed elliptical polarization with $\theta = 79.2°$ with $\epsilon = 0.2$.   □

7.3.2 Orthogonality Between Constant θ and ϵ Lines

It will be shown that the constant θ lines and the constant ϵ lines are orthogonal to each other. Using this fact, a constant ϵ line can be drawn from a constant θ line or vice versa. In the next section, the orthogonality relationship is used to construct custom-made Argand diagrams for specific problems.

The slope of the constant θ line is the derivative of Eq. (7.13) with respect to u,

$$2u + 2vv' + 2\cot 2\theta = 0 \tag{7.27}$$

where $v' = dv/du$. Multiplying Eq. (7.27) by u and subtracting Eq. (7.13) gives

$$v' = -\frac{u^2 - v^2 + 1}{2uv} \tag{7.28}$$

Similarly, the slope of the line of constant ϵ is, from Eq. (7.20),

$$2u + 2vv' - 2v' \csc 2\beta = 0 \tag{7.29}$$

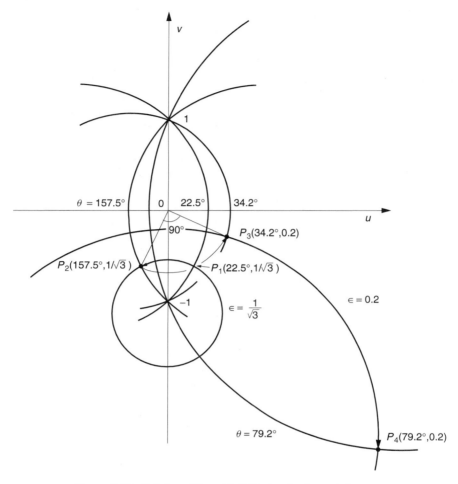

Figure 7.10   Solution of Example 7.5 using the Argand diagram.

Multiplying Eq. (7.29) by v and multiplying Eq. (7.20) by v', and then subtracting gives

$$v' = \frac{2uv}{u^2 - v^2 + 1} \tag{7.30}$$

From Eqs. (7.28) and (7.30), the slopes of the two curves are negative reciprocals and therefore the two curves are orthogonal to each other.

7.3.3   Solution Using a Custom-Made Argand Diagram

In using an already drawn Argand diagram, one frequently has to rely on interpolation between the lines, unless the point falls exactly on the line. What will be described here is a method of drawing specific constant θ and ϵ lines to solve a particular problem. The accuracy of a custom-made Argand diagram is higher than the accuracy of interpolating a ready-made diagram.

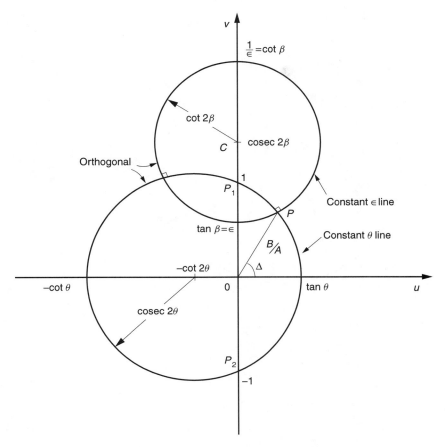

Figure 7.11 Curves for constant θ and ϵ.

First of all, information about the intersections with the axes is useful in drawing lines. The information so far obtained is summarized in Fig. 7.11.

Figure 7.12 illustrates how to draw the constant θ and ϵ lines that were used to solve Example 7.4 in Fig. 7.9. The circled numbers in the figure correspond to the step numbers below. First, the circles associated with the state of polarization of the incident light are drawn. The effect of the $\lambda/4$ plate is accounted for to obtain the final results.

① Find the point $(u, v) = (\tan 22.5°, 0)$ on the u axes. If 5 cm is taken as the unit length in the drawing, (u, v) is located at 2.1 cm horizontally from the origin.

② Draw the bisect of a line connecting $(\tan 22.5°, 0)$ and $(0, -1)$. Extend the bisect to find intersection C_1 with the u axis.

③ Centered at C_1, draw the constant θ circle passing through points $(\tan 22.5°, 0)$ and $(0, -1)$. This is the circle of $\theta = 22.5°$.

④ Next, the line of $\epsilon = 1/\sqrt{3}$ will be found. Find the point $(u, v) = (0, -1/\sqrt{3})$, which is -2.9 cm vertically from the origin from Eq. (7.25).

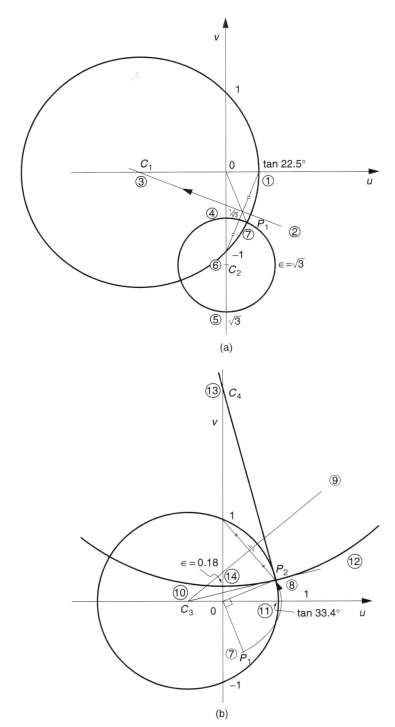

Figure 7.12   How to draw the constant θ and ϵ lines. (a) The first seven steps for drawing your own Argand diagram. (b) Steps from 7 to 13 for drawing your own Argand diagram.

⑤ Find the point $(u, v) = (0, -\sqrt{3})$, which is -8.7 cm vertically from the origin from Eq. (7.24).

⑥ Find the center $C_2(0, -\frac{1}{2}(1/\sqrt{3} + \sqrt{3}))$. C_2 is the circle of $\epsilon = 1/\sqrt{3}$.

⑦ The intersection P_1 between the circles centered at C_1 and C_2 is found. The phasor $\overline{0P_1}$ represents the component field ratio $0.69e^{-j67.5°}$ of the incident light.

⑧ Rotate point P_1 counterclockwise by $90°$ to point P_2 to account for the retardance $\Delta = 90°$ as shown in Fig. 7.12b.

⑨ Next, the value of θ at P_2 will be found. Find the bisect of the line connecting P_2 and the point $(0,1)$, and extend it to find the intersection C_3 with the u axis.

⑩ Centered at C_3 and passing through points P_2 and $(0,1)$, draw the constant θ circle.

⑪ The intersection of the circle centered at C_3 with the u axis is measured as 3.3 cm horizontally from the origin. The value of u at this point is 0.66; $\tan\theta = 0.66$ or $\theta = 33.4°$.

⑫ Finally, the circle of constant ϵ will be found. Draw the straight line $\overline{C_3P_2}$.

⑬ Draw the normal to the straight line $\overline{C_3P_2}$ from the point P_2, and this will locate the intersection C_4 with the v axis. Orthogonality between constant θ and ϵ lines is being invoked to construct the desired ϵ line from the $\theta = 33.4°$ line.

⑭ Centered at C_4, draw the constant ϵ circle passing through P_2. The intersection of the circle with the v axis is 0.9 cm from the origin, which is $v = 0.18$. Thus, $\epsilon = 0.18$.

The state of polarization of the emergent light is left-handed circularly polarized with $(\theta, \epsilon) = (33.4°, 0.18)$. The custom-made Argand diagram provides better accuracy than the results obtained in Example 7.4.

7.4 FROM ARGAND DIAGRAM TO POINCARÉ SPHERE

The Poincaré sphere is generated by back projecting the Argand diagram onto a unit diameter sphere. Figure 7.13 illustrates the relative orientation between the Poincaré sphere and the Argand diagram. The surface of the Poincaré sphere touches at the origin of the Argand diagram. The real axis u is back-projected onto the equator of the sphere and the imaginary axis v is back-projected onto the great circle passing through the north and south poles of the sphere. The plane of the *great circle* contains the center of the sphere. The diameter of the sphere being unity, points (0, 1) and (0, −1) on the Argand diagram back-project onto the north and south poles, respectively. Back-projections of other general points will be calculated using elementary analytic geometry in the next section.

7.4.1 Analytic Geometry of Back-Projection

As shown in Fig. 7.13, all the points on the Argand diagram are back-projected onto the single point $0'$, which is diametrically opposite to the tangent point 0. The concept of the Poincaré sphere boils down to finding the intersection of a straight line with the surface of a sphere.

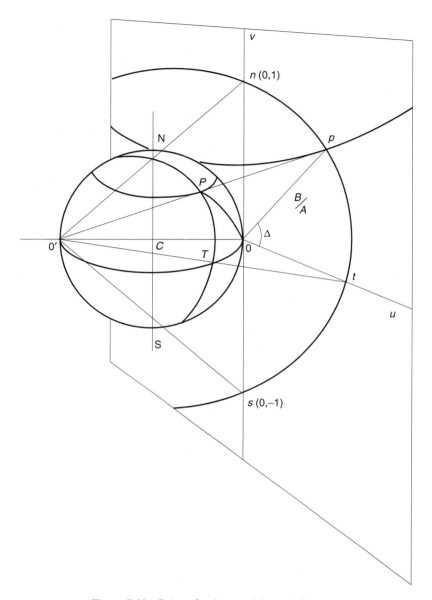

Figure 7.13 Poincaré sphere and Argand diagram.

Let us begin with the expression for a straight line passing through two specific points in space. Referring to Fig. 7.14, let the coordinates of these two points P_1 and P_2 be (a_1, b_1, c_1) and (a_2, b_2, c_2). If these two points are represented by position vectors $\mathbf{r}_1$ and $\mathbf{r}_2$, then the line segment $\overline{P_1P_2}$ is expressed by the vector $\mathbf{r}_2 - \mathbf{r}_1$. Let us pick another point P at (x, y, z). If all three points P_1, P_2, and P lie on the same straight line, then the line segments $\overline{P_1P}$ and $\overline{P_1P_2}$ must be parallel and share at least one point in common. The line segment $\overline{P_1P}$ is represented by the vector $\mathbf{r} - \mathbf{r}_1$. The line segments $\overline{P_1P}$ and $\overline{P_1P_2}$ share point P_1. Now, the condition that the two line segments

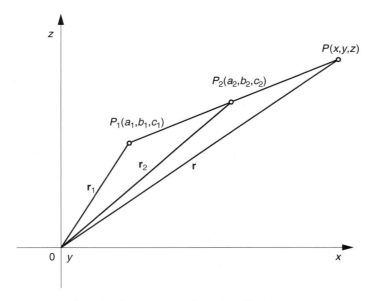

Figure 7.14 Geometry of a straight line in space.

are parallel is that the vector product is zero:

$$(\mathbf{r}_2 - \mathbf{r}_1) \times (\mathbf{r} - \mathbf{r}_1) = 0 \qquad (7.31)$$

Equation (7.31) leads to the following expression of a straight line passing through points $P_1, P_2,$ and P:

$$\frac{x - a_1}{a_2 - a_1} = \frac{y - b_1}{b_2 - b_1} = \frac{z - c_1}{c_2 - c_1} \qquad (7.32)$$

Both the sphere and the Argand diagram will be put into x, y, z coordinates, as shown in Fig. 7.15. Let the point where two surfaces touch be O. Let us pick the origin C of the x, y, z coordinates at the center of the Poincaré sphere, and let the radius of the Poincaré sphere be $\frac{1}{2}$. Let $P(x, y, z)$ be a point on the surface of the sphere. The choice of the senses of the coordinates should be noted. The positive x direction is from C to 0. The positive y direction is antiparallel to the u axis (indicated as $\overrightarrow{CW}$ in the figure) so that the direction of the positive z axis is vertically upward in the right-hand rectangular coordinate system. In this arrangement the Argand diagram stands vertically in the plane of $x = \frac{1}{2}$ with its imaginary axis parallel to the z axis. The point $0'$ diametrically opposite to point 0 is at $(-\frac{1}{2}, 0, 0)$. All back-projections converge to point $0'$.

The coordinates of a point p on the Argand diagram are, from Eq. (7.6),

$$(a_2, b_2, c_2) = \left(\tfrac{1}{2}, - \tan \alpha \cos \Delta, \tan \alpha \sin \Delta \right)$$

and the coordinates of point $0'$ are

$$(a_1, b_1, c_1) = \left(-\tfrac{1}{2}, 0, 0 \right)$$

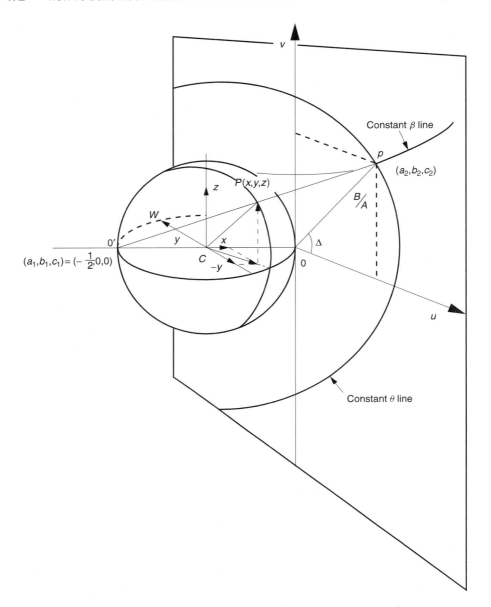

Figure 7.15   Projection of a point on the Argand diagram to the Poincaré sphere.

The expression for the straight line connecting $0'$ and p is, from Eq. (7.32),

$$\frac{x + \frac{1}{2}}{1} = \frac{y}{-\tan\alpha\cos\Delta} = \frac{z}{\tan\alpha\sin\Delta} \tag{7.33}$$

The expression for the unit diameter sphere is

$$x^2 + y^2 + z^2 = \tfrac{1}{4} \tag{7.34}$$

The solution of the simultaneous equations (7.33) and (7.34) gives the intersection of the line with the sphere. The calculation of the solution is simpler than it looks. First, the value of x will be found. Solving Eq. (7.33) for y and z in terms of x gives

$$y = - \left(x + \tfrac{1}{2}\right) \tan \alpha \cos \Delta \tag{7.35}$$

$$z = \left(x + \tfrac{1}{2}\right) \tan \alpha \sin \Delta \tag{7.36}$$

Insertion of Eqs. (7.35) and (7.36) into Eq. (7.34) gives

$$x^2 \cos^2 \alpha + \left(x + \tfrac{1}{2}\right)^2 \sin^2 \alpha = \tfrac{1}{4} \cos^2 \alpha$$

which can be rewritten in terms of $\cos 2\alpha$ as

$$4x^2 + 2x(1 - \cos 2\alpha) - \cos 2\alpha = 0 \tag{7.37}$$

which can be factored as

$$(2x - \cos 2\alpha)(2x + 1) = 0 \tag{7.38}$$

The intersections are at

$$2x = \cos 2\alpha \tag{7.39}$$

and

$$2x = -1 \tag{7.40}$$

Equation (7.39) gives the intersection P, and Eq. (7.40) gives the point $0'$ on the sphere. Using Eq. (6.125), Eq. (7.39) becomes

$$2x = \cos 2\beta \cos 2\theta \tag{7.41}$$

Next, y is obtained by inserting Eq. (7.39) into (7.35) and expressing $\cos 2\alpha$ in terms of $\cos \alpha$ as

$$2y = - \sin 2\alpha \cos \Delta \tag{7.42}$$

Like Eq. (7.41), Eq. (7.42) will be expressed in terms of β and θ. In Eq. (6.99), $\tan 2\alpha$ is first expressed as $\sin 2\alpha / \cos 2\alpha$ and then Eq. (6.125) is used as a substitution for $\cos 2\alpha$. The resulting relationship is

$$\cos 2\beta \sin 2\theta = \sin 2\alpha \cos \Delta \tag{7.43}$$

Insertion of Eq. (7.43) into (7.42) gives

$$2y = - \cos 2\beta \sin 2\theta \tag{7.44}$$

Finally, the value of z will be obtained from Eqs. (7.36) and (7.39):

$$2z = \sin 2\alpha \sin \Delta \tag{7.45}$$

Using Eq. (6.112),

$$2z = \begin{cases} \sin 2\beta & \text{for } \sin \Delta > 0 \\ - \sin 2\beta & \text{for } \sin \Delta < 0 \end{cases} \tag{7.46}$$

In summary, the coordinates of the projected point P on the sphere are

$$x = \tfrac{1}{2} \cos 2\beta \cos 2\theta \tag{7.41}$$

$$y = -\tfrac{1}{2} \cos 2\beta \sin 2\theta \tag{7.44}$$

$$z = \pm \tfrac{1}{2} \sin 2\beta \tag{7.46}$$

where the $+$ sign of z is for left-handed rotation while the $-$ sign of z is for right-handed rotation. The Poincaré sphere can now be drawn from these results.

7.4.2 Poincaré Sphere

Let us first express point $P(x, y, z)$ in terms of the latitude l and longitude k on the sphere, rather than (x, y, z) coordinates. The latitude l (angle of elevation) of an arbitrary point (x, y, z) on the sphere of radius $\tfrac{1}{2}$ is, from Fig. 7.16,

$$l = \sin^{-1} 2z \tag{7.47}$$

The z coordinate of P, back-projected from point p, which is the intersection point of the constant θ and ϵ lines on the Argand diagram, is given by Eq. (7.46). Inserting Eq. (7.46) into (7.47) gives

$$l = 2\beta \tag{7.48}$$

Thus, the latitude l of point P is 2β and is linearly proportional to β. The constant β (or ϵ) lines on the Poincaré sphere are equally spaced in β, unlike the unequal spacing of the corresponding lines in the Argand diagram.

Moreover, as point p on the Argand diagram shown in Fig. 7.16 moves along the constant β line, its height in the v direction varies. The corresponding point P on the Poincaré sphere, however, stays at the same height or at the same latitude l, and the constant β lines are cylindrically symmetric.

Next, the longitude k of an arbitrary point (x, y, z) on the sphere is, from Fig. 7.17,

$$k = \tan^{-1} \left(\frac{-y}{x} \right) \tag{7.49}$$

The x and y coordinates of P back-projected from point p are given by Eqs. (7.41) and (7.44), and they are inserted into Eq. (7.49) to obtain

$$k = 2\theta \tag{7.50}$$

The longitude k is 2θ and is linearly proportional to θ. The constant θ lines are also equally spaced in θ on the Poincaré sphere. Thus, both constant β lines and constant θ lines are equally spaced on the Poincaré sphere. These equal spacings together with the above-mentioned cylindrical symmetry make the Poincaré sphere strikingly more versatile than the Argand diagram.

The back-projection of the other two quantities B/A and Δ will now be considered. An arc that connects two points on a spherical surface with the shortest distance along the surface is called a *geodesic*. In Fig. 7.18, the projection of $\overline{0p}$ in the Argand diagram is the geodesic $\overline{0P}$ on the Poincaré sphere. Since the Argand diagram is the tangent plane 0, the angle Δ on the Argand diagram is preserved when it is back-projected onto the sphere, and the angle of the geodesic $\overline{0P}$ with respect to the equator is also Δ.

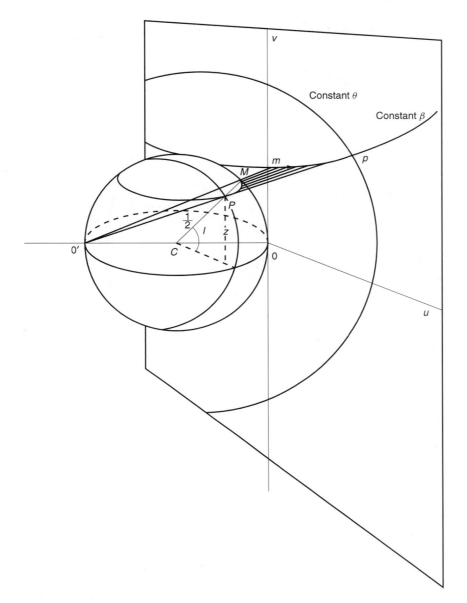

Figure 7.16 Projection of the constant ϵ curve onto the Poincaré sphere.

As shown in Fig. 7.18, $B/A (= \tan \alpha)$ is represented by $\overline{0p}$. Since $\overline{00'} = 1$,

$$\angle P0'0 = \alpha \tag{7.51}$$

$$\angle PC0 = 2\alpha \tag{7.52}$$

Thus, the Poincaré sphere provides a quick way of finding α from a given state of polarization. Referring to Fig. 7.19, it is interesting to note that, the radius of the sphere being $\frac{1}{2}$, geodesic $\overline{PT}$ is β, and geodesic $\overline{OP}$ is α.

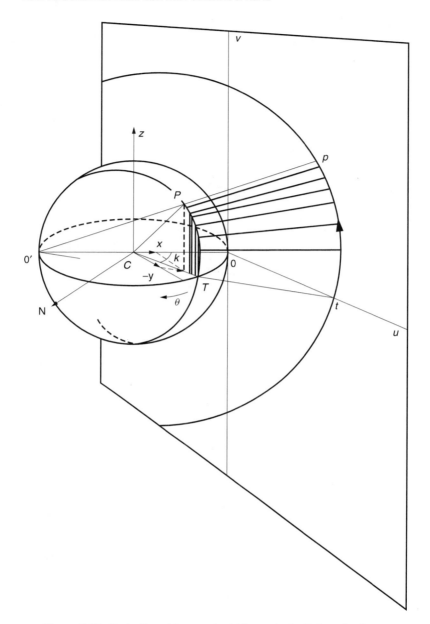

Figure 7.17   Projection of the constant θ line onto the Poincaré sphere.

Finally, the direction of the movement of the back-projected point on the sphere relative to that on the Argand diagram will be considered. Referring to Fig. 7.16, as the point moves from the left to the right, that is, from m to p on the Argand diagram, the projected point moves from the right to the left, that is, from M to P on the sphere, when observed from outside the sphere facing toward the center of the sphere. The points on the Argand diagram and on the sphere are more or less like mirror images of each other and are left and right reversed. This mirror image effect influences the

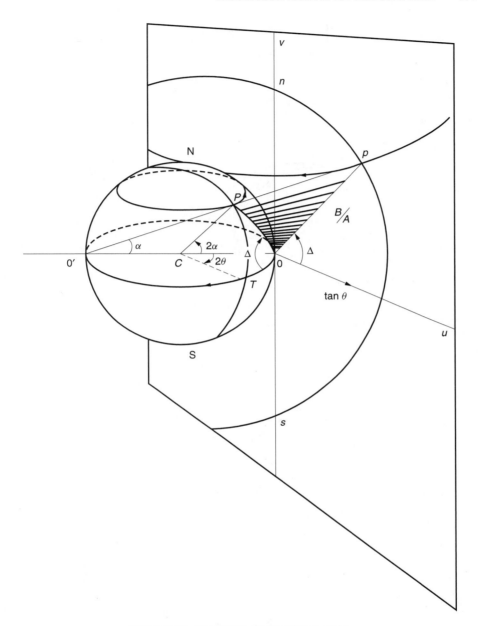

Figure 7.18 Directions of increasing Δ and θ.

direction of increasing Δ. The direction of increasing Δ is counterclockwise on the Argand diagram, whereas on the Poincaré sphere, the direction of increasing Δ is clockwise, as indicated when the sphere is viewed from the reader's vantage point in Fig. 7.18. Similarly, referring to Fig. 7.18, θ increases toward the right on the Argand diagram, and the direction of increasing θ is clockwise when the sphere is viewed from a point above the north pole. Whenever confusion about the direction arises, always go back to the Argand diagram, which is, after all, the basis for the Poincaré sphere.

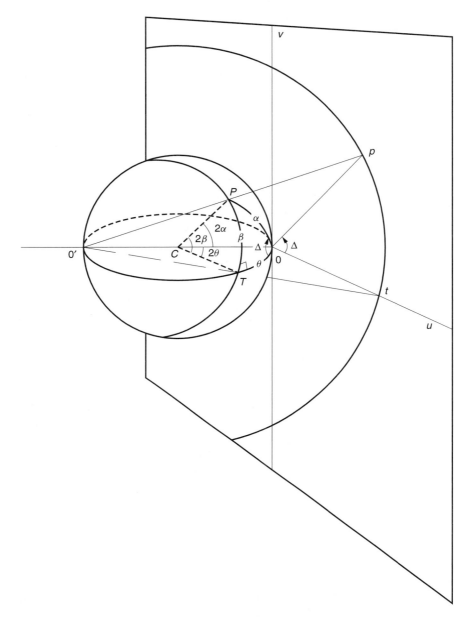

Figure 7.19 Geodesic triangle.

Figure 7.20 illustrates various states of polarization and their corresponding locations on the Poincaré sphere. The latitude lines are constant β (or ϵ) lines. Along the equator, $\epsilon = 0$ and the states of polarizations are linearly polarized waves with various azimuth angles. As higher latitudes are approached, the ellipticity increases and finally reaches unity at the poles. The states of polarization with β are represented by the latitude of 2β. The state of polarization $\beta = \pi/4$ (or $\epsilon = \tan\beta = 1$) is represented by the north pole of the sphere.

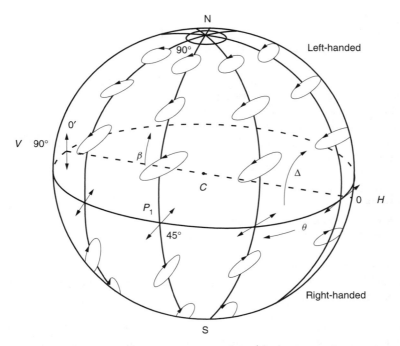

Figure 7.20 Poincaré sphere based on the convention of $e^{-j\omega t}$. The Poincaré sphere based on the $e^{j\omega t}$ convention is obtained by rotating in the plane of the page by 180°.

The longitudinal lines are constant θ (or azimuth) lines. The reference of the longitude is the longitudinal line passing through point 0. The state of polarization with θ is represented by the longitudinal line of 2θ. The state of polarization with $\theta = 90°$ is represented by the longitudinal line that passes through point 0′. The value of the azimuth increases in a clockwise sense when the sphere is viewed from a point above the north pole. The state of polarization with $\theta = 180°$ is represented by the longitude line that passes through point 0 again. Point 0 represents horizontal linear polarization and point 0′, vertical linear polarization. These points are often designated by points H and V. The azimuths of the ellipses along the same longitude stay the same. The states of polarization in the northern hemisphere are all left-handed rotation, while those in the southern hemisphere are right-handed rotation.

It is important to remember that the physical locations of the constant θ and β graduation lines are at 2θ and 2β, respectively, on the Poincaré sphere, while the retardance $\Delta°$ means a rotation of the point on the Poincaré sphere by Δ degrees, not 2Δ degrees.

Depending on which convention, $e^{-j\omega t}$ or $e^{j\omega t}$ is used, the same retardance can be expressed by either Eq. (6.8) or (6.10). The Poincaré sphere shown in Fig. 7.20 is based on the $e^{-j\omega t}$ convention of this book. The Poincaré sphere based on $e^{j\omega t}$, however, can be obtained by rotating the figure by 180° in the plane of the page.

7.5 POINCARÉ SPHERE SOLUTIONS FOR RETARDERS

Uses of the Poincaré sphere will be explained by a series of examples [4].

Example 7.6   Light linearly polarized at $\theta = 45°$ is incident onto a $\lambda/4$ plate whose fast axis is along the x axis. Find the state of polarization of the emergent light using both the Argand diagram and the Poincaré sphere.

Solution   The input light is characterized by $B/A = 1$ and $\epsilon = 0$ and is represented by point p_1 at $(u, v) = (1, 0)$ on the Argand diagram, as shown in Fig. 7.21. The $\lambda/4$ plate with its fast axis along the x axis rotates point p_1 by $90°$ to p_2 at $(u, v) = (0, 1)$, which represents left-handed circular polarization.

The same problem will be solved by using the Poincaré sphere. The back-projected point from point p_1 on the Argand diagram to the Poincaré sphere is P_1 at $(\theta, \epsilon) = (45°, 0)$ in either Fig. 7.20 or 7.21.

Geodesic $\overline{0P_1}$ represents B/A. With point 0 as the center of rotation, point P_1 is rotated by $90°$ in a clockwise direction due to the $\lambda/4$ plate. The final point P_2 is on the north pole. The answer is left-handed circularly polarized light. The answers agree with the result in Fig. 6.4.

The next example deals with a retarder whose fast axis is at an arbitrary angle and is along neither the x nor y axis. A significant advantage of the Poincaré sphere over the Argand diagram is seen.

Example 7.7   A linearly polarized light with azimuth $\theta = 40°$ is incident onto a $\lambda/4$ plate whose fast axis is oriented at $\Theta = 30°$. Find the state of polarization of the emergent light.

Solution   The orientation of the incident light and the $\lambda/4$ plate are represented by P_1 and R, respectively, in Fig. 7.22a. First, the coordinates have to be rotated by $30°$ so that the new x' axis lines up with the fast axis of the $\lambda/4$ plate. When the coordinates are rotated by $+30°$, the values of θ and Θ are decreased by $30°$ as shown in Example 7.3. After the rotation of the coordinate system, point P_1 and R are transferred to P_1' and R' at $\theta = 10°$ and $0°$, respectively, as shown in Fig. 7.22b. The $\lambda/4$ plate rotates P_1' to P_2' around R'. The final result is obtained by rotating the coordinates back by $-30°$ and the value of θ is increased by $30°$. As shown in Fig. 7.22c, the final point is P_2''. Approximate values of $\theta = 30°$, $\epsilon = 0.17$ can be read directly from the graduation on the Poincaré sphere.

An important feature of the Poincaré sphere is that rotation and rerotation of the coordinate system can be avoided because of the cylindrical symmetry of the Poincaré sphere with respect to the polar axis. Geodesic $\overline{P_1'P_2'}$, which was obtained following the rotation of the coordinates is the same as geodesic $\overline{P_1''P_2''}$. Rotation of the coordinates by $30°$ followed by rotation back by $-30°$ is not necessary. The same result can be obtained by rotating P_1'' to P_2'' by $90°$ around point R'' from the very beginning owing to the symmetry that the Poincaré sphere has. This is one of the major advantages of the Poincaré sphere. Compare this with the steps needed when the Argand diagram was used in Example 7.3.

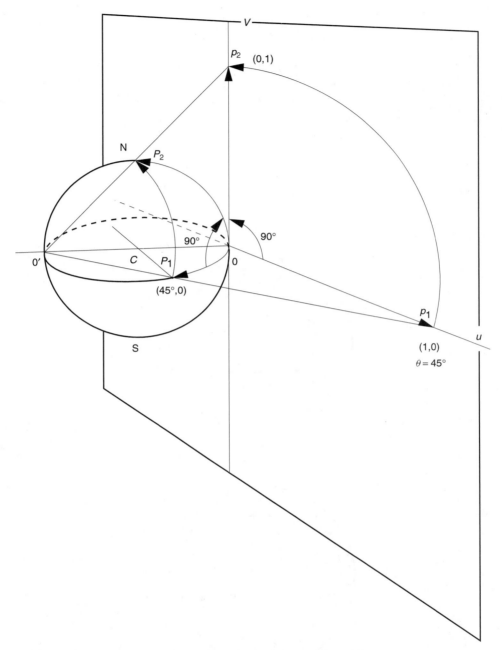

Figure 7.21 Linearly polarized light with $\theta = 45°$ is incident onto a $\lambda/4$ plate whose fast axis is along the x axis. The Poincaré sphere with Argand diagram is used to find the emergent wave.

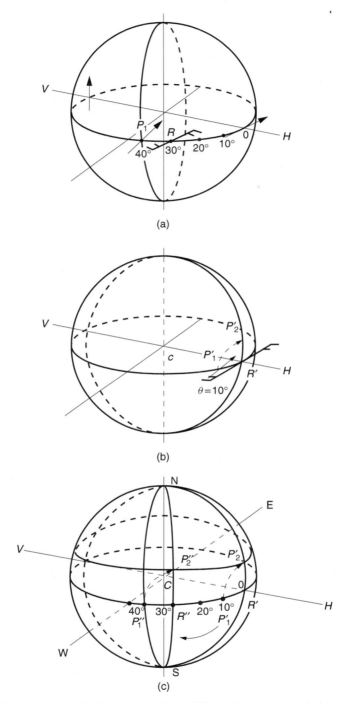

Figure 7.22   Linearly polarized light with azimuth $\theta = 40°$ is incident onto a $\lambda/4$ plate whose fast axis is oriented at $\Theta = 30°$. (a) State of polarization of the incident wave and the $\lambda/4$ plate are plotted on the Poincaré sphere. (b) Conversion of the state of polarization due to the $\lambda/4$ plate after rotation of the coordinates. (c) Similarity between the operations with and without the rotation of the coordinates.

A more accurate number can be obtained by noting the geometry. As already shown in Fig. 7.19, geodesic $\overline{P_2''R''}$ in Fig. 7.22c represents β. From the relationship $\overline{P_2''R''} = \overline{P_1''R''}$ combined with $\overline{P_1''R''} = (\pi/180°) \times 10$, β is found to be $\beta = (\pi/180°) \times 10$ rad and $\epsilon = \tan\beta = 0.176$. ☐

Example 7.8 Using the Poincaré sphere solve the following problems:

(a) Prove that a $\lambda/4$ plate can convert a left-handed or right-handed circularly polarized wave into a linearly polarized wave regardless of the orientation of the fast axis of the $\lambda/4$ plate.

(b) Identify the handedness of an incident circularly polarized wave by using a $\lambda/4$ plate of known fast axis orientation.

(c) Find the proportion of powers of each handedness when the incident light is a mixture of left-handed and right-handed circularly polarized waves.

(d) Consider the converse to part (a). Can a $\lambda/4$ plate convert a linearly polarized wave into a circularly polarized wave, regardless of the orientation of the fast axis of the $\lambda/4$ plate?

Solution

(a) As seen from Fig. 7.23, regardless of the orientation of the $\lambda/4$ plate, the 90° rotation from either pole brings the point onto the equator and the emergent light becomes linearly polarized. The direction of the emergent linear polarization, however, depends on the orientation of the fast axis of the $\lambda/4$ plate.

(b) As seen from Fig. 7.23, when the fast axis is oriented at Θ, the direction of the polarization of the emergent light is at $\theta = \Theta \pm 45°$. If the direction of polarization of the emergent light is at $\theta = \Theta + 45°$, the incident light has right-handed rotation, and if the direction is at $\theta = \Theta - 45°$, the incident light has left-handed rotation. In this manner, the handedness of the incident wave is identified.

(c) As shown in Fig. 7.23, mixed left and right circularly polarized waves are incident on a $\lambda/4$ plate with fast axis at $\Theta = 45°$. The north pole is brought to H and the south pole is brought to V. The emergent light is a combination of horizontal and vertical linear polarization, which can be separated using a polarization beam-splitter. Figure 7.24 illustrates the arrangement for measuring the ratio, between the two oppositely handed rotations.

(d) As illustrated in Fig. 7.25, if the orientation of the fast axis is other than $\pm 45°$ with respect to the direction of the linear polarization of the incident light, represented by point P_1, then it is not possible for the emergent state, represented by P_1', to reach the pole. Therefore, the converse to part (a) is not true. A $\lambda/4$ plate cannot convert a linearly polarized wave into a circularly polarized wave when the fast axis is in an arbitrary orientation. A circularly polarized wave is obtainable only when the fast axis is at $\pm 45°$ with respect to the direction of linear polarization of the incident light. ☐

Example 7.9 Fabricate your own Poincaré sphere by drawing the state of polarizations on flat one-eighth sectors that can later be taped together to form a sphere. Choose the length of $\sqrt{A^2 + B^2}$ to be $\pi/20$ of the radius of the Poincaré sphere. Draw the states of polarization at steps of $\Delta\theta = 22.5°$ and $\Delta\beta = 11.25°$.

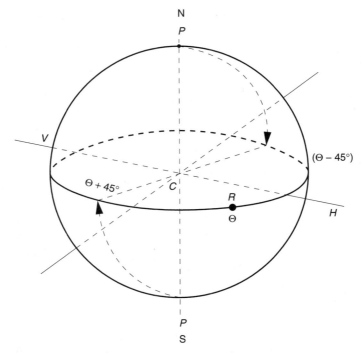

Figure 7.23   Regardless of the orientation of the fast axis, a $\lambda/4$ plate converts a circularly polarized wave into a linearly polarized wave.

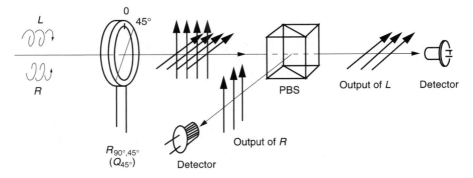

Figure 7.24   Determination of the ratio of powers of left- and right-handed circular polarizations.

Solution   The expressions for the lengths of the major and minor axes of the ellipses for a given β are obtained from Eqs. (6.100), (6.107) and (6.113), as

$$a = \sqrt{\frac{A^2 + B^2}{1 + \tan^2 \beta}} \tag{7.53}$$

$$b = a \tan \beta \tag{7.54}$$

Figure 7.26 shows the finished pattern.

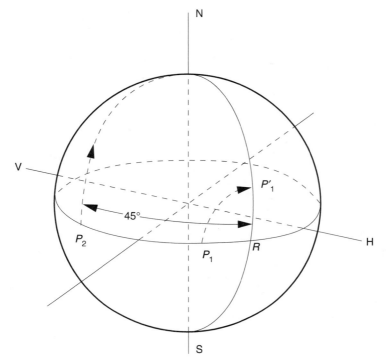

Figure 7.25 The difference in angle between the azimuth of the incident wave and the fast axis of the λ/4 plate has to be 45° to convert a linearly polarized wave into a circularly polarized wave.

Try to fabricate your own Poincaré sphere by photo copying the enlarged pattern. Firm materials such as acetate sheets for overhead projectors work best. Then cut and tape the sections to form a balloon. Such a balloon made out of thin colorful paper is called a Fusen and is a popular toy among Japanese children.   □

7.6 POINCARÉ SPHERE SOLUTIONS FOR POLARIZERS

The Poincaré sphere will be used to find the power transmittance k of a polarizer [4], which is the ratio of the polarizer's transmitted to incident power.

Consider light of an arbitrary state of polarization incident onto a polarizer whose transmission axis is along the x axis as shown in Fig. 7.27a. Of the total power of the incident light $A^2 + B^2$, only the component in the x direction transmits through the polarizer. The power transmittance k of an ideal polarizer ($k_1 = 1, k_2 = 0$) is

$$k = \frac{A^2}{A^2 + B^2} \qquad (7.55)$$

Equation (7.55) is true for any state of polarization as long as the direction of the transmission axis is along the x axis.

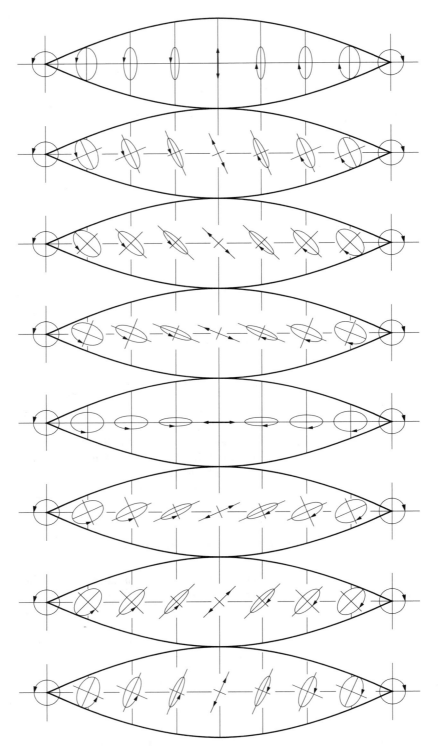

Figure 7.26 Pattern for making a Poincaré sphere.

Japanese children playing with a Fusen.

With Eq. (7.7), Eq. (7.55) becomes

$$k = \cos^2 \alpha \qquad (7.56)$$

Thus, once α of the incident light is found by the Poincaré sphere, k can be calculated from Eq. (7.56).

A method of finding α using the Poincaré sphere will be explained. First, the transmission axis of the polarizer is assumed to be along the x axis. Let the state of polarization of the incident light be (θ, β), as represented by point P_1 on the Poincaré sphere shown in Fig. 7.27b.

In order to find 2α, $\angle P_0 C P_1$ has to be found, where P_0 represents the azimuth of the polarizer transmission axis and is located at H for this case. The sphere is cut by a plane perpendicular to the $\overline{HV}$ axis and containing point $P_1(\theta, \beta)$. Point Q is any point on the circle made by the intersection of the perpendicular plane and the sphere, and $\angle P_0 C Q$ is always equal to $\angle P_0 C P_1$. Thus, when Q falls on the equator,

$$2\alpha = 2\theta' \quad \text{or} \quad \alpha = \theta'$$

where θ' is the value shown on the graduation line. Recall from Fig. 7.20 that the value shown on the graduation line, for instance, $\theta' = 45°$, is at the longitudinal angle of $90°$ from point 0 or H.

In short, in order to find α:

1. Draw the cross-sectional circle that contains point $P_1(\epsilon, \theta)$ of the incident light and is perpendicular to $\overline{VH}$.
2. Find the intersection of the cross-sectional circle with the equator.
3. The value of the graduation line of the intersection is the desired α.

A few interesting observations are (1) $\theta \neq \theta'$, unless $\theta = 45°$, and (2) $\angle C V P_1 = \angle C V Q = \alpha$, and this angle can also be used to find α.

Now consider the situation where the transmission axis of the polarizer is not along the x axis and the azimuth is Θ. This is represented by P_Θ in Fig. 7.27c. One way to solve this problem is to rotate the coordinates by Θ so that the x axis lines up with the transmission axis. The value α is found as in Fig. 7.27b. Once α is obtained, the coordinates are rotated back by $-\Theta$. A simpler approach is to use the symmetry of

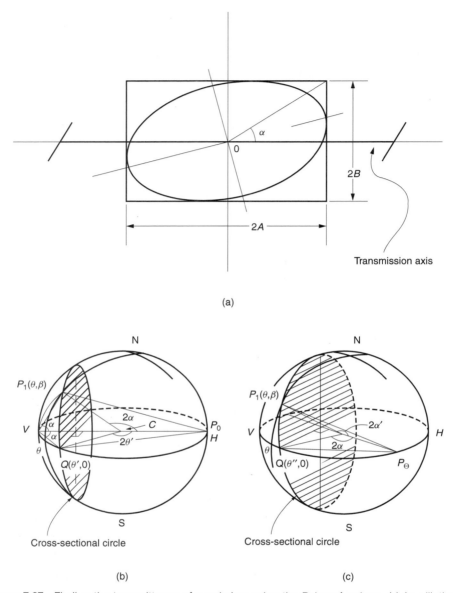

Figure 7.27 Finding the transmittance of a polarizer using the Poincaré sphere. (a) An elliptically polarized wave is incident onto a polarizer. (b) $\Theta = 0$. (c) $\Theta = \Theta$.

the sphere, as was done for the retarder, and eliminate the rotation and rerotation of coordinates. The procedure is the same as the first case except the plane cutting the sphere has to be perpendicular to $\overline{P_\Theta C}$ rather than $\overline{P_0 C}$, as illustrated in Fig. 7.27c. The required angle α is

$$\alpha = P_\Theta - \theta''$$ (7.57)

where values are those shown on the graduation line.

Example 7.10   Using the Poincaré sphere, find the transmittance through an ideal polarizer with the following configurations:

(a) Vertically polarized light is incident onto a polarizer whose transmission axis is horizontal.

(b) Linearly polarized light with azimuth 45° is incident onto a polarizer whose transmission axis is along the x axis.

(c) A circularly polarized wave with left-hand rotation is incident onto a polarizer whose transmission axis is at Θ.

(d) Left-handed elliptically polarized light with $\epsilon = 0.414$ and $\theta = 60°$ is incident onto a polarizer whose transmission axis is at $\Theta = 20°$.

Solution

(a) With Fig. 7.27b, the cross-sectional circle is tangent to the sphere at V, and the intersection with the equator is at the graduation line of $\theta' = 90°$, and thus $\alpha = 90°$.

$$k = \cos^2 90° = 0$$

(b) With Fig. 7.27b, the cross-sectional circle cuts the sphere into equal halves. The intersection with the equator is at the graduation line of $\theta = 45°$ and $\alpha = 45°$.

$$k = \cos^2 45° = 0.5$$

(c) The cross-sectional circle is through the N–S axis and perpendicular to $\overline{CP_\Theta}$ and

$$\alpha = \theta' - P'_\Theta = 45°$$
$$k = \cos^2 45° = 0.5$$

(d) Referring to Fig. 7.28, the cross-sectional circle containing the point at ($\theta = 60°$, $\epsilon = 0.414$) is drawn and is perpendicular to $\overline{CP_\Theta}$. The intersection with the equator is at the graduation line of

$$\theta' = 61°,$$
$$\alpha = \theta' - P_\Theta = 61° - 20° = 41°$$
$$k = 0.56$$

It should be noted that $\theta' \neq 60°$.   □

The analytical expression for k gives better accuracy. Rewriting Eq. (7.56) in terms of $\cos 2\alpha$, and using Eq. (6.125) gives

$$k = \tfrac{1}{2}(1 + \cos 2\theta \cos 2\beta) \tag{7.58}$$

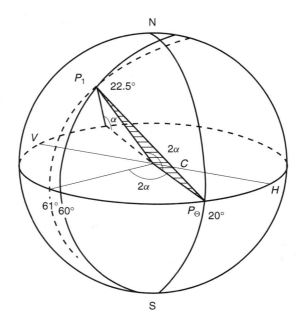

Figure 7.28   The Poincaré sphere is used to find the transmittance through a polarizer.

7.7   POINCARÉ SPHERE TRACES

For ease and accuracy, traces of the Poincaré sphere on a plane are often used [3]. Any convenient plane can be selected. The method will be explained using the example that a linearly polarized wave with $\theta = 50°$ is transmitted through a compensator whose fast axis is oriented at $\Theta = 30°$ with retardance $\Delta = 135°$.

Figure 7.29 explains how the traces are drawn [8]. The top drawing is the projection onto the horizontal plane. In this plane, the azimuth angles are clearly seen. Point P_1 corresponds to the azimuth of the incident light, and point R corresponds to the azimuth of the compensator.

The left bottom trace is the projection onto plane 1_F. Plane 1_F is the frontal plane perpendicular to radius RC. (Rather than projecting perpendicular to $\overline{VH}$, it is projected off the orthogonal direction $\overline{CR}$ and denoted as 1_F.) In this plane, the true angle Δ of the retardance can be seen, and from Δ, point P_2 for the emergent light is drawn in. The corresponding point P_2 in the horizontal plane can be obtained.

The Poincaré sphere is also projected onto the profile plane 1_P, which is the side-view plane perpendicular to both H and 1_F planes. Point P_2 in this plane is obtained by extensions from the corresponding points in the H and 1_F planes.

The essentials for finding (θ, β) of the emergent light are now in place. Referring to the projection of P_2 onto the H plane, $2\theta = 30°$ can be read directly from the diagram.

The angle P_2CP_1 in the 1_P plane is not the true 2β angle. In order to see the true angle, the sphere is rotated around the $\overline{NS}$ axis until $\overline{P_2'C}$ falls in a plane parallel to the 1_P plane. The height from the equator does not change by this rotation because the axis of rotation is $\overline{NS}$. The true angle is $2\beta = \angle P_2'CP_1 = 28°$.

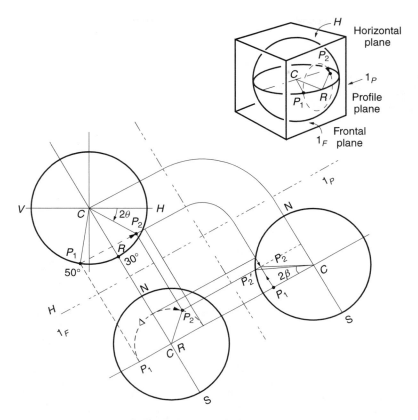

Figure 7.29   Poincaré sphere traces.

The emergent wave is therefore a left-handed elliptically polarized wave with $\theta = 15°$ and $\epsilon(= \tan \beta) = 0.25$.

Example 7.11   The retardance of a sample was measured by means of Senarmont's method in Section 6.4.3.3. The azimuth angle of the extinction axis (minor principal transmission axis) of the analyzer was at $\theta = 150°$. Find the retardance using the method of tracing onto the projected planes from the Poincaré sphere.

Solution   As shown in Fig. 6.21 in the last chapter, Senarmont's method determines the retardance of a sample by measuring the ellipticity of the emergent light when linearly polarized light is incident with its direction of polarization at $45°$ with respect to the horizontally oriented fast axis of the sample. As already proved (Problem 6.11), 2β of the emergent light equals the retardance Δ. 2β is measured by means of a $\lambda/4$ plate with $\Theta = 45°$. The light from the $\lambda/4$ plate emerges as linearly polarized light with $\theta = 45° + \beta$. θ is measured by means of an analyzer, and β is determined.

The operation is shown on the Poincaré sphere in Fig. 7.30. P_1 represents linearly polarized incident light with $\theta = 45°$. The sample, whose fast axis is along the x axis, rotates P_1 centered around R_x by Δ to P_2. Point P_2 is further rotated by the $\lambda/4$ plate centered around $R(\theta = 135°)$ by $90°$ to P_3, which represents a linearly polarized wave

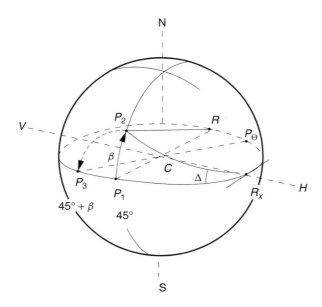

Figure 7.30   Senarmont's method represented on the Poincaré sphere.

with $\theta = 45° + \beta$. When the polarizer P_Θ is adjusted to $\theta = 135° + \beta$, null output is observed.

Now, the points on the Poincaré sphere in Fig. 7.30 will be transferred onto traces in Fig. 7.31. Projections are made in the horizontal, frontal, and profile planes. The trace in the profile plane shows the true retardance angle Δ. P_1 of the incident light is rotated around R_x by the retardance Δ of the sample. Point P_2 is further rotated by the $\lambda/4$ plate by 90° around R at $\theta = 135°$ to reach point P_3. The frontal projection shows the true angle of the rotation from P_2 to P_3.

Point P_Θ of the azimuth of the analyzer for extinction is diametrically opposite to point P_3. The true angle of P_Θ is seen in the horizontal plane.

Now in order to obtain the value of Δ from a given value of P_Θ, one has to follow the procedure backward. Keep in mind that the locations on the Poincaré sphere are graduated from $\theta = 0°$ to $180°$ and from $\beta = 0°$ to $45°$. As was shown in Fig. 7.20, the relevant angles on the projected planes are $2\theta, 2\beta$, and Δ. The amount of Δ is not doubled. A retardance of 90° is represented by a rotation of 90°.

Now, with the given value of $P_\Theta = 150°$, Δ will be obtained. The point graduated as $\Theta = 150°$ is labeled as P_Θ in the horizontal plane in Fig. 7.31. The graduation of the diagonally opposite point P_3 is $150° - 90° = 60°$. If one goes from the horizontal plane to the profile plane, angle $\angle P_1 C P_2$ indicates that the desired retardance is $\Delta = 30°$. □

The next example addresses (1) how to draw traces for a given state of polarization, (2) how to manipulate the operation of a $\lambda/4$ plate with an arbitrary orientation, and (3) how to read off the true value of the state of polarization from the trace.

Example 7.12   Light with $(\theta, \epsilon) = (77°, 0.34)$ is incident onto a $\lambda/4$ plate whose fast axis azimuth is $\Theta = 56°$.

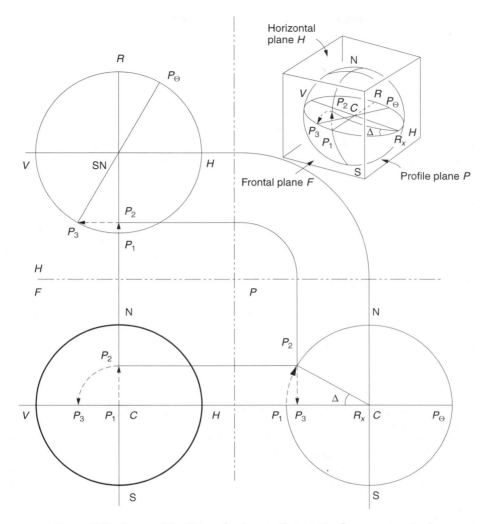

Figure 7.31 Traces of the Poincaré sphere to illustrate the Senarmont method.

(a) Find the state of polarization of the light emergent from the λ/4 plate.

(b) The above wave is further transmitted through an analyzer. The azimuth angle of the transmission axis of the analyzer is $\Theta = 15°$. Find the value of α and k for the wave emergent from the analyzer.

Solution Dividing the procedure into separate steps makes it easier to follow.

Step 1. Represent the orientation of the fast axis of the λ/4 plate.

Step 2. Represent the state of polarization of the incident light.

Step 3. Draw the operation of the λ/4 plate.

Step 4. Read the value of (θ, ϵ) from the traces.

In this example, many subscripts are needed and new notations involving $A, B, C, \ldots$ are introduced to represent the states of polarization, in place of $P_1, P_2, P_3, \ldots$ used in previous examples. For instance, the projections of point A in the horizontal, frontal, and profile planes are designated as A_H, A_F, and A_P.

Step 1. The azimuth of the fast axis is $\Theta = 56°$ and is represented by the point R_H on the equator in the horizontal plane. The true azimuth angle is seen in the horizontal plane at $2\Theta = 112°$, as shown in Fig. 7.32a. The intersection between the projection from R_H and the equatorial line in the frontal plane determines point R_F in the frontal plane. Extensions from R_H and R_F determine R_P in the profile plane.

Step 2. Figure 7.32b illustrates point A of the incident light with $\beta = \tan^{-1} \epsilon = 19°$ and $\theta = 77°$. In the frontal plane, the true latitude angle is seen. A straight horizontal line with $2\beta = 38°$ is the line for the graduation line of $\beta = 19°$. In the horizontal plane, the latitude 2β sweeps out a circle of radius r. The longitude angle 2θ is seen in the horizontal plane, and the longitude of $2\theta = 154°$ represents the graduation line of $\theta = 77°$. The intersection of the circle with radius r and the longitude determines A_H in the horizontal plane. The projection of A_H onto the frontal plane intersects at A_F. Points A_H and A_F determine point A_P in the profile plane.

Step 3. Figure 7.32c outlines the operation of the $\lambda/4$ plate. An off-orthogonal frontal plane 1_F, which is perpendicular to CR_H, is drawn. The $\lambda/4$ plate rotates A_1 to B_1 by 90°, and A_H moves perpendicular to CR_H to B_H in the horizontal plane. B_F in the frontal plane is obtained by the projection from B_H and the height b in the 1_F plane, because the true height of B from the equator is seen both in the 1_F plane and the frontal plane. Point B_P in the profile plane is determined from the projections of B_H and B_F. Neither geodesic $\overline{A_F B_F}$ nor $\overline{A_P B_P}$ is circular.

Step 4. The state of polarization of the emergent wave (θ, ϵ) is read from point B. The three-dimensional representation of point B is shown in the inset to Fig. 7.32b. From the horizontal projection in Fig. 7.32c, the azimuth is $2\theta = 65°$ or $\theta = 32.5°$.

In order to find 2β of B in the frontal plane in Fig. 7.32c, the sphere is rotated around NS so that the true angle $\angle B'_F CH$ in the frontal plane is $2\beta = 32°$ or $\epsilon = 0.29$.

The output from the analyzer is found from the drawing in Fig. 7.32d. The value of α is $21.5°$, and from Eq. (7.56), $k = 0.87$.   □

7.8   MOVEMENT OF A POINT ON THE POINCARÉ SPHERE

In principle, any given state of polarization can be converted into any other state of polarization by moving along lines of constant longitude and latitude. Hence, any general movement along the Poincaré sphere [4] can be treated by decomposing the movement into these two directions. Although this way of decomposition has a simple conceptual appeal, it is not necessarily the simplest from the viewpoint of implementation.

7.8.1   Movement Along a Line of Constant Longitude (or Constant θ Line)

As shown in Fig. 7.33, the state of polarization of an incident wave is represented by P_1 on the Poincaré sphere. P_1 is to be moved along a line of constant longitude, which can be accomplished with a compensator. When the azimuth Θ of the fast axis is set

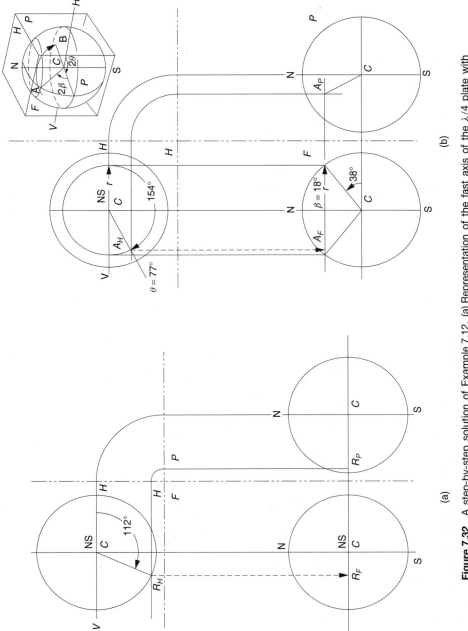

Figure 7.32 A step-by-step solution of Example 7.12. (a) Representation of the fast axis of the $\lambda/4$ plate with $\Theta = 56°$. (b) Representation of the incident light with $(\theta, \beta) = (77°, 19°)$. (c) Operation of the $\lambda/4$ plate. (d) Operation of a polarizer with $\Theta = 15°$ in the horizontal plane.

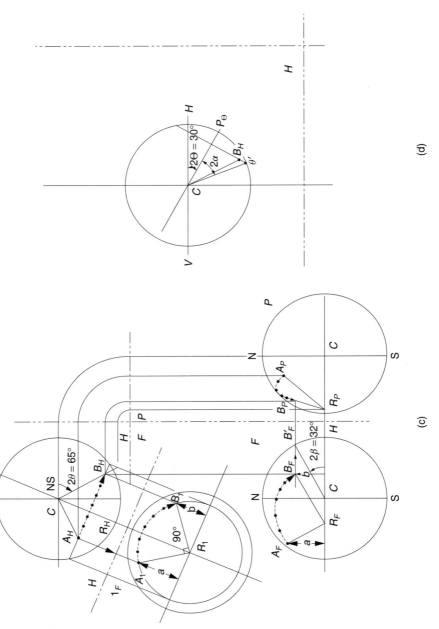

(c)

(d)

Figure 7.32 (Continued)

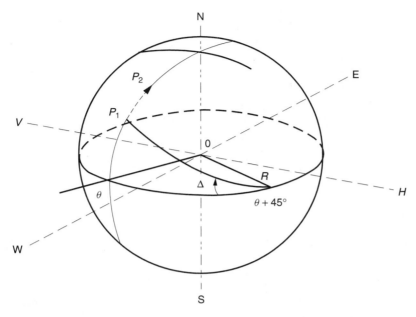

Figure 7.33 Moving the state of polarization along a line of constant longitude.

at 45° with respect to the azimuth θ of the incident elliptic light, point P_1 will move up or down the constant θ line as the retardance of the compensator is varied.

7.8.2 Movement Along a Line of Constant Latitude (or Constant β Line)

By using two half-waveplates, the state of polarization can be moved along a line of constant latitude. Let the initial state of polarization be represented by P_1 on the Poincaré sphere shown in Fig. 7.34. The first $\lambda/2$ plate rotates P_1 around R_1 by 180° to P_2. P_2 has the same ellipticity as P_1 but in the opposite direction of rotation with different azimuth. The second half-waveplate further rotates P_2 around R_2 to P_3. The ellipticity and the sense of rotation of P_3 are the same as that of the incident light P_1. Only the azimuth is changed from that of the incident light. If reversal of the handedness is acceptable, a solitary half-waveplate will suffice.

The last example of this chapter [9] is a comprehensive one and will serve as a good review of the material presented in Chapters 5, 6, and 7.

Example 7.13 Figure 7.35 shows the diagram of a TM–TE mode converter on a lithium niobate wafer. The direction of polarization of the TM mode is vertical inside a rectangular waveguide, as shown in Fig. 7.35, and that of the TE mode is horizontal inside the waveguide. In order to convert the direction of polarization inside the waveguide, a TM–TE mode converter is used. Chapters 9 and 10 are devoted to the topic of optical waveguides. In particular, TM and TE modes are discussed in Section 9.3 and mode converters are discussed in Section 10.9.

The converter in Fig. 7.35 consists of conversion retarder regions, where an external dc electric field is applied by the fingers of the interdigital electrodes, and modal retarder regions, which are located in between the conversion retarder regions and

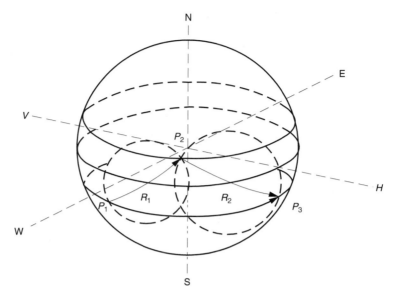

Figure 7.34   Moving the state of polarization along a line of constant latitude.

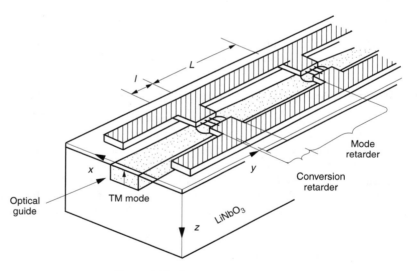

Figure 7.35   TM–TE mode converter.

whose function is based on the difference in the propagation constants of the TE and TM modes in the rectangular optical guide.

Use the Poincaré sphere to explain the principle of operation of the TM–TE mode converter. Assume the input light is in the TM mode.

Solution   The function of the conversion retarder and modal retarder will be described separately, and then they will be combined to explain the principle of operation of the TM–TE converter.

(a) Let us first consider the conversion retarder. Interdigital electrodes are deposited on a LiNbO$_3$ wafer to create a periodic external dc field ε_x. The electric lines of flux of the external field are periodic and parallel to the x axis. The external E field ε_x rotates the indicatrix by θ degrees (in the case of lithium niobate, $A < C$ and θ in Eq. (5.32) becomes a negative quantity). The fast axis of the conversion retarder is, therefore, at $\Theta = \pi/2 + \theta$ as shown in Fig. 7.36a. The retardance Δ_i of each electrode with length l is

$$\Delta_i = 2(\Delta N)kl \tag{7.59}$$

where ΔN is the change in the index of refraction due to the application of the external E field and k is the free space propagation constant. Each interdigital electrode creates a small retardance, but as shown next, the retardance is accumulated constructively and the incident point P_1 moves from point V to H.

(b) The modal retarder refers to the region where no external electric field is present. The propagation constants β_{TE} and β_{TM} for the TE and TM modes are quite different due to both birefringence and geometry. Lithium niobate is birefringent, and $n_0 > n_e$. The geometry of the cross section of the optical guide seen by the TM and TE modes is different. These regions are considered as retarders with their fast axis along the z direction ($\Theta = \pi/2$). The length L of the modal retarder is chosen such that the converted component constructively accumulates, as given by the condition

$$(\beta_{TM} - \beta_{TE})L = 2\pi n \tag{7.60}$$

where n is an integer.

(c) Finally, the principle of operation of the TM–TE converter will be explained using the Poincaré sphere in Fig. 7.36b. The incident light in the TM mode is represented by P_1 located at point V. The fast axis of the finger retarder is at $\Theta = \pi/2 + \theta$, which is represented by point R on the Poincaré sphere. The retardance of the first finger will rotate point P_1 by Δ_1 around R to P_2. Then the modal retarder region whose length is designed to provide 360° retardance rotates point P_2 by 360° around point V to P_3. The second finger electrode provides a retardance of Δ_2 and moves point P_3 to point P_4. Point P_4 is further rotated by 360° by the modal retarder. The same process is repeated, and the state of polarization moves toward H, or the direction of polarization of the TE mode.

The retardance of the modal retarder must be close to 360°, otherwise the point does not proceed toward H effectively. For instance, if the retardance of the first modal retarder region were 180°, point P_4 would have been at point P_1 again after the second finger electrode.   □

This example looked at the specific case of converting V to H polarization on the Poincaré sphere. Various other states of polarization are obtainable by adjusting the external electric field and the length of the modal converter.

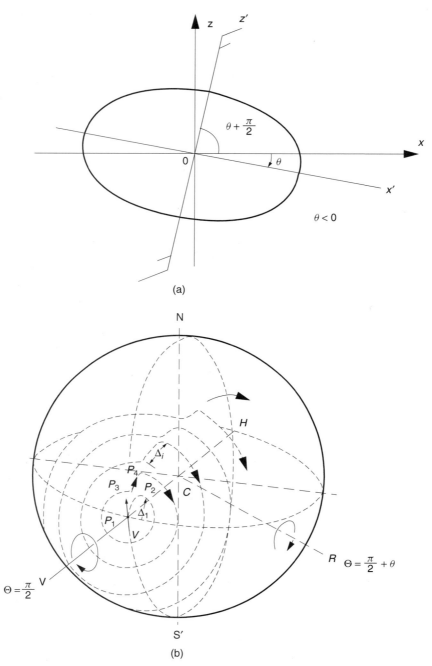

Figure 7.36 Solution of Example 7.13. (a) Indicatrix of the optical guide. (b) Movement of the state of polarization on the Poincaré sphere.

PROBLEMS

7.1   The Poincaré sphere is useful for quickly finding an approximate answer in the laboratory. Example 6.11 asked the calculation of the emergent wave from a $\lambda/4$ plate with $\Theta = 45°$ when the incident wave is a right-handed elliptically polarized wave with $a = \sqrt{3}$ V/m and $b = 1$ V/m and $\theta = 22.5°$. Draw a diagram of the operation on the Poincaré sphere.

7.2   **(a)**   With the fast axis of a $\lambda/4$ plate fixed along the x axis, use the Poincaré sphere to obtain the emergent states of polarization for incident linearly polarized waves with azimuth angles of $\theta = 15°$, $30°$, and $45°$.

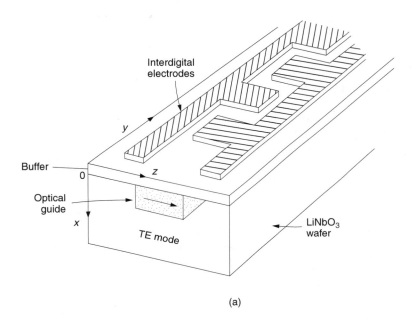

(a)

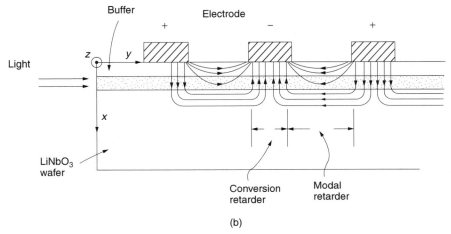

(b)

Figure P7.8   TE–TM converter on LiNbO$_3$ wafer. (a) Bird's eye view. (b) Profile view.

(b) With the azimuth of the incident linearly polarized wave fixed along the x axis, use the Poincaré sphere to obtain the states of polarization emergent from a $\lambda/4$ plate for three different azimuths of the fast axis at $\Theta = 15°$, $30°$, and $45°$.

7.3   The diagram in Fig. 6.35 illustrates how the glare from a radar screen is suppressed by an antiglare circularly polarizing sheet. Indicate the state of polarization of the light at each step of the explanation on the Poincaré sphere.

7.4   Problem 6.11 asked one to prove that when linearly polarized light with $B/A = 1$ is incident onto a retarder whose fast axis is along the x axis, 2β of the emergent light is equal to Δ of the retarder. Solve this problem using the Poincaré sphere.

7.5   Problem 6.2 asked for the state of polarization emergent from a retarder whose fast axis is oriented along the x axis with $\Delta = 315°$ when linearly polarized light is incident with $\theta = 63.4°$. Answer the same problem using Poincaré sphere traces.

7.6   Use Poincaré sphere traces to find the state of polarization of the emergent wave from a $\lambda/4$ plate. The state of polarization of the incident light is $(\theta, \epsilon) = (100°, 0.5)$ and the azimuth of the fast axis of the $\lambda/4$ plate is $\Theta = 35°$.

7.7   **(a)** Example 6.9 asked one to obtain θ and ϵ of the emergent light from a retarder with $\Delta = 38°$ and its fast axis along the x axis. The incident light was $E_x = 2.0$ V/m and $E_y = 3.1$ V/m. Verify the answer using the Poincaré sphere trace.

(b) The above system is followed by an analyzer with its transmission axis along the x axis. Find the emergent light power.

(c) Next, the transmission axis of the analyzer is rotated to $\Theta = 25°$. Find the new transmittance from the analyzer.

7.8   A TE–TM converter is shown in Fig. P7.8. The converter uses Y propagating X-cut lithium niobate. The external field ε_x is vertical and bipolar. Explain the operation of the mode converter on the Poincaré sphere. Assume the incident light is in the TE mode.

7.9   One of the methods of laser cooling is polarization gradient cooling. This method needs to create a spatial change of the state of polarization in the cooling laser

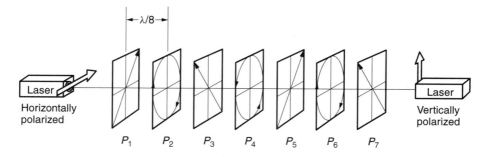

Figure P7.9   Polarization grating used for laser cooling.

beam. Such a laser beam is created by two oppositely propagating laser beams. Both laser beams are linearly polarized, but one is vertically polarized and the other is horizontally polarized. Such laser beams establish a spatial variation of the state of circular polarization, as shown in Fig. P7.9. Atoms moving along the laser beam are slowed down by the interaction with the magnetic sublevels of the ground state established by this circularly polarized laser beam [10]. Using the Poincaré sphere, verify the establishment of such a polarization gradient in the cooling laser beam.

REFERENCES

1. H. Poincaré, *Théorie Mathématique de la Lumière II*, Georges Carré Editeur, Paris, 1892, pp. 275–306.
2. W. A. Shurcliff and S. S. Ballard, *Polarized Light*, Book #7, Van Nostrand Momentum, Princeton, 1964.
3. Selected Reprints on "Polarized Light," published for the American Association of Physics Teachers by the American Institute of Physics, 1963.
4. H. Takasaki, "Representation of states of polarization," in *Crystal Optics*, Applied Physics Society of Japan (Ed.), Morikita Publishing Company, Tokyo, 1975, pp. 102–163.
5. R. M. A. Azzam and N. M. Bashara, *Ellipsometry and Polarized Light*, North Holland Publishing, New York, 1979.
6. D. S. Kliger, J. W. Lewis, and C. E. Randall, *Polarized Light in Optics and Spectroscopy*, Academic Press, Boston, 1990.
7. J. G. Ellison and A. S. Siddiqui, "Automatic matrix-based analysis method for extraction of optical fiber parameters from polarimetric optical time domain reflectometry data," *J. Lightwave Tech.*, **18**(9), 1226–1232 (2000).
8. G. F. Pearce, *Engineering Graphics and Descriptive Geometry in 3D*, McGraw-Hill Ryerson, Toronto, 1985.
9. N. G. Walker, G. R. Walker, and R. C. Booth, "Performance of lithium niobate polarisation convertor/tunable filter," *Electron. Lett.*, **24**(3), 268–270 (Mar. 1988).
10. R. Srinivasan, "Research news; 1997 Nobel Prize for Physics: Laser cooling and trapping of ions and atoms," *Curr. Sci.*, **76**(2), 183–189 (1999).

8

PHASE CONJUGATE OPTICS

This chapter deals with optical phase conjugation that can be generated in a medium with third order nonlinear susceptibility. The path of the phase conjugate wave retraces itself. This is analogous to playing a videotape backward, or in other words, a time-reversed videotape. The chapter begins with an illustration of the striking differences between an ordinary mirror and a phase conjugate mirror.

8.1 THE PHASE CONJUGATE MIRROR

The manner of reflection from a phase conjugate mirror is compared with that of a plain, ordinary mirror in Fig. 8.1 [1]. With the ordinary mirror in Fig. 8.1a, the reflected wave not only changes its direction in accordance with the orientation of the mirror, but also keeps on diverging if the incident light is diverging. On the other hand, the wave reflected from the phase conjugate mirror heads back to where it came from regardless of the orientation of the mirror. Furthermore, if the incident wave is a diverging wave, the reflected wave becomes a converging wave. The wave reflected from the phase conjugate mirror is called a phase conjugate wave.

The phase conjugate wave formed by an ordinary hologram provides greater insight into this phenomenon, as explained in the next section.

8.2 GENERATION OF A PHASE CONJUGATE WAVE USING A HOLOGRAM

A phase conjugate wave can be generated almost in real time if a special type of nonlinear crystal or gas is used. Even though an ordinary hologram cannot operate in real time, it is useful for explaining the principle of generating the phase conjugate wave [2,3]. Figure 8.2 shows the geometry for recording a hologram using photographic film. Let O be the field emanating from a point object o onto the photographic film, and R be the reference wave emanating from the reference point source r onto

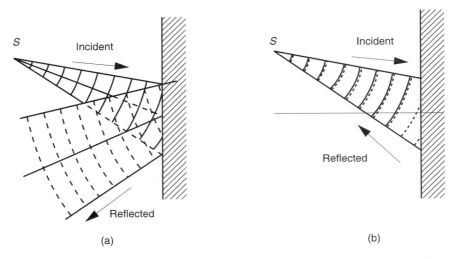

Figure 8.1   Difference between (a) a plain, ordinary mirror and (b) a phase conjugate mirror.

the same photographic film. The O and R waves form an interference fringe pattern. The exposed and developed photographic film is filled with hyperbolic-shaped, silver-grained, miniature mirrors as shown in Fig. 8.2. The transmittance of such a fringe pattern was given in Eq. (1.261) as

$$t = t_0 - \beta[|R|^2 + |O|^2 + OR^* + O^*R] \tag{8.1}$$

When this hologram is illuminated by R^*, the fourth term, $\beta O^* R$, in the square bracket generates the phase conjugate wave $|R|^2 O^*$. This is the shortest mathematical explanation of the generation of the phase conjugate wave O^*. Here, however, a graphical explanation is attempted because it can readily be used for explaining related phenomena.

Referring back to Fig. 8.2, a new point light source r' and a convex lens are arranged such that its path retraces the original reference wave R from the opposite direction. This light is R^*, the complex conjugate of the original R (we see in Section 1.5.4 that Eq. (1.162) is the complex conjugate of Eq. (1.161).), and the hologram is illuminated by R^*. The normals to the miniature mirrors are always oriented in the plane of the bisector of the angle between the object O and reference R beams. For instance, the surface of the small mirror at point c is in a plane that bisects $\angle ocr$ made by $\overline{oc}$ and $\overline{cr}$, and thus, $\alpha = \alpha'$ and R^* is reflected toward the object point o.

In conclusion, R^* (pump wave) from r' generates a wave that traces back the object wave O and converges to the source point. This wave is O^*, the phase conjugate of the O wave. The phase conjugate wave can be separated from the signal wave by means of a half-mirror (HM). It would seem that the O^* wave is the exact retrace of the O wave but the photons of the O^* wave come from the pump wave R^*, and *not from the original point source o*. The intensity of the reflected wave is controlled by the efficiency of the small mirrors and the intensity of the pump wave R^*. The intensity of the phase conjugate wave O^* can be even larger than that of the O wave and, as such, serves as a useful way of amplifying the intensity.

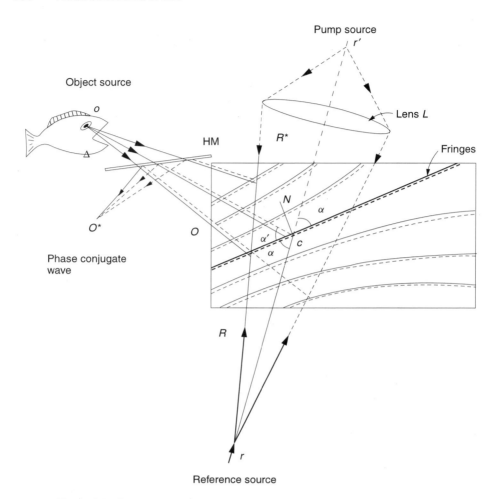

Figure 8.2 Phase conjugate wave generated by a hologram. HM is a half-mirror.

In the hologram in Fig. 8.2, the generation of the phase conjugate wave requires the participation of the following three waves: the object wave O, the reference wave R, and the reconstruction wave R^*. Including the phase conjugate wave O^*, a total of four waves are involved—hence the term *four-wave mixing* (FWM). Discussions on holography and phase conjugation generally use different terminology, as summarized in Table 8.1.

Table 8.1 Comparison of terminology

Wave in Fig. 8.2	Holography	Phase Conjugation
O	Object wave	Signal wave
R	Reference wave	First pump wave
R^*	Reconstruction wave	Second pump wave
O^*	Phase conjugate of the object wave	Phase conjugate wave

In the present example, all the waves are at the same frequency, which is said to be the degenerate case. In the nondegenerate case, a phase conjugate wave is generated by mixing waves of different frequencies.

8.3   EXPRESSIONS FOR PHASE CONJUGATE WAVES

The expression for the phase conjugate wave will be explained taking a spherical wave as an example. The expression for a diverging spherical wave is

$$E_s(r, t) = \text{Re } E(r, \omega)e^{-j\omega t} \tag{8.2}$$

where

$$E(r, \omega) = \frac{A_0}{r}e^{jkr+j\phi} \tag{8.3}$$

and where A_0 is a constant real number.

The phase conjugate wave $E_{\text{pc}}(r, t)$ of this signal wave is obtained by performing the conjugation operation on every term except the temporal term, namely,

$$E_{\text{pc}}(r, t) = \text{Re } [E^*(r, \omega)e^{-j\omega t}] \tag{8.4}$$

where

$$E^*(r, \omega) = \frac{A_0}{r}e^{-jkr-j\phi} \tag{8.5}$$

The term e^{-jkr}/r indicates a converging spherical wave (see Section 1.5.4). The phase conjugation converts a diverging spherical wave into a converging spherical wave, and vice versa.

The phase conjugate wave is sometimes called a time-reversed wave. If the sign of the temporal term in Eq. (8.2) is reversed, that is, $t \to -t$, then

$$E_{\text{pc}}(r, t) = \text{Re } \left(\frac{A_0}{r}e^{j\omega t+jkr+j\phi}\right) \tag{8.6}$$

which is identical to $E_{\text{pc}}(r, t)$ given by Eq. (8.4). This means that $E_{\text{pc}}(r, t)$ is obtained by reversing the time. $E_{\text{pc}}(r, t)$ is like a motion picture played backward.

It is important to understand the meaning of ϕ. Let us compare the phases between two diverging spherical waves with and without ϕ. From Eqs. (8.2) and (8.3), these spherical waves are given by

$$E_s(r, t) = \text{Re } \left(\frac{A_0}{r}e^{-j\omega t+jkr+j\phi}\right) \tag{8.7}$$

$$E_d(r, t) = \text{Re } \left(\frac{A_0}{r}e^{-j\omega t+jkr}\right) \tag{8.8}$$

The waves are observed at a particular radius $r = r_0$. The phase of $E_s(r, t)$ at $t = t$ is identical to the phase of $E_d(r, t)$ at $t = t - \phi/\omega$. This means that the phase of $E_s(r, t)$

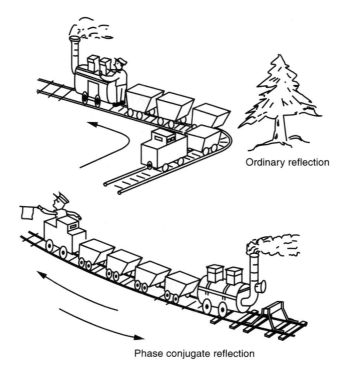

Ordinary reflection

Phase conjugate reflection

Comparison of ordinary reflection and phase conjugate reflection.

is equal to the phase that $E_d(r, t)$ had ϕ/ω seconds earlier. Thus, ϕ means that the phase of $E_s(r, t)$ is delayed from that of $E_d(r, t)$ by ϕ radians.

On the other hand, when the phase conjugate waves with and without ϕ are compared, the corresponding expressions are

$$E_{\mathrm{pc}}(r, t) = \mathrm{Re}\ \left(\frac{A_0}{r}e^{-j\omega t-jkr-j\phi}\right) \qquad (8.9)$$

$$E_c(r, t) = \mathrm{Re}\ \left(\frac{A_0}{r}e^{-j\omega t-jkr}\right) \qquad (8.10)$$

The phase of $E_{\mathrm{pc}}(r, t)$ at $t = t$ is equal to the phase that $E_c(r, t)$ will have at $t = t + \phi/\omega$. Hence, ϕ means that the phase of $E_{\mathrm{pc}}(r, t)$ is leading by ϕ radians.

In conclusion, the phase conjugate wave propagates in the reverse direction of the signal wave. The phase is also reversed and if the phase of the signal wave is *delayed* by ϕ radians, then that of the phase conjugate wave is *leading* by ϕ radians. This is just like a train reversing its direction. The trailing coach becomes the leading coach if the phase is compared to the location of the coach.

8.4   PHASE CONJUGATE MIRROR FOR RECOVERING PHASEFRONT DISTORTION

One of the most important applications of the phase conjugate mirror is for eliminating wavefront distortion incurred during light transmission through a turbulent atmosphere or dispersive optical fiber. The holographic principle will be used for explaining the recovery of the wavefront free from distortion.

First, the amount of fringe pattern shift due to the phase shift of the incident wave will be calculated. Figure 8.3a shows the fringe pattern formed by two counterpropagating plane waves:

$$E_1 = A_1 e^{-j\omega t + jkz} \tag{8.11}$$

$$E_2 = A_2 e^{-j\omega t - jkz} \tag{8.12}$$

If the amplitudes are equal, $A_1 = A_2 = A$, then the sum of $E_1 + E_2$ can be expressed as

$$E_1 + E_2 = 2Ae^{-j\omega t} \cos kz \tag{8.13}$$

The intensity peaks of Eq. (8.13) appear at every half-wavelength, as shown by the solid lines in Fig. 8.3a. Let us focus our attention on the particular peak at the center $z = 0$ in order to find how much the peak moves when one of the two waves shifts its phase. Let us say the phase of the forward wave E_1 is delayed by ϕ radians, and

$$E_1' = A_1 e^{-j\omega t + jkz + j\phi} \tag{8.14}$$

The interference pattern between E_1' and E_2 then becomes

$$E_1' + E_2 = 2Ae^{-j\omega t + j\phi/2} \cos(kz + \phi/2) \tag{8.15}$$

Peaks appear when the value inside the parentheses in Eq. (8.15) is zero, and the new location of the center peak is at

$$\Delta z = -\frac{1}{k}\frac{\phi}{2} \tag{8.16}$$

As a matter of fact, all peaks shift by Δz toward the left or toward the source of the delayed incident wave, as indicated by the dashed lines in Fig. 8.3a. This shift of the fringe pattern plays an important role in recovering the wavefront free from distortion.

Figure 8.3b explains how the wavefront disrupted by passing through a distorting medium is restored by means of the phase conjugate mirror. Let us say a plane wave whose wavefront is represented by the solid line in Fig. 8.3b(i) is incident from the left to the right. A rectangular shaped distorting medium whose refractive index is larger than that of the surrounding medium is placed in the way. The portion of the wavefront that has passed through the distorting medium is delayed, and the shape of the wavefront upon leaving the distorting medium becomes indented and resembles the letter C, as indicated by the solid line in Fig. 8.3b(ii). A hologram is generated from this distorted wavefront O and the reference wave R originating from point r. The recorded fringes in the hologram have protrusions shaped like the letter C in their pattern. The direction of the protrusion is toward the source of the incident wave as explained using Fig. 8.3a.

Next, consider the case when this hologram is illuminated by the pump wave R^* from the point source r'. Compare the ray that is reflected from the protruding

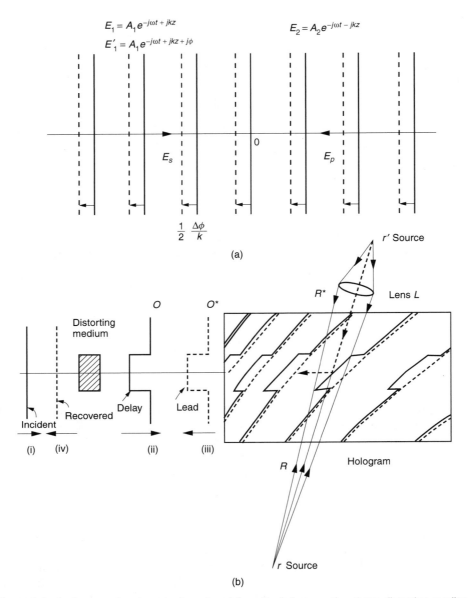

Figure 8.3 A phase conjugate wave is restored from the influence of a phase-distorting medium. (a) Fringe pattern formed by two counterpropagating plane waves. (b) Explanation of how the wavefront disrupted by passing through a distorting medium is restored by means of the phase conjugate mirror.

section of the small mirror and the ray reflected from the nonprotruding section. Not only does the ray reflected from the protruding section reach the mirror sooner, but also the point of reflection is shifted toward the left, and the wavefront O^* reflected from the hologram has a C-shaped dent as indicated by the dashed line in Fig. 8.3b(iii).

The shapes of the indentation in the O wave in Fig. 8.3b(ii) and the O^* wave in Fig. 8.3b(iii) are the same, but the difference is in their direction of propagation. They

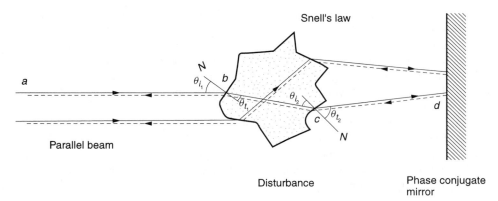

Figure 8.4 Recovery of the original parallel beam from a disturbed beam by means of a phase conjugate mirror. ⎯⎯, Signal beam. - - - -, Phase conjugate beam.

are propagating in opposite directions, and the indented part of the O wave is delayed, but that of the O^* wave is in the lead.

If the O^* wave continues to propagate to the left and passes through the same distorting medium again, only the distorted portion of the wavefront is delayed, and the wavefront of the emerging wave recovers from the distortion, as indicated by the dashed line in Fig. 8.3b(iv).

Next, let us consider the more general case where the distorting medium has an irregular shape, as shown in Fig. 8.4. The particular ray path *abcd* is examined. At point *d*, the nature of the phase conjugate mirror directs the reflected wave exactly toward the direction from which it came. Once this direction of the retrace is set at point *d*, the rest of the paths are solely determined by Snell's refraction law. Snell's law is reciprocal, which means that regardless of whether the ray goes from left to right or right to left, it takes the same path. Thus, the reflected ray takes the path of *dcba*, which is exactly the reversal of *abcd*, and the reflected wave becomes an undistorted parallel beam.

8.5 PHASE CONJUGATION IN REAL TIME

In the previous sections, in order to explain the method of generating a phase conjugate wave, a photographic film was used as the recording medium. For most applications, however, it is unrealistic to wait for the film to be developed. For real-time operation, the film has to be replaced by a more suitable recording medium. Third order nonlinear media are used. The refractive indices of such media change in real time when exposed to light [3,4].

The most commonly used materials are photorefractive crystals such as $BaTiO_3$, $LiNbO_3$, $LiTaO_3$, and $Bi_{12}SiO_{20}$ (BSO). These photorefractive crystals have a large nonlinear susceptibility, and the values of χ_{eff} are in the range of $10^{-20}-10^{-23} (V/m)^2$. The light intensity required to produce a noticeable effect can be as small as 1 mW/cm^2, which means lasers with output powers of the order of tens of milliwatts will suffice. The drawback with these crystals is the slow response time, which ranges from a few seconds to hours.

Nonlinear Kerr media such as glass, calcite, YAG, sapphire, benzene, liquid crystal, and semiconductors are an alternative to photorefractive crystals. Semiconductors like chromium-doped gallium arsenide (GaAs:Cr), iron-doped indium phosphide (InP:Fe), or titanium-doped indium phosphide (InP:Ti) change their energy band gap when illuminated by high-intensity light and hence change their refractive index. Their nonlinear susceptibility is low and $\chi_{\text{eff}} = 10^{-22} - 10^{-32} (\text{V/m})^2$, but the response time is as fast as $10^{-8} - 10^{-12}$ seconds.

Yet another possibility are materials that display either stimulated Brillouin scattering (SBS) or stimulated Raman scattering (SRS). Examples of such materials are gaseous methane (CH_4), carbon dioxide (CO_2), liquid carbon disulfide (CS_2), alcohol (C_2H_5OOH), and glass. The magnitude of χ_{eff} is $10^{-32} - 10^{-34} (\text{V/m})^2$ and the response time is $10^{-8} - 10^{-9}$ seconds.

8.6   PICTURE PROCESSING BY MEANS OF A PHASE CONJUGATE MIRROR

If the distorting medium in Fig. 8.3b is replaced by an inhomogeneous medium such as turbulent air, the system in Fig. 8.3b can be used immediately for correcting a distorted image.

Figure 8.5 shows an arrangement for compensating for the distortion caused by transmission of the signal light through an in homogeneous medium. Referring to Fig. 8.5, a light source illuminates the input mask, and the signal light from the input mask undergoes distortion as it passes through the turbulent air. The distorted signal is incident onto the phase conjugate mirror. The signal light reflected from the phase conjugate mirror reverses the sign of its phase. By going through the same inhomogeneous medium a second time, the distortion in the reversed phase is exactly canceled. The corrected wavefront reaches the image plane by way of the half-mirror. The location of the image plane is set such that the total distance between the input mask to the phase conjugate mirror is identical to that between the phase conjugate mirror and the image plane.

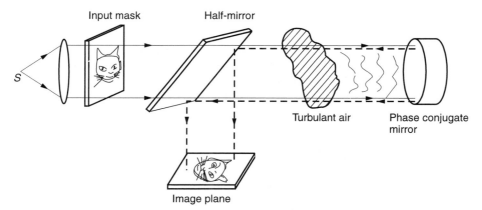

Figure 8.5 Arrangement designed to compensate for the wavefront distortion incurred during transmission through turbulent air.

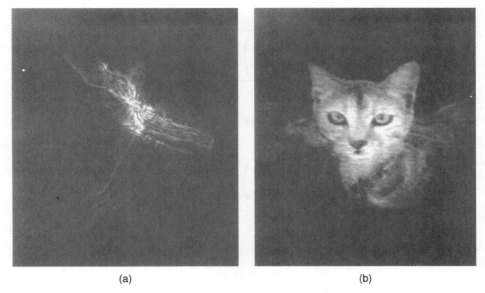

(a) (b)

Figure 8.6 Image restoration by means of a phase conjugate mirror. (a) Image through a distorted sheet of glass. (b) Image restored by means of a phase conjugate mirror. (Courtesy of J. Feinberg [5].)

It is important to realize that the wavefront has to retrace the same inhomogeneity. For this to be true, the air turbulence has to be stationary for the duration of the round trip of the signal light through the air turbulence.

Figure 8.6 shows the result of an experiment to demonstrate the effectiveness of compensation using such an arrangement as shown in Fig. 8.5 [5,6]. A sheet of surface-distorted glass was used instead of turbulent air. Figure 8.6a shows the image obtained using an ordinary mirror in place of the phase conjugate mirror. Figure 8.6b shows the restored image of the cat obtained using the phase conjugate mirror.

The arrangement shown in Fig. 8.5 has another application. By removing the inhomogeneous medium it can be used as a photolithography machine. The image of the input mask can be projected onto a substrate in the image plane. The system not only does away with imaging lenses but also avoids direct contact of the input mask with the substrate.

8.7 DISTORTION-FREE AMPLIFICATION OF LASER LIGHT BY MEANS OF A PHASE CONJUGATE MIRROR

The same principle for compensating distortion caused by inhomogeneity in Fig. 8.5 can be used to construct a light amplifier whose output light is free from distortion [7]. Problems of inhomogeneity normally occur within a high-power semiconductor laser amplifier. This can be compensated using the arrangement shown in Fig. 8.7, which is nothing but a modification of Fig. 8.5.

Figure 8.7a shows a semiconductor laser without compensation. The inhomogeneity of the amplifier generates wavefront distortion in the output light. In Fig. 8.7b, a phase conjugate mirror is placed in the amplified output light. The reflected phase conjugate

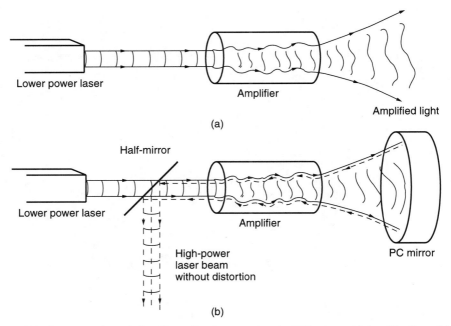

Figure 8.7 Compensation of distortion with phase conjugate (PC) mirror. (a) Amplification of light with distortion. (b) Amplification of light without distortion.

wave retraces through the inhomogeneity in the amplifier and takes the path of the original laser beam without distortion. The distortion-free amplified output exits by way of the half-mirror.

8.8   SELF-TRACKING OF A LASER BEAM

By nature, the phase conjugate wave retraces the path to the original source. Making use of this property, optical tracking or self-targeting systems can be realized [1,7].

Figures 8.8 −8.10 show examples of such systems.

Figure 8.8 shows an arrangement for directing a high-intensity laser beam to a point target. The target is illuminated by a laser. A portion of the light scattered by the target is intercepted by the optical amplifier and is amplified. The amplified output is incident onto the phase conjugate mirror. The reflected phase conjugate wave enters the optical amplifier again. The output from the amplifier is not only amplified twice but also converges to the point target. In Fig. 8.9, several self-targeting systems are combined to achieve super-high-intensity light concentrated on a single target with the goal of initiating thermal fusion of the pellet target. As long as the depth of the phase conjugate mirror is longer than the longest path differences among the targeting systems, the pulses from each system coincide at the pellet target and provide super-high-intensity light to the target.

Figure 8.10 shows another example of a tracking system [8]. This time, however, the arrangement is slightly different. Site A sends out a pilot light S through a turbulent medium to Site B. The pilot light contains information about the turbulence. At Site B,

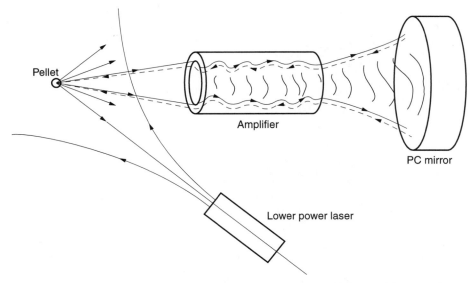

Figure 8.8 Optical self-targeting by means of a phase conjugate (PC) mirror. (After V. V. Shkunov and B. Ya Zel'dovich [7].)

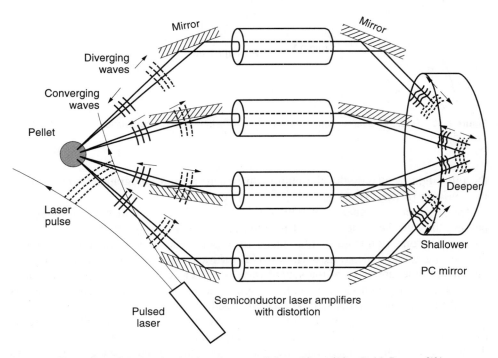

Figure 8.9 Synchronized pulses from parallel amplifiers. (After D. M. Pepper [3].)

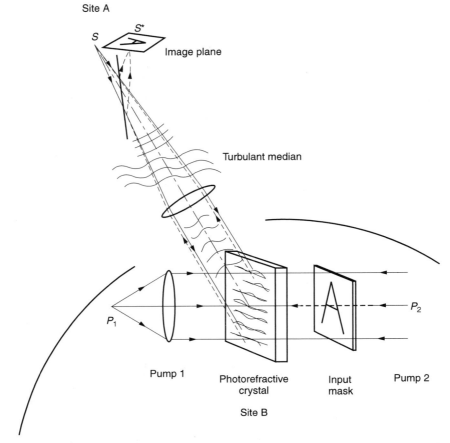

Figure 8.10 Tracking source with one-way transmission through turbulence. (After B. Fischer et al. [8].)

the pilot light is mixed with the pump wave P_1 to form a holographic fringe pattern in a photorefractive crystal.

The information signal, a picture of the letter A, is to be sent back from Site B, passing through the same turbulent medium to Site A. Pump beam P_2 propagating in the opposite direction from P_1 illuminates the input mask of the letter A and then illuminates the holographic fringes. The wave S^* diffracted from the holographic fringes is the phase conjugate wave of the source wave S that has come through the turbulence. S^* goes through the turbulence and the original wavefront is recovered and propagates toward Site A. The letter A will be imaged at Site A.

In the previous arrangements, the signal wave had to go through the turbulence twice, but what is unique about the present arrangement is that the signal wave goes through the turbulence once and the pump wave goes through once. It is a more practical configuration for transmitting information over a distance.

An interesting modification is that if the mask of the letter A is replaced by a fast-speed electronic shutter, then pump wave P_2 is temporally modulated and a free-space optical communication link immune to turbulence is established.

Another application of the self-tracking capability of the phase conjugate wave is the adaptive fiber coupler. The adaptive fiber coupler is a coupler that does not need critical alignment between the two connecting fibers [9]. Figure 8.11a shows the geometry of the coupling. A photorefractive crystal such as barium titanate ($BaTiO_3$) is placed inside an optical resonator formed by a pair of ordinary mirrors. Mirror M_1 is placed on one side of the crystal. A partially reflecting mirror M_2 has been deposited on the facet of fiber 2. M_2 is arranged to be parallel to M_1 so as to form an optical resonator. Light incident from fiber 1 is scattered by impurities in the crystal. The lightwave scattered in the direction perpendicular to mirrors M_1 and M_2 bounces back and forth between M_1 and M_2. Let the wave going horizontally from left to right be the pump wave P_1 and that going from right to left be pump wave P_2.

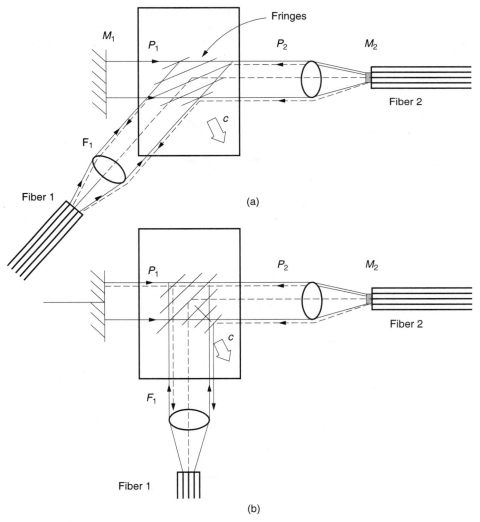

Figure 8.11 Adaptive fiber coupler. (a) Coupling between fibers 1 and 2. (b) When fiber 1 is moved, coupling adaptively continues.

Note that there is no external pump wave in this configuration. There is a special reason for using the photorefractive crystal as the phase conjugate element. For instance, when a nonlinear Kerr medium such as calcite is used, the recorded fringe pattern is an exact replica of the interference pattern of the incident lightwaves. However, when a photorefractive medium like $BaTiO_3$ is used, the recorded fringe pattern is shifted from that of the interference pattern of the incident lightwaves because, as mentioned in Section 5.7.2, the change in refractive index is proportional to the spatial derivative of the light intensity rather than the light intensity itself. Because of this fringe pattern shift, when two beams of equal intensity cross in a photorefractive medium as shown in Fig. 8.12a, the two outputs are uneven and the power at d is larger than at b. The energy is pulled toward the direction of the crystal axis c of the crystal. This is called the two-wave mixing gain of a photorefractive material.

With the configuration shown in Fig. 8.11, the light energy is preferentially bent toward M_2. This method of generating a conjugate wave without external pump waves is called self-pumped phase conjugation (SPPC).

Figures 8.12b and 8.12c show a few more SPPC configurations of the conjugate mirrors. In Fig. 8.12b, the walls of the crystal replace the external mirrors. Figure 8.12c makes use of total internal reflection at the crystal walls [10,11]. The SPPC configurations have significant practical value.

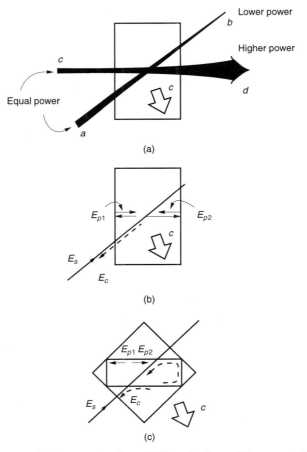

Figure 8.12 Self-pumped phase conjugation. (a) The direction of the energy transfer. (b) SPPC without external cavity. (c) SPPC with total internal reflection.

The incident light F_1 from fiber 1 and pump wave P_1 start forming holographic fringes. The direction of the fringes is such that the incident wave F_1 from fiber 1 is directed toward the input facet of fiber 2 and the connection is made between fibers 1 and 2.

If a misalignment of fiber 1 takes place, as shown in Fig. 8.11b, the direction of the fringe pattern in the crystal rotates such that F_1 is still reflected toward fiber 2. There is, however, a decrease in the energy transfer into the optical resonator. As the bisect between F_1 and P_1 moves away from the crystal axis, the diffraction efficiency from the photorefractive crystal decreases.

It should be noted that no external pump light is necessary in this coupler. The light F_1 from fiber 1 is transferred to the pump waves, and light energy pours into the optical resonator from the light F_1 of fiber 1.

8.9 PICTURE PROCESSING

By combining a pair of phase conjugate mirrors and a multiexposed hologram, an associative memory system such as shown in Fig. 8.13 can be constructed. The system can identify which one of a collection of memorized pictures best fits the interrogating obscure picture [12].

Let us say that the memorized pictures are of a cat, a dog, and a monkey. In memorizing these animal pictures, the angle of incidence of the reference beam is changed each time the input picture is exposed to the photographic film. Let us say the incident angle of reference beam R_1 used for recording the cat O_1 is at $10°$ from the normal to the photographic film. Reference beam R_2 used for recording the dog O_2 is at $20°$, and reference beam R_3 used for the monkey O_3 is at $30°$. After all three exposures are completed, the photographic film is developed and placed in the system shown in Fig. 8.13b.

The operation of the system will be explained with a picture of a cat as the interrogating picture. The transmittance t of the multiexposed hologram is

$$t = \beta \sum_{i=1}^{3} (R_i + O_i)(R_i^* + O_i^*)$$

$$= \beta \sum_{i=1}^{3} (|R_i|^2 + |O_i|^2 + R_i^* O_i + R_i O_i^*) \tag{8.17}$$

where β is a constant characterizing the photographic film. Only the fourth term,

$$t_4 = \beta \sum_{i=1}^{3} R_i O_i^* \tag{8.18}$$

is of concern. When the hologram is illuminated by the light pattern O_1 of the cat, the output light from the hologram is

$$t_4 O_1 = \beta(R_1 O_1^* O_1 + R_2 O_2^* O_1 + R_3 O_3^* O_1) \tag{8.19}$$

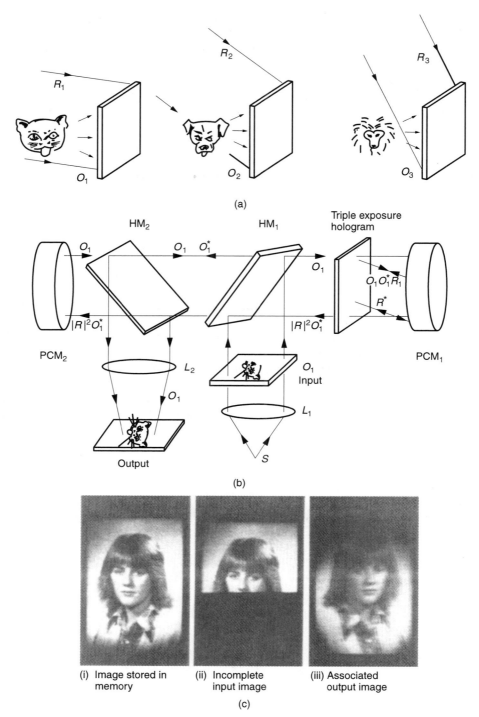

(a)

(b)

(i) Image stored in memory

(ii) Incomplete input image

(iii) Associated output image

(c)

Figure 8.13 Associative memory system. (a) Multi-exposed hologram. Three exposures are made on the same hologram. At each exposure, the direction of the reference beam is changed. (After D. M. Pepper [3].) (b) Interrogation of the input image. (c) With a portion of the portrait as the input, the image of the entire face is recovered. (Courtesy of Y. Owechko et al. [12].)

The first term predominates in Eq. (8.19), because not only do the amplitude distributions of O_1 and O_1^* match, but also the phase angle of $O_1^* O_1$ is constant (zero) throughout the pattern. The output $t_4 O_1$ from the hologram is essentially R_1 with reduced intensity.

Beam R_1 emergent from the hologram is now reflected by the right-hand-side phase conjugate mirror and the reflected beam becomes R_1^*. The hologram is now reilluminated by R_1^*. The contribution of the fourth term in Eq. (8.17) to the light transmitted through the hologram is

$$t_4 R_1^* = \beta(O_1^* R_1 R_1^* + O_2^* R_2 R_1^* + O_3^* R_3 R_1^*) \tag{8.20}$$

The contribution of the first term of Eq. (8.20) is predominant because the phase angle of $R_1 R_1^*$ is exactly zero, while that of $R_2 R_1^*$ is $10°$ and that of $R_3 R_1^*$ is $20°$ and propagates off axis. The emergent beam from the hologram becomes O_1^*.

O_1^* is further converted into O_1 by the phase conjugate mirror on the left-hand side and finally reaches the output image plane by way of the half-mirror HM_2. The image of the cat is formed by means of the imaging lens L_2. The beam that passes through the half-mirror HM_2 will repeat the same process to enhance the sensitivity of the system.

The sensitivity of the system can be improved significantly if a photographic film with a thick emulsion is used for fabricating the hologram. The sensitivity of the brightness of the reconstructed image to the angle of incidence of the reconstructing beam is enhanced due to the increased sizes of the miniature mirrors in the fringe pattern in the emulsion [2]. In fact, if the photographic film is replaced by a volume holographic material such as a $BaTiO_3$ crystal, a significant improvement in sensitivity and flexibility is achieved.

The power of discrimination can be adjusted by the threshold level of the phase conjugate mirror, and even a picture of the cat's brother or a faded imperfect input image can still be interrogated. Such a system is useful for interrogating handwritten letters or for fingerprint detection.

Figure 8.13c gives a similar demonstration for a portrait. Using only a portion of the portrait as input, the entire portrait is generated as a result of the interrogation.

8.10 THEORY OF PHASE CONJUGATE OPTICS

The theory of phase conjugate optics will be presented. Even though the level of treatment is elementary, it is still useful for solving practical problems.

8.10.1 Maxwell's Equations in a Nonlinear Medium

Maxwell's equations are the starting point for the quantitative representation of nonlinear phenomena [13–19]. Maxwell's equations are repeated here for convenience:

$$\nabla \times \mathbf{E} = -\frac{\partial \mathbf{B}}{\partial t} \tag{8.21}$$

$$\nabla \times \mathbf{H} = \mathbf{J} + \frac{\partial \mathbf{D}}{\partial t} \tag{8.22}$$

$$\nabla \cdot \mathbf{D} = \rho \tag{8.23}$$

$$\nabla \cdot \mathbf{B} = 0 \tag{8.24}$$

where

$$\mathbf{D} = \epsilon_0 \mathbf{E} + \mathbf{P} \tag{8.25}$$

$$\mathbf{B} = \mu_0 \mathbf{H} + \mathbf{M} \tag{8.26}$$

$\mathbf{P}$ is the induced electric polarization and is the focus of attention in this chapter. The medium is assumed to be nonconducting and nonmagnetic. This assumption leads to

$$\mathbf{J} = \rho = \mathbf{M} = 0 \tag{8.27}$$

From Eqs. (8.21), (8.22), (8.25), and (8.27), the following expression is obtained:

$$\nabla \times \nabla \times \mathbf{E} + \frac{1}{c^2} \frac{\partial^2 \mathbf{E}}{\partial t^2} = -\mu_0 \frac{\partial^2 \mathbf{P}}{\partial t^2} \tag{8.28}$$

where $c^2 = (\epsilon_0 \mu_0)^{-1}$. The identities involving differential operators that will be used to simplify Eq. (8.28) are

$$\nabla \times \nabla \times \mathbf{E} = \nabla(\nabla \cdot \mathbf{E}) - \nabla^2 \mathbf{E} \tag{8.29}$$

$$\nabla \cdot \mathbf{D} = \mathbf{E} \cdot \nabla \epsilon + \epsilon \nabla \cdot \mathbf{E} \tag{8.30}$$

If the spatial variation $\nabla \epsilon$ is negligible, then Eqs. (8.23), (8.27), and (8.30), lead to

$$\nabla \cdot \mathbf{E} = 0 \tag{8.31}$$

With Eqs. (8.29) and (8.31), Eq. (8.28) becomes

$$\nabla^2 \mathbf{E} - \frac{1}{c^2} \frac{\partial^2 \mathbf{E}}{\partial t^2} = \mu_0 \frac{\partial^2 \mathbf{P}}{\partial t^2} \tag{8.32}$$

Equation (8.32) can be interpreted as the wave equation of $\mathbf{E}$ whose source of excitation is $\mu_0 \partial^2 \mathbf{P} / \partial t^2$. However, the electric polarization $\mathbf{P}$ is induced by $\mathbf{E}$; and $\mathbf{P}$ is

$$\mathbf{P} = \epsilon_0 (\chi^{(1)} \cdot \mathbf{E} + \chi^{(2)} : \mathbf{EE} + \chi^{(3)} \vdots \mathbf{EEE} + \cdots) \tag{8.33}$$

The first term in Eq. (8.33) is proportional to $\mathbf{E}$, while the rest of the terms are proportional to higher orders of $\mathbf{E}$. The former is called the linear part $\mathbf{P}_\mathrm{L}$; and the latter, the nonlinear part $\mathbf{P}_\mathrm{NL}$ of the induced electric polarization

$$\mathbf{P} = \mathbf{P}_\mathrm{L} + \mathbf{P}_\mathrm{NL} \tag{8.34}$$

where

$$\mathbf{P}_\mathrm{L} = \epsilon_0 \chi^{(1)} \cdot \mathbf{E} \tag{8.35}$$

$$\mathbf{P}_\mathrm{NL} = \epsilon_0 (\chi^{(2)} : \mathbf{EE} + \chi^{(3)} \vdots \mathbf{EEE} + \cdots) \tag{8.36}$$

$\chi^{(i)}$ is the ith order optical susceptibility and is a tensor of rank $i + 1$. A nonlinear dielectric medium is characterized by $\mathbf{P}_\mathrm{NL}$.

Assuming a sinusoidal time dependence and substituting for $\mathbf{P_L}$ from Eq. (8.35) and $\mathbf{P_{NL}}$ from Eq. (8.36), the wave equation Eq. (8.32) becomes

$$\nabla^2 \mathbf{E} + \left(\frac{\omega}{c}\right)^2 \mathbf{E} = -\mu_0 \epsilon_0 \omega^2 \chi^{(1)} \cdot \mathbf{E} + \mu_0 \frac{\partial^2 \mathbf{P_{NL}}}{\partial t^2} \tag{8.37}$$

Noting that

$$\epsilon_r = 1 + \chi^{(1)} \tag{8.38}$$

Eq. (8.37) can be rewritten as

$$\nabla^2 \mathbf{E} + k_0^2 \epsilon_r \cdot \mathbf{E} = \mu_0 \frac{\partial^2 \mathbf{P_{NL}}}{\partial t^2} \tag{8.39}$$

with

$$k_0^2 = \omega^2 \mu_0 \epsilon_0 \tag{8.40}$$

8.10.2   Nonlinear Optical Susceptibilities $\chi^{(2)}$ and $\chi^{(3)}$

Susceptibilities with i larger than 3 are hardly used, so that only the properties of $\chi^{(2)}$ and $\chi^{(3)}$ are investigated here. Certain materials such as glass or NaCl have zero $\chi^{(2)}$ but have nonzero $\chi^{(3)}$. For experiments to be performed based solely on $\chi^{(3)}$, such materials are attractive because there are no second order nonlinearities, which might complicate the results. A slight detour will be taken to explain why some materials have zero $\chi^{(2)}$ but nonzero $\chi^{(3)}$.

First of all, for simplicity, let us choose

$$\mathbf{E} = (E \cos \omega t)\hat{\mathbf{x}} \tag{8.41}$$

Since only the $\hat{\mathbf{x}}$ component is considered, Eq. (8.41) can be treated as a scalar quantity. The second nonlinearity gives

$$P_{NL}^{(2)} = \tfrac{1}{2} \chi^{(2)} E^2 (1 + \cos 2\omega t) \tag{8.42}$$

and becomes the expression for second harmonic generation (SHG). On the other hand, the third nonlinearity gives

$$P_{NL}^{(3)} = \tfrac{1}{4} \chi^{(3)} E^3 (3 \cos \omega t + \cos 3\omega t) \tag{8.43}$$

and generates the third order higher harmonic.

Generation of the higher harmonics will be examined graphically in order to find out why crystals with inversion symmetry do not display the second order nonlinearity [19].

Figure 8.14a shows a one-dimensional model of a crystal with inversion symmetry. It possesses an inversion symmetry with respect to any one of the ions. Let us say that with respect to the c axis $\oplus$ and $\ominus$ charges are symmetrically distributed. Consider an instant that the electric field $\mathbf{E}$ of the light is in the positive x direction. The positive charges move to the right, and the negative charges move to the left, as shown in

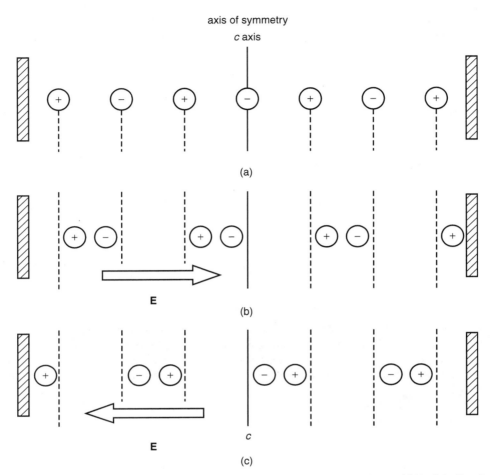

Figure 8.14 Induced electrical polarization in a crystal with inversion symmetry. (a) No light **E** = 0. (b) Light is on with **E** = $E\hat{\mathbf{x}}$. (c) Light is on with **E** = $-E\hat{\mathbf{x}}$.

Fig. 8.14b. At the next instant, the direction of **E** is reversed, and each charge moves in the opposite direction and the distribution of the charges becomes like the one shown in Fig. 8.14c. The distribution of the charges that **E** sees is the same for both instances, and the amounts of polarization are the same, except for the reversal of the outermost charges that determine the polarization polarity.

Figure 8.15a shows a plot of P_{NL} with respect to time. Even though the shape is distorted from a sinusoidal curve due to the nonlinearity, the shape of the curve for positive values of polarization in the range $0 < t < T/2$ is identical to that for negative values of polarization in the range $T/2 < t < T$ except for its sign, where T is the period of the fundamental frequency.

If the second harmonic is present, the shapes in the first half and the second half cannot be identical, as will be illustrated using Fig. 8.15b in which the curves of the fundamental and the second harmonic are plotted. In the region $0 < t < T/2$, the signs of the peaks of the fundamental and second harmonic are both positive; while in the

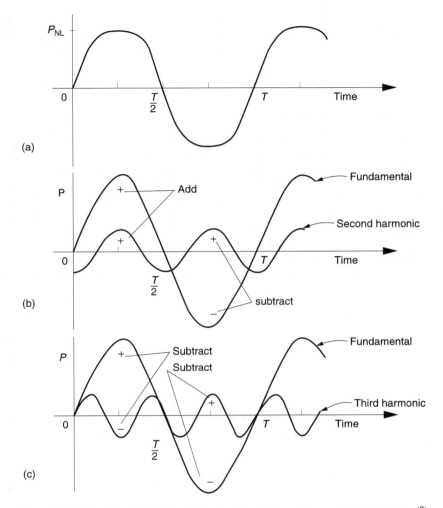

P_{NL}

0 $\dfrac{T}{2}$ T Time

(a)

P + Add Fundamental Second harmonic 0 $\dfrac{T}{2}$ T Time + + − subtract

(b)

P + Subtract Subtract Fundamental Third harmonic 0 − + $\dfrac{T}{2}$ T Time −

(c)

Figure 8.15 Graphical illustration that a crystal with inversion symmetry has zero $\chi^{(2)}$.

region $T/2 < t < T$, the peaks of the fundamental and second harmonic have opposite signs. As long as the second harmonic is added to the fundamental, the response curve cannot have the same shape in the regions $0 < t < T/2$ and $T/2 < t < T$, as shown in Fig. 8.15a for a crystal with inverse symmetry. Thus, the second order optical susceptibility $\chi^{(2)}$ has to be zero in a crystal with inversion symmetry.

The same crystal, however, can have a third order nonlinearity. Figure 8.15c shows the plot of the fundamental and the third harmonic. In the region $0 < t < T/2$, the center peak of the third harmonic is negative while that of the fundamental is positive. The resultant is the difference between these two peaks. In the region $T/2 < t < T$, the center peak of the third harmonic is positive while that of the fundamental is negative. The resultant is again the difference between the two peaks, and the shape in the region $0 < t < T/2$ becomes identical to that in $T/2 < t < T$. Thus, a crystal with inversion symmetry can support the third order nonlinearity.

In summary, a crystal with inversion symmetry cannot support the second order nonlinearity but can support the third order nonlinearity. Crystals that do not possess inversion symmetry can, in principle, support simultaneously second and third order nonlinearities.

In the next section, we return to solving the nonlinear Maxwell's equation.

8.10.3 Coupled Wave Equations

The first step toward solving the nonlinear Maxwell's equation, Eq. (8.39), is to find an expression for $\mathbf{P}_{NL}$ in Eq. (8.39). In the general case of four-wave mixing, $\mathbf{P}_{NL}$ is generated from a combination of four incident waves of different frequencies.

The analytic signal is one of the most common ways of solving differential equations in electrical engineering. A sinusoidal function, say, $\cos \omega t$, is replaced by the exponential $e^{j\omega t}$ and the differential equations are solved. The final answer is obtained by taking only the real part of the solution. This method does not necessarily work for solving problems in nonlinear optics; therefore, $\cos \omega t = \frac{1}{2}(e^{j\omega t} + \text{c.c.})$ will be used. See the boxed note and Appendix B of Volume 1.

All incident waves are assumed to be plane waves and are made up of four waves:

$$\mathbf{E}(\mathbf{r}, t) = \sum_{j=1}^{4} \hat{\mathbf{a}}_j |A_j(\mathbf{r}, \omega_j)| \cos(-\omega_j t + \mathbf{k}_j \cdot \mathbf{r} + \phi_j) \tag{8.44}$$

where $\hat{\mathbf{a}}_j$ are the unit vectors of the direction of polarization. The exponential expression that is exactly equivalent to Eq. (8.44) is

$$\mathbf{E}(\mathbf{r}, t) = \frac{1}{2} \sum_{j=1}^{4} \hat{\mathbf{a}}_j [E_j(\mathbf{r}, \omega_j) e^{-j\omega_j t} + \text{c.c.}] \tag{8.45}$$

where

$$E_j(\mathbf{r}, \omega_j) = A_j(\mathbf{r}, \omega_j) e^{j\mathbf{k}_j \cdot \mathbf{r}} \tag{8.46}$$

$$A_j(\mathbf{r}, \omega_j) = |A_j(\mathbf{r}, \omega_j)| e^{j\phi_j} \tag{8.47}$$

or simply

$$\begin{aligned} E_j &= E_j(\mathbf{r}, \omega_j) \\ A_j &= A_j(\mathbf{r}, \omega_j) \end{aligned} \tag{8.48}$$

For instance, it is well known that if two signals with frequencies ω_1 and ω_2 are put into a nonlinear element, the output contains both the upper beat frequency $\omega_1 + \omega_2$ and the lower beat frequency $\omega_1 - \omega_2$. If the analytic signal method is used for this case,

$$\text{Re } (e^{j\omega_1 t} + e^{j\omega_2 t})^2 = \cos 2\omega_1 t + \cos 2\omega_2 t + \cos(\omega_1 + \omega_2)t$$

the lower beat frequency component is missing. A more detailed explanation can be found in Appendix B of Volume 1.

Similarly, the induced electric polarization is expressed as

$$\mathbf{P}_{NL}(\mathbf{r}, t) = 1/2 \sum_{j=1}^{4} \hat{\mathbf{b}}_j [P_{NL_j}(\mathbf{r}, \omega_j) e^{-j\omega_j t} + \text{c.c.}] \qquad (8.49)$$

$$P_{NL_j}(\mathbf{r}, \omega_j) = B_j(\mathbf{r}, \omega_j) e^{j\mathbf{k}_j \cdot \mathbf{r}} \qquad (8.50)$$

or simply

$$P_{NL_j} = P_{NL_j}(\mathbf{r}, \omega_j) \qquad (8.51)$$

Next, the actual values of $\mathbf{P}_{NL}$ will be calculated for a Kerr medium whose susceptibility is predominantly the third order $\chi^{(3)}$.

$$\mathbf{P}_{NL} = \epsilon_0 \chi^{(3)} : \mathbf{EEE} \qquad (8.52)$$

We assume that all $\mathbf{E}$'s are nothing but the waves polarized in the x direction. $\mathbf{E}$, however, consists of waves of four different frequencies, ω_1, ω_2, ω_3, and ω_4. All four frequencies are assumed to be in the same frequency range, say, in the visible or infrared region. Their propagation directions and wavelengths are specified by the complex propagation constants $\mathbf{k}_1, \mathbf{k}_2, \mathbf{k}_3$, and $\mathbf{k}_4$.

Inserting Eq. (8.45) into (8.52) gives

$$\mathbf{P}_{NL} = \frac{\hat{\mathbf{x}} \epsilon_0 \chi_{xxxx}}{8} (E_1 e^{-j\omega_1 t} + E_1^* e^{j\omega_1 t} + E_2 e^{-j\omega_2 t} + E_2^* e^{j\omega_2 t}$$

$$+ E_3 e^{-j\omega_3 t} + E_3^* e^{j\omega_3 t} + E_4 e^{-j\omega_4 t} + E_4^* e^{j\omega_4 t})^3 \qquad (8.53)$$

As far as the subscript of χ_{xxxx} is concerned, the first subscript indicates the direction of polarization of the wave emergent from the nonlinear medium and the next three subscripts indicate the directions of polarization of the incident waves. In the present case, all are assumed in the $\hat{\mathbf{x}}$ direction. Manipulation of the cubic in Eq. (8.53) no doubt generates many beat frequencies. The manipulation is shown in Appendix C. We assume that frequencies associated with the third power, such as $3\omega_1$, $3\omega_2$, $\omega_1 + \omega_2 + \omega_4$, and $2\omega_2 + \omega_1$, are all out of the range of interest and are discarded, but all other terms are kept. Terms with frequencies such as $\omega_1 + \omega_2 - \omega_3$, $\omega_1 + \omega_2 - \omega_4$ or $2\omega_2 - \omega_3$ are of particular interest because they are all in the visible and infrared region.

In order that a significant exchange of energy take place among the frequency components, the generated beat frequency components have to be recycled to participate over and over again in the beating process. For instance, the beating among ω_2, ω_3 and ω_4 creates the original frequency ω_1 if the condition

$$\omega_1 = \omega_3 + \omega_4 - \omega_2 \qquad (8.54)$$

is satisfied. This ω_1 frequency component again participates in the beating and creates component ω_2 in accordance with Eq. (8.54), that is, $\omega_2 = \omega_3 + \omega_4 - \omega_1$. These cyclic conversions among the four frequencies are essential for four-wave mixing. If Eq. (8.54) is satisfied, the four frequencies are said to be commensurate. Among the terms in

Appendix C of Volume 1, the following is the set of equations that are commensurate with each other [18].

$$\mathbf{P}_{NL} = \tfrac{1}{2}\hat{\mathbf{x}}[P_{NL}(\omega_1)e^{-j\omega_1 t} + P_{NL}(\omega_2)e^{-j\omega_2 t}$$

$$+ P_{NL}(\omega_3)e^{-j\omega_3 t} + P_{NL}(\omega_4)e^{-j\omega_4 t} + \text{c.c.}] \tag{8.55}$$

$$P_{NL}(\omega_1) = \chi_{\text{eff}}(Q_1 E_1 + 2E_3 E_4 E_2^*) \tag{8.56}$$

$$P_{NL}(\omega_2) = \chi_{\text{eff}}(Q_2 E_2 + 2E_3 E_4 E_1^*) \tag{8.57}$$

$$P_{NL}(\omega_3) = \chi_{\text{eff}}(Q_3 E_3 + 2E_1 E_2 E_4^*) \tag{8.58}$$

$$P_{NL}(\omega_4) = \chi_{\text{eff}}(Q_4 E_4 + 2E_1 E_2 E_3^*) \tag{8.59}$$

where

$$Q_1 = Q - |E_1|^2 \tag{8.60}$$

$$Q_2 = Q - |E_2|^2 \tag{8.61}$$

$$Q_3 = Q - |E_3|^2 \tag{8.62}$$

$$Q_4 = Q - |E_4|^2 \tag{8.63}$$

$$Q = 2(|E_1|^2 + |E_2|^2 + |E_3|^2 + |E_4|^2) \tag{8.64}$$

$$\chi_{\text{eff}} = \frac{3\epsilon_0}{4}\chi_{xxxx} \tag{8.65}$$

Four frequency components are separately associated with the nonlinear wave equation, Eq. (8.39). $P_{NL}(\omega_1)$ is the source of excitation of E_1, and $P_{NL}(\omega_2)$ is the source of excitation of E_2, and so on. From Eq. (8.39) and Eqs. (8.56)–(8.59), the following simultaneous differential equations are generated:

$$(\nabla^2 + k_1^2)E_1 = -\mu_0\omega_1^2\chi_{\text{eff}}(Q_1 E_1 + 2E_3 E_4 E_2^*) \tag{8.66}$$

$$(\nabla^2 + k_2^2)E_2 = -\mu_0\omega_2^2\chi_{\text{eff}}(Q_2 E_2 + 2E_3 E_4 E_1^*) \tag{8.67}$$

$$(\nabla^2 + k_3^2)E_3 = -\mu_0\omega_3^2\chi_{\text{eff}}(Q_3 E_3 + 2E_1 E_2 E_4^*) \tag{8.68}$$

$$(\nabla^2 + k_4^2)E_4 = -\mu_0\omega_4^2\chi_{\text{eff}}(Q_4 E_4 + 2E_1 E_2 E_3^*) \tag{8.69}$$

Equations (8.66)–(8.69) are called the coupled wave equations. It should be noted that it is the condition of Eq. (8.54) that allows $E_3 E_4 E_2^*$ to participate in the generation of E_1. Inserting Eq. (8.46) into Eqs. (8.66)–(8.69), the required vector propagation constants are found. For instance, with Eq. (8.66), the induced polarization (2nd term) has a vector propagation constant of $\mathbf{k}_3 + \mathbf{k}_4 - \mathbf{k}_2$ and the excited E_1 field (1st term) has the propagation constant $\mathbf{k}_1$. An important condition for the maximum transfer of energy of the induced polarization $\mathbf{P}_{NL}$ into the electric field $\mathbf{E}_1$ is that both waves propagate in phase throughout their paths. The condition of maximum coupling, therefore, is

$$\mathbf{k}_1 = \mathbf{k}_3 + \mathbf{k}_4 - \mathbf{k}_2 \tag{8.70}$$

Similarly, the maximum transfer of energy with Eq. (8.67) is $\mathbf{k}_2 = \mathbf{k}_3 + \mathbf{k}_4 - \mathbf{k}_1$. This equation is exactly identical to Eq. (8.70). Similar conditions generated by Eqs. (8.68) and (8.69) also satisfy Eq. (8.70).

In addition to the previously mentioned frequency condition, Eq. (8.54), the phase matching condition, Eq. (8.70), has to be satisfied simultaneously for the maximum energy coupling.*

In summary, the coupled wave equations govern the exchange of energy among the four different frequency components. In the next section, solutions will be found with an approximation imposed on the coupled wave equation.

8.10.4 Solutions with Bohr's Approximation

Assumptions and approximations are imposed on the coupled wave equations, Eqs. (8.66)–(8.69), to find the differential equations for the amplitudes. The first assumption is that all waves are propagating in the z direction and

$$E_j = \mathbf{A}_j(z)e^{js_jk_jz} \tag{8.71}$$

with $k_j = \omega_j\sqrt{\mu_0\epsilon_0\epsilon_r}$, where $s_j = +1$ when the jth wave propagates in the positive z direction, and $s_j = -1$ when the jth wave propagates in the negative z direction. Inserting Eq. (8.71) into the left hand side of Eq. (8.39) gives

$$(\nabla^2 + k_j^2)E_j = \left(j2s_jk_j\frac{dA_j}{dz} + \frac{d^2A_j}{dz^2}\right)e^{js_jk_jz} \tag{8.72}$$

The second assumption is that the variation of $A_j(z)$ with respect to z is so slow that its second derivative can be ignored compared to other terms. This approximation is called Bohr's approximation or the slowly varying envelope approximation. With these approximations, Eq. (8.39) finally becomes

$$\frac{dA_j}{dz} = js_j\frac{\omega_j}{2}\sqrt{\frac{\mu_0}{\epsilon_0\epsilon_r}}P_{NL}(\omega_j)e^{-js_jk_jz} \tag{8.73}$$

Insertion of Eqs. (8.56)–(8.59) into Eq. (8.73) results in the following set of equations:

$$\frac{dA_1}{dz} = js_1K_1(Q_1A_1 + 2A_3A_4A_2^*e^{j(s_3k_3+s_4k_4-s_2k_2-s_1k_1)z}) \tag{8.74}$$

$$\frac{dA_2}{dz} = js_2K_2(Q_2A_2 + 2A_3A_4A_1^*e^{j(s_3k_3+s_4k_4-s_1k_1-s_2k_2)z}) \tag{8.75}$$

---

* In quantum mechanics, the four-photon collision has to satisfy both the conservation of energy,

$$\hbar\omega_1 = \hbar\omega_3 + \hbar\omega_4 - \hbar\omega_2$$

and the conservation of momentum

$$\hbar\mathbf{k}_1 = \hbar\mathbf{k}_3 + \hbar\mathbf{k}_4 - \hbar\mathbf{k}_2$$

where $\hbar = h/2\pi$ and h is Planck's constant.

$$\frac{dA_3}{dz} = js_3 K_3 (Q_3 A_3 + 2A_1 A_2 A_4^* e^{j(s_1 k_1 + s_2 k_2 - s_4 k_4 - s_3 k_3)z}) \tag{8.76}$$

$$\frac{dA_4}{dz} = js_4 K_4 (Q_4 A_4 + 2A_1 A_2 A_3^* e^{j(s_1 k_1 + s_2 k_2 - s_3 k_3 - s_4 k_4)z}) \tag{8.77}$$

$$K_j = \frac{\omega_j}{2} \sqrt{\frac{\mu_0}{\epsilon_0 \epsilon_r}} \chi_{\text{eff}} \tag{8.78}$$

Example 8.1   Figure 8.16 shows an arrangement for generating a phase conjugate wave using the principle of holography [20]. Explain the operation using the coupled wave equations.

Solution   All frequencies used in the hologram are the same and

$$\omega_1 = \omega_2 = \omega_3 = \omega_4 = \omega \tag{8.79}$$

This satisfies the frequency condition in Eq. (8.54). Let E_3 and E_4 be the signal wave E_s and phase conjugate wave E_c, respectively. Let the propagation directions of these two waves be along the z axis as shown in Fig. 8.16, with $s_3 = +1$ for E_3 and $s_4 = -1$ for E_4.

$$E_3 = E_s = A_s e^{jk_s z}$$
$$E_4 = E_c = A_c e^{-jk_c z} \tag{8.80}$$

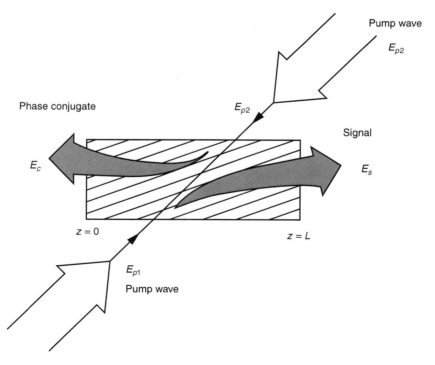

Figure 8.16   Generating a phase conjugate wave from a hologram.

Let the two pump waves E_{p_1} and E_{p_2} be represented by

$$E_1 = E_{p_1} = A_p e^{j\mathbf{k}_{p_1} \cdot \mathbf{r}}$$
$$E_2 = E_{p_2} = A_p e^{j\mathbf{k}_{p_2} \cdot \mathbf{r}} \tag{8.81}$$

Their amplitudes are assumed equal. Furthermore, the amplitudes of the pump waves are assumed to be so large that the depletion of the energy into either the signal or phase conjugate waves is negligible, and the amplitudes can be considered not only constant with respect to distance but also

$$|A_p|^2 >> |A_s|^2$$
$$|A_p|^2 >> |A_c|^2 \tag{8.82}$$

The signal and conjugate waves propagate in opposite directions, so that

$$\mathbf{k}_3 + \mathbf{k}_4 = 0 \tag{8.83}$$

and

$$s_3 = +1$$
$$s_4 = -1 \tag{8.84}$$

and in order to satisfy Eq. (8.70),

$$\mathbf{k}_{p_1} + \mathbf{k}_{p_2} = 0 \tag{8.85}$$

With Eq. (8.80)–(8.85), the coupled wave equations, Eqs. (8.76) and (8.77), become

$$\frac{dA_s}{dz} = jK\{2|A_p|^2 A_s + A_p^2 A_c^*\}$$
$$\frac{dA_c}{dz} = -jK\{2|A_p|^2 A_c + A_p^2 A_s^*\} \tag{8.86}$$

where

$$K = \omega \sqrt{\frac{\mu_0}{\epsilon_0 \epsilon_r}} \chi_{\text{eff}} \tag{8.87}$$

In order to remove the first term from the right-hand side of Eq. (8.86), the amplitude and phase factors of A_s and A_c are explicitly written as

$$A_s = A_{s0} e^{+j\beta_{\text{NL}} z}$$
$$A_c = A_{c0} e^{-j\beta_{\text{NL}} z}$$
$$A_{s0} = A_{s0}(z)$$
$$A_{c0} = A_{c0}(z) \tag{8.88}$$

where

$$\beta_{NL} = 2K|A_p|^2 \tag{8.89}$$

Inserting Eq. (8.88) into (8.86) gives

$$\frac{dA_{s0}}{dz} = jKA_p^2 A_{c0}^* \tag{8.90}$$

$$\frac{dA_{c0}}{dz} = -jKA_p^2 A_{s0}^* \tag{8.91}$$

Taking the derivative of Eq. (8.90) and inserting Eq. (8.91) gives

$$\frac{d^2 A_{c0}}{dz^2} + K^2|A_p|^4 A_{c0} = 0 \tag{8.92}$$

and similarly,

$$\frac{d^2 A_{s0}}{dz^2} + K^2|A_p|^4 A_{s0} = 0 \tag{8.93}$$

The general solution of Eq. (8.92) is

$$A_{c0} = A\cos K|A_p|^2 z + B\sin K|A_p|^2 z \tag{8.94}$$

From Eqs. (8.91) and (8.94), A_{s0}^* is expressed as

$$A_{s0}^* = j\frac{A_p^*}{A_p}(-A\sin K|A_p|^2 z + B\cos K|A_p|^2 z) \tag{8.95}$$

The integration constants A and B are determined from the boundary conditions:

$$A_{c0}(L) = 0 \quad \text{at } z = L \tag{8.96}$$

$$A_s(0) = A_{s0}(0) \quad \text{at } z = 0 \tag{8.97}$$

From Eqs. (8.94) and (8.96), the integration constant A is

$$A = -B\tan K|A_p|^2 L \tag{8.98}$$

From Eqs. (8.95) and (8.97), the integration constant B is

$$B = \frac{A_p}{jA_p^*}A_{s0}^*(0) \tag{8.99}$$

Inserting these constants into Eq. (8.94) and using Eq. (8.88) gives

$$A_c(z) = j\frac{A_p}{A_p^*}A_s^*(0)e^{-j\beta_{NL}z}\frac{\sin K|A_p|^2(L-z)}{\cos K|A_p|^2 L} \tag{8.100}$$

and similarly, inserting Eqs. (8.98) and (8.99) into Eq. (8.95) and using Eq. (8.88) gives

$$A_s(z) = A_s(0)e^{j\beta_{\mathrm{NL}}z}\frac{\cos K|A_p|^2(L-z)}{\cos K|A_p|^2 L} \tag{8.101}$$

Now let us interpret the calculated results. From Eq. (8.100), the magnitude of $A_c(z)$ increases with distance from the back surface at $z = L$. Referring to Fig. 8.16, the pump wave E_{p_2} is depleted into the phase conjugate wave by the deflection from the fringes established by E_{p_1} and E_s and is accumulated toward the front surface at $z = 0$. Similarly, the signal wave $A_s(z)$ grows from the front surface to the back surface $z = L$ by the depletion of the pump wave E_{p_1} into E_s. $\square$

8.11 THE GAIN OF FORWARD FOUR-WAVE MIXING

The geometry shown in Fig. 8.16 is one example that satisfies both the frequency and phase matching conditions. In this geometry, not only the two pump waves are counterpropagating but also the signal and phase conjugate waves are counterpropagating. In order to meet the phase matching condition of Eq. (8.70), each side of

$$\mathbf{k}_1 + \mathbf{k}_2 = \mathbf{k}_3 + \mathbf{k}_4 \tag{8.102}$$

was set individually to zero. The frequency condition, Eq. (8.54),

$$\omega_1 + \omega_2 = \omega_3 + \omega_4 \tag{8.103}$$

was met by letting all the frequencies be the same.

Another geometry will be investigated here. This time, all the waves are copropagating in the forward direction [16,21]. It is certainly possible to meet the condition of Eq. (8.102) by choosing identical $\mathbf{k}$'s and choosing identical frequencies to meet the condition of Eq. (8.103). In the earlier counterpropagating case, a half-mirror was good enough to separate the phase conjugate and signal waves. This is not possible in the copropagating case. A remedy for this is the use of multiple frequencies that meet the frequency condition of Eq. (8.103). One way this can be done is to set the average value of ω_3 and ω_4 equal to the average value of ω_1 and ω_2, as shown in Fig. 8.17a.

A special case of Fig. 8.17a is shown in Fig. 8.17b. That special case occurs when ω_1 and ω_2 are identical, and the four-wave mixing is semidegenerate. This arrangement, when ω_1 (or ω_2) is taken as a pump wave, necessitates only one pump wave and simplifies the implementation. Figure 8.18 shows the implementation. The signal and pump waves are fed into a dispersion-shifted fiber and the outputs are the phase conjugate, signal, and pump waves, among which the phase conjugate wave is selected by means of an optical filter. The core glass of the dispersion-shifted fiber is used as the $\chi^{(3)}$ nonlinear medium.

Next, the output powers of the signal and phase conjugate waves are calculated using the coupled wave equations, Eqs. (8.74)–(8.77). The procedure is quite similar to

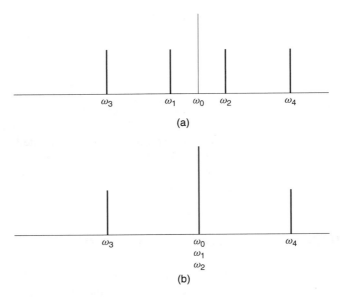

(a)

(b)

Figure 8.17 Spectra of forward four-wave mixing. (a) Nondegenerate case. (b) Semidegenerate case.

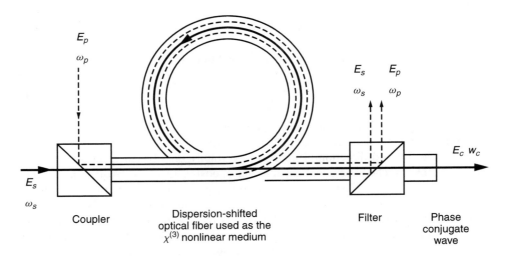

Figure 8.18 Forward four-wave mixing.

the reflection type presented in Example 8.1, and emphasis is placed on pointing out the differences, as well as the significance of the phase matching condition imposed on $\mathbf{k}_j$ by Eq. (8.70). Referring to Fig. 8.18, let us denote

$$E_1 = E_2 = A_p e^{jk_p z} \tag{8.104}$$

$$E_3 = E_s = A_s e^{jk_s z} \tag{8.105}$$

$$E_4 = E_{pc} = A_c e^{jk_c z} \tag{8.106}$$

In this case, all waves are propagating in the forward direction, and

$$s_j = 1 \tag{8.107}$$

With Eq. (8.82), the coupled wave equations, Eqs. (8.74)–(8.77), become

$$\frac{dA_p}{dz} \approx j\frac{3}{2}K|A_p|^2 A_p \tag{8.108}$$

$$\frac{dA_s}{dz} = jK(2|A_p|^2 A_s + A_p^2 A_c^* e^{j\Delta kz}) \tag{8.109}$$

$$\frac{dA_c}{dz} = jK(2|A_p|^2 A_c + A_p^2 A_s^* e^{j\Delta kz}) \tag{8.110}$$

$$\Delta k = k_1 + k_2 - k_3 - k_4 \tag{8.111}$$

In order to remove the first terms from both Eqs. (8.109) and (8.110), the amplitude and phase factors of A_s and A_c are explicitly written as

$$A_s = A_{s0}e^{j\beta_{NL}z} \tag{8.112}$$

$$A_c = A_{c0}e^{j\beta_{NL}z} \tag{8.113}$$

$$\beta_{NL} = 2k|A_p|^2$$

Inserting the expressions for A_s and A_c in Eqs. (8.112) and (8.113) into Eqs. (8.109) and (8.110) gives

$$\frac{dA_{s0}}{dz} = jKA_p^2 A_{c0}^* e^{j(\Delta k - 2\beta_{NL})z} \tag{8.114}$$

$$\frac{dA_{c0}}{dz} = jKA_p^2 A_{s0}^* e^{j(\Delta k - 2\beta_{NL})z} \tag{8.115}$$

The procedure for solving the differential equations starts with Eq. (8.108). The solution of Eq. (8.108) with the boundary condition $A_p = A_p(0)$ at $z = 0$ is

$$A_p = A_p(0)e^{j\beta_p z} \tag{8.116}$$

where

$$\beta_p = \frac{3}{2}K|A_p|^2 \tag{8.117}$$

It should be noted that both β_p and β_{NL} are a function of the intensity $|A_p|^2$ and are nonlinear with the pump field.

Inserting Eq. (8.116) into Eqs. (8.114) and (8.115) gives

$$\frac{dA_{s0}}{dz} = jKA_p^2(0)A_{c0}^* e^{j\Gamma z} \tag{8.118}$$

$$\frac{dA_{c0}^*}{dz} = -jKA_p^2(0)A_{s0}e^{-j\Gamma z} \tag{8.119}$$

where

$$\Gamma = 2(\beta_p - \beta_{NL}) + \Delta k \tag{8.120}$$

With Eqs. (8.89) and (8.117), Eq. (8.120) is further rewritten as

$$\Gamma = \Delta k - KA_p^2(0) \tag{8.121}$$

Assumed solutions

$$A_{s0} = [Ae^{gz} + Be^{-gz}]e^{j(\Gamma/2)z} \tag{8.122}$$

$$A_{c0}^* = [Ce^{gz} + De^{-gz}]e^{-j(\Gamma/2)z} \tag{8.123}$$

are put into Eqs. (8.118) and (8.119). The prime target of this calculation is to obtain the value of the gain g.

Let

$$KA_p^2(0) = a \tag{8.124}$$

Inserting Eqs. (8.122) and (8.123) into Eq. (8.118) gives

$$[(g + j\Gamma/2)A - jaC]e^{(g+j\Gamma/2)z} + [(-g + j\Gamma/2)B - jaD]e^{(-g+j\Gamma/2)z} = 0 \tag{8.125}$$

For Eq. (8.125) to be satisfied for any value of z, the values in the square brackets have to vanish:

$$(g + j\Gamma/2)A - jaC = 0 \tag{8.126}$$

$$(-g + j\Gamma/2)B - jaD = 0 \tag{8.127}$$

Similarly, inserting Eqs. (8.122) and (8.123) into Eq. (8.119) gives

$$jaA + (g - j\Gamma/2)C = 0 \tag{8.128}$$

$$jaB - (g + j\Gamma/2)D = 0 \tag{8.129}$$

Equations (8.126)–(8.129) are rearranged in a matrix form as

$$\begin{vmatrix} (g + j\Gamma/2) & 0 & -ja & 0 \\ 0 & (-g + j\Gamma/2) & 0 & -ja \\ ja & 0 & (g - j\Gamma/2) & 0 \\ 0 & ja & 0 & -(g + j\Gamma/2) \end{vmatrix} \begin{vmatrix} A \\ B \\ C \\ D \end{vmatrix} = 0 \tag{8.130}$$

For nonzero A, B, C, and D to exist, the determinant of Eq. (8.130) has to vanish. The value of the determinant is

$$[g^2 + (\Gamma/2)^2 - a^2]^2 = 0 \tag{8.131}$$

and finally,

$$g = \pm\sqrt{a^2 - (\Gamma/2)^2} \tag{8.132}$$

Putting back the parameters from Eqs. (8.121) and (8.124) gives

$$g = \sqrt{(2KA_p^2(0))^2 - (\Delta k - KA_p^2(0))^2}/2 \qquad (8.133)$$

Thus, for a given value of A_p, the gain g becomes maximum when

$$\Delta k - KA_p^2(0) = 0$$

That is, when the combination of the linear and nonlinear phase factors becomes zero rather than Δk alone becomes zero.

If $A_{c0} = 0$, at $z = 0$, from Eq. (8.123),

$$C = -D \qquad (8.134)$$

and with Eqs. (8.113) and (8.134), Eq. (8.123) becomes

$$A_c(z) = 2C \sinh(gz) e^{j(\beta_{NL} + \Gamma/2)z} \qquad (8.135)$$

The amplitude of the conjugate wave increases with the hyperbolic sine of the distance.

8.12 PULSE BROADENING COMPENSATION BY FORWARD FOUR-WAVE MIXING

Pulse broadening in an optical fiber limits the transmission capability of fiber optic communication. A method for narrowing a broadened light pulse is by means of four-wave mixing [13,22–24]. The principle is exactly the same as that illustrated in Fig. 8.5, where the distorted wave is reflected by a phase conjugator and retraces the time history of the distortion up to the original waveform as it goes back through the distorting medium again.

When applying this principle to fiber optic communication systems, a phase conjugator is placed in the middle of the transmission cable. The signal wave propagates down the first half of the fiber cable, and the phase conjugate wave is funneled into the second half of the fiber cable. An assumption has to be made that both halves of the fiber have the same physical properties and the same length.

Figure 8.19 shows the scheme for pulse broadening compensation by means of semidegenerate forward four-wave mixing with $\omega_1 = \omega_2 = \omega_p$, $\omega_3 = \omega_s$ and $\omega_4 = \omega_c$. Referring to Fig. 8.19, the transmitter light pulse is fed into a single mode optical fiber of length L_1. At L_1, the pulse enters the phase conjugator. The phase conjugator utilizes the nonlinear property of the core glass of a dispersion-shifted fiber. The broadband nature of the dispersion-shifted fiber allows for easy phase matching among the signal, phase conjugate, and pump waves. The pump wave is added to the signal by means of a beam combiner to drive the phase conjugator. Both the pump and the signal waves are removed at the exit of the phase conjugator by means of an optical filter. Only the phase conjugate wave is fed into the other half of the single mode fiber. The phase conjugate wave travels a distance L_2 to the receiver. If $L_1 = L_2$, then the transmitted pulse will be recovered when the phase conjugate wave reaches the receiver.

Now let us analyze the compensation process in more detail. Referring again to Fig. 8.19, a light pulse $E_1(z, t)$ is launched into a single mode fiber at $z = 0$. The

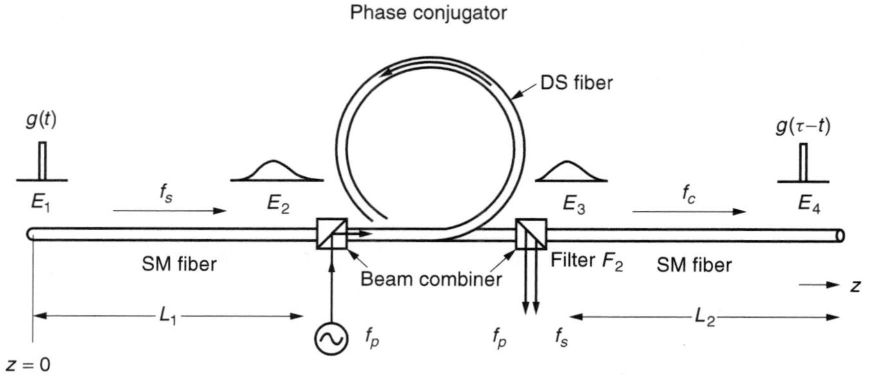

Phase conjugator

Figure 8.19 Pulse broadening compensation by four-wave mixing. DS fiber, dispersion-shifted fiber; SM fiber, single mode fiber.

carrier frequency f_s of the light pulse is modulated by an envelope function $g(t)$:

$$E_1(0, t) = g(t) \cos 2\pi f_s t \tag{8.136}$$

The frequency spectra of the input light is obtained by the Fourier transform as

$$\mathcal{F}(E_1) = \tfrac{1}{2}[G(f - f_s) + G(f + f_s)] \tag{8.137}$$

where

$$\mathcal{F}\{g(t)\} = G(f) \tag{8.138}$$

The Fourier transform $G(f)$ of the envelope is shifted by f_s to the right and by $-f_s$ to the left in the frequency domain. The narrower the width of the input pulse in time, the greater the spread of the spectra in the frequency domain.

Each frequency component in this spectra propagates at its own phase velocity and reaches the receiver. Unless each frequency component propagates at the same velocity, the relative phase relationship is upset and the received pulse becomes distorted.

First, the behavior of a single frequency wave as it propagates to its destination is analyzed. Once the behavior of one frequency component is known, the received pulse shape is obtained by integrating over frequency.

Let the chosen frequency be $f = f_s + \eta$, which is η away from f_s. From Eq. (8.137), this frequency component has an amplitude of $\tfrac{1}{2}G(\eta)$. The propagation of this frequency component through the first half L_1 of a long fiber is

$$dE_2 = \tfrac{1}{2}G(\eta)\, d\eta e^{-j2\pi(f_s+\eta)t + j\beta(f_s+\eta)L_1} + \text{c.c.} \tag{8.139}$$

The value of the propagation constant β at $f_s + \eta$ can be approximated by the Taylor series expansion, η being usually at most one thousandth of f_s,

$$\beta(f_s + \eta) = \beta(f_s) + \beta'(f_s)\eta + \tfrac{1}{2}\beta''(f_s)\eta^2 + \cdots \tag{8.140}$$

Thus, the expression for the single frequency spectrum at the input to the phase conjugator is

$$dE_2 = \tfrac{1}{2}G(\eta)\, d\eta e^{-j2\pi(f_s+\eta)t} e^{+j\beta(f_s)L_1 + j\beta'(f_s)L_1\eta + \tfrac{1}{2}j\beta''(f_s)L_1\eta^2} + \text{c.c.} \tag{8.141}$$

The output dE_3 from the phase conjugator is the phase conjugate of the input except for the time factor,

$$dE_3(\eta) = \tfrac{1}{2}\sqrt{\eta_c}G^*(\eta)\,d\eta e^{-j2\pi(f_s+\eta)t}e^{-j\beta(f_s)L_1-j\beta'(f_s)L_1\eta-\frac{1}{2}j\beta''(f_s)L_1\eta^2} + \text{c.c.} \qquad (8.142)$$

where η_c is the conversion efficiency, which is determined by such parameters as the gain given by Eq. (8.133), fiber loss and beam combiner loss.

In the second half of the optical fiber, the phase conjugate wave propagates. In the degenerate case of Eq. (8.103), the frequency f_c of the phase conjugate wave is shifted to the other side of the pump frequency f_p, as shown in Fig. 8.17b, and

$$f_c = 2f_p - f_s \qquad (8.143)$$

With the input of $f_s + \eta$, the new shifted frequency f'_c is

$$f'_c = f_c - \eta \qquad (8.144)$$

The propagation constant in the second fiber at frequency $f_c - \eta$ is obtained from the Taylor series expansion

$$\beta(f_c - \eta) = \beta(f_c) - \beta'(f_c)\eta + \tfrac{1}{2}\beta''(f_c)\eta^2 + \cdots \qquad (8.145)$$

The signal reaching the receiver is therefore

$$dE_4(\eta) = \tfrac{1}{2}\sqrt{\eta_c}G^*(\eta)d\eta e^{j[-2\pi(f_c-\eta)t+\beta(f_c)L_2-\beta(f_s)L_1]}$$

$$\times e^{-j[\beta'(f_s)L_1+\beta'(f_c)L_2]\eta+\frac{1}{2}j[\beta''(f_c)L_2-\beta''(f_s)L_1]\eta^2} + \text{c.c.} \qquad (8.146)$$

where the f_s and f_p components have been filtered out by filter F_2.

Let us put

$$\phi = \beta(f_c)L_2 - \beta(f_s)L_1 \qquad (8.147)$$

$$\tau = \frac{1}{2\pi}[\beta'(f_c)L_2 + \beta'(f_s)L_1] \qquad (8.148)$$

$$\psi = \tfrac{1}{2}[\beta''(f_c)L_2 - \beta''(f_s)L_1] \qquad (8.149)$$

$$dE_4(\eta) = \tfrac{1}{2}\sqrt{\eta_c}G^*(\eta)\,d\eta e^{-j2\pi f_c t+j\phi+j2\pi(t-\tau)\eta+j\psi\eta^2} + \text{c.c.} \qquad (8.150)$$

The waveform of the received signal is obtained by integrating over frequency:

$$E_4 = \tfrac{1}{2}\sqrt{\eta_c}e^{-j2\pi f_c t+j\phi}\int G^*(\eta)e^{j2\pi(t-\tau)\eta+j\psi\eta^2}\,d\eta + \text{c.c.} \qquad (8.151)$$

First, let us deal with the case when

$$\psi = 0 \qquad (8.152)$$

Equation (8.151) is in the form of an inverse Fourier transform and

$$E_4(L_1 + L_2, t) = \tfrac{1}{2}\sqrt{\eta_c}e^{-j2\pi f_c t + j\phi}g^*(\tau - t) + \text{c.c.} \qquad (8.153)$$

If the envelope function g is assumed real, the final result is

$$E_4(L_1 + L_2, t) = \sqrt{\eta_c}g(\tau - t)\cos(-2\pi f_c t + \phi) \qquad (8.154)$$

Equation (8.154) shows that the envelope function of the received pulse is exactly the same as that of the transmitted pulse except that $g(\tau - t)$ is time reversed. The original envelope $g(0)$ reappears τ seconds later. Thus, τ is the total transmission time of the envelope from $z = 0$ to $z = L_1 + L_2$.

Next, the case when $\psi \neq 0$ is considered. Equation (8.151) becomes

$$E_4 = e^{-j2\pi f_c t + j\phi}g(\tau - t) * \mathcal{F}^{-1}\{e^{-j\Psi\eta^2}\} \qquad (8.155)$$

The pulse shape is now convolved with $\mathcal{F}^{-1}\{e^{-j\psi\eta^2}\}$, creating a distortion in the received pulse. The distortionless condition, however, can be achieved from Eq. (8.149) by setting

$$\beta''(f_c)L_2 = \beta''(f_s)L_1 \qquad (8.156)$$

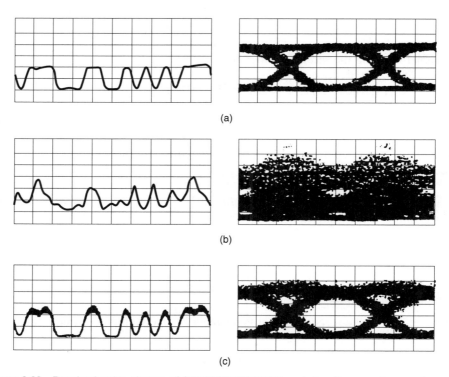

(a)

(b)

(c)

Figure 8.20 Received pulse shapes of (1110001100101010) coded patterns and eye-patterns for a 10-Gb/s intensity modulated signal at $P_1 = P_2 = +5$ dBm. (a) A 5-m transmission without OPC. (b) A 200-km transmission without OPC. (c) A 200-km transmission with OPC at the midpoint. (Scale units of pulse shapes in left column: V, 50 mV/div.; H, 200 ps/div. Scale units of pulse shapes in right column: V, 25 mV/div.; H, 20 ps/div. (After S. Watanabe et al. [22].)

One way of obtaining this condition is to use a fiber with the same length for both halves of the transmission cable.

Pulse broadening compensation with 10 Gb/s pulse modulated light is demonstrated in Fig. 8.20 [22]. Figure 8.20a shows the input signal. Figure 8.20b shows the same signal after 200-km transmission without the optical phase conjugator (OPC) and Fig. 8.20c shows the result when the phase conjugator is inserted in the middle of the fiber transmission. Figure 8.20 confirms the effectiveness of pulse broadening compensation in an optical fiber by means of four-wave mixing.

PROBLEMS

8.1 In the text, the distorting medium in Fig. 8.4 was assumed to be free of temporal variations: that is, the temporal variations either did not exist or were so slow that they could be taken as constant for the duration of the experiment. Consider a distorting medium in which temporal fluctuations cannot be ignored. For simplicity, assume that the fluctuation is sinusoidal with time and is expressed as $\Phi(t) = \Phi \cos \omega t$. What are the distances L between the distorting medium and the phase conjugate mirror (Fig. P8.1) that make the best distortion-free image and the worst distorted image?

8.2 Consider a crystal whose one-dimensional charge distribution is as shown in Fig. P8.2.

   (a) Does this crystal have inversion symmetry?

   (b) Draw the redistributed charges when exposed to the **E** field of an incident light wave.

   (c) Does such a crystal possess a second order nonlinearity?

8.3 Does a crystal with inversion symmetry have a fourth order nonlinearity?

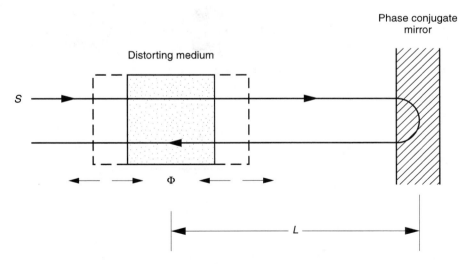

Figure P8.1 Temporally varying distorting medium.

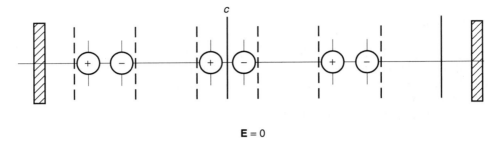

$\mathbf{E} = 0$

Figure P8.2 Does this crystal have nonzero $\chi^{(2)}$?

8.4 Three lightwaves having adjacent frequencies are incident onto an optical fiber (Fig. P8.4). Find the frequency spectra generated in the fiber due to the third order nonlinear effect. All incident waves are assumed to be polarized in the x direction [25].

8.5 Assuming the degenerate case, if the directions of $\mathbf{k}_1$ and $\mathbf{k}_2$ are set as shown in Fig. P8.5, find the directions of $\mathbf{k}_3$ and $\mathbf{k}_4$ that sustain four-wave mixing.

8.6 Draw all possible fringe patterns in a medium when four waves are incident as shown in Fig. P8.6.

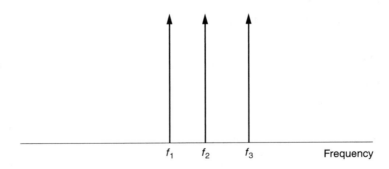

Figure P8.4 Spectra of light incident onto an optical fiber having a third order nonlinearity.

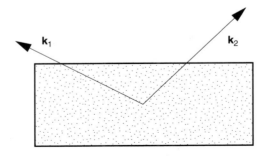

Figure P8.5 Finding the condition of four-wave mixing.

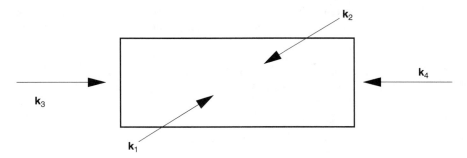

Figure P8.6 A configuration for a degenerate type of four-wave mixing.

8.7 In the text, the case of forward four-wave mixing was dealt with, but the attenuation in the dispersion-shifted fiber was not taken into consideration. With an amplitude attenuation constant α, Eqs. (8.108)–(8.111) become

$$\frac{dA_p}{dz} = (-\alpha + j\tfrac{3}{2}K|A_p|^2)A_p$$

$$\frac{dA_s}{dz} = (-\alpha + j2K|A_p|^2)A_s + jKA_p^2 A_c^* e^{j\Delta kz}$$

$$\frac{dA_c}{dz} = (-\alpha + j2K|A_p|^2)A_c + jKA_p^2 A_s^* e^{j\Delta kz}$$

Find the differential equations with attenuation in the fiber.

REFERENCES

1. D. M. Pepper, "Applications of optical phase conjugation," *Sci. Am.*, 74–83 (Jan. 1986).
2. H. Kogelnik, "Reconstructing response and efficiency of hologram gratings," *Proceedings of the Symposium on Modern Optics*, Polytechnic Press, Brooklyn, NY, 1967, pp. 605–617.
3. D. M. Pepper, "Nonlinear optical phase conjugation," *Opt. Eng.*, **21**(2), 156–183 (1982).
4. R. W. Hellwarth, "Third-order optical susceptibilities of liquids and solids," *Prog. Quantum Electron.*, **5**, 1–68 (1977).
5. J. Feinberg, "Self-pumped, continuous-wave phase conjugator using internal reflection," *Opt. Lett.*, **7**(10), 486–488 (1982).
6. D. M. Bloom and G. C. Bjorklund, "Conjugate wave-front generation and image reconstruction for four-wave mixing," *Appl. Phys. Lett.*, **31**(9), 592–594 (1977).
7. V. V. Shkunov and B. Ya. Zel'dovich, "Optical phase conjugation," *Sci. Am.*, 54–59 (Dec. 1985).
8. B. Fischer, M. Cronin-Golomb, J. O. White, and A. Yariv, "Real-time phase conjugate window for one-way optical field imaging through a distortion," *Appl. Phys. Lett.*, **41**(2), 141–143 (1982).
9. A. Chiou, P. Yeh, C.-X. Yang, and C. Gu, "Experimental demonstration of photorefractive resonator for adaptive fault-tolerant coupling," *Opt. Photonics News*, 20–21 (Dec. 1995).
10. P. Yeh, "Photorefractive phase conjugators," *Proc. IEEE*, **80**(3), 436–450 (1992).

11. M. Cronin-Golomb, B. Fischer, J. O. White, and A. Yariv, "Theory and applications of four-wave mixing in photorefractive media," *IEEE J. Quautum Electron.*, **QE-20**(4), 12–30 (1984).

12. Y. Owechko, G. J. Dunning, E. Marom, and B. H. Soffer, "Holographic associative memory with nonlinearities in the correlation domain," *Appl. Opt.* **26**(10), 1900–1910 (1987).

13. A. Yariv, *Optical Electronics*, 4th ed., Saunders, New York, 1991.

14. Y. R. Shen, *Principles of Nonlinear Optics*, Wiley, New York, 1984.

15. R. A. Fisher (Ed.), *Optical Phase Conjugation*, Academic Press, New York, 1983.

16. G. P. Agrawal, *Applications of Nonlinear Fiber Optics*, Academic Press, San Diego, CA, 2001.

17. N. Bloembergen, *Nonlinear Optics*, 4th ed., World Scientific, River Edge, NJ, 1996.

18. B. E. A. Saleh and M. C. Teich, *Fundamentals of Photonics*, Wiley, New York, 1991.

19. G. C. Baldwin, *An Introduction to Nonlinear Optics*, Plenum Press, New York, 1974.

20. A. Yariv, D. Fekete, and D. M. Pepper, "Compensation for channel dispersion by nonlinear optical phase conjugation," *Opt. Lett.*, **4**(2), 52–54 (1979).

21. R. H. Stolen and J. E. Bjorkholm, "Parametric amplification and frequency conversion in optical fibers," *IEEE J. Quantum Electron.*, **QE 18**(7), 1062–1072 (1982).

22. S. Watanabe, G. Ishikawa, T. Naito, and T. Chikama, "Generation of optical phase-conjugate waves and compensation for pulse shape distortion in a single-mode fiber," *J. Lightwave Technol.*, **12**(12), 2139–2146 (1994).

23. N. Shibata, R. P. Braun, and R. G. Waarts, "Phase-mismatch dependence of efficiency of wave generation through four-wave mixing in a single-mode optical fiber," *IEEE J. Quantum Electron.*, **QE-23**(7), 1205–1210 (1987).

24. S. Murata, A. Tomita, J. Shimizu, and A. Suzuki, "THz optical-frequency conversion of 1 Gb/s-signals using highly nondegenerate four-wave mixing in an InGaAsP semiconductor laser," *IEEE Photonics Technol. Lett.*, **3**(11), 1021–1023 (1991).

25. M. W. Maeda, W. B. Sessa, W. I. Way, A. Yi-Yan, L. Curtis, R. Spicer, and R. I. Laming, "The effect of four-wave mixing in fibers on optical frequency-division multiplexed systems," *J. Lightwave Technol.*, **8**(9), 1402–1408 (1990).

DERIVATION OF THE FRESNEL–KIRCHHOFF DIFFRACTION FORMULA FROM THE RAYLEIGH–SOMMERFELD DIFFRACTION FORMULA

The Rayleigh–Sommerfeld diffraction formula uses the Fourier transform of the input field, but Fresnel–Kirchhoff's integral equation uses the input field directly to find the diffraction field. This appendix shows how the latter is derived from the former formula (Kazuo Tanaka, private communication).

By combining Eq. (1.177) with (1.178), an expression for the diffraction pattern can be obtained directly from the input field as

$$E(x_i, y_i, z_i) = \iint \left(\iint E(x_0, y_0, 0) e^{-j2\pi f_x x_0 - j2\pi f_y y_0} dx_0 \, dy_0 \right)$$
$$\times e^{j2\pi \sqrt{f_s^2 - f_x^2 - f_y^2} z_i} e^{j2\pi f_x x_i + j2\pi f_y y_i} \, df_x \, df_y \tag{A.1}$$

Reversing the order of integration gives

$$E(x_i, y_i, z_i) = \iint dx_0 \, dy_0 E(x_0, y_0, 0) \iint e^{j2\pi \sqrt{f_s^2 - f_x^2 - f_y^2} z_i}$$
$$\times e^{j2\pi f_x(x_i - x_0) + j2\pi f_y(y_i - y_0)} \, df_x \, df_y \tag{A.2}$$

The integration Eq. (A.2) can be performed using Weyl's expansion theorem [1], which expresses a spherical wavefront in integral form as

$$\frac{e^{j2\pi f_s r}}{r} = j \iint \frac{e^{j2\pi f_x(x_i - x_0) + j2\pi f_y(y_i - y_0) + j2\pi f_z z}}{f_z} \, df_x \, df_y \tag{A.3}$$

where

$$r^2 = (x_i - x_0)^2 + (y_i - y_0)^2 + z_i^2 \tag{A.4}$$

545

The factor $1/f_z$ in Eq. (A.3) must be removed in order to use it in the integral of Eq. (A.2). This can be accomplished by differentiating both sides of Eq. (A.3) with respect to z. The result is

$$-\frac{1}{2\pi}\frac{\partial}{\partial z}\left(\frac{e^{j2\pi f_s r}}{r}\right) = \iint e^{2\pi f_x(x_i-x_0)+j2\pi f_y(y_i-y_0)+j2\pi f_z z}\, df_x\, df_y \tag{A.5}$$

Inserting Eq. (A.5) into (A.2) gives

$$E(x_i, y_i, z_i) = -\frac{1}{2\pi}\iint E(x_0, y_0, 0)\frac{\partial}{\partial z}\left(\frac{e^{j2\pi f_s r}}{r}\right)dx_0\, dy_0 \tag{A.6}$$

The derivative with respect to z in Eq. (A.6) is first performed as

$$\frac{\partial}{\partial z}\left(\frac{e^{j2\pi f_s r}}{r}\right) = \left(-\frac{1}{r^2}e^{j2\pi f_s r} + j2\pi f_s\frac{e^{j2\pi f_s r}}{r}\right)\frac{dr}{dz} \tag{A.7}$$

For a large value of r, the second term on the right-hand side of Eq. (A.7) dominates, and

$$\frac{\partial}{\partial z}\left(\frac{e^{j2\pi f_s r}}{r}\right) \doteq j2\pi f_s\frac{e^{j2\pi f_s r}}{r}\frac{z}{r} \tag{A.8}$$

where z/r was obtained from the derivative of Eq. (A.4).

From the para-axial approximation, namely, in the region where

$$z^2 \gg (x_i - x_0)^2 + (y_i - y_0)^2 \tag{A.9}$$

the ratio $z/r \cong 1$. Thus Eq. (A.8) becomes

$$\frac{\partial}{\partial z}\left(\frac{e^{j2\pi f_s r}}{r}\right) \doteq j2\pi f_s\frac{e^{j2\pi f_s r}}{r} \tag{A.10}$$

From this result, and by substituting $f_s = 1/\lambda$, Eq. (A.6) becomes

$$E(x_i, y_i, z_i) = \frac{1}{j\lambda}\iint E(x_0, y_0, 0)\frac{e^{j(2\pi/\lambda)r}}{r}\, dx_0\, dy_0 \tag{A.11}$$

This equations is known as the *Fresnel–Kirchhoff integral* of diffraction, which represents the diffraction pattern for a given input field.

The final equation, the Fresnel–Kirchhoff integral (Eq. (A.11)), is identical to Eq. (1.28), which was derived earlier without rigorous proof. The analysis here has proved that the constant K given by Eq. (1.29) is true.

REFERENCE

1. E. Wolf and M. Nieto-Vesperinas, "Analyticity of the angular spectrum amplitude of scattered fields and some of its consequences," *J. Opt. Soc. Am. A* **2**(6), 886–890 (1985).

WHY THE ANALYTIC SIGNAL METHOD IS NOT APPLICABLE TO THE NONLINEAR SYSTEM

Let us review the common approach to solving an *RL* network by the method of the analytic signal (phasor). A rule that electrical engineering students use is to replace $V \cos \omega t$ by $V e^{j\omega t}$, but not by $\frac{1}{2}(V e^{j\omega t} + \text{c.c.})$, which is the mathematically exact equivalent, and to "*take the real part*" of the final solution instead of both real and imaginary parts of the solution.

Method I   This is how an electrical engineering student solves the problem of the *RL* circuit shown in Fig. B.1. The differential equation for the analytic signal $V e^{j\omega t}$ is

$$L \frac{di_a}{dt} + R i_a = V e^{j\omega t} \tag{B.1}$$

The method of undetermined coefficients is used. Let an assumed solution be

$$i_a(t) = I e^{j\omega t} \tag{B.2}$$

Inserting Eq. (B.2) into (B.1) gives

$$(j\omega L + R)I = V \tag{B.3}$$

$$I = \frac{V}{j\omega L + R} \tag{B.4}$$

$$= \frac{V}{\sqrt{(\omega L)^2 + R^2}} e^{-j\phi} \tag{B.5}$$

where

$$\phi = \tan^{-1}\left(\frac{\omega L}{R}\right)$$

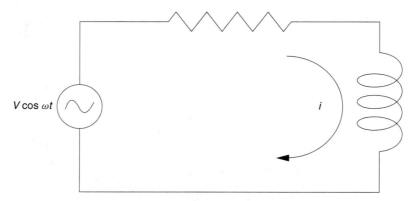

Figure B.1 *RL* circuit driven by $V \cos \omega t$.

From Eqs. (B.2) and (B.5),

$$i_a(t) = \frac{V}{\sqrt{(\omega L)^2 + R^2}} e^{j(\omega t - \phi)} \tag{B.6}$$

The final step is to take the real part of Eq. (B.6). The answer is

$$i(t) = \mathrm{Re}\, i_a(t) = \frac{V}{\sqrt{(\omega L)^2 + R^2}} \cos(\omega t - \phi) \tag{B.7}$$

Method II The same problem will be solved by driving with

$$v(t) = V \cos \omega t \tag{B.8}$$

Equation (B.8) is rewritten as

$$v(t) = \frac{V}{2}(e^{j\omega t} + e^{-j\omega t}) \tag{B.9}$$

Now, since the differential equation (B.1) is linear, the law of superposition holds true. The solution will be obtained by adding i_1 when driven by $(V/2)e^{j\omega t}$ and i_2 when driven by $(V/2)e^{-j\omega t}$.

The value of i_1 is immediately obtained by replacing V by $V/2$ in Eq. (B.6) as

$$i_1(t) = \frac{V}{2} \frac{1}{\sqrt{(\omega L)^2 + R^2}} e^{j(\omega t - \phi)} \tag{B.10}$$

and

$$i_1(t) = \tfrac{1}{2} i_a \tag{B.11}$$

Next, the circuit is driven by $(V/2)e^{-j\omega t}$. An assumed solution

$$i_2(t) = I e^{-j\omega t} \tag{B.12}$$

is put into Eq. (B.1):

$$I = \frac{V}{2} \left(\frac{1}{-j\omega L + R} \right) \tag{B.13}$$

Inserting Eq. (B.13) into (B.12) gives

$$i_2(t) = \frac{V}{2} \frac{1}{\sqrt{(\omega L)^2 + R^2}} e^{-j(\omega t - \phi)} \tag{B.14}$$

and

$$i_2(t) = \tfrac{1}{2} i_a^* \tag{B.15}$$

Using the law of superposition, the solution of Eq. (B.9) is

$$i(t) = i_1(t) + i_2(t) = \frac{V}{\sqrt{(\omega L)^2 + R^2}} \cos(\omega t - \phi) \tag{B.16}$$

Thus, it has been proved that Methods I and II provide the same answer.

Let's examine the key points that made the two answers the same. The law of superposition gave

$$i(t) = i_1(t) + i_2(t) \tag{B.17}$$

From Eqs. (B.11) and (B.15),

$$i(t) = \tfrac{1}{2}[i_a(t) + i_a^*(t)] \tag{B.18}$$

Since Re $z = \frac{1}{2}(z + z^*)$ the operation of Eq. (B.18) is identical with the operation of "taking the real part of $i_a(t)$." Thus,

$$i(t) = \text{Re } i_a(t) \tag{B.19}$$

The equality of the solutions by the two methods cannot be realized if the law of superposition expressed by Eq. (B.17) is not true. The law of superposition, however, is realized only when the differential equation is linear, as explained below. This leads to the conclusion that the method of analytic signal cannot be used for solving a nonlinear differential equation, and expressions such as Eq. (B.8) or (B.9) have to be used to express the driving voltage.

Finally, we give a proof that the law of superposition is realized only when the differential equation is linear. The proof is made using the same differential equation. When the circuit is driven by v_1, the solution i_1 has to satisfy

$$L\frac{di_1}{dt} + Ri_1 = v_1 \tag{B.20}$$

whereas when driven by v_2, the solution i_2 has to satisfy

$$L\frac{di_2}{dt} + Ri_2 = v_2 \tag{B.21}$$

Now, let the solution be i when the circuit is excited by $v_1 + v_2$:

$$L\frac{di}{dt} + Ri = v_1 + v_2 \tag{B.22}$$

Inserting Eqs. (B.20) and (B.21) into the right-hand side of Eq. (B.22) gives

$$L\frac{di}{dt} + Ri = L\frac{d}{dt}(i_1 + i_2) + R(i_1 + i_2) \tag{B.23}$$

Comparing both sides of Eq. (B.23), we see that

$$i = i_1 + i_2 \tag{B.24}$$

is the solution when the circuit is driven by $v_1 + v_2$.

In other words, when i_1 is the response of stimulus v_1 and i_2 is the response of stimulus v_2, the response of the two stimuli together, $v_1 + v_2$, is the addition of the two responses, $i_1 + i_2$. This is the law of superposition, which holds true only when the differential equation is linear; meaning that L and R, which are the coefficients of di/dt and i in Eq. (B.1), are not a function of i.

Let us now examine the case of the nonlinear differential equation. As an example of a nonlinear differential equation, let us consider the case when the value of the resistance R is changed due to the generated heat. Such a system may be represented approximately by the differential equation

$$L\frac{di}{dt} + Ri^3 = v \tag{B.25}$$

When the circuit is driven by v_1, the current i_1 has to satisfy

$$L\frac{di_1}{dt} + Ri_1^3 = v_1 \tag{B.26}$$

Similarly, when it is driven by v_2, i_2 has to satisfy

$$L\frac{di_2}{dt} + Ri_2^3 = v_2 \tag{B.27}$$

Next, when v_1 and v_2 are applied simultaneously,

$$L\frac{di}{dt} + Ri^3 = v_1 + v_2 \tag{B.28}$$

Inserting Eqs. (B.25) and (B.26) into the right-hand side of Eq. (B.28) gives

$$L\frac{di}{dt} + Ri^3 = \frac{d}{dt}(i_1 + i_2) + R(i_1 + i_2)^3 - 3R(i_1^2 i_2 + i_1 i_2^2) \tag{B.29}$$

Comparing both sides of Eq. (B.29), we see that

$$i = i_1 + i_2$$

is no longer the solution of Eq. (B.28). The law of superposition no longer holds true with a nonlinear differential equation.

APPENDIX C

DERIVATION OF $\mathbf{P}_{\mathrm{NL}}$

In order to shorten the descriptions, let us put

$$a = E_1 e^{-j\omega_1 t}$$
$$b = E_2 e^{-j\omega_2 t}$$
$$c = E_3 e^{-j\omega_3 t}$$
$$d = E_4 e^{-j\omega_4 t}$$

(C.1)

Equation (8.53) becomes

$$\mathbf{P}_{\mathrm{NL}} = \hat{\mathbf{x}}\epsilon_0 \frac{\chi_{xxxx}}{8}(a + a^* + b + b^* + c + c^* + d + d^*)^3$$

(C.2)

Putting

$$q = a + b + c + d$$

(C.3)

$$P_{\mathrm{NL}} = \epsilon_0 \frac{\chi_{xxxx}}{8}(q + q^*)^3$$

(C.4)

Note that

$$(q + q^*)^3 = q^3 + 3q^2 q^* + \text{c.c.}$$

(C.5)

Frequencies associated with q^3 are too high and are out of the range of interest. The q^3 terms will be discarded.

$$P_{\mathrm{NL}} = \tfrac{3}{8}\epsilon_0 \chi_{xxxx}(q^2 q^* + \text{c.c.})$$

(C.6)

Discarding q^3 makes a substantial reduction in the number of calculations. The last step is inserting Eq. (C.3) into Eq. (C.6) and performing the multiplication. The result is

$$q^2 q^* + \text{c.c.} = a(|a|^2 + 2|b|^2 + 2|c|^2 + 2|d|^2) + b(2|a|^2 + |b|^2 + 2|c|^2 + 2|d|^2)$$

$$+ c(2|a|^2 + 2|b|^2 + |c|^2 + 2|d|^2) + d(2|a|^2 + 2|b|^2 + 2|c|^2 + |d|^2)$$
$$+ 2a^*(bc + cd + db) + 2b^*(ac + cd + da) + 2c^*(ab + bd + da)$$
$$+ 2d^*(ab + bc + ca) + a^*(b^2 + c^2 + d^2) + b^*(a^2 + c^2 + d^2)$$
$$+ c^*(a^2 + b^2 + d^2) + d^*(a^2 + b^2 + c^2) + \text{c.c.} \tag{C.7}$$

These terms generate a variety of beat frequencies. From Eq. (C.1), terms such as a^*bc, a^*cd, and a^*db create frequency components of $\omega_2 + \omega_3 - \omega_1, \omega_3 + \omega_4 - \omega_1$, and $\omega_2 + \omega_4 - \omega_1$, respectively. Moreover, for the set of equations that are commensurate with each other,

$$\omega_4 = \omega_1 + \omega_2 - \omega_3 \tag{C.8}$$

and these frequency components become $2\omega_2 - \omega_4, \omega_2$, and $2\omega_2 - \omega_3$, respectively. Note, in particular, that b^*cd, a^*cd, d^*ab, and c^*ab become $\omega_1, \omega_2, \omega_3$, and ω_4, respectively.

Rewriting Eq. (C.7) using Eqs. (C.1) and (C.8) reduces Eq. (C.6) to

$$\mathbf{P}_{NL} = \tfrac{1}{2}\hat{\mathbf{x}}[P_{NL}(\omega_1)e^{j\omega_1 t} + P_{NL}(\omega_2)e^{j\omega_2 t} + P_{NL}(\omega_3)e^{j\omega_3 t} + P_{NL}(\omega_4)e^{j\omega_4 t}$$
$$+ P_{NL}(2\omega_1 - \omega_2) + P_{NL}(2\omega_1 - \omega_3) + P_{NL}(2\omega_1 - \omega_4)$$
$$+ P_{NL}(2\omega_2 - \omega_1) + P_{NL}(2\omega_1 - \omega_3) + P_{NL}(2\omega_2 - \omega_4)$$
$$+ P_{NL}(2\omega_3 - \omega_1) + P_{NL}(2\omega_3 - \omega_2) + P_{NL}(2\omega_3 - \omega_4)$$
$$+ P_{NL}(2\omega_4 - \omega_1) + P_{NL}(2\omega_4 - \omega_2) + P_{NL}(2\omega_4 - \omega_3) + \text{c.c.}]$$

where

$$P_{NL}(\omega_1) = \chi_{\text{eff}}[(|E_1|^2 + 2|E_2|^2 + 2|E_3|^2 + 2|E_4|^2)E_1 + 2E_3E_4E_2^*]$$
$$P_{NL}(\omega_2) = \chi_{\text{eff}}[(2|E_1|^2 + |E_2|^2 + 2|E_3|^2 + 2|E_4|^2)E_2 + 2E_3E_4E_1^*]$$
$$P_{NL}(\omega_3) = \chi_{\text{eff}}[(2|E_1|^2 + 2|E_2|^2 + |E_3|^2 + 2|E_4|^2)E_3 + 2E_1E_2E_4^*]$$
$$P_{NL}(\omega_4) = \chi_{\text{eff}}[(2|E_1|^2 + 2|E_2|^2 + 2|E_3|^2 + |E_4|^2)E_4 + 2E_1E_2E_3^*]$$
$$P_{NL}(2\omega_1 - \omega_2) = \chi_{\text{eff}}E_1^2E_2^*$$
$$P_{NL}(2\omega_1 - \omega_3) = \chi_{\text{eff}}(E_1^2E_3^* + 2E_1E_4E_2^*)$$
$$P_{NL}(2\omega_1 - \omega_4) = \chi_{\text{eff}}(E_1^2E_4^* + 2E_1E_3E_2^*)$$

$$P_{NL}(2\omega_2 - \omega_1) = \chi_{\text{eff}}E_2^2E_1^*$$
$$P_{NL}(2\omega_2 - \omega_3) = \chi_{\text{eff}}(E_2^2E_3^* + 2E_2E_4E_1^*)$$
$$P_{NL}(2\omega_2 - \omega_4) = \chi_{\text{eff}}(E_2^2E_4^* + 2E_2E_3E_1^*)$$

$$P_{NL}(2\omega_3 - \omega_1) = \chi_{\text{eff}}(E_3^2E_1^* + 2E_2E_3E_4^*)$$
$$P_{NL}(2\omega_3 - \omega_2) = \chi_{\text{eff}}(E_3^2E_2^* + 2E_1E_3E_4^*)$$
$$P_{NL}(2\omega_3 - \omega_4) = \chi_{\text{eff}}E_3^2E_4^*$$

$$P_{NL}(2\omega_4 - \omega_1) = \chi_{eff}(E_4^2 E_1^* + 2E_2 E_4 E_3^*)$$

$$P_{NL}(2\omega_4 - \omega_2) = \chi_{eff}(E_4^2 E_2^* + 2E_1 E_4 E_3^*)$$

$$P_{NL}(2\omega_4 - \omega_3) = \chi_{eff} E_4^2 E_3^*$$

$$\chi_{eff} = \frac{3\epsilon_0}{4} \chi_{xxxx}$$

ANSWERS TO PROBLEMS

Chapter 1

1.1 (a)

$$\mathbf{k} \cdot \mathbf{r} = 67.32$$

$$2\pi f = 2.44 \times 10^{15}$$

$$\mathbf{k} \cdot \mathbf{r} = \frac{2\pi}{\lambda}(x \sin\phi \cos\theta + y \sin\phi \sin\theta + z \cos\phi)$$

Hence, $\lambda = 0.497$ μm, where λ is the wavelength in the medium.

(b) The wavelength λ_a in air is found from

$$2\pi f = 2\pi \frac{c}{\lambda_a}$$

and $\lambda_a = 0.77$ μm. Thus,

$$n = \frac{\lambda_a}{\lambda} = \frac{0.772}{0.497} = 1.55$$

1.2 (a) We have

$$f_z = \sqrt{f_s^2 - f_x^2 - f_y^2} = \sqrt{\left(\frac{1}{0.84}\right)^2 - 0.6^2 - 0.8^2}$$

$$= 0.645 \quad \text{line/μm}$$

From Eq. (1.20)

$$e_x = \frac{f_x}{f_s} = 0.84 \times 0.6 = 0.504$$

$$e_y = \frac{f_y}{f_s} = 0.84 \times 0.8 = 0.672$$

$$e_z = \frac{f_z}{f_s} = (0.84)(0.65) = 0.55$$

$$e_x = \sin\phi \cos\theta = 0.504$$

$$e_y = \sin \phi \sin \theta = 0.672$$

$$e_z = \cos \phi = 0.545$$

$$\phi = 57° \quad \text{and} \quad \theta = 53°$$

(b) The angle θ between $\hat{\mathbf{e}}$ and $\hat{\mathbf{e}}_{\hat{\mathbf{l}}}$ (see Fig. A1.2) is

$$\cos \theta = \hat{\mathbf{e}} \cdot \hat{\mathbf{e}}_{\hat{\mathbf{l}}}$$

Hence, $\lambda_l \hat{\mathbf{e}} \cdot \hat{\mathbf{e}}_l = \lambda_s$.

$$0.84 = \lambda_l (\tfrac{3}{5}\mathbf{i} + \tfrac{4}{5}\mathbf{j}) \cdot (0.504\mathbf{i} + 0.672\mathbf{j} + 0.55\mathbf{k}) = 0.84\lambda_l$$

$$f_l = \frac{1}{\lambda_l} = 1 \text{ line/μm}$$

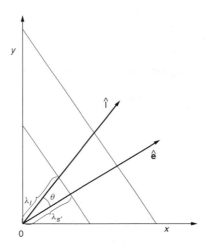

Figure A1.2 Wavelength measured along a line in the direction $\hat{\mathbf{l}}$.

1.3 The Fourier transform of the input function is

$$\mathcal{F}\{t(x_0, y_0)\} = \iint_{-\infty}^{\infty} \left(1 + \frac{e^{j2 f_{x0} x_0} + e^{-j2\pi f_{x0} x_0}}{2} \right) e^{-j2\pi f_x x_0} dx_0 \, dy_0$$

$$= \left[\delta(f_x) + \tfrac{1}{2}\delta(f_x - f_{x0}) + \tfrac{1}{2}\delta(f_x + f_{x0}) \right] \delta(f_y)$$

From Eq. (1.36), the diffraction pattern is

$$E(x_i, y_i) = \frac{e^{jk[z_i + (x_i^2 + y_i^2)/2z_i]}}{j\lambda z_i} \left[\delta\left(\frac{x_i}{\lambda z_i} \right) + \frac{1}{2}\delta\left(\frac{x_i}{\lambda z_i} - f_{x0} \right) \right.$$

$$\left. + \frac{1}{2}\delta\left(\frac{x_i}{\lambda z_i} + f_{x0} \right) \right] \delta\left(\frac{y_i}{\lambda z_i} \right)$$

The result is three peaks, as shown in Fig. A1.3.

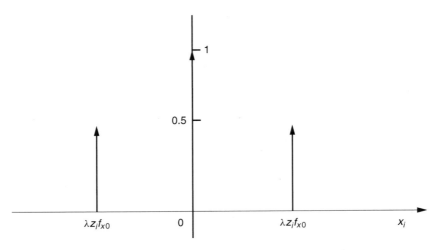

Figure A1.3 Diffraction pattern from the sinusoidal transmittance.

1.4 The current along the dipole antenna is expressed as

$$I(x_0) = I_0 \cos\left(\frac{2\pi}{\lambda}x_0\right) \Pi\left(\frac{2x_0}{\lambda}\right)$$

$$\mathcal{F}\{I(x_0)\} = \frac{\lambda}{4}I_0\left[\delta\left(f - \frac{1}{\lambda}\right) + \delta\left(f + \frac{1}{\lambda}\right)\right] * \text{sinc}\left(\frac{\lambda}{2}f\right)$$

$$= \frac{\lambda}{4}I_0\left[\text{sinc}\,\frac{\lambda}{2}\left(f - \frac{1}{\lambda}\right) + \text{sinc}\,\frac{\lambda}{2}\left(f + \frac{1}{\lambda}\right)\right]$$

$$= \frac{\lambda}{2\pi}I_0\left[\frac{\sin\left(\frac{\pi}{2}\lambda f - \frac{\pi}{2}\right)}{(\lambda f - 1)} + \frac{\sin\left(\frac{\pi}{2}\lambda f + \frac{\pi}{2}\right)}{(\lambda f + 1)}\right]$$

The radiation pattern of the antenna is

$$E_\theta = \frac{60}{j\lambda}\mathcal{F}\{I(x_0)\}_{f=(\cos\theta)/\lambda}\sin\theta$$

A few things should be noted about the result. There exist some differences in convention for expressing the same quantities (see Fig. A1.4). In antenna theory, the elevation angle θ is measured from the antenna axis rather than θ' from the normal to the antenna axis, and $f = (\cos\theta)/\lambda$ rather than $f = (\sin\theta)/\lambda$ is used. Also, the field E_θ of the θ component is used rather than the E_x component, and a factor $\sin\theta$ has to be multiplied. The final result is

$$E_\theta = -j60I_0\frac{\cos\left(\frac{\pi}{2}\cos\theta\right)}{\sin\theta}$$

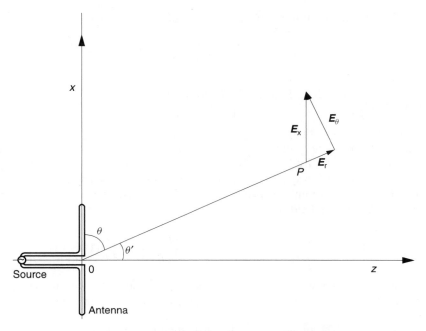

Figure A1.4   Field from a half-wave dipole antenna.

1.5

$$A - 2, \quad B - 1, \quad C - 4, \quad D - 3$$

1.6

$$A - 4, \quad B - 1, \quad C - 2, \quad D - 3$$

1.7 (a) The diffraction pattern at $z = z_i$ of the finite-sized lens is

$$E(x_i, y_i) = \frac{e^{jk[z_i + (x_i^2 + y_i^2)/2z_i]}}{j\lambda z_i} \mathcal{F}\left\{ \underbrace{\Pi\left(\frac{x}{a}\right) \Pi\left(\frac{y}{a}\right)}_{\text{Aperture}} \right.$$

$$\left. \times \underbrace{e^{-jk[(x^2+y^2)/2f_0]}}_{\text{Lens}} \underbrace{e^{jk[(x^2+y^2)/2z_i]}}_{\substack{\text{Part of point spread} \\ \text{function}}} \right\}_{f_x = x_i/\lambda z_i, \ f_y = y_i/\lambda z_i}$$

At the focal plane $z_i = f_0$ the last two factors cancel and

$$E(x_i, y_i) = \frac{e^{jk[f_0 + (x_i^2 + y_i^2)/2f_0]}}{j\lambda f_0} \operatorname{sinc}\left(a\frac{x_i}{\lambda f_0}\right) \operatorname{sinc}\left(a\frac{y_i}{\lambda f_0}\right)$$

The diffraction-limited main lobe spot size (zero crossing) is

$$\left(\frac{2\lambda f_0}{a}\right) \times \left(\frac{2\lambda f_0}{a}\right)$$

(b) $a = \dfrac{2\lambda f_0}{\Delta x} = \dfrac{2(0.555 \times 10^{-6}) + (50 \times 10^{-3})}{(1 \times 10^{-6})} = 55.5$ mm

1.8 Figure P1.8 shows the geometry for taking a picture with a pinhole camera. For simplicity, only one point P_0 at $(l_0, 0, d_1)$ on the object is considered and the object is expressed by $\delta(x_0 - l_0)\delta(y_0)$. The aperture of the camera is in the x–y plane, which is d_1 away from the object. The aperture function is $\Pi(x/a)\Pi(y/a)$. The field that is illuminating the aperture is a spherical wave emanating from $(l_0, 0, d_1)$, and the field on the aperture is

$$E(x, y) = \frac{E_0}{j\lambda d_1} \exp\left[jk\left(d_1 + \frac{(x - l_0)^2 + y^2}{2d_1}\right)\right] \Pi\left(\frac{x}{a}\right) \Pi\left(\frac{y}{a}\right) \tag{1}$$

Before calculating the diffraction pattern of the field within the aperture, projected onto the film of the camera, some assumptions are made.

Assumption 1. The pinhole is made small such that the term with $x^2 + y^2$ in the exponent can be ignored, and $E(x, y)$ can be approximated as

$$E(x, y) \doteq \frac{E_0}{j\lambda d_1} e^{jk[d_1 + l_0^2/2d_1 - l_0 x/d_1]} \tag{2}$$

Using Eq. (1.38) again, the image on the film is calculated as

$$E(x_i, y_i) = \frac{-E_0}{\lambda^2 d_1 d_2} \exp\left[jk\left(d_1 + d_2 + \frac{l_0^2}{2d_1} + \frac{x_i^2 + y_i^2}{2d_2}\right)\right] \mathcal{F}\left\{e^{-jkl_0 x/d_1}\right.$$

$$\times \Pi\left(\frac{x}{a}\right) \Pi\left(\frac{y}{a}\right) e^{jk(x^2+y^2)/2d_2}\Bigg\}_{f_x=x_i/\lambda d_2,\ f_y=y_i/\lambda d_2} \tag{3}$$

where d_2 is the length between the aperture and the film of the pinhole camera. An additional assumption is made.

Assumption 2. The limits of the Fourier transform integral of Eq. (3) are from $-a$ to a. The value of the exponent in the fourth term inside the Fourier transform is at most $ka^2/2d_2$, and the size $a \times a$ of the pinhole will be chosen so that this value is much smaller than unity.

With Assumption 2, Eq. (3) can be approximated as

$$E(x_i) = -\frac{E_0 a^2}{\lambda^2 d_1 d_2} \exp\left[jk\left(d_1 + d_2 + \frac{l_0^2}{d_1} + \frac{x_i^2 + y_i^2}{2d_2}\right)\right]$$

$$\times \text{ sinc}\left[a\left(\frac{l_0}{\lambda d_1} - \frac{x_i}{\lambda d_2}\right)\right] \text{sinc}\left(\frac{a y_i}{\lambda d_2}\right) \tag{4}$$

Thus, the image formed on the film is a sinc function centered at $(-l_0 d_2/d_1, 0)$ with the size of the main lobe of $2\lambda d_2/a$. Because of Assumption 2, this no longer holds true for larger values of a.

1.9 The output from lens L_1 is

$$E_1(x, y, f_1) = \frac{e^{jkf_1}}{j\lambda f_1} G\left(\frac{x}{\lambda f_1}, \frac{y}{\lambda f_1}\right)$$

which is the input to lens L_2. The output from lens L_2 is

$$E_2(x_i, y_i) = \frac{e^{jk(f_1+f_2)}}{j\lambda f_2} \mathcal{F}\left\{\frac{1}{j\lambda f_1} G\left(\frac{x}{\lambda f_1}, \frac{y}{\lambda f_1}\right)\right\}_{f_x=x_i/\lambda f_2, f_y=y_i/\lambda f_2}$$

$$= \frac{e^{jk(f_1+f_2)}}{j\lambda f_2} \frac{(\lambda f_1)^2}{j\lambda f_1} g\left(-\frac{\lambda f_1}{\lambda f_2}x_i, -\frac{\lambda f_1}{\lambda f_2}y_i\right)$$

$$E_2(x_i, y_i) = -\frac{f_1}{f_2} e^{jk(f_1+f_2)} g\left(-\frac{f_1}{f_2}x_i, -\frac{f_1}{f_2}y_i\right)$$

The output image is therefore f_2/f_1 times the input image and is inverted. The amplitude of the output image is $-f_2/f_1$ times that of the input image. The minus sign in the amplitude means the phase of the output light is reversed from that of the input, but human eyes, which are sensitive only to intensity, cannot recognize this. Recall that the magnified image from a single lens expressed by Eq. (1.151) has a quadratic phase factor. This quadratic phase factor can be eliminated with the two-lens arrangement.

1.10 **(a)** The case of the opaque dot: The field distribution at the back focal plane of lens L_2 is

$$\frac{e^{j2kf}}{j\lambda f_2}\left[\delta\left(\frac{x}{\lambda f_2}\right)\delta\left(\frac{y}{\lambda f_2}\right) + j\Phi\left(\frac{x}{\lambda f_2}, \frac{y}{\lambda f_2}\right)\right]$$

where

$$\Phi = \mathcal{F}\{\phi\}$$

With the opaque dot, the first term is removed. The second term is further Fourier transformed by L_3, and the field on the screen is

$$-e^{j4kf}\left(\frac{f_2}{f_3}\right) j\phi\left(-\frac{f_2}{f_3}x_i, -\frac{f_2}{f_3}y_i\right)$$

and the intensity pattern is

$$I_a(x_i, y_i) = \left(\frac{f_2}{f_3}\right)^2 \phi^2\left(-\frac{f_2}{f_3}x_i, -\frac{f_2}{f_3}y_i\right)$$

(b) The case of $\pi/2$-radian phase plate: The field distribution just after the phase plate in the back focal plane (x, y) of L_2 is

$$\frac{e^{j2kf}}{j\lambda f_2}\left[e^{j\pi/2}\delta\left(\frac{x}{\lambda f_2}\right)\delta\left(\frac{y}{\lambda f_2}\right) + j\Phi\left(\frac{x}{\lambda f_2}, \frac{y}{\lambda f_2}\right)\right]$$

The field on the screen after the second lens L_3 is

$$-e^{j4kf}\frac{f_2}{f_3}\left[j + j\phi\left(-\frac{f_2}{f_3}x_i, -\frac{f_2}{f_3}y_i\right)\right]$$

The intensity distribution is

$$I_b(x_i\, y_i) = \left(\frac{f_2}{f_3}\right)^2\left[1 + \phi\left(-\frac{f_2}{f_3}x_i, -\frac{f_2}{f_3}y_i\right)\right]^2$$

$$\simeq \left(\frac{f_2}{f_3}\right)^2\left[1 + 2\phi\left(-\frac{f_2}{f_3}x_i, -\frac{f_2}{f_3}y_i\right)\right]$$

(c) Comparing I_a and I_b: While case (a) provides ϕ^2, case (b) provides $1 + 2\phi$. The result of case (b) is linear with ϕ and a more truthful representation and more sensitive when $|\phi|^2 < 1$.

1.11 If the card of encryption is $n(x, y)$, that of decryption has to be $n^{-1}(-x, -y)$. The phase distribution of a convex lens is

$$n(x, y) = e^{-jk(x^2+y^2)/2f_0}$$

Thus

$$n^{-1}(-x, -y) = e^{jk(x^2+y^2)/2f_0}$$

which is the transmission coefficient of a concave lens of focal length f_0.

1.12

$$g(r, \theta) = \begin{cases} 1 & b < r < a,\ \dfrac{\pi}{c} \leq \theta \leq \pi - \dfrac{\pi}{c},\ \text{and}\ \pi + \dfrac{\pi}{c} \leq \theta \leq 2\pi - \dfrac{\pi}{c} \\ 0 & \text{elsewhere} \end{cases}$$

$$g_n(r) = \frac{1}{2\pi}\int_{\pi/c}^{\pi-\pi/c} g(r, \theta)e^{-jn\theta}\, d\theta + \frac{1}{2\pi}\int_{\pi+\pi/c}^{2\pi-\pi/c} g(r, \theta)e^{-jn\theta}\, d\theta$$

$$= \frac{-1}{n\pi}[1 + (-1)^n]\sin\left(\frac{n\pi}{c}\right) \tag{1}$$

$$g_n(r) = \begin{cases} 1 - \dfrac{2}{c} & n = 0 \\ -\dfrac{2}{n\pi}\sin\left(n\dfrac{\pi}{c}\right) & n = \text{even} \\ 0 & n = \text{odd} \end{cases}$$

The Fourier–Hankel transform is calculated from Eq. (1.61) and Eq. (1):

$$G(\rho, \phi) = 2\pi \left(1 - \frac{2}{c}\right) \int_0^a r J_0(2\pi \rho r)\, dr$$

$$- \sum_{\substack{n = -\infty \\ n = \text{even} \\ \text{except } n = 0}} (-j)^n e^{jn\phi} \left(\frac{4}{n}\right) \sin\left(\frac{n\pi}{c}\right) \int_b^a r J_n(2\pi \rho r)\, dr$$

With the relationship

$$J_{-n}(x) = (-1)^n J_n(x)$$

$G(\rho, \phi)$ becomes

$$G(\rho, \phi) = 2\pi \left(1 - \frac{2}{c}\right) \int_0^a r J_0(2\pi \rho r)\, dr$$

$$+ \sum_{m=1}^{\infty} (-1)^{m+1} \frac{4j}{m} \sin(2m\phi) \sin\left(2m\frac{\pi}{c}\right) \int_b^a J_{2m}(2\pi \rho r)\, dr$$

or

$$G(\rho, \phi) = \left(1 - \frac{2}{c}\right) \left(\frac{a}{\rho} J_1(2\pi \rho a) - \frac{b}{\rho} J_1(2\pi \rho b)\right)$$

$$+ \sum_{m=1}^{\infty} (-1)^{m+1} \frac{4j}{m} \sin(2m\phi) \sin\left(2m\frac{\pi}{c}\right) \int_b^a J_{2m}(2\pi \rho r)\, dr$$

For large c and small b, $G(\rho, \phi)$ approaches the value of a complete circular aperture.

1.13 The transmittance of the hologram is

$$t(x, y) = |O|^2 + |R|^2 + OR^* + O^*R \tag{1}$$

where t_0 and β are suppressed. According to the geometry for fabricating the hologram, the expressions for O and R are

$$O(x, y) = \frac{A}{j\lambda d_0} e^{jk[d_0 + (x^2 + y^2)/2d_0]} \tag{2}$$

$$R(x, y) = R_0 e^{-jkx \sin\theta} \tag{3}$$

According to the geometry for reconstructing the image, we can express the reconstruction beam P as

$$P(x, y) = P_0 e^{jkz} \tag{4}$$

Reconstruction of the virtual image is, from the third term of Eq. (1),

$$E_3(x_i, y_i) = AR_0P_0 \frac{e^{jk[z_i+d_0+(x_i^2+y_i^2)/2z_i]}}{j\lambda z_i} \frac{1}{j\lambda d_0}$$
$$\times \mathcal{F}\left\{e^{jk(x^2+y^2)/2D} \cdot e^{jkx\sin\theta}\right\}_{f_x=x_i/\lambda z_i, f_y=y_i/\lambda z_i} \tag{5}$$

where

$$\frac{1}{D} = \frac{1}{d_0} + \frac{1}{z_i} \tag{6}$$

Using Eqs. (1.43) and (1.110), the result of the Fourier transform is

$$E_3(x_i, y_i) = AR_0P_0 \frac{e^{jk[z_i+d_0+(x_i^2+y_i^2)/2z_i]}}{j\lambda(z_i+d_0)}$$
$$\cdot \left\{e^{-j\pi\lambda D[(f_x-\sin\theta/\lambda)^2+f_y^2]}\right\}_{f_x=x_i/\lambda z_i, f_y=y_i/\lambda z_i} \tag{7}$$

Insertion of $f_x = x_i/\lambda z_i$, $f_y = y_i/\lambda z_i$ provides

$$E_3(x_i, y_i) = AR_0P_0 \frac{e^{jk(z_i+d_0)}}{j\lambda(z_i+d_0)} \cdot \exp\left\{j\frac{k}{2}\left[\frac{x_i^2+y_i^2}{z_i}\left(1-\frac{D}{z_i}\right)\right.\right.$$
$$\left.\left. +2\frac{D}{z_i}x_i\sin\theta - D\sin^2\theta\right]\right\} \tag{8}$$

$$E_3(x_i, y_i) = \frac{AR_0P_0}{j\lambda(z_i+d_0)} \exp\left[jk\left((z_i+d_0) + \frac{1}{2(z_i+d_0)}[(x_i+d_0\sin\theta)^2\right.\right.$$
$$\left.\left. +y_i^2 - d_0(d_0+z_i)\sin^2\theta]\right)\right] \tag{9}$$

The last term in the exponent is an aberration term that is small when either $z_i \doteq -d_0$ or $\theta \doteq 0$. The virtual image is at a distance $z = z_i + d_0$ from the observer and is laterally shifted by $-d_0\sin\theta$ in the x_i direction.

Next, the real image is considered using the fourth term of Eq. (1):

$$E_4(x_i, y_i) = AR_0P_0 \frac{e^{jk[z_i-d_0+(x_i^2+y_i^2)/2z_i]}}{j\lambda z_i} \frac{1}{j\lambda d_0}$$
$$\times \mathcal{F}\left\{e^{jk(x^2+y^2)/2D'} \cdot e^{-jkx\sin\theta}\right\}_{f_x=x_i/\lambda z_i, f_y=y_i/\lambda z_i} \tag{10}$$

where

$$\frac{1}{D'} = \frac{1}{z_i} - \frac{1}{d_0} \tag{11}$$

Replacing D by D' and $jkx \sin \theta$ by $-jkx \sin \theta$ in Eq. (8), $E_4(x_3, y_3)$ is obtained directly as

$$E_4(x_4, y_4) = \frac{AR_0P_0}{\lambda(z_i - d_0)} \exp\left[jk \left((z_i - d_0) + \frac{1}{2(z_i - d_0)}[(x_i + d_0 \sin \theta)^2 \right. \right.$$
$$\left. \left. + y_i^2 - d_0(d_0 - z_i) \sin^2 \theta] \right) \right] \tag{12}$$

The results are summarized in Fig. A1.13.

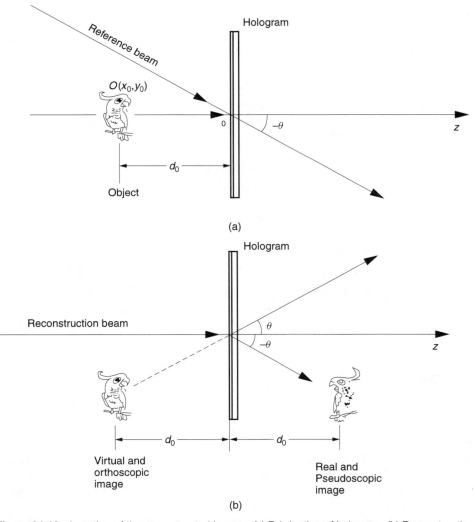

(a)

(b)

Figure A1.13 Location of the reconstructed images. (a) Fabrication of hologram. (b) Reconstructing the images.

Chapter 2

2.1 From Eqs. (2.2), (2.31), and (2.32),

$$\eta_1 H_1 \cos\theta_1 - \eta_1 H_3 \cos\theta_3 = \eta_2 H_2 \cos\theta_2 \tag{1}$$

$$H_1 + H_3 = H_2 \tag{2}$$

Equation (1) + Eq. (2) ($\eta_1 \cos\theta_3$) gives

$$\frac{H_2}{H_1} = \frac{2\eta_1 \cos\theta_1}{\eta_1 \cos\theta_1 + \eta_2 \cos\theta_2}$$

$$t_{\|H} = \frac{2n_2 \cos\theta_1}{n_2 \cos\theta_1 + n_1 \cos\theta_2} \neq t_{\|E}$$

Similarly, Eq. (1) − Eq. (2) ($\eta_2 \cos\theta_2$) gives

$$r_{\|H} = \frac{n_2 \cos\theta_1 - n_1 \cos\theta_2}{n_2 \cos\theta_1 + n_1 \cos\theta_2} = r_{\|E}$$

2.2 Snell's law changes Eq. (2.35) to

$$r_\| = \frac{E_3}{E_1} = \frac{\sin\theta_1 \cos\theta_1 - \sin\theta_2 \cos\theta_2}{\sin\theta_1 \cos\theta_1 + \sin\theta_2 \cos\theta_2}$$

which can be further rewritten as

$$\begin{aligned}
r_\| &= \frac{\sin\theta_1 \cos\theta_1 (\cos^2\theta_2 + \sin^2\theta_2) - \sin\theta_2 \cos\theta_2 (\cos^2\theta_1 + \sin^2\theta_1)}{\sin\theta_1 \cos\theta_1 (\cos^2\theta_2 + \sin^2\theta_2) + \sin\theta_2 \cos\theta_2 (\cos^2\theta_1 + \sin^2\theta_1)} \\
&= \frac{(\cos\theta_1 \cos\theta_2 - \sin\theta_1 \sin\theta_2)(\sin\theta_1 \cos\theta_2 - \cos\theta_1 \sin\theta_2)}{(\sin\theta_1 \sin\theta_2 + \cos\theta_1 \cos\theta_2)(\cos\theta_1 \sin\theta_2 + \sin\theta_1 \cos\theta_2)} \\
&= \frac{\cos(\theta_1 + \theta_2) \sin(\theta_1 - \theta_2)}{\sin(\theta_1 + \theta_2) \cos(\theta_1 - \theta_2)}
\end{aligned}$$

Thus,

$$r_\| = \frac{\tan(\theta_1 - \theta_2)}{\tan(\theta_1 + \theta_2)}$$

2.3 In the case of perpendicular polarization, E fields are parallel to the interface. They are identical with the tangential component of the field and $r_\perp + 1 = t_\perp$.

In the case of parallel polarization, the E field is not identical with the tangential component. Only a fraction of the E field is tangential to the interface. The tangential components satisfy the continuity condition while the E field itself does not. Hence, $r_\| + 1 \neq t_\|$ is true except in the case of normal incidence.

2.4   **(a)** From Eqs. (2.42) and (2.56), $R_\perp$ is

$$R_\perp = r_\perp^2 = \frac{\sin^2(\theta_1 - \theta_2)}{\sin^2(\theta_1 + \theta_2)}$$

$$= \frac{(\sin\theta_1\cos\theta_2 - \cos\theta_1\sin\theta_2)^2}{\sin^2(\theta_1 + \theta_2)}$$

From Eqs. (2.44) and (2.55), $T_\perp$ is

$$T_\perp = \frac{n_2\cos\theta_2}{n_1\cos\theta_1} \cdot \frac{4\cos^2\theta_1\sin^2\theta_2}{\sin^2(\theta_1 + \theta_2)}$$

With Snell's law, $T_\perp$ is rewritten as

$$T_\perp = \frac{4\sin\theta_1\cos\theta_1\sin\theta_2\cos\theta_2}{\sin^2(\theta_1 + \theta_2)}$$

and hence,

$$R_\perp + T_\perp = 1$$

(b) From Eqs. (2.43) and (2.56), $R_\parallel$ is

$$R_\parallel = r_\parallel^2 = \frac{\tan^2(\theta_1 - \theta_2)}{\tan^2(\theta_1 + \theta_2)}$$

$$= [\cos(\theta_1 + \theta_2)\sin(\theta_1 - \theta_2)]^2[\sin(\theta_1 + \theta_2)\cos(\theta_1 - \theta_2)]^{-2}$$

Thus,

$$R_\parallel = \frac{(\sin\theta_1\cos\theta_1 - \sin\theta_2\cos\theta_2)^2}{[\sin(\theta_1 + \theta_2)\cos(\theta_1 - \theta_2)]^2}$$

From Eqs. (2.45) and (2.55), $T_\parallel$ is

$$T_\parallel = \frac{n_2\cos\theta_2}{n_1\cos\theta_1}\left(\frac{2\cos\theta_1\sin\theta_2}{\sin(\theta_1 + \theta_2)\cos(\theta_1 - \theta_2)}\right)^2$$

With Snell's law, $T_\parallel$ is rewritten as

$$T_\parallel = \frac{4\sin\theta_1\cos\theta_1\sin\theta_2\cos\theta_2}{[\sin(\theta_1 + \theta_2)\cos(\theta_1 - \theta_2)]^2}$$

$$R_\parallel + T_\parallel = \frac{(\sin\theta_1\cos\theta_1 + \sin\theta_2\cos\theta_2)^2}{[\sin(\theta_1 + \theta_2)\cos(\theta_1 - \theta_2)]^2}$$

$$= \frac{\{\frac{1}{2}(\sin 2\theta_1 + \sin 2\theta_2)\}^2}{[\sin(\theta_1 + \theta_2)\cos(\theta_1 - \theta_2)]^2}$$

$$= \frac{[\sin(\theta_1 + \theta_2)\cos(\theta_1 - \theta_2)]^2}{[\sin(\theta_1 + \theta_2)\cos(\theta_1 - \theta_2)]^2} = 1$$

2.5   The condition for $r_\perp$ to be zero is $\theta_1 = \theta_2$, but due to Snell's law if $n_1 \neq n_2$, θ_1 can never be equal to θ_2.

2.6   When the angle of incidence is 57°, the light that is polarized in the plane of incidence is at Brewster's angle. Only the light that is polarized perpendicular to the plane of incidence (or perpendicular to the page) is reflected by G_1. The light then reaches G_2 resulting in l_0.

   By rotating G_2 by 90° this light polarized perpendicular to the plane of incidence also satisfies Brewster's angle and is not reflected by G_2.

2.7

$$n_2 \sin \theta_1 = \sin \theta_2$$

$$n_x = n_2 \sin(90° - \theta_1) = n_2 \cos \theta_1$$

Eliminate θ_1 from the above two equations to obtain

$$n_x = \sqrt{n_2^2 - \sin^2 \theta_2}$$

Pulfrich's refractometer is widely used for measuring the index of refraction of fluid.

2.8

$$(n_1 k)^2 \cos^2 \theta_1 - (n_2 k)^2 \cos^2 \theta_2 = \left(n_1^2 - n_2^2\right) k^2$$

$$(n_1 k)^2 (1 - \sin^2 \theta_1) - (n_2 k)^2 (1 - \sin^2 \theta_2) = \left(n_1^2 - n_2^2\right) k^2$$

Thus,

$$k \left(n_1^2 \sin^2 \theta_1 - n_2^2 \sin^2 \theta_2\right) = 0$$

Chapter 3

3.1   The concentric fringe rings on the screen in Fig. 3.10 become blurred. The $x - z$ plane cross section of the cone shape is shown in Fig. A3.1.

3.2   Note from Eq. (3.66) that the measured quantity Δd_1 is augmented by m_1 times and the accuracy of the method in Section 3.4.3 is m_1 times higher than the method proposed in this problem.

3.3   From Eq. (3.70), $\Delta\lambda/\lambda$ is

$$\frac{\Delta\lambda}{\lambda} = \frac{\Delta d_2 - \Delta d_1}{d_2 - d_1} = \frac{0.1441 - 0.0829}{300}$$

$$= 2.040 \times 10^{-4}$$

The frequency of the He–Ne laser light is

$$v = \frac{c}{\lambda} = \frac{3 \times 10^{14}}{0.6328} = 4.741 \times 10^{14} \text{ Hz}$$

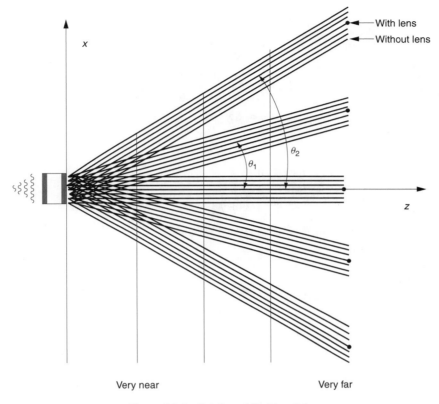

Figure A3.1 Solution of Problem 3.1.

Since

$$\frac{\Delta \nu}{\nu} = \frac{\Delta \lambda}{\lambda}$$

the modulation frequency is

$$\Delta \nu = \frac{\Delta \lambda}{\lambda} \nu = (2.040 \times 10^{-4})(4.741 \times 10^{14})$$

$$= 96.71 \text{ GHz}$$

3.4   (a)

$$d_{m+1} - d_m = \frac{\lambda}{2}$$

$$\lambda = 2(d_{m+1} - d_m) = 2(0.42) = 0.84 \ \mu m$$

(b)

$$m \frac{\lambda}{2} = d$$

$$m = 2 \frac{d}{\lambda} = \frac{2(420)}{0.84} = 1000$$

(c)

$$\Delta \nu = \frac{d_1}{d_2} \Delta \nu_{FSR}$$

$$= \frac{d_1}{d_2} \frac{c}{2d} = \frac{(0.1)(3 \times 10^{14})}{(0.42)(2)(420)}$$

$$= 85 \text{ GHz}$$

3.5

$$r = f \sin \theta_i$$

$$\theta_i = \sin^{-1} \left(\frac{r}{f} \right) = \sin^{-1} \left(\frac{5.29}{50} \right) = 6.07°$$

$$\sin \theta_i = n_2 \sin \theta = 1.05 \sin \theta$$

$$\theta = 5.78°$$

Note that

$$2n_2 d = m\lambda$$

$$2n_2 d \cos \theta = (m - 1)\lambda$$

$$\therefore \lambda = 2n_2 d(1 - \cos \theta)$$

$$= 2(1.05)60(1 - \cos 5.78°)$$

$$= 0.64 \text{ μm}$$

3.6   The output intensity from the Fabry–Pérot cavity is, from Eq. (3.30) with $A = 0$,

$$I_t = I_0 \frac{1}{1 + M \sin^2 \phi/2}$$

where

$$\phi = \frac{4\pi \nu}{c} d$$

At the resonance length

$$L = m\lambda/2$$

When the length of the cavity is enlongated by ΔL from the resonance length,

$$d = m\lambda/2 + \Delta L$$

ΔL that reduces the output to one-half of the value at the resonance is

$$1 = M \sin^2 \left[\frac{1}{2} \left(\frac{4\pi \nu}{c} \right) \left(m \frac{\lambda}{2} + \Delta L \right) \right]$$

$$= M \sin^2 \left(\frac{\pi}{\lambda} \Delta L \right)$$

$$\Delta L \doteq \frac{\lambda}{\pi\sqrt{M}} = \frac{\lambda}{2}\frac{1}{F}$$

$$\frac{\Delta L}{L} = \frac{\lambda}{2FL} = \frac{1.064 \times 10^{-6}}{2(516)(300)} = 3.4 \times 10^{-12}$$

3.7 Using the parameters in Fig. A.3.7, $z_0/f = 0.5$. From Table 3.2 or from Figure A.3.7, $M = 1.8$, $W_1 = 0.18$ mm, $d_1 = 18$ cm, $z_1 = 160$ mm, and $\theta_1 = 0.063°$.

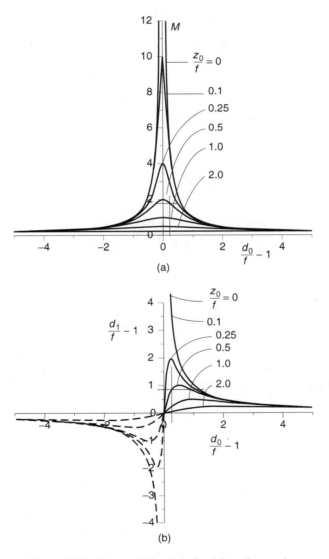

Figure A3.7 Values of M and $d_1/f - 1$ from the graphs.

3.8   **(a)** Using Eqs. (3.118) and (3.119), $(d_1 - f)$ in Eq. (3.117) is converted into $(d_0 - f)$, and $(d_0 - f)$ in the same equation is converted into $(d_1 - f)$; then the expression for d_0 in terms of f, d_1 and z_1 is

$$d_0 - f = \frac{f^2(d_1 - f)}{(d_1 - f)^2 + z_1^2}$$

$$= \frac{10^2(3.6 \times 10^6)}{(3.6 \times 10^6)^2 + \left[\left(\dfrac{\pi}{0.63 \times 10^{-6}}\right)(1)^2\right]^2}$$

$$= 9.5 \ \mu m$$

$$d_0 = 10 \ m + 9.5 \ \mu m$$

From Eqs. (3.116) and (3.118),

$$W_0 = 1\sqrt{\frac{9.5 \times 10^{-6}}{3.6 \times 10^6}}$$

$$W_0 = 1.62 \ \mu m$$

(b) When the distance is large compared to the beam waist, Eq. (3.106) and Fig. 3.25 give

$$\lim_{z \to \infty} \tan \theta = \frac{2}{kW_0}$$

$$\frac{2}{kW_0} \simeq \frac{1}{3.6 \times 10^6}$$

$$W_0 \simeq \frac{3.6 \times 10^6}{\left(\dfrac{\pi}{0.63 \times 10^{-6}}\right)}$$

$$W_0 \simeq 72 \ cm$$

3.9   If the beam is unfolded with respect to the plane mirror, then the problem becomes the same as the two concave mirror cavity with spacing $2\,d$ in place of d. From Eqs. (3.151) and (3.153),

$$W_1^2 = \frac{2R_c}{k\sqrt{\dfrac{R_c}{d} - 1}}$$

$$W_0^2 = \frac{2\,d}{k}\sqrt{\frac{R_c}{d} - 1}$$

The same result can be obtained from the boundary conditions directly. Referring to Fig. P3.9 the phase of the waist matches that of the flat mirror. The flat mirror is placed at $z = 0$ or at the waist and the concave mirror is placed at $z = d$. At resonance, the radius of curvature R of the beam has to be matched with the radius of curvature R_c of the mirror. The phase of the waist matches the flat mirror. Equation (3.100) gives the relationship among them. Insert W_0^2 of Eq. (3.100) into (3.96) to obtain

$$1 + \left(\frac{2R_c}{kW_1^2} \right)^2 = \frac{R_c}{d}$$

which leads to

$$W_1^2 = \frac{2R_c}{k\sqrt{\dfrac{R_c}{d} - 1}}$$

3.10 The propagation constant α is, from Eq. (3.171),

$$\alpha = k \sin \theta$$

The value of θ is determined by the size of the annular slit and the focal length as

$$\tan \theta = \frac{a}{f}$$

If θ is small, then $\sin \theta \doteq \tan \theta$ and from the above two equations

$$\alpha = 2\pi a / \lambda f$$

The first zero of the main lobe appears when

$$\alpha \rho = 2.4$$

and

$$\rho = 0.38 \lambda f / a$$

Thus, the focal length f of the lens is

$$f = \frac{\rho a}{0.38 \lambda} = \frac{(60 \times 10^{-6})(2.5 \times 10^{-3})}{(0.38)(0.63 \times 10^{-6})}$$

$$= 62.7 \text{ cm}$$

The radius R of the lens is, from Eq. (3.173),

$$R = z_{max} \frac{a}{f} = (1) \frac{2.5 \times 10^{-3}}{(0.627)}$$

$$= 4.0 \text{ mm}$$

3.11 On the white side of the fin, a photon bounced off the surface changes its momentum from p to $-p$ and the amount of change in momentum is $2p$, while on the black side, the photon is absorbed and the change of the momentum is from p to 0 and the amount of change is p. It should rotate toward the black side of the fin. Crook's radiometer rotates in the opposite direction. It is propelled by the heat expansion of the gas on the black side.

3.12   When a photon collides, the momentum is changed from p to $-p$. The change of momentum is $2p$. From Eq. (3.199), the pushing force is

$$F = 2qn_1 \frac{P}{c}$$

where q is the fraction of light effectively reflected back.

(a) $F = 2(0.07)(1.43)\dfrac{1}{3 \times 10^8} = 6.67 \times 10^{-10}$ newtons

(b) The mass of the sphere is

$$m = \tfrac{4}{3}\pi r^3 = \tfrac{4}{3}\pi(0.5145 \times 10^{-4} \text{ cm})^3 \times 10^{-3}$$
$$= 5.7 \times 10^{-16} \text{ kg}$$

The acceleration is

$$\frac{dv}{dt} = \frac{F}{m} = \frac{6.67 \times 10^{-10}}{5.7 \times 10^{-16}} = 1.17 \times 10^6 \text{ m/s}^2$$

(c) The gravitational acceleration is

$$g = 9.8 \text{ m/s}^2$$

The ratio is

$$\frac{dv/dt}{g} = 1.2 \times 10^5 \text{ times}$$

3.13   As shown in Fig. A3.13, the direction is toward the focused beam.

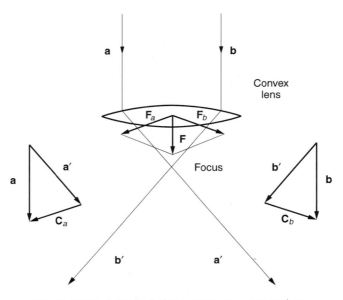

Figure A3.13   Laser radiation pressure on a convex lens.

Chapter 4

4.1   Inserting Eq. (4.61) into (4.62) gives

$$k_x^2 \left(\frac{1}{n_o^2} - \frac{1}{n_e^2} \right) = 0$$

(a) $k_x = 0$ together with the original assumption $k_y = 0$ means that the propagation vector $\mathbf{k}$ is along the z axis (or the optic axis).

(b) $n_o = n_e$ means that the medium is isotropic.

4.2

$$\begin{bmatrix} k_0^2 n_\alpha^2 - k_y^2 - k_z^2 & k_x k_y & k_x k_z \\ k_y k_x & k_0^2 n_\beta^2 - k_x^2 - k_z^2 & k_y k_z \\ k_z k_x & k_z k_y & k_0^2 n_\gamma^2 - k_x^2 - k_y^2 \end{bmatrix} \begin{bmatrix} E_x \\ E_y \\ E_z \end{bmatrix} = 0$$

where $k_0 = \omega/c$.

4.3   (a) The answer is given in Fig. A4.3a.

(b) The answer is given in Fig. A4.3b. It should be realized that the critical angle from the left is different from that from the right. There are two intersections

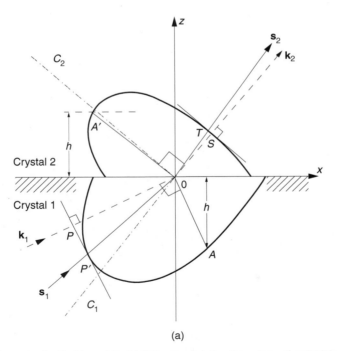

(a)

Figure A4.3   Answers to Problem 4.3. (a) Refraction at the boundary of uniaxial media. (b) The condition for total internal reflection at the boundary between uniaxial media. (c) Direction of the reflected ray.

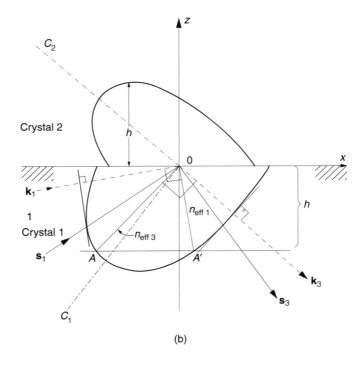

(b)

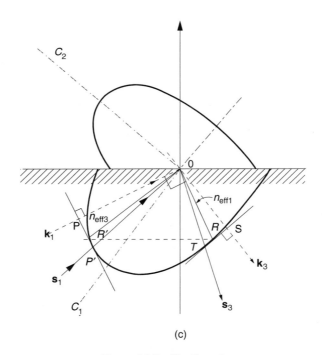

(c)

Figure A4.3 (*Continued*)

A and A' with the indicatrix, which are not symmetric with respect to the normal to the boundary, and the critical angle from the left side is different from that from the right.

(c) Referring to Fig. A4.3c, draw tangent $\overline{PP'}$ at P'. The line normal to $\overline{PP'}$ is $\mathbf{k}_1$. The line $\overline{OR}$, which is perpendicular to $\mathbf{k}_1$, represents n_{eff_1}. From R, line $\overline{RR'}$ is drawn parallel to the boundary until it intersects with the indicatrix at R'. $\overline{OR'}$ represents n_{eff_3}. The normal to $\overline{OR'}$ is $\mathbf{k}_3$. (Thus far, this discussion duplicates the discussion connected with Fig. 4.17.) The point of contact T to the indicatrix that the tangent $\overline{ST}$ makes determines the direction $\mathbf{s}_3$ of the reflected light ray.

4.4 The answer will be given by referring to Fig. A4.4. The critical angle ϕ_c^a between air and the LiNbO$_3$ interface is first calculated.

The point where the horizontal line a–a' at a distance $h = 1$ from the boundary intersects the ellipse determines n_{eff_3} at the critical angle. Equation (4.101) is used for calculating $n_{\mathrm{eff}_3} = k/k_0$, but keep in mind that angle θ in Eq. (4.101) is the angle with respect to the optic axis of crystal. The optic axis of our crystal is horizontal. Put $\theta = 90° + \theta_c$ and $h = n_{\mathrm{eff}_3} \sin \theta_c$.

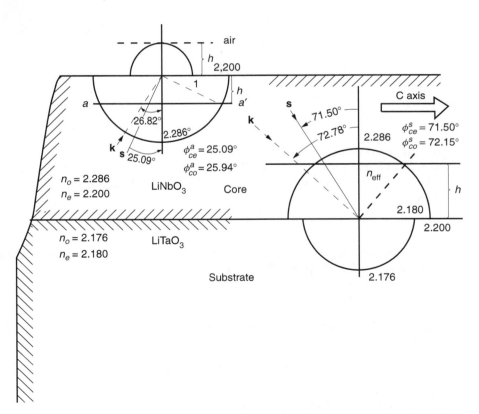

Figure A4.4 Examination of critical angles at the boundaries of an anisotropic optical guide.

$$h = \frac{\sin \theta_c}{\left[\left(\dfrac{\cos \theta_c}{n_e}\right)^2 + \left(\dfrac{\sin \theta_c}{n_o}\right)^2\right]^{1/2}}$$

which can be solved for $\sin \theta_c$ as

$$\sin \theta_c = \frac{h/n_e}{\left[1 + h^2\left(\dfrac{1}{n_e^2} - \dfrac{1}{n_o^2}\right)\right]^{1/2}}$$

Note that this expression is different from Eq. (4.112). Equation (4.112) applies when the crystal axis is normal to the interface. Inserting the given values, θ_c of the wavenormal at the critical angle is calculated as

$$\theta_c = 26.82°$$

The direction ϕ of the light ray of the e-wave is obtained from Eq. (4.56). Again, the angles in Eq. (4.56) are with respect to the optic axis, which means $\theta = 26.82° + 90°$ and

$$\phi_{ce}^a = 25.09°$$

where ϕ_{ce}^a is referred to the normal to the interface. The critical angle of the o-wave is simply

$$n_o \sin \phi_{co}^a = 1$$

and

$$\phi_{co}^a = 25.94°$$

The difference in critical angle between the o- and e-waves is 0.88°.
Next, the critical angle between LiNbO$_3$ and LiTaO$_3$ is calculated. The maximum value of h associated with the interface is $h = 2.174$. Hence, θ_c from the above expression is

$$\theta_c = 72.78°$$

Once again, the angles in Eq. (4.56) are with respect to the optic axis, which means that $\theta = 72.78° + 90°$. From Eq. (4.56), we find

$$\phi_{ce}^s = 71.50°$$

where ϕ_{ce}^s is referred to the normal to the interface. The critical angle of the o-wave is

$$\sin \phi_{co}^s = \frac{2.176}{2.286}, \qquad \phi_{co}^s = 72.15°$$

The difference in the critical angles between the o- and e-waves is 0.65°. The answers are summarized in Fig. A4.4.

4.5 **(a)** As shown in Fig. 4.12 as well as in Fig. A4.5a, by the time the wavefront of **D** reaches point P, the wavefront of **E** reaches point P'; thus,

$$v = u \cos \gamma \tag{1}$$

where v and u are the phase and ray velocities of the e-wave, respectively. γ is the angle between **s** and **k**, which is the same as that between **E** and **D**.

The left-hand side of Eq. (4.82) is the projection of **E** onto **D** and is rewritten as

$$E \cos \gamma = \left(\frac{v}{c}\right)^2 \frac{D}{\epsilon_0} \tag{2}$$

Inserting Eq. (1) into the right-hand side of Eq. (2) gives

$$\frac{D}{\epsilon_0 E} \cos \gamma = \left(\frac{c}{u}\right)^2 \tag{3}$$

With the relationship

$$\cos \gamma = \frac{\mathbf{E} \cdot \mathbf{D}}{ED} \tag{4}$$

Equation (4) is written as

$$\frac{n_o^2 E_x^2 + n_e^2 E_z^2}{E^2} = \left(\frac{c}{u}\right)^2 \tag{5}$$

where use was made of Eq. (3).

Note from Fig. A4.5a that

$$\frac{E_x}{E} = -\cos\phi, \qquad \frac{E_z}{E} = \sin\phi \tag{6}$$

$$\frac{\cos^2 \phi}{v_o^2} + \frac{\sin^2 \phi}{v_e^2} = \frac{1}{u^2} \tag{7}$$

where

$$v_o = \frac{c}{n_o}, \qquad v_e = \frac{c}{n_e} \tag{8}$$

Equation (7) is the expression for the ray velocity surface (suppressing the y components) and is called Huygens' wavelet ellipsoid. It provides the ray velocity for a given direction. The ray velocity for $\phi = 0$ or along the z direction is v_o, and that for $\phi = 90°$ or along the x direction is v_e. In other words, when the wave propagates along the z axis, **D** is in the x direction and sees the index of refraction n_o; and when the wave propagates along the x axis, **D** is in the z direction and sees the index of refraction n_e. The index of refraction is determined by what **D** sees.

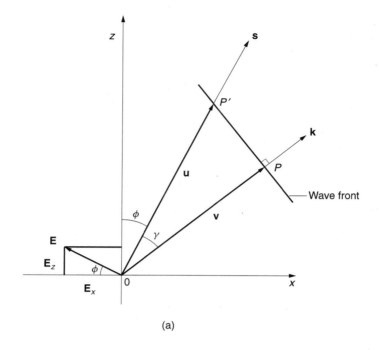

(a)

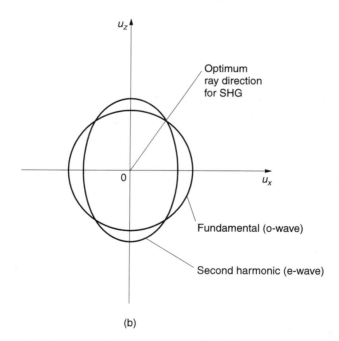

(b)

Figure A4.5 (a) Phase velocity **v** and ray velocity **u** of the e-wave. (b) Ray velocity matching for SHG.

Equation (7) immediately provides an expression for the wavelet spread in terms of spatial coordinates. Multiplying Eq. (7) by the distance r from the origin to the point of observation gives

$$\left(\frac{x}{v_e}\right)^2 + \left(\frac{z}{v_o}\right)^2 = t^2 \tag{9}$$

Equation (9) is the expression for an ellipse whose x axis is $v_e t$ and z axis is $v_o t$, both of which expand with t.

(b) As shown in Fig. A4.5b, the ray velocity diagram of the fundamental wave (o-wave) is overlayed with that of the second harmonic (e-wave). The intersection of the two diagrams is the optimum ray velocity direction for the SHG.

4.6 As the direction of the propagation is tilted from the z axis in the $x - z$ plane, the length of the major axis $\overline{OP}$ of the elliptic cross section moves along the circumference of the ellipse made with $y = 0$ in Eq. (4.84). $\overline{OP}$ of such an ellipse can be found by replacing

$$x = r \cos \theta$$
$$z = r \sin \theta$$
$$y = 0$$

in Eq. (4.84):

$$\left(\frac{\cos \theta}{n_\alpha}\right)^2 + \left(\frac{\sin \theta}{n_\gamma}\right)^2 = \frac{1}{r^2}$$

The angle θ is obtained by inserting

$$r = n_\beta$$

A method for experimentally finding the optical axes can be found in Ref. 15 in Chapter 4.

Chapter 5

5.1 It is seen from Eq. (5.10) that with $\varepsilon = \varepsilon_z$ one can take advantage of the large r_{33} of lithium niobate. The equation of the indicatrix, Eq. (5.4), becomes

$$\left(\frac{1}{n_o^2} + r_{13}\varepsilon_z\right)x^2 + \left(\frac{1}{n_o^2} + r_{13}\varepsilon_z\right)y^2 + \left(\frac{1}{n_e^2} + r_{33}\varepsilon_z\right)z^2 = 1$$

which can be approximated as

$$\frac{x^2}{(n_o - \frac{1}{2}r_{13}n_o^3\varepsilon_z)^2} + \frac{y^2}{\left(n_o - \frac{1}{2}r_{13}n_o^3\varepsilon_z\right)^2} + \frac{z^2}{\left(n_e - \frac{1}{2}r_{33}n_e^3\varepsilon_z\right)^2} = 1$$

Maximum retardation is achieved when light propagates in the $x - y$ plane. The amount of retardation is similar for any propagation direction in the $x - y$ plane.

For $k = k_y$,

$$\Delta = \frac{2\pi}{\lambda}(n_e - n_o)d + \frac{\pi}{\lambda}\left(n_o^3 r_{13} - n_e^3 r_{33}\right)\frac{V}{h}d$$

The first term is independent of the applied dc field and is called the *natural birefringence*, whereas the second term is the *induced birefringence*.

5.2 The linear portion of the I versus Δ curve is located at the point where $dI/d\Delta$ is constant. Therefore, the good biasing point is where $d^2I/d^2\Delta = 0$ is satisfied. Insertion of Eq. (5.29) into this equation leads to

$$\Delta_b = \frac{\pi}{2}$$

The retardance Δ is the sum of the bias term Δ_b and the modulated term Δ_m:

$$\Delta = \frac{\pi}{2} + \Delta_m$$

With the assumption that Δ_m is much smaller than $\pi/2$, and using the trigonometric relationship $\sin^2 A = \frac{1}{2}(1 - \cos 2A)$, Eq. (5.29) is approximated as

$$I = \frac{I_0}{2}(1 + \Delta_m)$$

where

$$\Delta_m = -\frac{2\pi}{\lambda}\left(r_{22}n_o^3\varepsilon_m \cos \omega_m t\right)h$$

5.3 From Eq. (5.10) and $\varepsilon = (\varepsilon_x, \varepsilon_y, 0)$, the expression of the indicatrix is obtained. Since the direction of the propagation is along the z direction, the cross-sectional ellipse is

$$\left(\frac{1}{n_o^2} - r_{22}\varepsilon_y\right)x^2 + \left(\frac{1}{n_o^2} + r_{22}\varepsilon_y\right)y^2 - 2r_{22}\varepsilon_x xy = 1$$

The major and minor axes of this expression are immediately obtained by using the results of Example 5.4. Insertion of

$$A = \frac{1}{n^2} - r_{22}\varepsilon_y$$

$$C = \frac{1}{n^2} + r_{22}\varepsilon_y$$

$$B = -r_{22}\varepsilon_x$$

into Eq. (5.35) gives

$$\tan 2\theta = \frac{\varepsilon_x}{\varepsilon_y}$$

Let Φ be the angle between the resultant electric field and the x axis. Then Φ is expressed in terms of ε_x and ε_y as $\cot \Phi = \varepsilon_x/\varepsilon_y$. Using the identity

$$\cot \Phi = \tan\left(\frac{\pi}{2} - \Phi\right)$$

θ becomes

$$\theta = \frac{\pi}{4} - \frac{\Phi}{2}$$

If the external field is a rotating field at $\Phi = \Omega t$, the axis of the ellipse rotates in the opposite direction at one-half of the angular velocity Ω:

$$\theta = \frac{\pi}{4} - \frac{\Omega}{2}t$$

This fact is used for building a frequency shifter.

5.4   The two beams are

$$R = R_0 e^{j(k_x x + k_z z + \psi)}$$

$$S = S_0 e^{j(-k_x x + k_z z)}$$

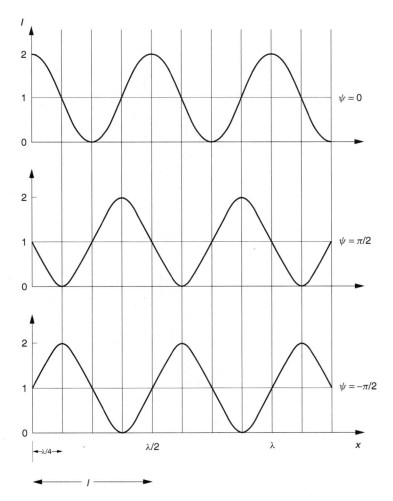

Figure A5.4   Intensity distributions $I = 2[1 + \cos(2k_x x + \psi)]$ for $\psi = 0, \pi/2, -\pi/2$ radians.

With $R_0 = S_0 = 1$, the fields on the $z = 0$ plane are found:

$$R + S = e^{j\psi/2}(e^{j(k_x x + \psi/2)} + e^{-j(k_x x + \psi/2)})$$

$$= 2e^{j\psi/2}\cos(k_x x + \psi/2)$$

$$|R + S|^2 = 4\cos^2(k_x x + \psi/2)$$

$$I = 2[1 + \cos(2k_x x + \psi)]$$

The intensity distributions for $\psi = 0, \pi/2, -\pi/2$ are shown in Fig. A5.4.

5.5 Extending the vector diagram in Fig. 5.21 to the case where $\psi = -\pi/2$, one finds that energy transfers from $S(x)$ to $R(x)$.

5.6 See Fig. A5.6.

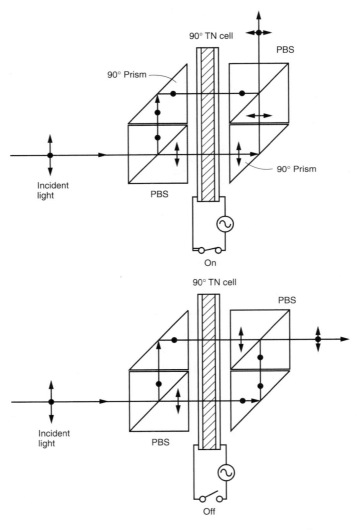

Figure A5.6 Light paths when the switch is on and off.

Chapter 6

6.1 Let

$$E_- = e^{-j\omega t + j\beta z}$$

$$E_+ = e^{j\omega t - j\beta z}$$

Let the plane of observation be $z = 0$. The phasor circles C_1 and C_2 of E_- rotate clockwise, while those of E_+ rotate counterclockwise. The 90° phase delay of the retarder is represented by

$$E_{-y} = e^{-j\omega t + j90°}$$

$$E_{+y} = e^{j\omega t - j90°}$$

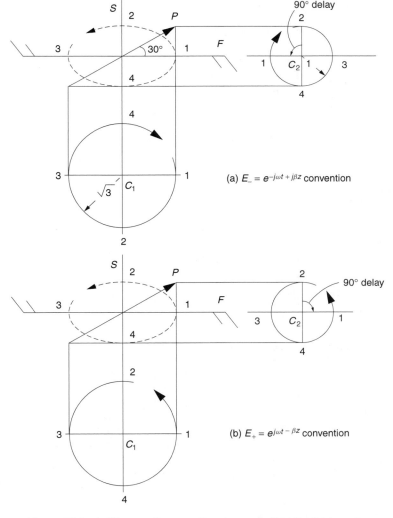

Figure A6.1 A difference in convention does not affect the final results.

The ellipses using these two different conventions are drawn in Fig. A6.1. The same final results are obtained.

6.2 The solution is obtained by the circle diagram as shown in Fig A6.2. From the diagram

$$\theta_2 = 68° \quad \text{and} \quad \epsilon = 0.31$$

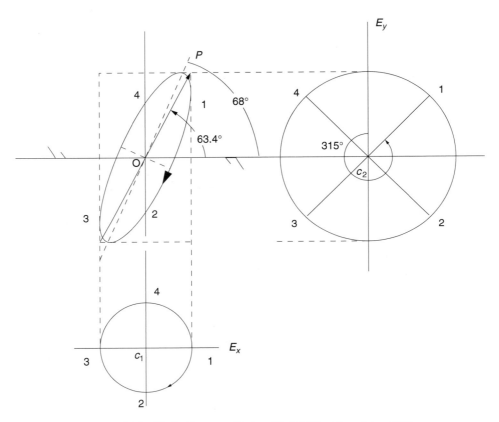

Figure A6.2 Circle diagram with $\theta_1 = 63.4°(B/A = 2)$ and $\Delta = 315°$.

6.3 The results are summarized in Fig. A6.3. As expected from the answer of Example 6.1, the major or minor axis is always along the fast axis, which is a characteristic of the combination of a linearly polarized wave and the quarter-waveplate.

When E_x and E_y of the incident linear polarization have the same sign, the result is left-handed rotation or counterclockwise rotation, whereas when E_x and E_y have opposite signs, the result is right-handed or clockwise rotation, as summarized in Fig. A6.3b.

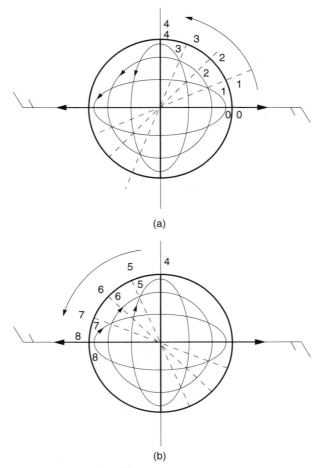

Figure A6.3 Transitions of the state of polarization as the azimuth of the incident linearly polarized wave is rotated. The fast axis of the $\lambda/4$ plate is along the x direction.

6.4 Referring to Fig. A6.4, point 1 on the major axis of the ellipse corresponds to the 1's on the circles C_1 and C_2. The phasor of C_2 is delayed from that of C_1 by 90°.

6.5 The result by the circle diagram method is shown in the dotted line in Fig. A6.5. The change of the sense of rotation makes a significant change in the state of polarization of the emergent light.

6.6 For $k_1 = 1$ and $k_2 = 0$, the direction of polarization of the emergent light from the stacked polarizers is always in the azimuth direction of the last polarizer.

(a) $\theta = \theta_2$.

(b) $\theta = \theta_1$.

6.7 The horseshoe crab sees linear polarization when it is oriented in a north–south direction. When it is oriented in other directions, the polarization is elliptical.

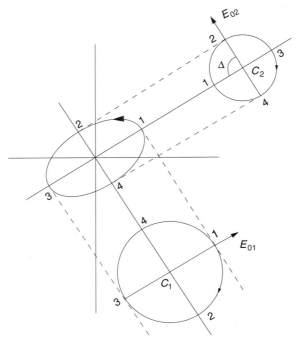

Figure A6.4 When an elliptically polarized wave is decomposed into parallel and perpendicular components, there is a 90° phase difference between the components.

To face north (or south) it turns its body until linear polarization is sensed. This can be seen from Fig. 6.33 by making these substitutions: the observer is the horseshoe crab, the light source is a ray of sunlight, and the scattering center is a particle in the water. The direction of linear polarization is perpendicular to the plane containing the horseshoe crab, the ray of sunlight, and the scattering center. The horseshoe crab sees vertically linearly polarized sunlight.

6.8 Because of the optical activity, the direction of polarization rotates. Wherever the light is horizontally polarized, no scattered light propagates horizontally because the direction of the **E** field becomes parallel to the direction of propagation. This happens every 180° of rotation. If the rotary power of quartz is $[\theta]_{0.63\ \mu m}^{20°C} \doteq 19.5$ deg/mm, the period of modulation is $L = 180/19.5 = 9.2$ mm.

6.9 Since $d_1 > d_2$, the retardance is dominated by the anisotropy of the crystal with d_1. Since the emerging light is left-handed, the polarization direction of the incident light has to be to the left of the fast axis. In other words, the fast axis is to the right of the polarization direction of the incident wave, which means the fast axis is along the optic axis in crystal d_1. The refractive index in the direction of the optic axis is n_e, while the refractive index in the direction perpendicular to the optic axis (y direction) is n_o. Hence,

$$n_e < n_o$$

and the crystal is classified as negative, as mentioned in Chapter 4.

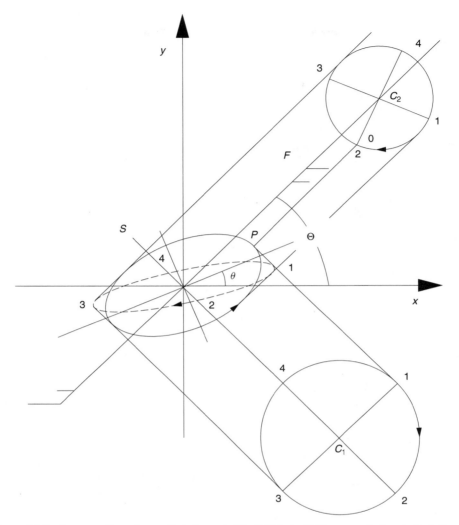

Figure A6.5 Same configuration as that shown in Fig. 6.10 but with the opposite handed rotation of the incident wave.

6.10 Insert the retarder in question between a pair of crossed polarizers. This configuration is illustrated in Fig. 6.38 with the retarder as the inserted optical element. As a precaution, if the incident wave happens to be linearly polarized, check that at least some of the incident light is getting through the first polarizer. The situation where no light passes through the first polarizer is to be avoided.

Null output from the second polarizer is obtained only when the fast and slow axes of the retarder match the direction of the polarizer principal axes. Determining which axis is the fast axis, and determining the actual value of the retardance can be done by Senarmont's method (see Section 6.4.3.3).

6.11 First, the solution is obtained by circle diagrams. Partition the retardance in equal proportions between the E_x and E_y component fields. If $\Theta = 0$, delay E_y

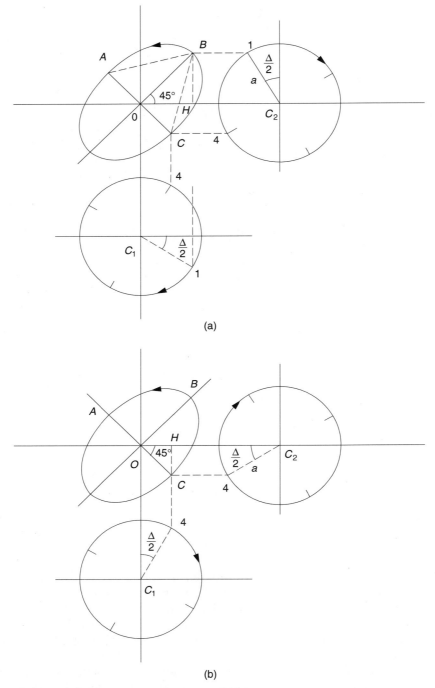

(a)

(b)

Figure A6.11 Circle diagrams. (a) Point 1 on phasors representing point B on the ellipse. (b) Point 4 on phasors representing point C on the ellipse.

by $\Delta/2$ while advancing E_x by $\Delta/2$, and draw circle diagrams as shown in Fig. A6.11(a). Points 1 on C_1 and C_2 are extended to point B on the ellipse. Point 4 on C_1 and C_2 is extended to point C on the ellipse. If we find the lengths $\overline{OB}$ and $\overline{OC}$ on the ellipse in terms of Δ, the angle $\angle ABC$ can be found in terms of Δ. From the diagram in Fig. A6.11a, we have

$$\overline{BH} = a\cos\frac{\Delta}{2}$$

As we learned from Fig. 6.4, the major and minor axes are at either $\theta = 45°$ or $\theta = 135°$, when $B/A = 1$, and

$$\overline{OB} = \sqrt{2}\, a\cos\frac{\Delta}{2}$$

From the diagram in Fig. A6.11b, we have

$$\overline{CH} = a\sin\frac{\Delta}{2}$$

$$\overline{OC} = \sqrt{2}\, a\sin\frac{\Delta}{2}$$

From the above two values, $\tan\beta$ is expressed as

$$\tan\beta = \frac{\overline{OC}}{\overline{OB}} = \tan\frac{\Delta}{2}$$

Finally,

$$2\beta = \Delta$$

Next, the result is obtained simply using Eq. (6.127) with $\theta = 45°$. The answer is $2\beta = \Delta$.

6.12 **(a)** Note from Eqs. (6.95) and (6.98) that

$$\tan 2\alpha = \frac{2AB}{A^2 - B^2}$$

Use of Eq. (6.99) to find $\cos 2\theta$ to put into Eq. (6.123) leads to

$$(a^2 - b^2)\sin 2\theta = 2AB\cos\Delta$$

(b) Applying the identity

$$\cos 2\alpha = \frac{1 - \tan^2\alpha}{1 + \tan^2\alpha}$$

to Eq. (6.91) gives

$$\cos 2\alpha = \frac{A^2 - B^2}{A^2 + B^2}$$

Using Eq. (6.123) to find a substitute for the numerator in the above expression gives the result

$$(a^2 - b^2)\cos 2\theta = (A^2 + B^2)\cos 2\alpha$$

Chapter 7

7.1   The diagram is shown in Fig. A7.1. The emergent wave is a right-handed elliptically polarized wave with

$$\epsilon = 0.2 \quad \text{and} \quad \theta = 80°$$

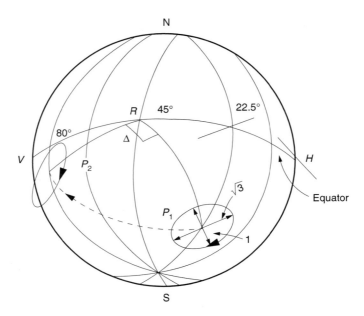

Figure A7.1   Solution of Problem 7.1.

7.2   The answers are shown in Fig. A7.2.

7.3   Figure A7.3 shows the geometry of the antiglare screen. Incident white light is first converted into a vertically polarized wave, which can be represented as P_1 on the Poincaré sphere. The azimuth Θ of the fast axis of the $\lambda/4$ sheet is 45°, which is represented by R on the Poincaré sphere. The incident point P_1 is rotated by 90° around R to P_2 at the north pole. Thus, the incident wave to the radar screen is a left-handed circularly polarized wave.

The radar screen reflects the incident light. The direction of rotation of the reflected circularly polarized wave is the same as that of the incident wave on the radar screen, but the direction of the propagation reverses and the reflected light looking toward the radar screen from the operator is right-handed. It is represented by the south pole on the Poincaré sphere.

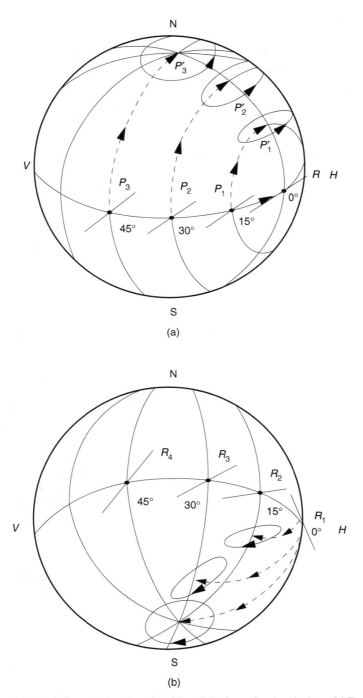

Figure A7.2 (a) Three different azimuths of incident light for a fixed λ/4 plate. (b) Three different azimuths of the fast axis of λ/4 plate for a fixed incident light.

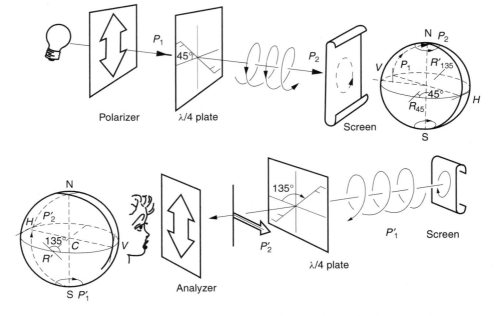

Figure A7.3 Polarization stages of an antiglare sheet plotted on the Poincaré sphere.

The azimuth of the fast axis of the same $\lambda/4$ sheet now appears at $\Theta = 135°$. The reflected light P'_1 is rotated by $90°$ around R' at $\theta = 135°$ to generate a horizontally linearly polarized wave, and the reflected light is blocked by the analyzer.

Light originating from the radar screen is attenuated but reaches the operator.

7.4 Point P_1 of the incident light with $\theta = 45°$ is at the midpoint between points H and V, and P_1 appears on the circumference of the circle made by the projection of the Poincaré sphere along the HV axis, as shown in Fig. A7.4.

Point P_2 of the emergent light with β is at the latitude of 2β on the Poincaré sphere.

$$\angle P_1 C P_2 = \Delta \quad \text{and} \quad \Delta = 2\beta$$

7.5 Projections onto the horizontal, frontal, and profile planes are made. The order of drawing the points in Fig. A7.5 is:

- P_1 at $2\theta_1 = 126.8°$ in the H plane.

- P_1 at $\beta = 0$ in the F plane.

- P_1 in the P plane.

- P_2 from P_1 by $\Delta = 315°$ in the P plane.

- P_2 in the H plane to find $2\theta = 137°$

- P'_2 rotated from P_2 around NS to obtain the true angle $2\beta = 34°$.

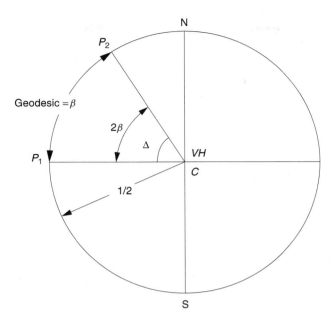

Figure A7.4 Poincaré sphere viewed along the *HV* axis.

7.6 Follow the steps explained in Example 7.12. Figure A7.6 shows the key operations. The state of the polarization of the emergent light is left-handed elliptical polarization with $(\theta, \epsilon) = (158°, 0.23)$.

7.7 **(a)** The procedure is similar to that of Problem 7.5. The answers of $\theta = 60°$ and $\epsilon = 0.31$ with left-handedness are verified from Fig. A7.7a.

(b) The transmittance of the analyzer is $k = \cos^2 \angle HCP_2/2$ in the upper right figure in Fig. A7.7a. The true angle is obtained when P_2 is rotated around the axis CH to the equator, so that $2\alpha = 114°$ and $\alpha = 57°$. The transmitted power intensity P_t is

$$P_t = \frac{\eta_0}{2}(E_x^2 + E_y^2)\cos^2 57° = 2.0\eta_0 \text{ W/m}^2$$

(c) The answer is in Fig. A7.7b:

$$2\alpha = 74° \quad \text{and} \quad k = \cos^2 \alpha = 0.64$$

7.8 The operation is quite similar to that of Example 7.13. The major difference is the bipolar nature of the external field ε_x. As shown in Fig. A7.8a the solid line indicates the shape of the indicatrix when the external field is $+\varepsilon_x$. The indicatrix is rotated as shown by the dotted line when the external field is changed to $-\varepsilon_x$. The direction of the fast axis rotates by 2θ, when the polarity of ε_x is reversed. Thus, the azimuth Θ of the fast axis is alternately at R and R' on the Poincaré sphere. The amount of retardance Δ is the same for both.

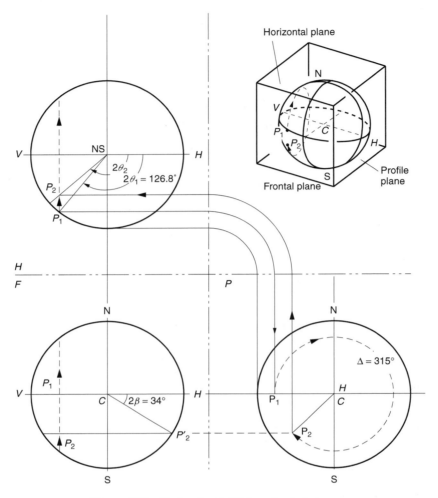

Figure A7.5   Finding θ and β Poincaré sphere traces.

With this TE–TM converter, the retardance of the modal retarder is 180° so that the spiral locus on the Poincaré sphere moves toward point V as the light passes through the interdigital electrodes, as shown in Fig. A7.8b.

The incident light P_1 is first rotated around R by Δ to P_2 by the first conversion retarder. The first modal retarder rotates point P_2 by 180° around the HV axis to P_3. The fast axis of the second conversion retarder, however, is moved to $+\theta$ and P_3 is rotated around R' rather than around R by Δ to P_4. The second modal retarder rotates by 180° from P_4 to P_5 around point H.

The same procedure repeats, and the point moves toward point V.

7.9   Let the two oppositely propagating beams be represented by

$$E_x = Ae^{j\beta z - j\omega t}$$
$$E_y = Be^{-j\beta z - j\omega t} \tag{1}$$

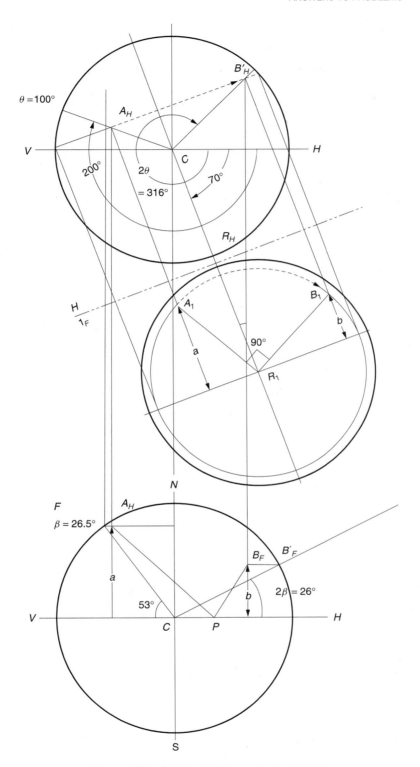

Figure A7.6   Solution by Poincaré sphere traces.

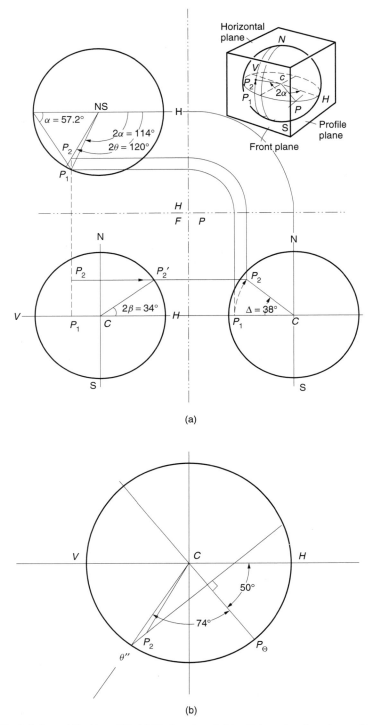

Figure A7.7 Solution of Problem 7.7. (a) Finding θ and β by Poincaré sphere trace. (b) Finding the transmittance of an analyzer with $\Theta = 25°$.

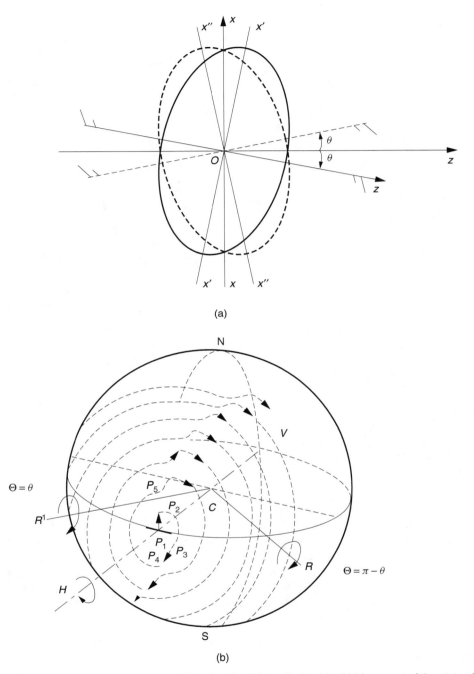

(a)

(b)

Figure A7.8 Solution of Problem 7.8. (a) Indicatrix of the optical guide. (b) Movement of the state of polarization on the Poincaré sphere.

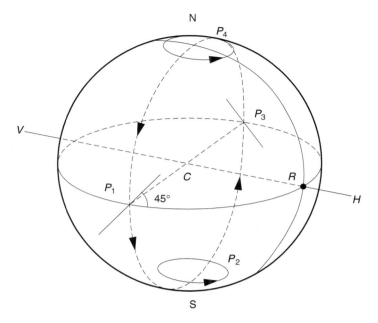

Figure A7.9 Polarization grating for laser cooling.

The component field ratio is

$$\frac{E_y}{E_x} = e^{-j2\beta z} \tag{2}$$

where $A = B$ (or $\alpha = 45°$) is assumed. Equation (2) is equivalent to a retarder whose fast axis is along the x axis and whose retardance is

$$\Delta = -2\beta z \tag{3}$$

The distribution of the state of polarization along the z axis is found using the Poincaré sphere in Fig. A7.9. P_1 represents linearly polarized light with $\theta = 45°$ at $z = 0$. As z is increased P_1 rotates by $-2\beta z$ (counterclockwise) on the Poincaré sphere with point H as its center of rotation. The state of polarization changes in the following order: linearly polarized at $\theta = 45°$, right-handed circularly polarized, linearly polarized at $\theta = 135°$, left-hand circularly polarized, and so on, at every $\Delta z = \lambda/8$.

Chapter 8

8.1 There is a time lapse $\tau = 2L/c$ between the two entries of the incident and reflected waves into the distorting medium. The distortion-free image is recovered when

$$\Phi(t) = \Phi(t - \tau)$$

$\Delta \Phi = \Phi(t) - \Phi(t - \tau)$ can be used as a criterion for the quality of the recovered image:

$$\Delta \Phi = \Phi_0[\cos \omega t - \cos \omega(t - \tau)]$$
$$= \tfrac{1}{2}\Phi_0 \sin(\omega\tau/2) \cos \omega(t - \tau/2)]$$

If $\Delta \Phi$ is zero, the best image is recovered; that is, when

$$\omega\frac{\tau}{2} = n\pi \quad \text{or} \quad \frac{2L}{c} = nT$$

where T is the period of the fluctuation. This happens when the return-trip time matches the fluctuation period and the wave sees the same fluctuation on both trips. The worst case is when

$$\frac{2L}{c} = \frac{2n + 1}{2}T$$

The wave that saw the maximum Φ on the initial trip sees the minimum on the return trip, and the worst recovery is made.

8.2 **(a)** Although the atoms are symmetric with respect to position, they are not symmetric with respect to the polarity of the charges; therefore, the crystal does not possess inversion symmetry.

(b) The redistributed charges are shown in Figs. A8.2a and A8.2b for $\mathbf{E} = E\hat{\mathbf{x}}$ and $\mathbf{E} = -E\hat{\mathbf{x}}$. The amount of polarization is different when the direction of $\mathbf{E}$ is reversed, and the polarization with respect to time is as shown in Fig. A8.2c.

(c) Since the curve for positive values of P_{NL} is different from that of negative P_{NL}, the second order nonlinearity can exist. The addition of the fundamental and the second harmonic in Fig. A8.2d conforms with that in Fig. A8.2c.

8.3 The curve of P_{NL} for the crystal with inversion symmetry should have the same shape for positive P_{NL} and negative P_{NL}, as indicated in Fig. A8.3a. If the fundamental and the fourth order harmonic in Fig. A8.3b are added, the result is as shown in Fig. A8.3c. The shape of positive P_{NL} is not the same as that of negative P_{NL} and cannot conform with the curve in Fig. A8.3a. Thus, $\chi^{(4)} = 0$.

8.4

$$\mathbf{P_{NL}} = \hat{\mathbf{x}}\epsilon_0\chi_{xxxx}[A_1 \cos(k_1z - \omega t + \phi_1)$$
$$+ A_2 \cos(k_2z - \omega t + \phi_2) + A_3 \cos(k_3z - \omega t + \phi_3)]^3 \tag{1}$$

Put

$$a = A_1 e^{j(-\omega_1 t + k_1 z + \phi_1)}$$
$$b = A_2 e^{j(-\omega_2 t + k_2 z + \phi_2)} \tag{2}$$
$$c = A_3 e^{j(-\omega_3 t + k_3 z + \phi_3)}$$

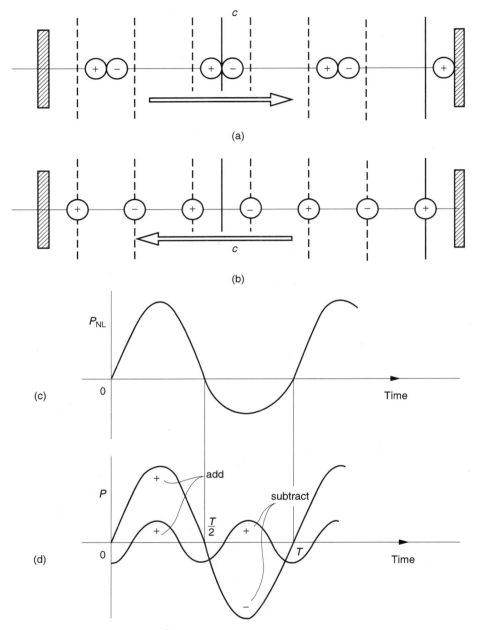

Figure A8.2 Proof of nonzero $\chi^{(2)}$. (a) Charge distribution for $E = E\hat{\mathbf{x}}$. (b) Charge distribution for $E = -E\hat{\mathbf{x}}$. (c) P_{NL} as a function of time. (d) Addition of fundamental and second harmonic.

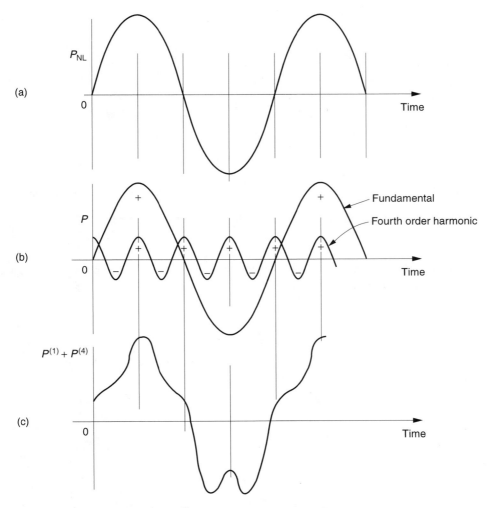

Figure A8.3 $\chi^{(4)} = 0$ for a crystal with inversion symmetry.

Then, Eq. (1) becomes

$$\mathbf{P}_{NL} = \hat{\mathbf{x}}\epsilon_0 \chi_{xxxx} \tfrac{1}{8}(d + d^*)^3 \tag{3}$$

where

$$d = a + b + c \tag{4}$$

which is reduced to

$$P_{NL} = \frac{\epsilon_0}{8}\chi_{xxxx}(d^3 + 3d^2 d^* + \text{c.c.}) \tag{5}$$

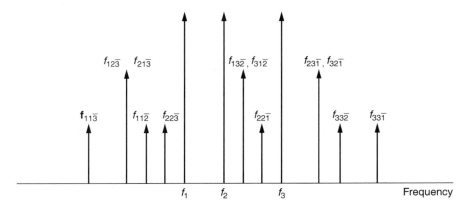

Figure A8.4 Spectra generated due to $\chi^{(3)}$ of the optical fiber. For instance, $f_{23\bar{1}} = f_2 + f_3 - f_1$.

Frequencies associated with d^3 are too high and are out of the range of interest. Discarding these terms gives

$$P_{NL} = \frac{3\epsilon_0}{8} \chi_{xxxx} (d^2 d^* + \text{c.c.}) \qquad (6)$$

Insertion of Eq. (4) into Eq. (6) generates many terms and the calculated result is

$$
\begin{aligned}
\frac{8P_{NL}}{3\epsilon_0 \chi_{xxxx}} = \; & a(|a|^2 + 2|b|^2 + 2|c|^2) \\
& + b(2|a|^2 + |b|^2 + 2|c|^2) \\
& + c(2|a|^2 + 2|b|^2 + |c|^2) \\
& + b^2 a^* + c^2 a^* + a^2 b^* + c^2 b^* + a^2 c^* + b^2 c^* \\
& + bca^* + cba^* + acb^* + cab^* + abc^* + bac^* + \text{c.c.} \qquad (7)
\end{aligned}
$$

For instance, frequency component of $b^2 a^*$ is, from Eq. (2), $2\omega_2 - \omega_1$ and that of bca^* is $\omega_2 + \omega_3 - \omega_1$. There are 15 frequencies, of which 3 are redundant in the last 6 terms, for a total of 12 different frequencies, as shown in Fig. A8.4.

After going through the fiber, the three input frequencies have generated 12 output frequencies. Stimulated Brillouin scattering in the glass is the main contributor to the third order nonlinearity. This, sometimes, does harm and creates serious crosstalk problems among highly wavelength division multiplexed channels and prevents the use of high-intensity light in this type of optical communication system [25].

8.5 In order to sustain four-wave mixing, both the frequency (Eq. (8.54)) and phase (Eq. (8.70)) matching conditions must be met. Since all frequencies are identical, Eq. (8.54) is satisfied. Since the magnitudes of all vector propagation constants are identical, the tips of the propagation vectors lie on the circumference of the same circle. The vector $\mathbf{k}_1 + \mathbf{k}_2$ has to be identical to $\mathbf{k}_3 + \mathbf{k}_4$. As shown in

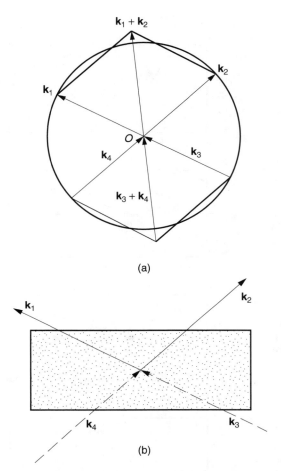

Figure A8.5 Conditions to sustain four-wave mixing. (a) Phasor diagram of **k** vectors. (b) Directions of **k**₃ and **k**₄.

Fig. A8.5, the only possible combination is for $\mathbf{k}_3$ to be identical to $\mathbf{k}_1$ (or $\mathbf{k}_2$), and for $\mathbf{k}_4$ to be identical to $\mathbf{k}_2$ (or $\mathbf{k}_1$).

8.6   There are six combinations, but two are redundant. There are four different patterns altogether, as shown in Fig A8.6.

8.7   The first equation is rewritten as

$$\frac{dA_p}{dz} = \gamma A_p$$

with $\gamma = j\beta_p - \alpha$. The solution of this equation is

$$A_p = A_p(0)e^{\gamma z}$$

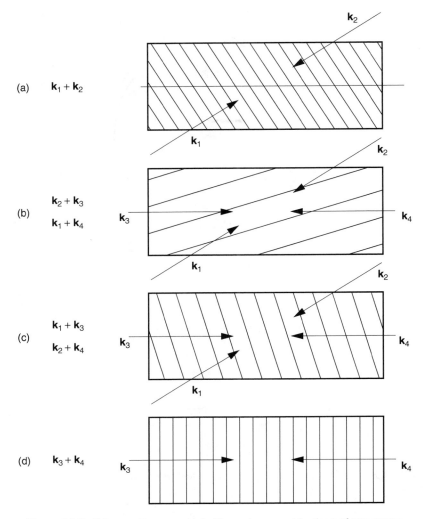

Figure A8.6 Fringe patterns generated by various combinations of two waves.

Next, the other two equations are simplified by letting

$$A_s = A_{s0}e^{\gamma z}$$

$$A_c = A_{c0}e^{\gamma z}$$

The result is

$$\frac{dA_{s0}}{dz} = jKA_p^2(0)A_{c0}^*$$

$$\frac{dA_{c0}}{dz} = jKA_p^2(0)A_{s0}^*$$

INDEX

A (Einstein's *A* coefficient), 842, 900
A mode in a nonlinear layer, 1048
Absorption indicatrix, 400
Acceptor
  action in semiconductors, 1153–1155
  atoms, 1153
  energy level, 1153
Accommodation in eye vision, 93
Acetone $(CH_3)_2CO$, 239
Acoustic wave
  imaging, 95
  surface, 324
Acoustooptic modulator (AOM), 100, 324
  surface acoustic wave (SAW), 324
  used for holographic video displays, 100
Active layer of laser diodes, 904
  gain of, 905
  plasma effect of, 944
Adaptive fiber coupler, 517
ADP, *see* Ammonium dihydrogen phosphate
AFM (atomic force microscope), 160, 161
Ahrens polarizing prisms, 404, 405
Airy pattern, 28
  calculation of, 1172
Alcohol C_2H_5OOH, stimulated Brillouin
    scattering of, 512
Al_2O_3 (aluminum oxide), 681, 894
$Al_xGa_{1-x}As$ (gallium arsenide doped with
    aluminum) laser, 895, 949
Alkali metals as cathode materials, 796
Aluminum oxide (Al_2O_3), 687, 894
AM (amplitude modulation), 919, 1086, 1087
Amino acids, optical activity of, 412
Ammonium dihydrogen phosphate (ADP)
  elastooptic properties of, 322
  electrooptic properties of, 305, 316, 1018
Ammonium fluoride (NH_4F), 158
Amplified signal power, 847
Amplified spontaneous emission (ASE) noise, 838,
    847, 856, 864, 866, 868, 870

Amplitude distribution
  in coupled slab guides, 646
  in slab optical guides, 615, 616
Amplitude modulation (AM), 919, 1086, 1087
Amplitude modulators, 312, 679, 1018
Amplitude shift keying (ASK), 812, 818, 831, 832,
    1089, 1095
Analog modulation, 1083, 1134, 1137
Analytic signal, 526, 547, 1022
  proof of not being applicable to nonlinear cases,
    547, 1022
Angle modulation, 1088, 1092
Angular frequency, 1
  convention of, 3, 368
Anisotropic media, 263
Annihilation of negative carriers, 898, 1155
Anomalous dispersion region, 1056, 1066
Antenna radiation pattern, 13, 95
  calculated by Fourier optics, 103, 556
  visualized by microwave holography, 96
Antiglare sheet, 409
  analyzed using Poincaré sphere, 502
Antiglare TV camera, 419
Antireflection (AR) coating, 803, 838, 954
Anti-Stokes radiation, 816
AOM, *see* Acoustooptic modulator
APC (automatic power control), 952
APD detectors, *see* Avalanche photodiode
Aperture functions
  of rectangles, 20
  of triangles, 21
  of circles, 25
  of delta functions, 28
  of shah functions, 30
  of SNOM probes, 114, 158
  numerical (NA), 151, 693, 694, 701, 703, 706,
    770
Apodization of radiation patterns, 22
AR (antireflection) coating, 803, 838, 954
Ar_2 (argon) for excimer lasers, 894

Combined index for Elements of Photonics, Volumes I and II. Volume I: pp. 1–604; Volume II: pp: 605–1197.

I.1